Environmental Science and Engineering

The ultimate goal of this series is to contribute to the protection of our environment, which calls for both profound research and the ongoing development of solutions and measurements by experts in the field. Accordingly, the series promotes not only a deeper understanding of environmental processes and the evaluation of management strategies, but also design and technology aimed at improving environmental quality. Books focusing on the former are published in the subseries Environmental Science, those focusing on the latter in the subseries Environmental Engineering.

Dongfei Han · Mukesh Khare · Siyue Li
Editors

Environmental Crisis: Pollution and Governance

Selected Papers from 2024 4th International Conference on Environmental Pollution and Governance (ICEPG 2024)

Set 1

Springer

Editors
Dongfei Han
Department of Environmental Sciences
Suzhou University of Science
and Technology
Suzhou, China

Mukesh Khare
Indian Institute of Technology (IIT) Delhi
New Delhi, Delhi, India

Siyue Li
Wuhan Institute of Technology
Wuhan, Hubei, China

ISSN 1863-5520 ISSN 1863-5539 (electronic)
Environmental Science and Engineering
ISBN 978-3-031-86831-3 ISBN 978-3-031-86829-0 (eBook)
https://doi.org/10.1007/978-3-031-86829-0

This Springer imprint is published by the registered company Springer Nature Switzerland AG
The registered company address is: Gewerbestrasse 11, 6330 Cham, Switzerland

Preface

The 2024 4th International Conference on Environmental Pollution and Governance (ICEPG 2024) was held from October 25–27 in Zhengzhou, China. As environmental issues remain an important global problem, this conference has become an important platform for exchanging ideas and research results. More than 150 attendees from around the world gathered together to share their insights and engage in productive discussions.

Keynote Speakers brought invaluable perspectives to ICEPG 2024, enriching the dialogue with their expertise and innovative ideas. Among them were Professor Huijun Wang, from the Institute of Atmospheric Physics at the Chinese Academy of Sciences; Professor Shijie Han, from Henan University; and Professor Junguo Liu, from North China University of Water Resources and Electric Power. Apart from the keynote speeches, oral and poster presentation parts were also held and displayed by various scholars, leading to a warm atmosphere of academic discussion.

The conference proceedings is a valuable record of the conference which presents a selection of papers submitted to the conference from universities, research institutes, and industries. All the papers have been through a rigorous peer-reviewed process in order to meet international publication standards. The papers highlight the collaborative efforts among researchers, policymakers, and industry leaders, reflecting the interdisciplinary nature of tackling environmental challenges. As readers explore these works, they will gain valuable insights into cutting-edge solutions and methodologies that hold the potential to transform our approach to environmental stewardship in the years to come.

Gratitude extends to every participant, speaker, and organizer who contributed to making ICEPG 2024 a resounding success. Their enthusiasm and commitment to enhance environmental research and policy have created a lasting impact. It is hoped that the proceedings will not only commemorate the achievements of the conference but also inspire ongoing efforts to improve environmental governance for a sustainable future.

The Committee of ICEPG 2024

Committee Member

Conference General Chair

Huijun Wang	Institute of Atmospheric Physics, Chinese Academy of Sciences, China

Technical Program Committee Chairs

Junguo Liu	North China University of Water Resources and Electric Power, China
Baoyu Gao	Shandong University, China

Publication Chairs

Mukesh Khare	Indian Institute of Technology (IIT) Delhi, India
Siyue Li	Wuhan Institute of Technology, China
Dongfei Han	Suzhou University of Science and Technology, China

Technical Program Committee

Chunhui Li	Beijing Normal University, China
Shiliang Liu	Beijing Normal University, China
Rui Du	University of Chinese Academy of Sciences, China
Wenjing Lu	Tsinghua University, China
Hong Cui	Shanxi Xinhua Chemical Defense Equipment Research Institute, China
Snengr. Meichao Zhao	Shenyang Academy of Environmental Sciences, China
Xia Li	Inner Mongolia University of Science & Technology, China
Assoc. Qinzhong Feng	University of Chinese Academy of Sciences, China
Zhengguang He	Zhengzhou University, China

Wenyan He	Xi'an University of Science and Technology, China
Xiaojuan Wang	Northwest A&F University, China
Rongkui Su	Central South University of Forestry Science and Technology, China
Yu Shen	Nanjing Forestry University, China
Junfeng Wan	Zhengzhou University, China
Lingjun Kong	Guangzhou University, China
Wanjun Wang	Guangdong University of Technology, China
Tiecheng Wang	Northwest A&F University, China
Lijuan Feng	Zhejiang Ocean University, China

Contents

Water and Wastewater Management

Hydrochemical Characteristics and Causes of Shallow Groundwater in Construction Waste Landfills

Jie Liu[1,2](✉), Haiming Li[1], Ye Li[2], Yizhang Zhang[2,3], and Haojie Lu[2,3]

[1] Tianjin University of Science and Technology, Tianjin 300222, China
1519607775@qq.com

[2] Research Institute for Environmental Innovation (Tianjin Binhai), Tianjin 300457, China

[3] Chinese Research Academy of Environmental Sciences, Beijing 100012, China

Abstract. By collecting groundwater samples, And Piper trilinear diagram, Gibbs diagram, ion-terminal element diagram, ion ratio and PMF model were comprehensively applied to study the chemical characteristics of groundwater and the main ion sources in a construction waste landfill in Xi'an. The results show that the pH of shallow groundwater in this study area spanned from 7.37 to 8.98, the TDS ranged from 172 to 1590 mg-L-1, and the groundwater was neutral to weakly alkaline fresh water. This cations were dominated by Ca^{2+} and the anions by HCO_3^- and SO_4^{2-}, and the water chemistry types showed partitioning, with Ca-HCO_3 type dominating, followed by Ca-SO_4 type. Construction waste landfill, agricultural activities and water rock action are the major factors affecting the shallow groundwater chemistry in the study area, contributing 41.8%, 33.7% and 24.5%, respectively.

Keywords: Construction waste · Shallow groundwater · Hydrochemical characteristics · PMF

1 Introduction

China's rapid urbanization has led to significant construction waste from building demolition, new construction, and alterations. Construction waste production increased from 610 million tons in 2018 to 2,385 million tons in 2020 [1]. Predominantly disposed of in simple landfills and open piles, this waste generates leachate containing pollutants like heavy metals, sulfates, and calcium ions, which can seep into and contaminate groundwater [2].

The quality of shallow groundwater is vital for the ecological environment and human health. Understanding the hydrochemical characteristics of groundwater in construction waste landfills is essential for tracking pollutant migration and informing pollution prevention strategies. This research supports the development of effective waste management policies and environmental protection measures.

D. Han et al. (Eds.): ICEPG 2024, ESE, pp. 3–10, 2025.
https://doi.org/10.1007/978-3-031-86829-0_1

2 Material and Methods

2.1 Subject of Research

The study area is an informal construction waste landfill in Shaanxi, covering approximately 110,000 square meters (Fig. 1). Constructed in 2005 and decommissioned in 2021, the landfill mainly contained slag, with small amounts of tile and concrete, deposited in sand pits using improvised methods.

Situated in the warm temperate Dalu monsoon climate zone, the area receives an average annual rainfall of about 504.7 mm. The groundwater belongs to the loose rock pore water of the Quaternary system, with shallow groundwater depths ranging from 1.15 to 13.65 m. The lithology is relatively homogeneous, characterized by good permeability and dominated by Quaternary loess, Upper Pleistocene ice-accumulated loess, and pebble gravel.

2.2 Sample Collection and Determination

In November 2022, thirty-four shallow groundwater sampling wells were installed in the study area, taking into account hydrogeologic conditions and landfill characteristics. Field tests for pH and total hardness (TH) were performed using a portable water quality meter. Major ions (Na^+, K^+, Mg^{2+}, Ca^{2+}) were analyzed with an ICP AES-600, $SO_4{}^{2-}$ and $HCO_3{}^-$ were measured using ion chromatography (CIC-D120), and total dissolved solids (TDS) were determined by the drying method. Quality control included three blanks and standards for each batch, with 10% of samples measured in parallel to ensure accuracy within $\pm$ 5%.

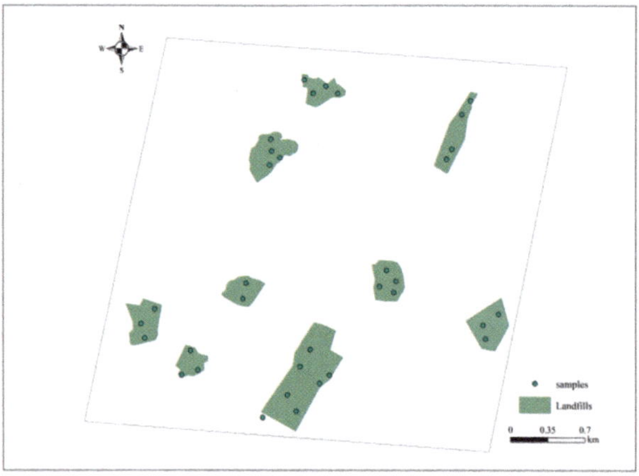

Fig. 1. The sampling point distribution

3 Results and Discussion

3.1 Hydrochemical Characterization

Table 1 shows the concentrations of the main hydrochemical components of the shallow groundwater in the study area. The pH of the groundwater ranged from 7.37 to 8.98, with a mean value of 7.82. The TDS ranged from 172 to 1,590 mg/L, with a mean value of 571.50 mg/L, indicating that the groundwater is neutral to soft alkaline freshwater (TDS $< 1{,}000$ mg/L) [3]. The total hardness (TH) of the shallow groundwater ranged from 131 to 1,070 mg/L, with an average value of 406.61 mg/L, classifying it as hard water (300 mg/L $<$ TH $<$ 450 mg/L). Cations were dominated by Ca^{2+}, characterized by the order $Ca^{2+} > Na^{+} > Mg^{2+} > K^{+}$, while anions were dominated by HCO_3^- and SO_4^{2-}, characterized by the order $HCO_3^- > SO_4^{2-} > NO_3^- > Cl^-$. The coefficient of variation for TDS and ion concentrations in the groundwater within the study area ranges from 0.03 to 1.02, indicating significant variability and considerable spatial heterogeneity. The Piper trilinear diagram is commonly used to characterize the major ion content and chemical composition type of water [4]. According to the Piper diagram (Fig. 2) and using the Shukarev classification method, the primary hydrochemical type is HCO_3-Ca. The study area is divided into three zones based on the milliequivalent percentage of SO_4^{2-}: 20%–50%, 50%–70%, and 80%–100%. The hydrochemical type in Zone I is HCO_3-Ca, with the background point located in Zone I, representing the main hydrochemical type of natural groundwater. Zones II and III are characterized by SO_4-Ca type water. The milliequivalent percentage of SO_4^{2-} in groundwater is less than 30%, gradually increasing from the background point towards Zones II and III.

3.2 Analysis of Hydrochemical Genesis

The Gibbs diagram (Fig. 3) can qualitatively determine the influence of natural processes such as rock weathering, atmospheric precipitation, and evaporation concentration [5]. Almost all groundwater samples in the study area are located in the rock weathering control zone, indicating that the chemical composition of shallow groundwater in this area is mainly affected by water-rock interaction [6]. The groundwater sample points are all far from the atmospheric precipitation control zone, indicating that atmospheric precipitation has little influence on the chemical composition of shallow groundwater. The end-member method can further reveal the impact of rock weathering and mineral dissolution on major ions in groundwater. As shown in Fig. 4, most of the sampling points in the study area fall between the carbonate rock and silicate rock end-members, indicating that the ion composition of shallow groundwater in the study area is primarily influenced by the weathering of silicate and carbonate rocks.

The ratio of milligram equivalent concentrations of ($Na^{+} + K^{+} - Cl^{-}$) to ($Ca^{2+} + Mg^{2+} - SO_4^{2-} - HCO_3^{-}$) can reflect the existence of cation adsorption in groundwater in the study area [7]. When the ratio is close to –1, it indicates the presence of cation adsorption. Most of the sampling points in the shallow groundwater in the study area showed a linear negative correlation, with a correlation coefficient of 0.65 and a slope of –1.58, indicating the existence of cation adsorption. The chlor-alkali index (CAI-1 and CAI-2) characterizes the type and intensity of cation exchange in groundwater. When

Table 1. Statistics of inorganic water chemistry parameters in the study area

Item	Minimum	Maximum	Average	Standard Deviation	Coefficient of Variation
pH	7.37	8.98	7.82	0.30	0.03
TH / ($mg{\cdot}L^{-1}$)	131	1070	406.61	200.13	0.49
TDS / ($mg{\cdot}L^{-1}$)	172	1590	571.50	322.83	0.56
K^+ / ($mg{\cdot}L^{-1}$)	2.35	27.5	5.49	4.83	0.87
Na^+ / ($mg{\cdot}L^{-1}$)	6.7	113	28.26	29.04	1.02
Ca^{2+} / ($mg{\cdot}L^{-1}$)	46	333	128.00	62.39	0.48
Mg^{2+} / ($mg{\cdot}L^{-1}$)	5.9	59.3	20.45	12.29	0.60
Cl^- / ($mg{\cdot}L^{-1}$)	7.2	138	34.68	32.28	0.93
$SO_4{}^{2-}$ / ($mg{\cdot}L^{-1}$)	42	802	165.09	182.67	0.42
$NO_3{}^-$ / ($mg{\cdot}L^{-1}$)	2	167	70.15	48.40	0.69
HCO_3- / ($mg{\cdot}L-1$)	46	565	243.56	105.08	0.43

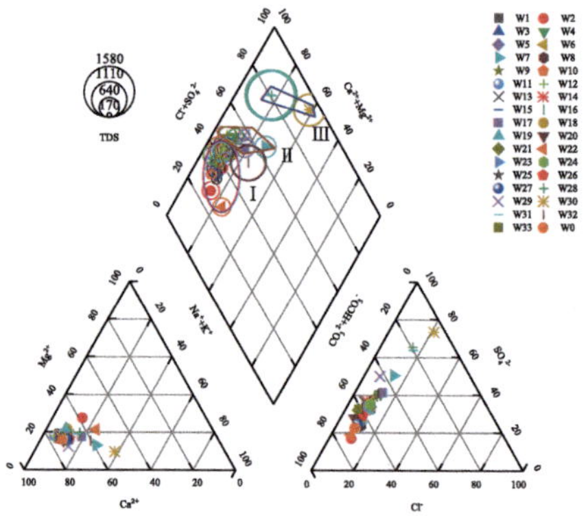

Fig. 2. Piper three-line diagram

both CAI-1 and CAI-2 are negative, it indicates that Ca^{2+} and Mg^{2+} in groundwater are exchanged with Na^+ and K^+ in the aquifer medium. Conversely, when they are positive,

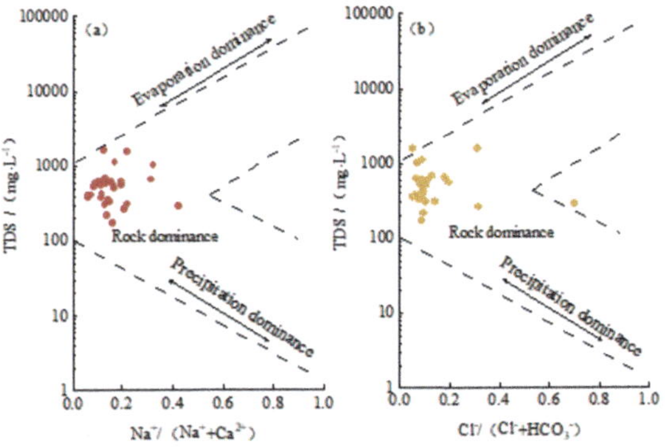

Fig. 3. Gibbs diagram

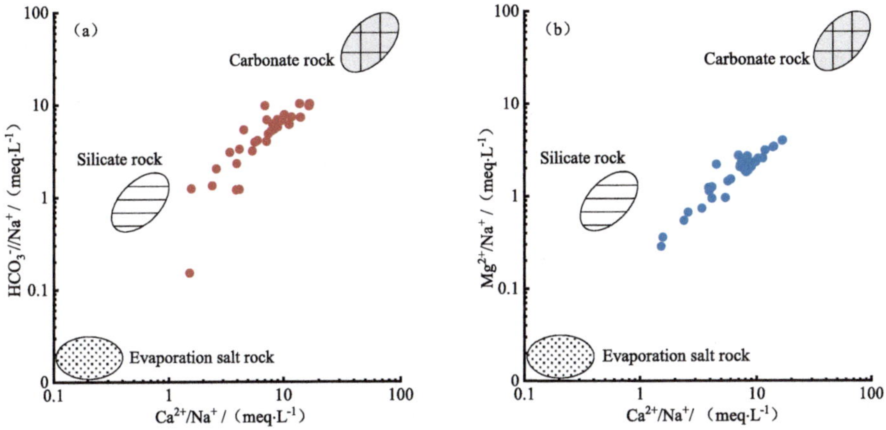

Fig. 4. Ionic end element diagram

Na^+ and K^+ in groundwater are exchanged with Ca^{2+} and Mg^{2+} in the aquifer medium. The stronger the alternating adsorption of cations, the greater the absolute values of CAI-1 and CAI-2. The CAI-1 and CAI-2 values of most of the sampling points of shallow groundwater in the study area were less than 0, showing a trend of moving away from the intersection of the 0 value line (Fig. 5), indicating strong displacement of Na^+ and K^+ by Ca^{2+} and Mg^{2+} in the aquifer medium.

3.3 PMF Quantitative Analysis

The qualitative analysis of ion sources indicates that the shallow groundwater chemical components in the study area originate from multiple sources, including water-rock interactions, construction waste landfills, and agricultural activities. The PMF model was used to quantitatively analyze the sources of chemical components in the shallow groundwater of the study area [8]· Ten inorganic components were selected, and the model was run 100 times under a random seed model. Three source factors were

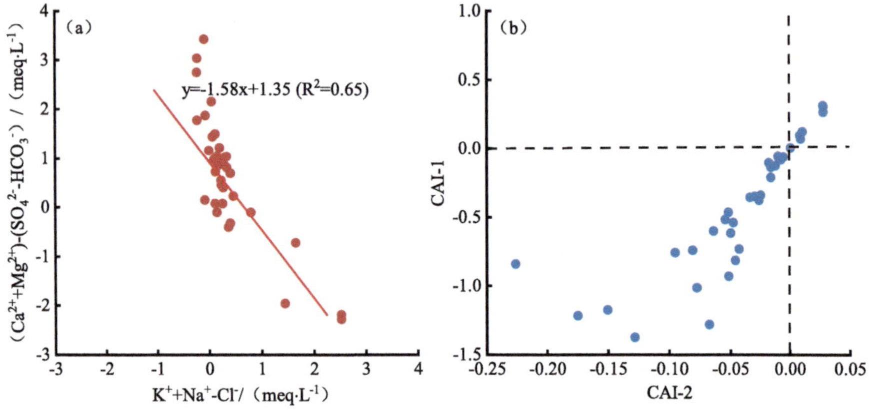

Fig. 5. Alternating cation adsorption and chlor-alkali index

monitored, with a Qrobusy/Qtur value of 0.97, which is close to 1, indicating a good correlation between the predicted and observed concentrations. The standardized residuals of all indicators were normally distributed between –3 and 3, indicating that the model simulation was satisfactory [9]. The contribution rates of various groundwater indicators are shown in Fig. 6. The characteristic components of source factor F1 are mainly HCO_3^- (43.5%) and Ca^{2+} (43.3%), followed by Mg^{2+} (31.4%). The main chemical type of groundwater in the district is HCO_3-Ca, and the HCO_3^-, Ca^{2+}, and Mg^{2+} in natural groundwater are primarily controlled by water-rock interactions. The groundwater in the typical district in the piedmont plain, adjacent to the Qinling Mountains to the east, which contain abundant carbonate minerals such as dolomite and calcite [10]. The aquifer medium gradually becomes finer from the piedmont inclined plain to the alluvial plain, resulting in poorer groundwater flow conditions, longer flow times, and gradual dissolution of carbonate rocks under water-rock interactions.Therefore, factor F1 is attributed to water-rock interactions, with a contribution rate of 24.5%.The characteristic components of source factor F2 are mainly NO_3^- (75.3%), followed by total hardness (TH) (43.4%) and total dissolved solids (TDS) (39.9%). In agricultural areas, the application of agricultural fertilizers (including inorganic fertilizers, organic fertilizers, and manure) and the decomposition of organic nitrogen in the soil can increase the concentration of nitrate (NO_3^-) in groundwater. Therefore, factor F2 is attributed to agricultural activities, with a contribution rate of 33.73%.The characteristic components of source factor F3 are mainly K^+ (86.5%) and Na^+ (70.0%), followed by SO_4^{2-} (54.9%). The spatial variability coefficients of Na^+ and K^+ in shallow groundwater in the study area are relatively high (Table 1), indicating potential influence from human activities. In addition to natural inputs from rock salt dissolution, K^+, Na^+, and $SO4^{2-}$ may also come from construction materials or waste such as cement, concrete, lime, discarded bricks, decorative stones, and gypsum. Therefore, factor F3 is likely attributed to construction waste landfill, with a contribution rate of 41.77%.

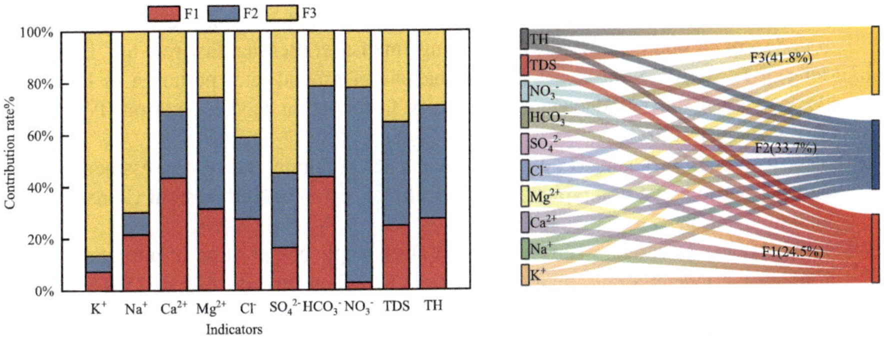

Fig. 6. Factor contributions of chemical species calculate

4 Conclusions

The pH of shallow groundwater between 7.37 and 8.98, and the TDS ranged from 172 to 1590 $mg{\cdot}L^{-1}$, indicating that the groundwater as a whole is neutral to weakly alkaline fresh water. The dominant cations were $Ca2^{+}$, and the most advantageous anions were found to be HCO_3^- and SO_42^-. The coefficients of variation for Na^+, Cl^-, K^+, $Mg2^+$, and NO_3^- were large due to human activities. The water chemical types showed zonation, with HCO_3-Ca type being the main type, followed by SO_4-Ca type water. The chemical composition of shallow groundwater in the study area is influenced by water-rock interactions, evaporation concentration, cation exchange adsorption, and human factors. The quantitative analysis results of PMF showed that the contribution rates of water-rock interactions, agricultural activities, and construction waste landfill to the chemical composition of shallow groundwater in the study area were 24.5%, 33.73%, and 41.77%, respectively. Construction waste landfill was identified as the main factor affecting the chemical composition of shallow groundwater in the study area.

References

1. Cao, Y.H., Wang, S.J., Wang, Y., et al.: Present situation and future development trend of comprehensive utilization of construction waste in China. China Build. Mater. **09**, 118–121 (2021)
2. López, A., Lobo, A.: Emissions of C&D refuse in landfills: a European case. Waste Manage. **34**(8), 1446–1454 (2014)
3. Wang, D., et al.: Chemical characteristics and controlling factors of shallow groundwater in the lower reaches of Changhua River Basin, Hainan Island. China. Water **15**, 3508 (2023)
4. Brown, S.G., Lee, T., Norris, G.A., et al.: Receptor modeling of near-roadway aerosol mass spectrometer data in Las Vegas, Nevada, with EPA PMF. Atmos. Chem. Phys. **12**(1), 309–325 (2012)
5. Chen, K., Wang, J., Yu, B., et al.: Critical evaluation of construction and demolition waste and associated environmental impacts: a scientometric analysis. J. Clean. Prod. **287**, 125071 (2021)
6. Liu, Y., Li, M.Q., Zhang, Y., et al.: Hydrogeochemical characteristics and origins factors of groundwater in in the Changbai Mountain Region via inverse hydrogeochemical modeling and unsupervised machine learning. Water **2024**, 16 (1853)

7. Hu, Y.X., Zhou, R.X.L., Song, W., et al.: Analysis on hydrochemical evolution of shallow groundwater east of Yong Ding River in Feng Tai district, Beijing. Environ. Sci. 1–18 (2023)
8. Meng, R.F., Meng, S.R., et al.: Sourer analysis of groundwater pollution in Hutuo river alluvial-pluvial fan based on positive matrix factorization model. Environ. Pollut. Cont. **43**(05), 586–591 (2021)
9. Chen, R., Teng, Y., Chen, H., et al.: Groundwater pollution and risk assessment based on source apportionment in a typical cold agricultural region in Northeastern China. Sci. Total Environ. 696 (2019)
10. Nikolenko, O., Jurado, A., Borges, A.V., et al.: Isotopic composition of nitrogen species in groundwater under agricultural areas: a review. Sci. Total Environ. **621**, 1415–1432 (2018)

Pollution Status and Cause Analysis of a Reservoir in Shenzhen

Chenjia Li[1,2], Haojie Lu[2,3], Yizhang Zhang[3](✉), Tianlong Deng[1], Yueling Liu[2], Yue Wang[2], Hongzhe Zhu[2], and Chen Chen[2]

[1] Tianjin University of Science and Technology, Tianjin 300457, China
[2] Research Institute for Environmental Innovation (Tianjin Binhai), Tianjin 300457, China
[3] Chinese Research Academy of Environmental Sciences, Beijing 100021, China
770286240@qq.com

Abstract. In this study, a reservoir in Shenzhen was taken as the object. In September, 2021, the water in the reservoir area, the surrounding surface water (pits, ditches, catchments, etc.) and the shallow groundwater in the muck landfill area where the groundwater flows upstream were sampled and analyzed. Combined with the hydrochemical characteristics analysis of surface-groundwater, isotope radon (^{222}Rn) tracer, Pearson correlation analysis and principal component analysis, the pollution status and causes of Tl, Fe, Mn and Iodide in the reservoir were deeply analyzed.The results show that Iodide, Mn and Tl are the main pollution indexes of the reservoir. The hydrochemical composition of reservoir water, surrounding surface water and groundwater in residue landfill area is similar and has certain connectivity. Both surface runoff and groundwater recharge can carry pollutants into the reservoir. Iodide comes from the same source as Mn, mainly from the recharge of groundwater to the reservoir; Tl mainly comes from the leakage of 1# pumping station and water collection tank.

Keywords: Reservoir · Cause analysis of pollution · Radon Isotope · Correlation Analysis

1 Introduction

Prevention and control of water pollution in lakes and reservoirs is one of the key tasks of ecoenvironmental protection in China. According to "China Water Resources Bulletin in 2018", among the 1,129 reservoirs investigated, there are 986 reservoirs with grade I-III water quality, 114 reservoirs with grade IV-V water quality, and 29 reservoirs with grade V water quality worse. The main pollutants are total phosphorus, permanganate index, BOD_5 biochemical oxygen demand and heavy metals [1]. It can be seen that the prevention and control of water pollution in lakes and reservoirs has a long way to go.

A reservoir in Shenzhen (Fig. 1) has a total storage capacity of 17.54 million m^3 and a maximum water depth of 38.42 m. It is a medium-sized reservoir and an important water storage reservoir for municipal water source projects. The detection data from 2014 to 2020 show that Tl, Fe, Mn and Iodide in the reservoir area exceed the standard

D. Han et al. (Eds.): ICEPG 2024, ESE, pp. 11–21, 2025.
https://doi.org/10.1007/978-3-031-86829-0_2

all the year round. As a reservoir within the built-up city, the surrounding environment is complex and human activities are frequent, which makes it difficult to accurately identify its pollution source and carry out targeted pollution prevention and control work. Therefore, it is urgent to analyze and study the pollution status and causes of reservoirs for key pollutants. Therefore, based on the sampling analysis of surface water, groundwater and sediment around the reservoir in September, 2021, combined with the hydrochemical composition analysis of surface and groundwater, isotope radon (^{222}Rn) tracing, Pearson correlation analysis and principal component analysis, this study deeply analyzed the characteristics and causes of reservoir pollution, in order to provide support for reservoir pollution prevention and ecological restoration.

2 Research Method

2.1 Study Area

The reservoir was built in 1977, with a total storage capacity of 1754 m^3. Its basin belongs to subtropical maritime climate, with abundant rainfall. The average rainfall in the reservoir area for many years is 1750 mm, with long sunshine time and mild climate. In the study area, the terrain is high in the northwest and low in the southeast, and there is a muck receiving yard in the northwest, and a river flows into the reservoir in the north. On October 3, 2008, the muck receiving yard began to fill the subway spoil, and it was closed on August 30, 2013. In April, 2014, the yard was closed and afforested, occupying an area of about 7.98 km^2, with an average receiving depth of 20.2 m below the current surface and a maximum receiving depth of 66.9 m below the current surface. The main landfill components are construction muck, construction waste and a small amount of domestic waste. The upstream river flows through the high-speed railway headquarters, landfill areas, residential areas and charging stations, and finally into the reservoir (Fig. 1). There are two types of groundwater in the study area: Quaternary loose rock pore water (plain fill) and granite weathering fissure water. The pore water of Quaternary loose rocks is affected by topography, and the runoff path is short. Generally speaking, it shows a trend of runoff from north to south and discharge from the periphery of the reservoir to the reservoir.

2.2 Sample Collection and Analysis

According to the key areas of concern in the study area, in September 2021, sampling site was selected to collect and detect samples in the muck receiving field, the pumping station in the receiving field, the overflow area of seepage and drainage, the motor vehicle maintenance station, the gully region and other areas with obvious pollution traces. At least 3 samples are collected at each point, and the average of the test results is taken for analysis. G1–15 are groundwater sampling site. W1–15 is the sampling site of surface water around the reservoir, and W2–1–6 is the sampling site of water sample in the reservoir area (Fig. 2). The water samples in the reservoir area are collected according to the upper layer (0–1 m), middle layer (5–8 m) and lower layer (9–15 m), filtered on site and transported back to the laboratory for processing and testing.

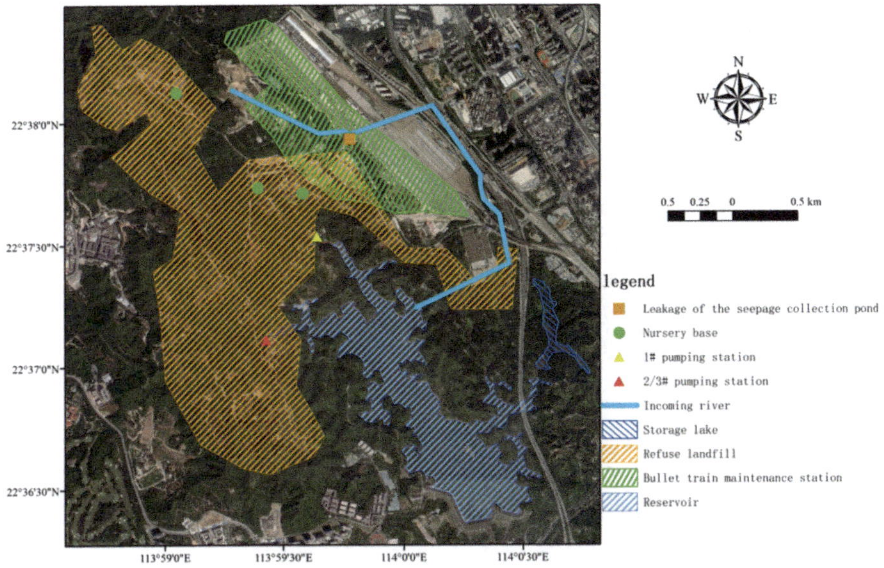

Fig. 1. Overview of the study area

The main test indexes are Tl, Fe, Mn and Iodide. Tl, Fe and Mn in water were determined according to Determination of 32 Elements in Water Quality by Inductively Coupled Plasma Emission Spectrometry (HJ 776–2015), and Iodide was determined according to Determination of Iodide in Water Quality by Ion Chromatography (HJ 778–2015). In order to assist the cause analysis of reservoir water pollution, Na^+, K^+, Mg^{2+}, Ca^{2+}, Cl^-, NO_3^-, SO_4^{2-} and HCO_3^- in water were also analyzed. Cl^-, NO_3^- and SO_4^{2-} are determined according to Determination of Inorganic Anions in Water Quality (F^-, Cl^-, NO_2^-, Br^-, NO_3^-, PO_4^{3-}, SO_3^{2-} and SO_4^{2-}) by Ion Chromatography (HJ 84–2016), and HCO_3^- is determined according to Underground water quality inspection Titration. Na^+, K^+, Mg^{2+}, Ca^{2+} refer to Determination of 32 Elements in Water Quality by Inductively Coupled Plasma Emission Spectrometry (HJ 776–2015).

2.3 Pearson Correlation Analysis

Pearson correlation coefficient can analyze the degree of correlation between two variables, which has been applied in many fields and is an effective statistical analysis method of correlation. In this study, Pearson correlation coefficient method is used to analyze the correlation of main ions in surface water to infer the relationship between ions in surface water, and its calculation formula is as follows:

$$r = \frac{N\sum x_i y_i - \sum x_i \sum y_i}{\sqrt{N\sum x_i^2 - \left(\sum x_i\right)^2}\sqrt{N\sum y_i^2 - \left(\sum y_i\right)^2}} \quad (1)$$

where: x and y are variables; R is the correlation coefficient with the range of [–1,1]. The closer the absolute value of R is to 1, the stronger the correlation between X and Y.

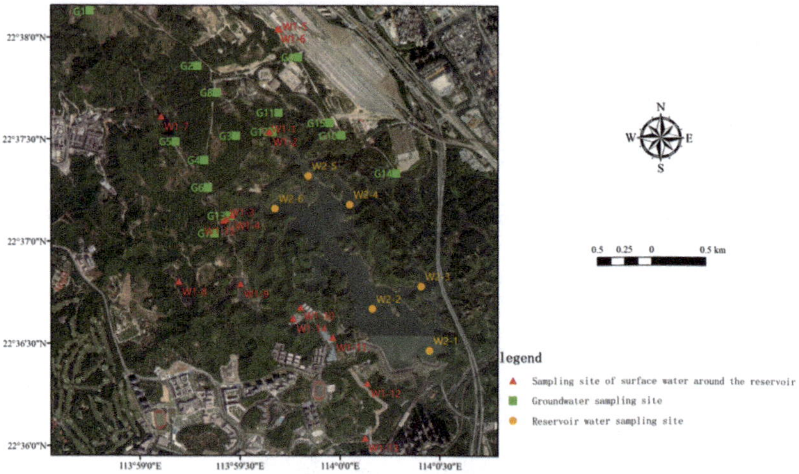

Fig. 2. Layout of sampling site in the study area

2.4 Principal Component Analysis(PCA)

Principal component analysis (PCA) is a multivariate statistical analysis method that uses projection method to reduce the dimension of data, also known as principal component analysis [2]. Usually, several new variables, so-called principal components, which are less than the original variables and can explain the variables in most data, are selected and used to explain the comprehensive indicators of data. The data were preliminarily processed by Excel, and the principal component analysis of pollutants was carried out by SPSS software, during which the data were standardized to zero mean and unit variance. Before evaluation, KMO test and Bartlett's sphericity test are usually needed to illustrate the feasibility of PCA. According to the number of principal components and the load factor, it can be revealed that there are several sources of pollutants. The calculation formula is as follows [3]:

$$(A_Z)_{jk} = \sum_{j=1}^{p} W_j \cdot Z_k \tag{2}$$

$$Z_k = \frac{c_k - \overline{c}}{\sigma} \tag{3}$$

where: Z_k is the standardized pollutant concentration at monitoring point k, mg/L; j is the principal component serial number obtained by PCA process; $(A_Z)_{jk}$ as the main component score; W_j is the factor coefficient of the j-th principal component; c_k is the pollutant concentration of monitoring point k; $\overline{c}$ is the arithmetic average of pollutant concentration; σ is the standard deviation of pollutant concentration.

3 Results and Discussion

3.1 Water Environment Quality in the Study Area

Figure 3 shows the concentration of pollutants in the upper, middle and lower layers of water at six sampling points in the reservoir. It can be seen from the figure that the main pollutants detected in the water in the reservoir area are Tl, Mn and Iodide, with the concentration of Iodide ranging from 0.131 to 0.174 mg/L, Mn ranging from 0.022 to 0.323 mg/L, and Tl ranging from 0.20×10^{-4} to 0.50×10^{-4} mg/L. As the level of the reservoir protection zone is Grade I, Class II standard in Surface Water Quality Standard (GB3838-2002) is adopted for evaluation. The evaluation results show that Iodide in the three layers of water at three sampling points in the southeast of the reservoir exceeded the standard, and in the three sampling points in the northwest (near the refuse landfill field), the middle and two layers of water exceeded the standard; The three sampling points of Mn in the southeast of the reservoir are mainly the lower water exceeding the standard, while the three sampling points in the northwest are not exceeding the standard; Tl did not exceed the standard. Four pollutants were not detected in the bottom water at three sampling points in the northwest.

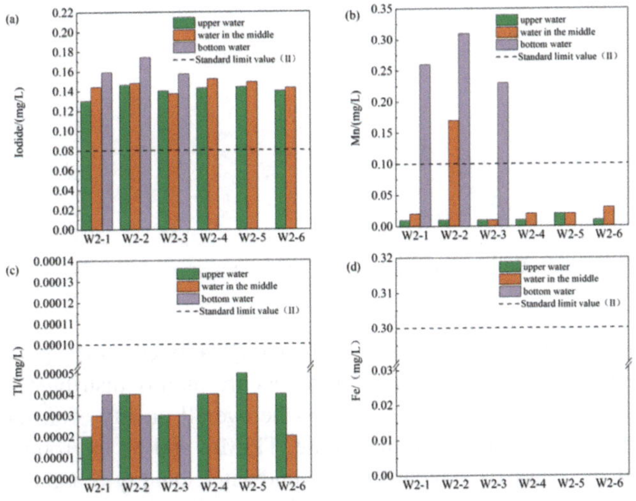

Fig. 3. Contents of pollutants in the reservoir

The surface water around the reservoir is mainly pits, ditches and catchments. Through sampling and testing, it is found that Iodide, Tl, Fe and Mn in the surface water around the reservoir exceed Class III standard of Surface Water Quality Standard (GB 3838-2002) in many places (Fig. 4). Among them, the content of Mn is the highest, the concentration range is 0.01–2.2 mg/L, the average value is 0.3934 mg/L, and the over-standard rate is 33.33%. The concentration of Iodide ranged from 0.04 to 1.17 mg/L, with an average of 0.284 mg/L, and the over-standard rate was 66.67%. The detected concentrations of Fe and Tl are relatively low, with the average value of

0.164 mg/L in the range of 0.03–0.5 mg/L for Fe and 0.00002–0.00033 mg/L for Tl, with the average value of 0.00012mg/L, and the over-standard rates of 20.00% and 53.33% respectively. Among them, W1–1 and W1–2 are in the catchment area of rainwater runoff and groundwater seepage in the valley downstream of the muck yard, and W1–5 and W1–6 are respectively located in the drainage area of the bullet train maintenance station and in the tributary of the storage (near the water discharge outlet of the 1# pumping station). The concentrations of Iodide, Mn and Tl in the four sampling points are generally high, and the water in the four areas will flow into the reservoir, which will lead to the deterioration of the water quality of the reservoir.

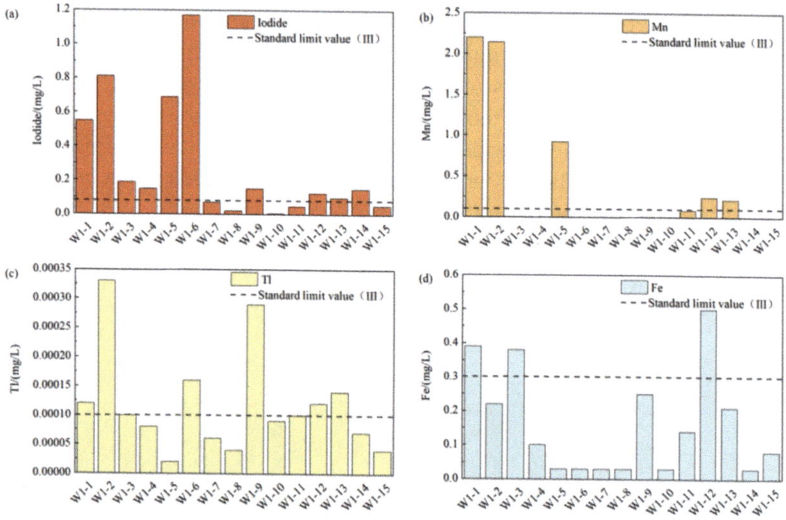

Fig. 4. Pollutant content of surface water around the reservoir

The groundwater around the reservoir is mainly sampled for the pore water of quaternary loose rocks, that is, the sampling points are mainly distributed in the residue receiving field. As can be seen from Fig. 5, the overall quality of groundwater in the residue receiving site is poor, and the Iodide, Tl, Mn, and Fe in groundwater at many points exceed the class III standard of Groundwater Quality Standard (GB/T14848-2017), with the exceeding rates being 73.33%, 66.67%, 60.00%, and 20.00%, respectively. Due to the difference of the main landfills in each area of the waste receiving site, the areas where the pollutants exceed the standard are located are different.

3.2 Water Chemical Composition of the Study Area

The water chemical composition of the study area is shown in Fig. 6. Comparing reservoir water, surrounding surface water and groundwater, the cation characteristics of the three are the same, but the anions are slightly different. The content of SO_4^{2-} in the surrounding surface water and groundwater is the highest, while the content of HCO_3^- in the reservoir water is the highest. The average value of Iodide, Tl and Mn in the

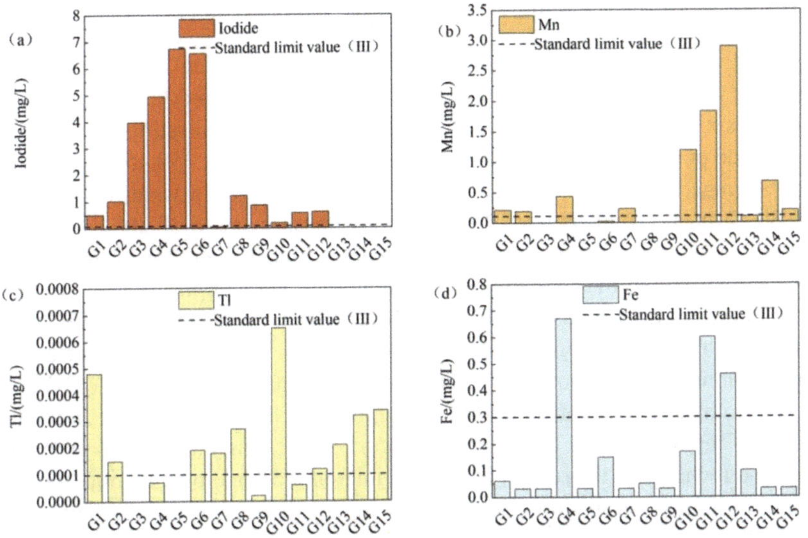

Fig. 5. Groundwater pollutant content

water under the slag receiving site was higher than that of the surface water around the reservoir and the reservoir water, indicating that the sewage pollution level of the slag receiving site was higher than that of the reservoir and the surface water around the reservoir. The coefficient of variation of Iodide and Mn in all water bodies is high. The coefficient of variation of Iodide and Mn in surface water and groundwater around the reservoir is 123.55%, 192.91%, 134.25% and 155.83%, respectively. The coefficient of variation of Mn in reservoir water is 141.52%, indicating that Iodide and Mn have strong spatial variability. It is caused by human activities.

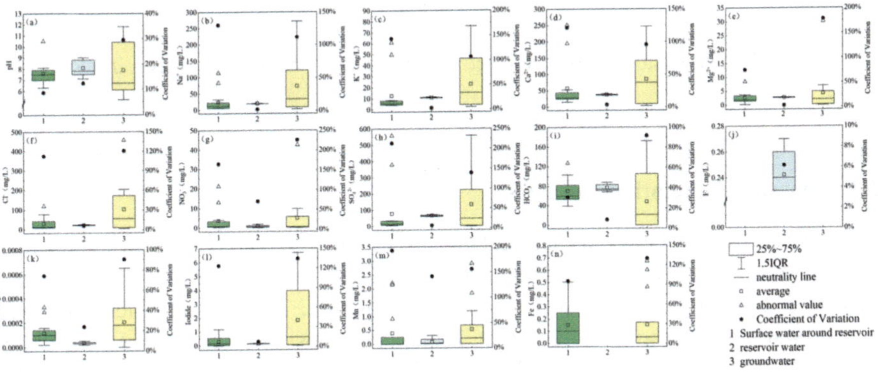

Fig. 6. Box diagram of each element content in reservoir water samples in the study area

3.3 Tracer Analysis of Isotope ^{222}Rn

The isotope ^{222}Rn was used to analyze whether groundwater leaked into the reservoir in the muck yard, and the radon concentration in the surface, middle and bottom water samples of six sampling points was tested (Fig. 7).The concentration of ^{222}Rn in the water at three points in the north of W2–4, W2–5 and W2–6 is 330.45–3416.9 Bq/m^3, with an average of 1621 Bq/m^3. The concentration of ^{222}Rn in water bodies of W2–1 and W2–2 in the south is 0–422.81 Bq/m^3, with an average of 208.8 Bq/m^3. The concentration of ^{222}Rn in the north is significantly higher than that in the south, and the spatial distribution of radon concentration in six water samples is high in the north and low in the south. Because the half-life of 222Rn is short, when the surface water contains a high concentration of ^{222}Rn (> 1000 bq/m^3), it shows that the surface water has an additional recharge source [4]. The concentrations of W3–20 and W3–21 ^{222}Rn are all more than 1000 Bq/m^3, which are significantly higher than those in the center and south of the reservoir (276.68–422.81 Bq/m^3), indicating that there is groundwater leakage and recharge to the reservoir near the inlet of 2/3# pumping station, the inlet of 1# pumping station and the junction where the inlet tributaries flow into the reservoir.

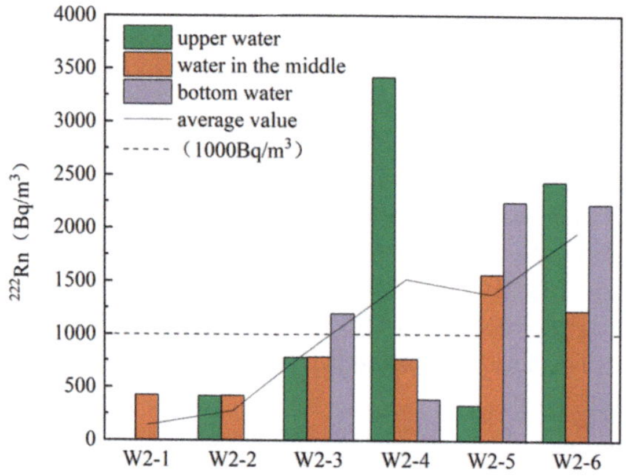

Fig. 7. Histogram of radon activity at different points in the reservoir

3.4 Pearson Correlation Analysis

When the correlation between chemical indexes in water is significant, it shows that there may be similar sources among chemical indexes [5]. In this study, SPSS25 software was used to analyze the Pearson correlation of Iodide, Tl, Mn and other substances in reservoir water. Results As shown in Fig. 8, Tl is positively correlated with iodide, negatively correlated with Mn, and negatively correlated with conventional ions such as Na^+, K^+, Mg^+, $SO_4{}^{2-}$, $NO_3{}^-$, Cl^-. Mn is positively correlated with iodide and negatively correlated with Tl. This shows that iodide has the same source as Mn, and has no significant correlation with Tl and some conventional ions, but there are some differences in sources.

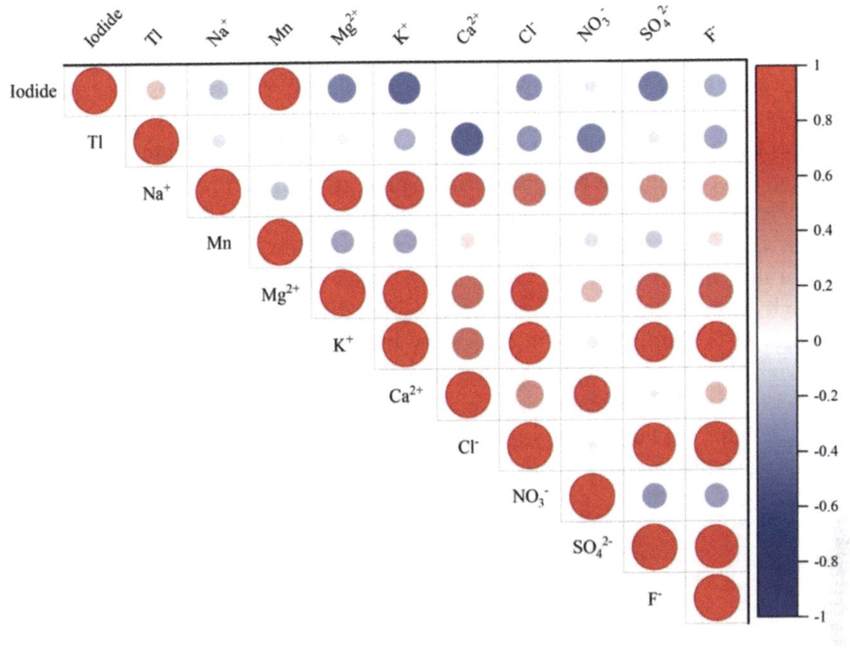

Fig. 8. Thermal map of correlation analysis of various elements in reservoir water samples

3.5 Principal Component Analysis(PCA)

In order to further analyze the sources of Mn, Tl and Iodide in the reservoir area, the principal component analysis of three pollutants was carried out. After inspection, the KMO value of the three pollutants is 0.516, and the corresponding probability of Bartlett spherical test is 0.000, which meets the statistical requirements, and the research data is suitable for principal component analysis. Through principal component analysis, four components with eigenvalues greater than 1 are obtained (Table 1), and the cumulative variance contribution rate of the four principal components is 90.807%, which shows that the extracted principal components cover enough information of the original variables [6].

PC1 shows that Na^+, K^+, Mg^{2+}, Ca^{2+}, Cl^-, $SO_4{}^{2-}$ and F^- have higher positive loads. Correlation analysis shows that Na +, K + and Mg2 + are significantly correlated, while Cl-, SO42- 2-and F- are significantly correlated, and the coefficient of variation of these ions is small, indicating that these ions are mainly from natural sources, which may be related to rock weathering, salt dissolution and hydrodynamic conditions in waters [7]. PC2 shows that Ca^{2+} and $NO_3{}^-$ have higher normal loads, which indicates that Ca^{2+} has other inputs besides natural sources. The coefficient of variation of $NO_3{}^-$ is 70.85%, with moderate variability, indicating that $NO_3{}^-$ in reservoir water is the input caused by human activities. The test data show that the concentrations of Ca^{2+} and NO_3- in the sampling point W2–4 of the inflow tributary and W1–5 and W1–6 near the seepage collection pond around the reservoir (north side) are high, indicating that Ca^{2+} and $NO_3{}^-$ in the reservoir are related to the leakage of the seepage collection pond. PC3 shows that iodide and Mn have high positive loads, and the coefficient of variation and Pearson correlation analysis show that the two substances are mainly input by human activities and come from the same source. The results of ^{222}Rn tracing show that there is groundwater leakage and recharge to the reservoir near the inlet of 2/3# pumping station, the inlet of 1# pumping station and the junction where the inlet tributaries flow into the reservoir. The concentration of Iodide and Mn in groundwater is much higher than that in the reservoir, and the high-value area of iodide and Mn in the northern part of the reservoir is mainly concentrated in the lower water body, so the iodide and Mn in the reservoir may mainly come from the recharge input of groundwater. PC4 shows that Tl has a high positive load, and the coefficient of variation and Pearson correlation analysis show that Tl is mainly the input caused by human activities, and it is different from iodide and Mn sources. The high-value area of Tl in the reservoir is mainly located in the upper layer of W2–5, and the leakage of 1# pumping station near this area is serious, indicating that Tl in the reservoir mainly comes from the high-concentration Tl leakage of pumping station.

Table 1. Factor analysis of elements in reservoir

element	PC1	PC2	PC3	PC4
pH	0.391	–0.156	–0.71	0.272
Iodide	–0.553	0.27	0.61	0.454
Tl	–0.278	–0.479	–0.267	0.769
Mn	–0.303	0.218	0.845	0.285
Na^+	0.735	0.42	–0.123	0.382
K^+	0.97	–0.033	0.034	0.059
Mg^{2+}	0.887	0.102	–0.066	0.266
Ca^{2+}	0.507	0.768	0.071	0.037
Cl^-	0.888	–0.141	0.377	–0.055
NO_3^-	0.17	0.875	–0.282	–0.044
SO_4^{2-}	0.783	–0.503	0.253	–0.032
F^-	0.784	–0.311	0.5	–0.086
eigenvalue	5.2	2.287	2.224	1.187
Contribution rate/%	43.33	19.055	18.53	9.893
Cumulative contribution rate/%	43.33	62.384	80.914	90.807

4 Conclusion

Iodide, Mn and Tl are the main pollution indexes of the reservoir, and iodide and Mn in some points exceed the Class II standard in Surface Water Quality Standard (GB3838-2002). The points of iodide, Tl, Mn and Fe in surface water bodies such as pits, ditches and water catchments around the reservoir that exceed Class III standard of Surface Water Quality Standard (GB 3838-2002) are 66.67%, 20.00%, 33.33% and 53.33% respectively. The percentage of points in groundwater exceeding Class III Standard of Groundwater Quality Standard (GB/T 14848–2017) is 73.33%, 66.67%, 60.00% and 20.00% respectively, and the overall water environment in the study area is poor. The hydrochemical composition of water in reservoir area, surrounding surface water and groundwater in residue landfill area is similar, and the spatial variability of iodide and Mn in surrounding surface water and groundwater in residue landfill area is extremely strong, which is significantly influenced by human activities such as residue landfill, construction waste landfill and the construction of motor train maintenance station. Iodide has the same source as Mn, but it is different from Tl and some conventional ions (such as Na^+, K^+, Mg^{2+}, Cl^-, $SO_4{}^{2-}$, F^-, etc.). Iodide and Mn mainly come from the leakage of groundwater to the reservoir; Tl mainly comes from the leakage of 1# pumping station and water collection tank.

References

1. Shen, L.Y., Zhu, Y.P., Chen, L.J.: Present situation and countermeasures of pollution treatment of reservoirs in China. J. Anhui Agric. Sci. **48**(11), 88–92 (2020)
2. Li, X., Liu, L., Wang, Y., et al.: Heavy metal contamination of urban soil in an old industrial city (Shenyang) in Northeast China. Geoderma **192**, 50–58 (2013)
3. Hou, X.K., Zhang, K., ping zhou, D., et al.: Pollution source apportionment of Tuohe River based on absolute principal component score-multiple linear regression. Res. Environ. Sci. **34**(10), 2350–2357 (2021)
4. Zhongbo, Y., minjuan, L., Yunchen, L., et al.: Study on hydraulic exchange of river water and groundwater based on radon isotope. J. Hohai Univ. (Nat. Sci.) **48**(01), 8–13 (2020)
5. qi, S., guangjun, Z., Weiran, L., et al.: Quality and suitability of surface irrigation water in Panjin City. J. Irrig. Drainage **42**(12), 165–171 (2023)
6. Lin, F., Yang, N., Qiang, Z., et al.: Air quality assessment based on five kinds of air quality indexes in Tianjin. J. Environ. Health **35**(09), 803–808 (2018)
7. Xu Mingming, Y., Chenglong, J.J., et al.: Source identification of chemical compounds in groundwater surrounding the landfill sites dumping construction and demolition waste. Geol. Bull. China **41**(12), 2125–2137 (2022)

Applicability Analysis of Base Flow Segmentation Method in the Lena River Basin

Yin Tang[1], Shaoyan Lv[2(✉)], Haiming Li[1], and Han Xiao[3]

[1] Institute of Geographic Sciences and Natural Resources Research, Chinese Academy of Sciences, Key Laboratory of Water Cycle and Related Land Surface Processes, Beijing 100101, China

[2] College of Marine and Environmental Sciences, Laboratory of Coastal Groundwater Utilization & Protection, Tianjin University of Science & Technology, Tianjin 300457, China
shaoyan3333@163.com

[3] State Key Laboratory of Groundwater Pollution Simulation and Control for Environmental Protection, Chinese Research Academy of Environmental Sciences, Beijing 100012, China

Abstract. Basic flow is the main source of runoff during the dry season in the Lena River Basin, and plays an important role in water resource planning and ecological environment protection. A stable basic flow segmentation method is particularly crucial for basic flow research. This study compared 9 basic flow segmentation methods, including digital filtering, sliding minimum method, and time step method, to select the optimal method suitable for the Lena River Basin, in order to determine the basic flow. The results showed that the Boughton Chapman filtering method had the best application effect in the Lena River Basin.

Keywords: Basic flow segmentation · Lena River · Basic flow index

1 Introduction

The total runoff can be divided into two parts: surface runoff and base flow. Surface runoff, including surface runoff, is an important factor in floods and soil erosion. The base flow comes from groundwater and other shallow groundwater, which can delay flowing into rivers [1]. As an important component of runoff, base flow has significant utilization value, especially in cold regions where groundwater is the main source of runoff [2]. However, a unified solution has not yet been formed when solving the problem of base flow segmentation, and there are significant differences in the results of base flow segmentation obtained by different methods.

The Lena River Basin in the Arctic is the second largest river flowing into the Arctic Ocean, and it is a key area of research on climate change and environmental response in the Arctic Circle [3–5]. This study will use three types of nine basic flow segmentation methods for applicability analysis, which will help to select the appropriate basic flow segmentation method for the Lena River basin and is of great significance for in-depth management of water resources in the basin.

D. Han et al. (Eds.): ICEPG 2024, ESE, pp. 22–28, 2025.
https://doi.org/10.1007/978-3-031-86829-0_3

2 Research Area and Data Sources

This study takes the Lena River Basin as the research area (Fig. 1), with a total length of about 4400 km. It originates from the Baikal Mountains in central and southern Siberia, flows northward into the Laptev Sea, and enters the Arctic Ocean. The basin area is 2.46 million km^2.

This study obtained the measured daily runoff data from the Stolb hydrological station in the Lena River Basin over a period of 40 years (1975–2014), which was systematically released by the Russian hydrological and meteorological department, as the basic data for the study.

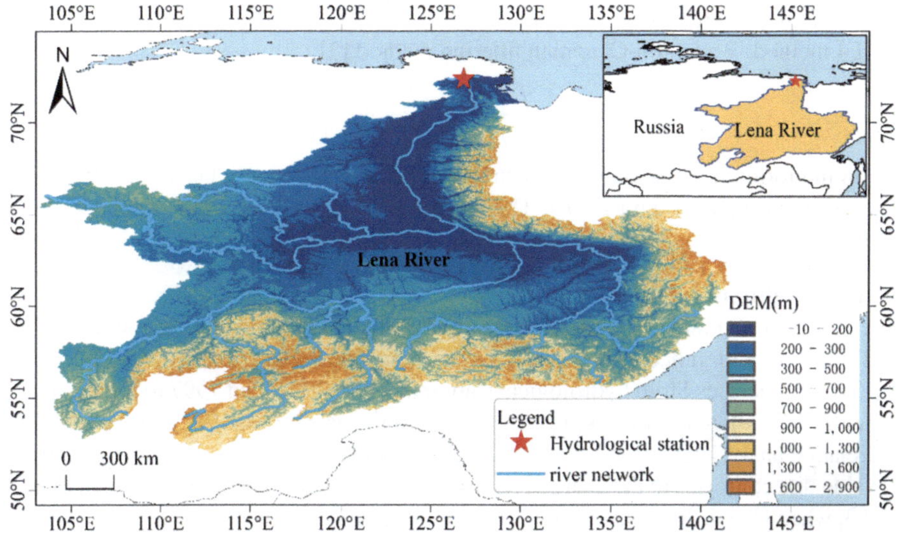

Fig. 1. Overview of the study area

3 Research Method

3.1 Basic Flow Segmentation Method

Digital Filtering Method

F1 method: Lyne Hollick filtering method [2, 6].

$$Q_{dt} = f_1 Q_{d(t-1)} + \frac{(1 + f_1)}{2}\left[Q_t - Q_{(t-1)}\right] \tag{1}$$

$$Q_{bt} = Q_t - Q_{dt} \tag{2}$$

In the formula, Q_{dt} and $Q_{d(t-1)}$ represent the surface runoff at time t and time t − 1, respectively; Q_t and $Q_{(t-1)}$ represent the runoff at time t and time t − 1, respectively; Q_{bt} is the base flow rate at time t; f_1 is the filtering parameter, usually taken as 0 925.

F2 method: Chapman filtering method [2]

$$Q_{dt} = \frac{3f_1 - 1}{3 - f_1} Q_{d(t-1)} + \frac{2}{3 - f_1} \left[Q_t - f_1 Q_{(t-1)} \right] \tag{3}$$

$$Q_{bt} = Q_t - Q_{dt} \tag{4}$$

The meaning of the parameters in the formula is the same as above.

F3 method: Chapman-Maxwell filtering method [2]

$$Q_{bt} = \frac{f_1}{2 - f_1} Q_{b(t-1)} + \frac{1 - f_1}{2 - f_1} Q_t \tag{5}$$

In the formula, f_1 is the filtering parameter, usually taken as 0 95; The meanings of other parameters are the same as above.

F4 method: Boughton-Chapman filtering method [2]

$$Q_{bt} = \frac{f_1}{1 + f_2} Q_{b(t-1)} + \frac{f_2}{1 + f_2} Q_t \tag{6}$$

In the formula, f_1. is the water withdrawal parameter, usually taken as 0.95; f_2 is a fixed value, usually taken as 0.15; The meanings of other parameters are the same as above.

Sliding Minimum Method

The BFI method mainly includes the standard BFI (B1) method and the improved BFI (B2) method, with inflection point test factors generally set at 0.9 and 0.97915.

This article uses Stolb hydrological station in 1996, 2004, and 1997 to represent dry years (P = 75%), normal years (P = 50%), and high years (P $\doteq$ 25%), respectively. The correlation analysis between N value and base flow index (BFI) is carried out to find the N value that conforms to the characteristics of the Lena River basin. As shown in Figs. 2 and 3, when N $\geq$ 7, In the B1 and B2 methods, there is not much change in the slope of the base flow index at each level year, and it can be considered that the base flow index tends to stabilize. Therefore, N = 7 is selected, that is, 7 days is taken as a time period for base flow segmentation.

Time Step Method

The time step method, also known as the HYSEP method, mainly includes three calculation methods: local minimum method (H1), fixed step method (H2), and sliding step method (H3) [7].

Fixed interval method [2]: Within the selected time interval, use the minimum flow rate within that time interval as the base flow for any day during that time period. Sliding interval method [2]: Take the minimum flow rate before and after a certain day (2N-1)/2d as the base flow for that day, and calculate the base flow for each day accordingly. Local minimum method [2]: Select the minimum flow rate before and after the center point within a time step (2N-1)/2d as the base flow value for the center point within adjacent time steps, and then use linear interpolation to obtain the base flow for the time period outside the center point of the step. All three methods use the endpoint of this calculation as the starting point for the next time, and repeat the above process to calculate the base flow segmentation process.

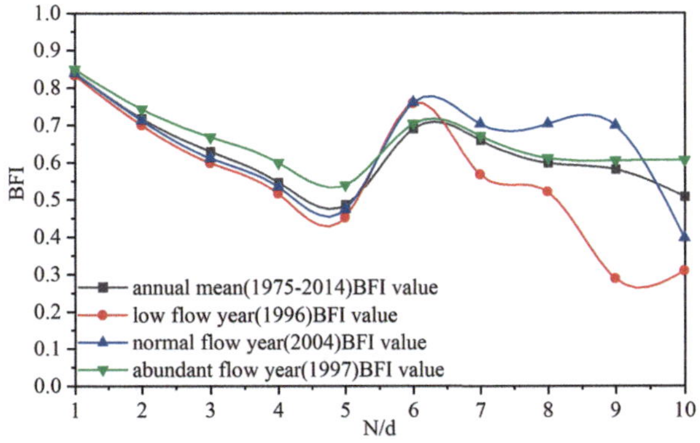

Fig. 2. The relationship between B1 method base flow index and N value

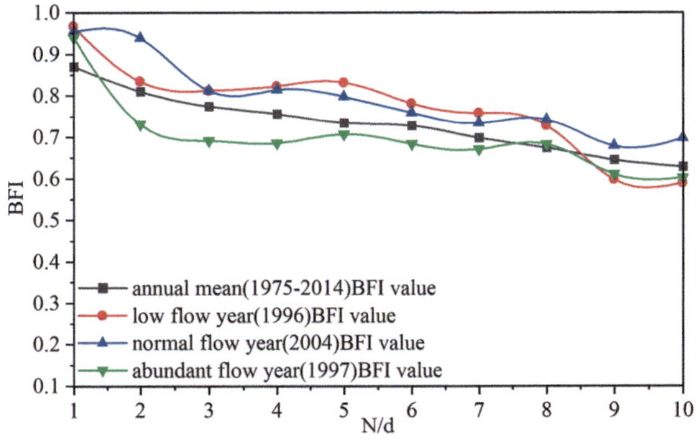

Fig. 3. The relationship between B2 method base flow index and N value

3.2 Evaluation Indicators

The product of low water index and annual total flow is considered as the observed value of annual base flow [7], and the specific formula is as follows:

$$Q_{oi} = \frac{Q_{90}}{Q_{50}} Q_i \tag{7}$$

In the formula, Q_{oi} is the observed annual base flow rate; Q_{90} and Q_{50} represent daily runoff greater than or equal to 90% and 50%, respectively, Q_{90}/Q_{50} is the annual low traffic index; Q_i is the total annual runoff.

Comparing the observed fundamental flow with the estimated fundamental flow separation results, this study used two evaluation metrics, Nash Sutcliffe efficiency coefficient (NSE) and mean relative error (MRE), to evaluate the segmentation results of nine fundamental flows, The larger the NSE, The smaller the MRE, the better the segmentation result of the base flow.

4 Result

4.1 Comparison of Basic Flow Segmentation Results

Selecting the year of normal flow (2004) as the representative year, the basic flow segmentation results of 9 methods are shown in Fig. 4. There are significant differences in the baseline process lines obtained by the 9 baseline segmentation methods. After being regulated by the aeration zone, the process change of the base flow should show a relatively slow characteristic compared to the runoff process, and may exhibit a certain lag phenomenon in the water withdrawal process. The basic flow process lines segmented by the sliding minimum B1 and B2 methods, as well as the time step H1 and H3 methods, show significant inflection points, which is inconsistent with the lag effect that the underlying surface of the watershed should have during the rainfall convergence process. For the baseline obtained by the time step H2 method, there is a significant fluctuation in the flow rate. When the runoff reaches its peak, the baseline flow rate also reaches its peak and exhibits synchronous fluctuations with the runoff process. This situation is not consistent with the actual water withdrawal process. In contrast, the baseline obtained by using digital filtering method is smoother, without any sharp increase or decrease, and basically conforms to the watershed convergence law.

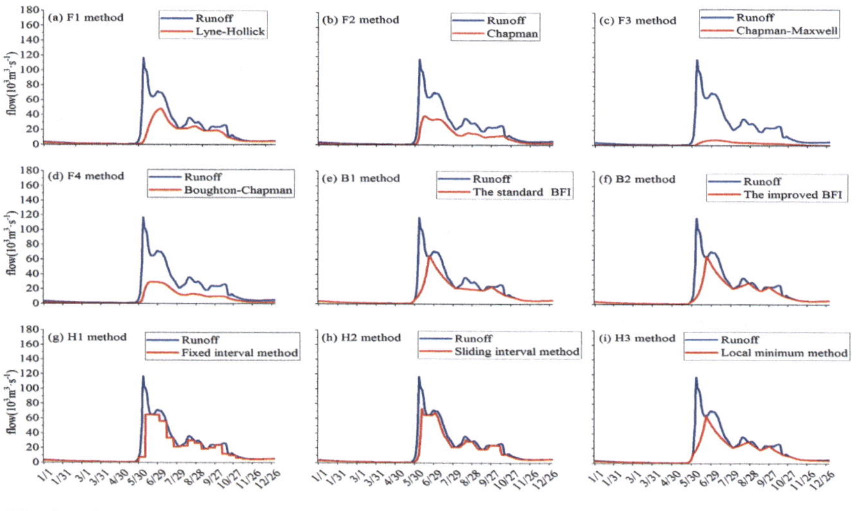

Fig. 4. The comparison of baseflow process in 2004 by the 9 baseflow separation methods

4.2 Comparative Analysis of Base Flow Index

The violin plot of the annual base flow index in the Lena River Basin is shown in Fig. 5, which shows that, F2 method F3 method The distribution of annual BFI data points obtained by the F4 method is relatively concentrated, among which the BFI obtained by the F3 method is much lower than the other two methods. F1 method The BFI calculated by the BFI method and the time step method are relatively large and have a discrete distribution, among which the B1 method B2 method The H3 method distribution is the most discrete.

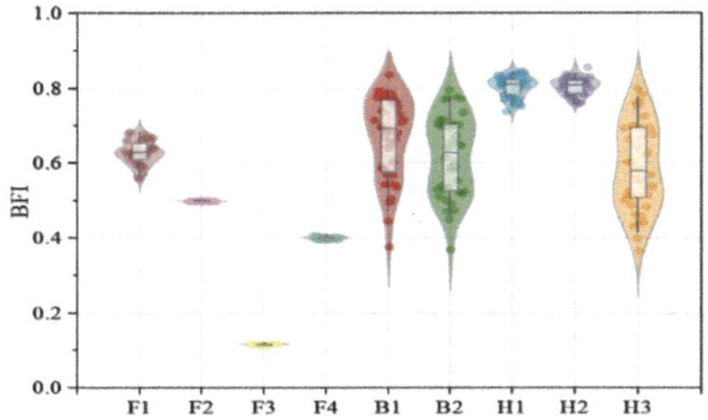

Fig. 5. The comparison of the interannual baseflow indices by the 9 baseflow separation methods

4.3 Evaluation of Basic Flow Segmentation Methods

As shown in Fig. 6, the determination coefficients R2 of F2, F3, and F4 in the digital filtering method reach 0.991 ~ 0.999, which is higher than the sliding minimum method (0.635 ~ 0.645) and the time step method (0.431 ~ 0.964), indicating that the digital filtering method is superior to the sliding minimum method and the time step method, and is more consistent with the basic flow process calculated by the low water index.

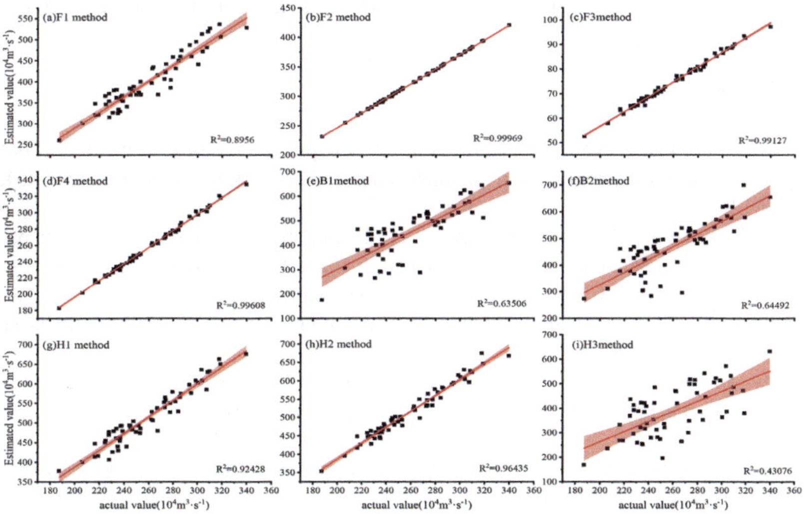

Fig. 6. Deterministic Coefficients of 9 base flow segmentation methods

From Table 1, it can be seen that except for the F4 method, the MRE of all other methods is greater than 30%, and the NSE is negative, indicating that the reliability of the annual base flow calculated by these methods is relatively low. In the digital filtering method, the evaluation result of F4 is relatively ideal, with an NSE of 0.991 and an MRE of 0.8%. Based on the comparative analysis of the base flow index and flow hydrograph, F4 can better reflect the rise and fall processes and hysteresis effects of the base flow. Therefore, it is selected as the optimal method for base flow segmentation in the Lena River Basin.

Table 1. Reliability verification results of 9 baseflow separation methods

	F1	F2	F3	F4	B1	B2	H1	H2	H3
NSE	-19.64	−2.72	-32.28	0.99	-41.34	-46.50	-63.39	-64.17	-21.47
MRE	0.55	0.24	0.71	0.01	0.75	0.81	0.98	0.99	0.49

5 Conclusions

The Boughton Chapman filtering method (F4 method) has the best application effect in the Lena River Basin among digital filtering methods. The average annual BFI is 0.40, with small standard deviation and coefficient of variation (0.003 and 0.008, respectively), making the separation results more reliable. Based on the estimated value calculated by the dry water index, the NSE value of the F4 method is closest to 1, which is 0.991, the MRE value is the smallest, and the R2 is 0.996. Therefore, using the F4 method for the division of base flow can better reflect the process of runoff retreat and lag effect.

References

1. Koskelo, A.I., Fisher, T.R., Utz, R.M., et al.: A new precipitation-based method of baseflow separation and event identification for small watersheds (<50 km^2). J. Hydrol. **450–451**, 267–78 (2012)
2. Wen, H., Zhanling, L.I., Zongxue, X.U., et al.: Suitability analysis of different baseflow separation methods in cold and arid watershed. J. China Hydrol. **38**(03), 21–28 (2018)
3. Gautier, E., Dépret, T., Cavero, J., et al.: Fifty-year dynamics of the Lena River islands (Russia): spatio-temporal pattern of large periglacial anabranching river and influence of climate change. Sci. Total Environ. **783**, 147020 (2021)
4. Ogneva, O., Mollenhauer, G., Juhls, B., et al.: Particulate organic matter in the Lena River and its delta: from the permafrost catchment to the Arctic Ocean. Biogeosciences **20**(7), 1423–41 (2023)
5. Sanders, T., Fiencke, C., Fuchs, M., et al.: Seasonal nitrogen fluxes of the Lena River Delta. Ambio **51**(2), 423–38 (2022)
6. Eckhardt, K.: A comparison of baseflow indices, which were calculated with seven different baseflow separation methods. J. Hydrol. **352**(1–2), 168–173 (2008)
7. Fang, L.I., Shaobo, J.I.N., Songbing, Z.O.U., et al.: Analysis of the applicability of base-flow separation methods in Baihe River basin, the source regions of Yellow River. J. Glaciol. Geocryol. **42**(2), 587–597 (2020)

Effect of Cascade Dam Construction on Distribution of Heavy Metals in River Sediment

Shuai Yan[1,2,4], Huijuan Yu[2,4], Tianlong Deng[1,4], Yizhang Zhang[2,3,4](✉), Changyuan Tang[2,4], Fang Li[3,4], Yanli Shi[2,4], and Zhenzhen Yu[2,4]

[1] Tianjin University of Science & Technology, Tianjin 300457, China
[2] Research Institute for Environmental Innovation (Binhai, Tianjin), Tianjin 300457, China
18048776801@163.com
[3] Chinese Research Academy of Environmental Sciences, Beijing 100012, China
[4] Tianjin Binhai New Area Eco-Environment Monitoring Center, Tianjin 300457, China

Abstract. The construction of dams alters the hydrodynamics of rivers, consequently leading to a significant influence on the deposition of pollutants within the rivers. To investigate the impact of cascade dam construction on the distribution of heavy metals in river sediments, samples were gathered from the Liujiang River. The analysis included the content of 10 heavy metals (Cd, Cr, Zn, Cu, Hg, Pb, As, Ni, Sb, and Al), as well as pH, particle size, and organic matter content of the sediments. The results showed an overall increasing and then decreasing trend of all heavy metals in the sediments of the Liujiang River from upstream to downstream. The highest concentrations of eight different heavy metals were found downstream of the three terrace dams. The levels of organic matter and ratio of clay particles were closely mirrored the spatial pattern of heavy metal distribution, exhibiting a strong correlation. Studies indicate that sediment heavy metal pollution is mainly affected by runoff contributions. The hydrodynamics are altered by the installation of cascade dams, resulting in the capture of sand particles by the dams and the subsequent downstream accumulation of heavy metals, together with clay particles.

Keywords: River · Sediment · Cascade dam · Heavy metals · Distributional features

1 Introduction

Cascade dam is a widely used method in various countries for managing water resources. It alters the river's flow dynamics, leading to reduced velocity upstream of the structure, gradual sediment deposition, and notable shifts in sediment flux downstream [1, 2]. The primary transporters of pollutants in rivers are sediment particles, thus the construction of dams could potentially alter the dispersion of contaminants within river sediments. With the rapid pace of urbanization and development, pollutant discharge rates are increasing, leading to a more prominent awareness of water environmental pollution

D. Han et al. (Eds.): ICEPG 2024, ESE, pp. 29–37, 2025.
https://doi.org/10.1007/978-3-031-86829-0_4

concerns. The presence of heavy metals is particularly alarming due to their potent toxicity, resistance to degradation, and propensity for bioaccumulation [3]. Heavy metals enter the food chain through the feeding activities of benthic organisms and zooplankton in river water ecosystems [4], posing a potential threat to river ecology and human health, thus becoming one of the key research issues in the water environment [5, 6].

Presently, researchers both domestically and internationally have carried out numerous investigations on the variables impacting the dispersion of toxic elements in stream deposits. Fang [7] discovered that the spatial patterns of Hg, Cd, Zn, Ni, Cr, Cu, Mn, and Pb in the sediment of the Ruxi River, devoid of dam interference, indicated that the levels of Hg, Cd, Zn, Ni, Cr, Cu, Mn, and Pb in the river segment influenced by human actions were greater compared to the standing water region, with the pristine river segment exhibiting the least concentration. In her research, Joanna Jaskula [8] analyzed the spatial distribution patterns of heavy metals in sediments found in the Oder and Vistula rivers in Poland. This findings revealed that reservoirs and lakes with specific hydrodynamic characteristics facilitate the accumulation of heavy metals in the reservoir or lake vicinity, leading to lower concentrations of heavy metals beneath the dam. Zhao [9] observed that the water and sediment management strategies implemented in the Xiaolangdi Water Conservancy Project on the Yellow River have caused a re-distribution of heavy metals within sediments along the downstream riverbanks. Bing [10] found that the areas with high heavy metal content in the sediments of the Three Gorges Reservoir in 2014 were distributed in the downstream areas. While in 2016 they were distributed in the middle and upper reaches, indicating that the regulation of periodic and anti-seasonal runoff of the reservoir affected the distribution of heavy metals in the river. The spatial pattern of heavy metal distribution in sediments within various aquatic systems is influenced by a variety of factors, including natural elements, human actions, dam building, and water management practices, leading to distinctive distribution patterns. The presence of dams significantly alters the deposition of heavy metals in riverbed sediments, while water management strategies can modify metal distribution both upstream and downstream of these structures. Therefore, the effects of dam construction on heavy metal distribution in rivers require further investigation. Specifically, the construction of cascading dams complicates the hydrodynamic conditions of the river, yet there is a scarcity of research on the impact of cascading dam construction on heavy metal distribution in river sediments, leaving their influence unclear.

This research focuses on Liujiang to exam the effects of the cascade dams on heavy metal deposition distribution in rivers. The analysis of Liujiang sediment includes studying particle size, heavy metal, and organic matter content to examine the spatial distribution patterns of heavy metals. By considering the correlation between sediment heavy metals and particle size and organic matter content, the study evaluates how cascade dam construction influences heavy metal distribution in sediment.

2 Material and Method

2.1 Study Area

The Liujiang River, situated in the middle portion of the Pearl River Basin, serves as the second largest tributary within the Xijiang River system. This particular investigation focuses on the Liujiang River along with its five primary tributaries, namely Xunjiang, Langxi, Beijiang, Longjiang, and Xinxu River. Within the study area, there are a total of five cascade hydropower stations situated along the Liujiang River, namely Maishi, Rong'an, Guding, Dapu, and Honghua (Fig. 1). To ensure adequate coverage, a total of 13 sampling points were strategically placed. In May and September of 2022, river sediment samples ranging from 0–10 cm were obtained utilizing a Van Veen mud miner. These samples were then carefully preserved in polyethylene self-sealing bags, and subsequently transported back to the laboratory under refrigerated conditions. The samples underwent freeze-drying and grinding processes before further analysis.

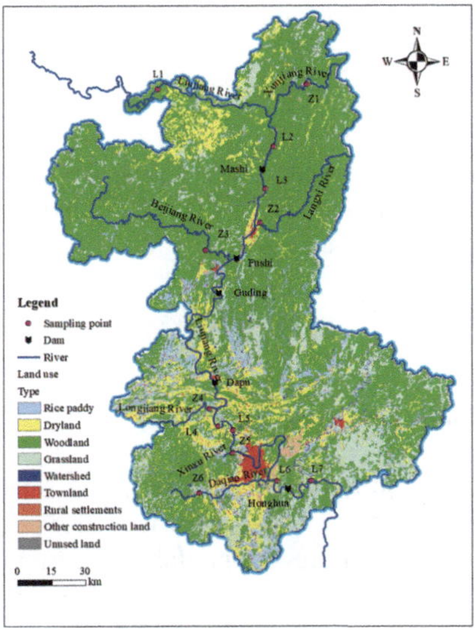

Fig. 1. Distribution of sampling sites.

2.2 Sample Analysis

Determination of Sediment Physicochemical Index.
According to sediment: deionized water = 1:2.5, 10 g ± 0.0005 g of sediment samples passed through a 10-mesh nylon sieve were taken in 50 mL polyethylene centrifuge tubes. Deionized water (25 mL) was added to the samples, which were then shaken at room temperature for 1 h, followed by a 30-min incubation period. The pH of the supernatant was measured using a Hach Parameter Meter (HQ30D).

Assessment of sediment particle size follows the guidelines outlined in Soil-Determination of Particle Size Distribution, employing both the pipette and hydrometer methods as specified in HJ 1068–2019. The determination of oxidation-reduction potential (ORP) follows the procedures outlined in the Marine Monitoring Specification, specifically Part 5: Sediment Analysis.

Approximately 1.0000 ± 0.0005 g of sediment, which had been sieved through a 10-mesh screen, was weighed into a ceramic crucible. The sample was then calcined in a muffle furnace at 550 °C for a duration of 4 h. After this, the crucible was transferred to a desiccator to cool to room temperature. The mass of the sample post-calcination was recorded, and the mass difference before and after burning was calculated.

Heavy Metal Analysis.
0.1000 ± 0.0005 g of sediment samples (≤ 200 μm) were taken in a polytetrafluoroethylene closed digestion tank, and the samples were digested by using aqua regia-HF until they were clarified and transparent, then the samples were fixed with distilled water into 50 mL volumetric flasks, and then the contents of the 10 heavy metals (Cd, Cr, Zn, Cu, Pb, As, Ni, Hg, Sb and Al) were deter-mined by ICP-MS.

2.3 Data Analysis

In this paper, the maps of sampling points were plotted using ArcGIS 10.8, and the heavy metal distribution graphs were drawn using origin 2018, and the correlation analysis was analyzed by using IBM SPSS Statistics 25.

3 Result

3.1 Physicochemical Index of Sediment

The percentage of clay (≤ 2 μm), silt (2 ~ 63 μm) and sand (63 ~ 2000 μm) in the sediments of Liujiang ranged from 1.5% to 16.1%, 23.8% to 76.1%, and 10.2% to 74.7%, respectively. Figure 2 illustrates the gradual increase in clay content in Liujiang sediments from L1 (1.5%) to L4 (16.1%), followed by a decrease at L5 (6.9%), and then another gradual increase from L5 to L7 (13.5%). The silt showed a trend that was basically the same as that of the clay particles, and correspondingly, the proportion of grit showed an opposite trend. The proportions of clay, silt and gravel in tributary sediments ranged from 2.8% ~ 28.1%, 5.9% ~ 89.9% and 4.8% ~ 91.3%, respectively.

The sediment pH in the Liujiang River and its tributaries was generally acidic, with levels ranging from 5.05 to 6.45 in the river sediments and from 4.26 to 6.08 in the tributaries. This acidity could potentially promote the leaching of heavy metals from the sediment.In the sediments of the Liujiang River, the organic matter content varied from 0.57 to 24.10 g/kg, averaging 12.08 g/kg, and exhibited a consistent increase in the downstream direction, except for L3 and L6. The organic matter content in the tributary sediments ranged from 0.79 to 66.50 g/kg, and the average content of 30.23 g/kg. Sediment organic matter content in the Langxi River (Z2), Xinxu River (Z5), and Daqiao River (Z6) was significantly higher than that of the Liujiang River.

The ORP of sediments in the Liujiang River ranged from –50 to 369 mV, with an overall decreasing and then increasing trend. The tributary sediment ORP ranged from –111 to 205 mV, with L5, Z2, Z5 and Z6 having more pronounced reductions compared to other sites.

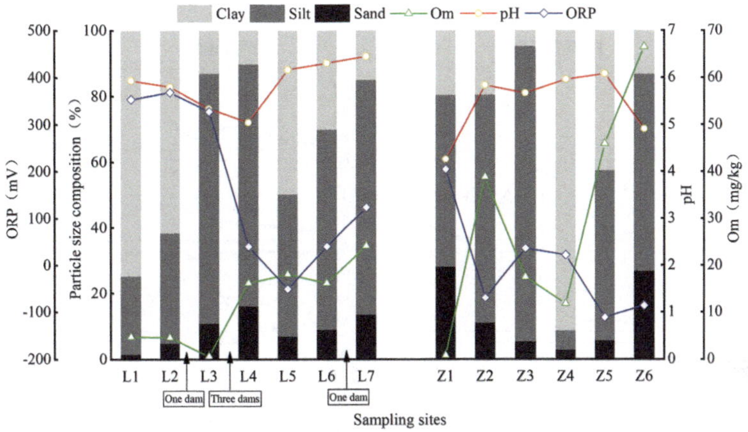

Fig. 2. Physicochemical indexes of sediments of Liujiang River (left) and its tributaries (right).

3.2 Spatial Distribution of Heavy Metals

The concentrations of Cd, Cr, Zn, Cu, Hg, Pb, As, Ni, Sb, and Al in the sediments of the Liujiang River were found to range between 0.09 and 4.36 mg/kg, 55.00 and 93.00 mg/kg, 88.00 and 239.00 mg/kg, 20.00 and 55.00 mg/kg, 0.05 and 0.69 mg/kg, 20.00 and 114.00 mg/kg, 3.57 and 14.60 mg/kg, 10.00 and 63.00 mg/kg, 0.13 and 6.61 mg/kg, and 2.60 and 77.30 g/kg, respectively. From Fig. 3, it is evident that the heavy metal content in the Liujiang River sediments exhibited a general pattern of increasing and then decreasing from upstream to downstream. Among them, Cd, As, Sb and Cr reached the maximum value at point L5, Zn, Hg, Pb and Al reached the maximum value at point L4, Cu and Ni reached the maximum value at point L7, L5 and L4 were the next highest values of Cu and Ni, respectively. It can be seen that L4 and L5 are the key points of heavy metal accumulation in Liujiang. The spatial distribution trend of Zn, Hg, Pb, Ni and Al was consistent, and all of them showed a gradual increase from L1 to L4, a gradual decrease from L4 to L6, and a trend of increasing again at L7. Both Cd and Cu showed a gradual increase from L1 to L5, experienced a sharp decline at L6, and then increased once more at L7. Conversely, As and Cr incrementally rose from L1 to L5 before gradually descending. The distribution patterns of Sb and other metallic elements exhibited more variability; nonetheless, they generally followed an upward trend.

The concentrations of Cd, Cr, Zn, Cu, Pb, As, and Ni in the sediment samples from various tributaries of the Liujiang River were found to be the most elevated at site Z2 (Langxi River), with values of 7.21, 111.00, 545.00, 51.00, 87.00, 17.50 and 49.00 mg/kg, respectively. This observation indicates that the Langxi River could potentially be a significant contributor of heavy metal pollutants compared to the other tributaries. On the other hand, Hg, Sb, and Al exhibited peak levels at locations Z6 (Daqiao River), Z4 (Longjiang River), and Z5 (Xinxu River), showing concentrations of 0.40 mg/kg, 1.06 mg/kg, and 47.7 g/kg, respectively.

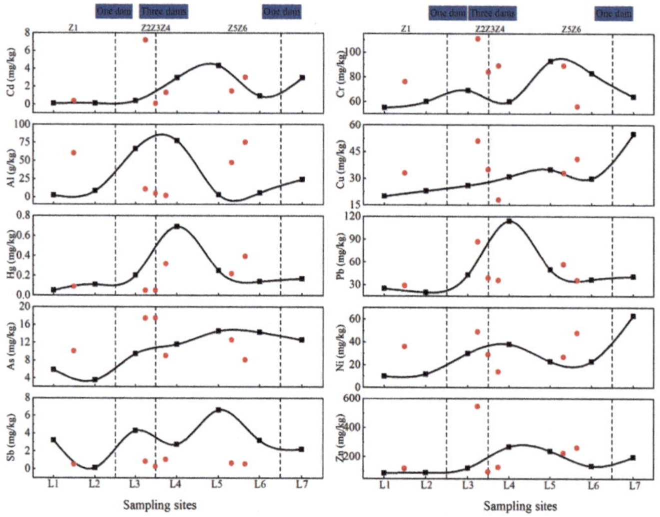

Fig. 3. Changes in heavy metal concentration in main streams and tributaries.

4 Discussion

The contribution of tributary rivers, sediment particle size makeup, and levels of organic substances frequently play a crucial role in determining how heavy metals are dispersed in river sediments. As shown in Fig. 3, there are fluctuations in the concentrations of heavy metals within the sediments of the Liujiang River and its associated tributary rivers. Cd, Zn, Pb, and Ni reach their highest levels at either L4 or L5, potentially influenced by the overflow from the Langxi River. Cr, Cu, and As levels peak at L5, likely due to the influx of water from the Xunjiang, Langxi, and Beijiang rivers. In contrast to the typical multi-year discharge of the Liujiang River, which totals 9.37 billion m^3, the Langxi River experiences a lower average multi-year discharge of 7.67 million m^3. Positioned between the river mouth and the points L4 and L5 are three dams named Fushi, Guding, and Dapu. Additionally, there are 4 dams and 2 dams separating the Xunjiang and Beijiang river mouths from the points L4 and L5, respectively. It is evident that runoff input primarily leads to the enrichment of heavy metals in river sediments, although it does not play a significant role in influencing the distribution of these metals. Notably, in the sediments of upstream tributaries, the concentrations of Hg and Sb are lower compared to those found in Liujiang sediments. Because Hg and Sb still demonstrate enrichment patterns in the L4 and L5 sections, corroborating this observation. Based on the 2022 land use data for the study region (Fig. 1), it is apparent that urban land use is predominantly located between L5 and L6. This area encompasses industries such as metal smelting and electroplating within the Liuzhou Steel and Liuzhou Automobile Industrial Parks, which are known to release heavy metals. Therefore, it can be concluded that the enrichment of heavy metals in L4 and L5 is not attributable to point source inputs.

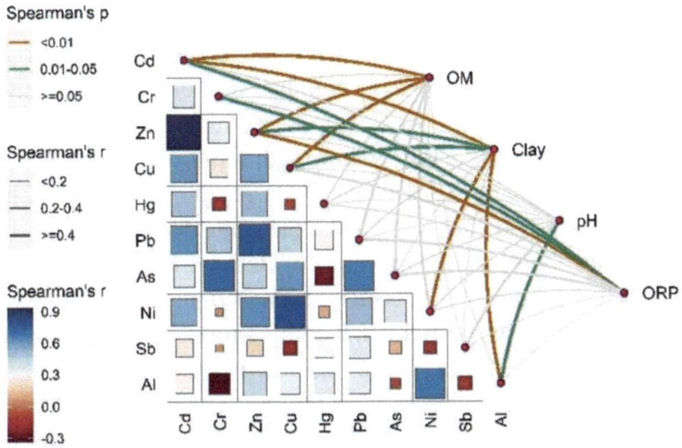

Fig. 4. Spearman correlation coefficient between heavy metals and Clay, pH, ORP and OM.

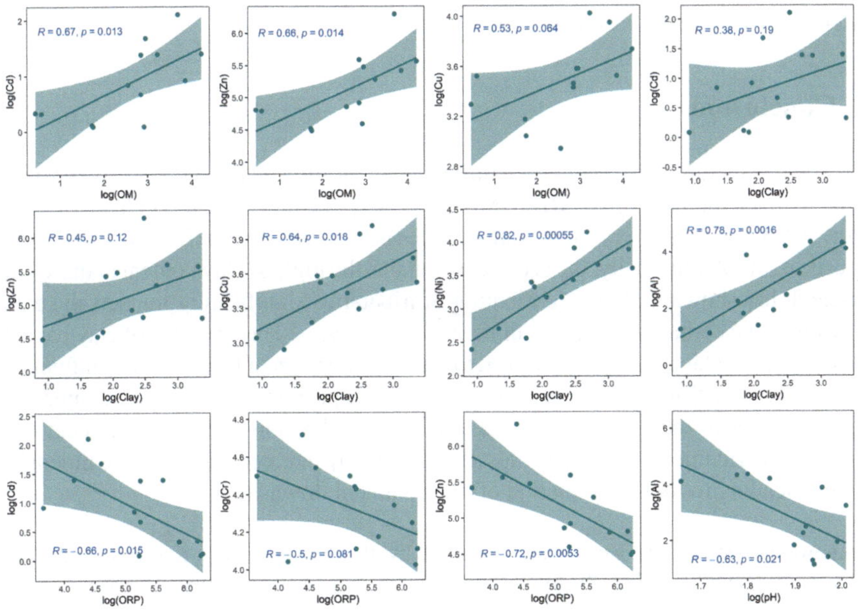

Fig. 5. Relationships between heavy metals and Clay, pH, ORP and OM.

Among the 10 heavy metals, only Cd, Zn, Cu, Ni, Al, and Cr exhibited a strong positive relationship with organic matter and clay particles, while showing an inverse correlation with ORP and pH (refer to Fig. 4). This could be due to the fact that the presence of Hg, Pb, As, and Sb is less influenced by these environmental factors. The decrease in pH and ORP may enhance the leaching of heavy metals from sediments, but the overall impact on heavy metal distribution is minor. Hence, the interplay between ORP and pH serves as a warning sign, indicating that sites with higher heavy metal concentrations are at a greater risk of heavy metal leaching. The strong correlation (0.38–0.82) between clay particles and heavy metals is more pronounced compared to the correlation between organic matter and heavy metals (0.38–0.67) as shown in Fig. 5. This suggests that the distribution of heavy metals is influenced more significantly by the distribution of clay particles than by organic matter. This trend has also been observed in research conducted on estuarine and nearshore sediments along the east and west coasts of India [11], as well as in the Pearl River Delta region of China [12]. These results support previous studies [13] that emphasize the considerable impact of clay content and organic matter levels on the distribution of heavy metals in the sediments of the Liujiang River and its tributaries. The construction of dams alters the hydrodynamics of rivers, impacting the sediment particle size composition. One dam is located between L2 and L3, three between L3 and L4, and one between L6 and L7. The clay content of Liujiang sediments gradually increases from L1 to L4, peaks at L4, decreases at L5, and then increases again. This may be attributed to the interception by the cascade dam, which traps the coarse particles in the runoff. Owing to the relatively slow sedimentation rate, the clay particles gradually migrate downstream [14], leading to a higher concentration of L4 clay particles. Most heavy metals peak at this location, and the opposite trend observed in sand particles further corroborates this conclusion.

5 Conclusion

All 10 heavy metals show a pattern of initial increase followed by a decrease downstream in the sediment of the Liujiang River. More precisely, the highest levels of 8 heavy metals (Cd, Zn, Pb, Ni, Zn, Hg, Pb, and Al) are detected at the downstream (L4 or L5) of the three cascade dams. Additionally, the sediment contains the highest clay content at these sites. Research has shown that the contribution of tributaries significantly impacts the extent of heavy metal contamination in sediment found in rivers. The presence of clay particles and organic substances is closely linked to the concentration of heavy metals, influencing their dispersion. Notably, clay particles have a more pronounced influence compared to organic matter in the distribution of heavy metals. The construction of gradient dams caused the enrichment of heavy metals downstream by intercepting coarse sediment step by step, leading to the accumulation of fine sediment downstream.

Acknowledgments. This work was supported by the National Key Research and Development Program(2022YFC3203404); the National Key Research and Development Program(2023YFC3205702).

References

1. Ji, D.B., Long, L.H., Xu, H., et al.: Advances in study on cumulative effects of construction of cascaded reservoirs on water environment. Adv. Sci. Technol. Water Resour. **37**(03), 3545–3551 (2017)
2. Li, H.N., Xia, P.H., Wu, S.L., et al.: Heavy metal distribution and plankton enrichment characteristics in cascade reservoirs from Beipan River. Sci. Technol. Eng. **18**(9), 156–162 (2018)
3. Zhang, G.Y., Wu, L.N., Ouyang, K.C., et al.: Pollution characteristics and risk assessment of heavy metals in soils along the upper reaches of the Duliu River. Carsologica Sinica **40**(3), 495–503 (2021)
4. Li, Y., Li, J.K., Xie, W.F., et al.: Assessment and source analysis of heavy metal pollution in sediments of the Yellow River Basin. Environ. Sci. Technol. **45**(7), 96–104 (2022)
5. Zhong, Y.H., Fu, C., Pan, J., et al.: Migration and transformation characteristics of heavy metals in sediments in the drawdown zone of Wanzhou section of the three Gorges Reservoir. Environ. Sci. Technol. **44**(12), 81–89 (2021)
6. Liang, G.Y., Wang, R., Sheng, G.D., et al.: Geochemical controls on the distribution and bioavailability of heavy metals in sediments from Yangtze Riverto the East China Sea: assessed by sequential extraction versus diffusive gradients in thin-films (DGT) technique. J. Hazard. Mater. **452**, 131253 (2023)
7. Fang, Z.Q., Wang, Y.M., Wang, X., et al.: Spatial distribution and risk assessment of heavy metals in sediments of the Ruxi Tributary of the Three Gorges Reservoir. Environ. Sci. **41**(03), 1338–1345 (2020)
8. Jaskula, J., Sojka, M.: Assessment of spatial distribution of sediment contamination with heavy metals in the two biggest rivers in Poland. CATENA **211**, 105959 (2022)
9. Zhao, Q., Ding, S., Hong, Z., et al.: Impacts of water-sediment regulation onspatial-temporal variations of heavy metals in riparian sediments along the middle and lower reaches of the Yellow River. Ecotoxicol. Environ. Saf. **227**, 112943 (2021)
10. Bing, H., Wu, Y., Zhou, J., et al.: Spatial variation of heavy metal contamination in the riparian sediments after two-year flow regulation in the Three Gorges Reservoir. Chin. Sci. Total Environ. **649**, 1004–1016 (2019)
11. Anbuselvan, N., Senthil Nathan, D., Sridharan, M.: Heavy metalassessment in surface sediments off coromandel coast of India: implication on marine pollution. Mar. Pollut. Bull. **131**, 712–726 (2018)
12. Zhao, G., Ye, S., Yuan, H., et al.: Surface sediment properties and heavy metal contamination assessment in river sediments of the Pearl River Delta. Chin. Mar. Pollut. Bull. **136**, 300–308 (2018)
13. Zhu, L., Wang, T.X., Liu, J.W., et al: Metal distribution in sediments of a drinkingwater reservoir: influence of reservoir morphometry and hydrodynamics. Environ. Sci. Pollut. Res. **26**(10), 9599–9609 (2019)
14. Yu, H.J., Xu, S.G., Li, W.J., et al.: Removal process of phosphorus during the settlement of particulates with runoff and its implication for reservoir management. Environ. Sci. Pollut. Res. **28**(36), 49588–49601 (2021)

Characteristics and Influencing Factors of Groundwater Chemistry in a Chemical Industry Park

Yuwen Dong[1](✉), Haiming Li[1], and Yanwei Wang[2]

[1] Tianjin University of Science and Technology, Tianjin 300222, China
954245022@qq.com

[2] Technical Center for Soil, Agricultural and Rural Ecology and Environment, Ministry of Ecology and Environment of the People's Republic China, Beijing 100012, China

Abstract. The groundwater chemical characteristics of a chemical industry park in Shandong Province were studied by collecting groundwater samples and using mathematical statistics, Piper three-line graph and Gibbs chart. The results show that HCO_3·Cl-Ca, Cl·SO_4-Ca and SO_4.-Cl-Ca are the main hydrochemical types of carbonate fissure karst water, and HCO_3·Cl-Ca and Cl· HCO_3-Ca are the main hydrochemical types of clastic pore fissure water. The anions are mainly Cl^-, $HCO_3{}^-$, $SO_4{}^{2-}$, and the cations are mainly Ca^{2+}. It is speculated that the ion change is mainly from the dissolution of carbonate sediments and gypsum sediments in the south to the dissolution of gypsum or sulfate sedimentary rocks in the north. Along the northeast to southwest direction, the hydrochemical type changes from Cl-HCO_3-Ca $\rightarrow$ Cl·SO_4-Ca type. The shallow groundwater in the park area is affected by water-rock interaction. The deep groundwater is affected by the water-rock interaction, and evaporation and concentration also contribute to it, but the water-rock interaction is dominant.

Keywords: chemical industry park · groundwater pollution · mathematical statistics · hydrochemical characteristics

1 Introduction

Due to the leakage of waste water and the unreasonable management of drainage channels in the chemical industrial park, pollutants can spread continuously through rainfall and surface infiltration, causing pollution to water bodies [1, 2]. The groundwater quality in the study area can not meet the requirements of Class III standard of "Groundwater Quality Standard" (GB/T14848–2017). The above standard of regional groundwater is mainly caused by agricultural and rural non-point source pollution, historical industrial pollution, regional geological conditions, groundwater exploitation and other reasons [3–5]. To investigate and determine the groundwater chemical characteristics of the chemical industry park and its surrounding areas, and to put forward prevention and control measures on this basis has become the primary task [4–6].

D. Han et al. (Eds.): ICEPG 2024, ESE, pp. 38–45, 2025.
https://doi.org/10.1007/978-3-031-86829-0_5

2 Materials and Methods

2.1 Research Object

Located in Shandong Province, the chemical industry park is situated in the western part of the province. It spans 22.33 km long. The region's geographical location and convenient transportation are some of the factors that make it an ideal location for research. Its industrial park has major transport facilities such as Huayang Railway special line with a transport volume of 200,000 t/a, providing strong support for logistics and transportation of enterprises.

The climate of this area is characterized by its warm continental climate and distinct seasons. There are four types of groundwater that can be found in this region: loose rock pore, carbonate karst, magmatic metamorphic and clastic fissures.

2.2 Sample Collection and Determination

A total of 53 sampling points of water-bearing group of karst water were arranged in the study area. The stable water level of groundwater is measured by electric water level meter before groundwater sampling. The quality of extracted groundwater was measured by multi-parameter water quality monitoring instrument. A turbidimeter is used to measure the turbidity of groundwater. When the extracted water level and water quality monitoring parameters (pH, conductivity, dissolved oxygen, water temperature, REDOX potential, turbidity) reach stability, the water sample is collected. The criterion is to collect water samples after at least 3 consecutive test readings meet the requirements. The sampling depth is below 0.5 m of groundwater surface. When sampling water, the container should be filled completely with the water sample, leaving no space in the upper part. In the study area, a total of 40 groundwater samples were collected, comprising 14 samples of loose rock pore water, 15 samples of carbonate rock fissure karst water, 5 samples of clastic rock pore fissure water, and 6 samples of magmatic metamorphic rock fissure water [7, 8].

3 Results and Discussion

3.1 Types and Distribution Characteristics of Groundwater Water Chemistry

The main chemical types of groundwater in the area are $HCO_3 \cdot SO_4 \cdot Cl\text{-}Ca$, $SO_4 \cdot Cl\text{-}Ca$, $SO_4 \cdot HCO_3\text{-}Ca$, $HCO_3 \cdot Cl \cdot SO_4\text{-}Ca$, with local distribution of $Cl \cdot HCO_3\text{-}Ca$ and $Cl \cdot SO_4\text{-}Ca$ [9, 10].

In this paper, the characteristics of groundwater chemical distribution in chemical industry park were studied by Piper three-line diagram.As can be seen from the figure:

As shown in Fig. 1, Most of the cations in the water samples in the study area are Ca^{2+} and Na^+, among which Ca^{2+} is the largest, followed by Na^+. The main anions are Cl^- and $SO_4{}^{2-}$. Groundwater sampling points are mainly concentrated in the lower left triangle region, and in the lower right triangle region and the diamond region, the distribution of groundwater sampling points can be found to be slightly dispersed, so the cation in the groundwater in this study area is relatively stable. However, $SO_4{}^{2-}$ also showed a

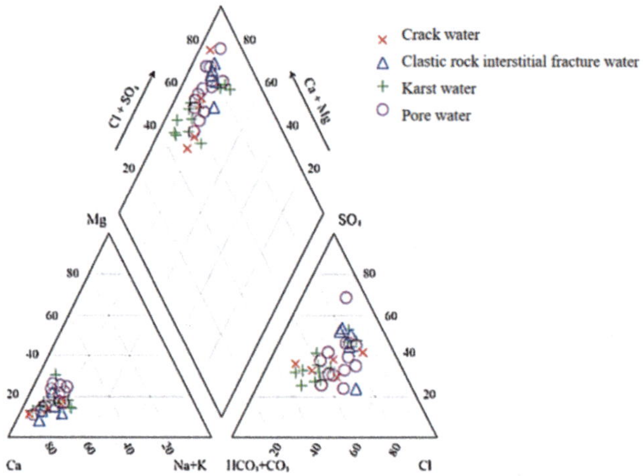

Fig. 1. Piper three-line diagram

significant increase in some water samples. According to the previous investigation of pollution sources in the study area, it was found that there were some polluted plots such as chemical plants, domestic waste and aquaculture farms in the chemical industry park, and the existence of these plots may lead to significant differences in anions in some groundwater.

Combined with the location of sampling points and the diamond diagram, it can be seen that the chemical type of groundwater collected in this time shows an obvious change law from $HCO_3 \cdot SO_4$-Ca type to SO_4·Cl-Ca type water along the direction of groundwater runoff, which initially shows that the water quality gradually deteriorates along the direction of groundwater runoff.

3. The distribution of fissure water in the figure is relatively dispersed, and there is no obvious change rule, showing a variety of hydrochemical types such as $HCO_3 \cdot SO_4$-Ca and Cl·SO_4-Ca, and the cations are mainly Ca^{2+} ions.

3.2 Gibbs Chart Analysis

With $Cl^-/(Cl^- + HCO_3^-)$ and $Na^+/(Na^+ + Ca^{2+})$ ratios as horizontal coordinates and TDS as vertical coordinates, Gibbs [10, 11] chart can be used to analyze the main natural mechanisms controlling groundwater chemistry. Soluble ions contained in water bodies come from a variety of sources, usually evaporation and concentration, water-rock interaction and atmospheric precipitation.

As can be seen from the Figs. 2 and 3, the $Na^+/(Na^+ + Ca^{2+})$ ratio of shallow groundwater samples is low, ranging from 0.1 to 0.4, a few samples are greater than 0.5, and the TDS of water samples is very high, and they are in the control zone of water-rock interaction and affected by obvious water-rock interaction. The $Cl^-/(Cl^- + HCO_3^-)$ratio of most samples is between 0.4 and 0.7, which is in the control zone of water-rock interaction and is also affected by evaporation and concentration. The ratio of $Cl^-/(Cl^- + HCO_3^-)$of some samples is around 0.8.

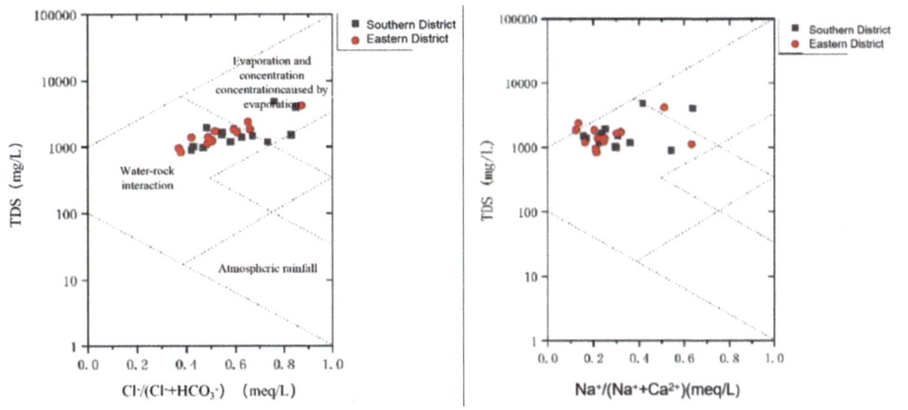

Fig. 2. Gibbs map of shallow groundwater

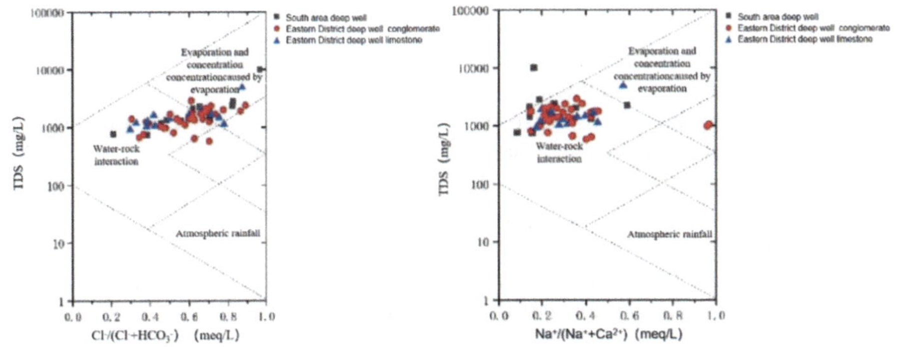

Fig. 3. Gibbs map of deep groundwater

As can be seen from the figure above, compared with shallow groundwater, the ratio of $Cl^-/(Cl^- + HCO_3^-)$ in deep groundwater is relatively high, ranging from 0.2–0.9, which is in the water-rock interaction area and affected by evaporation and concentration to a certain extent. $Na^+/(Na^+ + Ca^{2+})$ is between 0.1 and 0.5, most of the points are located in the water-rock interaction region and its boundary, and some of the points are relatively high outside the model.

In summary, Gibbs map analysis of shallow and deep groundwater shows that the shallow groundwater in the park area is affected by water-rock interaction. The deep groundwater is affected by the water-rock interaction, and evaporation and concentration also contribute to it, but the water-rock interaction is dominant.

4 Prediction of Groundwater Pollution Trend

4.1 Numerical Model of Groundwater

In this work, the groundwater in the working area is generalized to a heterogeneous and anisotropic stable three-dimensional groundwater flow system.

Its mathematical equation is as follows:

$$\begin{cases} 0 = \frac{\partial}{\partial x}\left(K_x \frac{\partial h}{\partial x}\right) + \frac{\partial}{\partial y}\left(K_y \frac{\partial h}{\partial y}\right) + \frac{\partial}{\partial z}\left(K_z \frac{\partial h}{\partial z}\right) + \varepsilon & x, y, z \in \Omega \\ 0 = K_x\left(\frac{\partial h}{\partial x}\right)^2 + K_y\left(\frac{\partial h}{\partial y}\right)^2 + K_z\left(\frac{\partial h}{\partial z}\right)^2 + p & x, y, z \in \Gamma_0 \\ h(x, y, z)|_{\Gamma_1} = h_0 & x, y, z \in \Gamma_1 \\ K_n \frac{\partial h}{\partial \vec{n}}\Big|_{\Gamma_2} = 0 & x, y, z \in \Gamma_2 \end{cases} \quad (1)$$

Ω is the seepage zone;

h is the water level elevation of aquifer (m);

K_x, K_y, K_z are the permeability coefficients (m/d) in x, y and z directions, respectively.

K_n is the permeability coefficient of the boundary surface in normal direction (m/d).

ε is the source and sink term (d-1) of the aquifer;

p is the evaporation and precipitation of the diving surface (d-1);

h_0 is the river level distribution (m);

Γ_0 is the upper boundary of the seepage zone, i.e. the free surface of groundwater.

Γ_1 is the boundary of constant water head in the seepage zone.

Γ_2 is the constant flow boundary in the seepage region.

ñ is the normal direction of the boundary surface.

In this solute transport simulation, based on the most unfavorable factors, the reaction and adsorption of pollutants in underground water bodies are not considered, and only the convective dispersion is considered.

$$\begin{cases} R\theta \frac{\partial c}{\partial t} = \frac{\partial}{\partial x_i}\left(\theta D_{ij} \frac{\partial c}{\partial x_j}\right) - \frac{\partial}{\partial x_i}(\theta v_i c) - W c_s - W c \\ c(x, y, z, t) = c_0(x, y, z)(x, y, z) \in \Omega, t = 0 \\ \theta D_{ij} \frac{\partial c}{\partial x_j}\Big|_{\Gamma_2} = f_i(x, y, z, t)(x, y, z) \in \Gamma_2, t \geq 0 \end{cases} \quad (2)$$

Formula:

R - hysteresis coefficient, dimensionless, R = 1 + ρ_b/θ K_d;

K_d - distribution coefficient, dm3/kg;

ρ_b- Medium density, 2 × 106 mg/(dm)^3 according to geotechnical test data;

θ - medium porosity, dimensionless, 0.5 according to soil test data;

c - concentration of component, mg/L;

$\bar{c}$- Solute concentration adsorbed by medium skeleton, mg/L;

t - time, d;

x, y, z - space position coordinates, m;

D_{ij} - hydrodynamic dispersion coefficient tensor, m2/d;

Vi - groundwater seepage velocity tensor, m/d;

W - Source and sink of water flow, 1/d;

c_s- Concentration of components, mg/L;
$c_0(x, y, z)$- Known concentration distribution;
Ω - Model simulation area.
Γ_2- flux boundary;
$f_i(x, y, z, t)$- Known dispersion flux function on the boundary.

4.2 Model Calibration

As shown in Fig. 4, According to the terrain and geomorphology, the groundwater flow field is generally consistent with the actual conditions. It can be observed from the diagram that the direction of flow and hydraulic gradient of groundwater are predominantly aligned.As shown in the figure below, the water level is about 86–102 m, the fitting effect is relatively ideal, the overall trend is basically the same, more in line with the actual conditions of the site, and can basically reflect the dynamic characteristics of the groundwater system in this area. Therefore, this model is used to predict the solute transport scenario of groundwater.

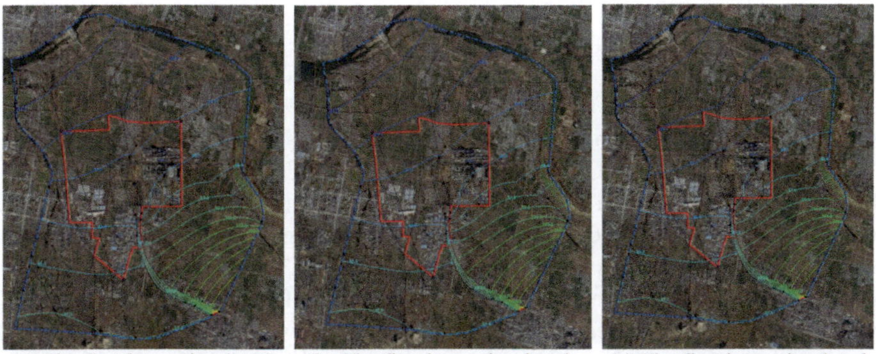

(a) The first layer simulated flow field diagram **(b) The first layer simulated flow field diagram** **(c) The first layer simulated flow field diagram**

Fig. 4. Simulated flow field diagram

4.3 Prediction Results of Groundwater Model

As shown in Fig. 5, Trichloromethane in groundwater of chemical industry park migrated from southeast to northwest in the 5th years. In the fifth year, the migration length of the pollution plume in the first aquifer was 540.8 m. The migration length of the pollution plume in the second aquifer is 520.1 m. The migration length of the pollution plume in the third aquifer is 505.4 m.

Considering that the simulation did not take into account the change of flow field caused by irrigation water intake downstream of the pollution plume, the groundwater flow in the park may be accelerated after water intake in the water source area, and the calculated migration distance may be conservative. At the same time, the increase of

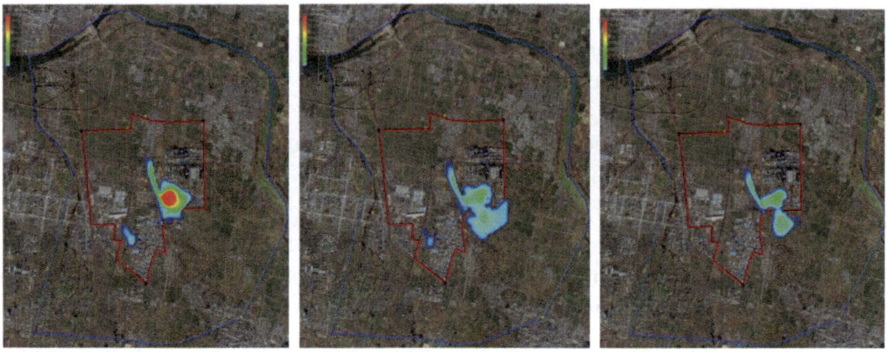

(a) In shallow groundwater (b) In middle groundwater (c) In deep groundwater

Fig. 5. Prediction of spatial distribution of trichloromethane after 5 years

rainfall infiltration will increase the migration speed of pollution plume, increase the pollution risk of downstream water source and surrounding irrigation Wells, and shorten the affected time of water source.

5 Conclusion

The main types of groundwater chemistry observed in the study area include HCO_3·SO_4·Cl-Ca, SO_4·Cl-Ca, SO_4· HCO_3-Ca, and HCO_3·Cl·SO_4-Ca, with localized distribution of Cl· HCO_3-Ca and Cl·SO_4-Ca. The predominant cations in the water samples from the study area are Ca^{2+} and Na^+, with Ca^{2+} being the most abundant, followed by Na^+. The primary anions detected are Cl^- and $SO_4{}^{2-}$.The chemical composition of groundwater in the study area exhibits a distinct pattern of change from HCO_3·SO_4-Ca type to SO_4·Cl-Ca type water with the direction of groundwater runoff. Shallow groundwater in the park area is influenced by water-rock interactions, while deep groundwater is primarily affected by water-rock interactions, with some contribution from evaporation and concentration processes. However, water-rock interactions are the dominant factor shaping the groundwater chemistry.Simulation results indicate that the pollution plume is expected to migrate towards the surrounding rivers, potentially impacting the water sources in the vicinity. Considering the recharge relationship between groundwater and rivers, this contamination could pose a health risk to human beings.

References

1. Yang, H., Xie, Y., Zhu, T., et al.: Reduced organic carbon content during the evolvement of calcareous soils in karst region. Forests **12**(2), 221 (2021)
2. Xie, H., Liang, Y.P., Li, J., et al.: Distribution characteristics and health risk assessment of metal elements in groundwater of Longzici spring area. Environmental Science **42**(9), 4257–4266 (2021)
3. Liu, Z., Li, S., Liu, B., et al.: Source-tracing of sulfate in groundwater of Xizhang Well Group in the Liuzheng water source area of Zibo City. Carsologica Sinica **42**(5), 1074–1084 (2023)

4. Yawei, F.E.N.G., Hongnian, C.H.E.N., Hua, B.U., Dewang, J.I.A.: Hydro-chemical genesis and isotope characteristics of Yangzhuang karst water system. Carsologica Sinica **38**(3), 394–403 (2019). https://doi.org/10.11932/karst20190309
5. Lu, M., Han, B., Wu, F., Sun, D., Zhang, Z.: Characteristics and genesis of high-fluorine groundwater in southwestern Shandong Province. Geology In China **41**(1), 294–302 (2014)
6. Yang, X., luo, W., Pei, J., You, J.: Analysis of variation characteristics of karst groundwater quality in Guiyang City. Carsologica Sinica **36**(5), 713–720 (2017). https://doi.org/10.11932/karst20170515
7. Mahaqi, A., Moheghy, M.A., Moheghi, M.M., et al.: Environmental hydrogeochemistry characteristics, controlling factors and groundwater quality assessment in Herat City, West Afghanistan. Water Resour. **47**(2), 325–335 (2020)
8. Brennan, S.T., Lowenstein, T.K.: The major-ion composition of silurian seawater. Geochim. Cosmochim. Acta **66**(15), 2683–2700 (2002)
9. Ramanathan, A.L.: Seasonal variation in the major ion chemistry of Pandoh Lake, Mandi District, Himachal Pradesh, India. Appl. Geochem. **22**(8), 1736–1747 (2007)
10. Gibbs, R.J.: Mechanisms controlling world water chemistry. Science **170**, 1088–1090 (1970)
11. Gibbs, R.J.: Mechanisms controlling world water chemistry: Evaporation-crystallization process. Science **172**, 870–872 (1971)

Application of PVDF/PVP Membrane in Chemical Synthesis Pharmaceutical Wastewater Treatment Project

Shan Wang[1](✉), Guibao Zhang[2], and Kanyu Zhou[3]

[1] Shandong Jiaotong University, Jinan 250300, China
18366181876@163.com
[2] Qilu Pharmaceutical Co., Ltd., Jinan 250000, China
[3] Jiangsu Peier Membrane Industry Co., Ltd., Yixing 214200, China

Abstract. The treatment of chemical synthesis pharmaceutical wastewater from a pharmaceutical plant was studied using the LSP-MBR combined process, with the effluent water quality reaching the Chemical Synthesis Pharmaceutical Industrial Wastewater Pollutant Discharge Standard. The membrane component used as the main unit of the MBR process was a PVDF/PVP flat membrane made by the non-solvent phase separation method. The changes in the pore size and porosity of the PVDF/PVP flat membrane was examined, the reasons for increased membrane fouling and decreased membrane flux were explored, and the cycle of membrane cleaning was determined.

Keywords: PVDF/PVP membrane · Chemical synthesis pharmaceutical wastewater · Membrane contamination

1 Introduction

The treatment of chemical synthesis pharmaceutical wastewater from a pharmaceutical plant was studied using the sludge reduction biological treatment process (LSP)-membrane bioreactor (MBR) combined process. The effluent water quality reached the Chemical Synthesis Pharmaceutical Industrial Wastewater Pollutant Discharge Standard (GB21904–208). The Polyvinylidene fluoride (PVDF)/ polyethylenepyrrolidone (PVP) membrane in MBR process was preparaed by nonsolvent Induce Phase Separation, N, N-Dimethylacetamide(DMAc) and N-Methylpyrrolidone (NMP) as the blending solvent, PVDF and PVP as pore-forming agent. Certain amount of PVP (K30) were added which could improve the flux and anti fouling ability of the membrane [1, 2]. PVDF and PVP have good chemical stability and can be applied to wastewater treatment with high organic pollutant concentrations.

D. Han et al. (Eds.): ICEPG 2024, ESE, pp. 46–56, 2025.
https://doi.org/10.1007/978-3-031-86829-0_6

2 Materials and Production Methods

2.1 A Subsection Sample

PVDF: 6020, Solvay (Shanghai) Co., LTD.

PVP: K30, Analytics pure, Us-Based ISP TECNOLOGIES INC.

DMAc: Analychem, Shanghai Sinopharm Group;

NMP: Analytically pure, Zhangjiagang Kerui Fine Chemical Co., Ltd;

Propylene glycol (PG): Analytically pure, Shanghai Sinopharm Group;

Nonwovens: 500 mm, Shanghai Tianlue Textile New Material Co., LTD.

2.2 Instruments and Equipment

Film scraper: DKN-500, Dalian Kona Sustainable Technology Development Co., LTD.

Integrity tester: IntegtestV4.0, Beijing Nuintec Biotechnology Co., LTD.

Digital viscosimeter: NDJ-8S, Shanghai Jingtian Electronic Instrument Co., LTD.

Contact Angle tester: OCA15EC, dataphysics, Germany.

2.3 Preparation Method

DMAc and NMP were mixed together and the solution was heated to 60–70°C, PG was added to solution and stirred well. PVDF was added to the solution in batches, and the solution was stirred at 70°C for 72 h. Then PVP K30 was added and stirred for 8–10 h. Solution was stood for defoaming. The membrane solution was dissolved and matured at a constant temperature of 50–65°C. And the film was scraped on the non-woven fabric at a speed of 3.5 m/min at room temperature of 20–22°C and humidity of 40%–50% to make membrane.

Membrane was cleaned, cut and moisturized by ultrasonic welding on ABS plate 1.78 m long and 0.49 m wide to make a membrane element with an area of 1.5 m^2.

3 Engineering Application

3.1 Designing the Water Quality and Quantity for Inlet and Effluent

The pharmaceutical residues and the first cleaning water were treated as hazardous waste. The other cleaning water, circulating water, laboratory wastewater, and factory domestic sewage were treated by LSP-MBR treatment system, with design water volume being 1000 m^3/d. The inlet and effluent water quality of the system design is shown in Table 1.

Table 1. Inlet and effluent quality diagram

Testing items	CODcr (mg/L)	BOD5 (mg/L)	TN (mg/L)	NH_3-N (mg/L)
Inlet water	1500–2500		50–80	30–40
Effluent water	<100	<20	<30	<20

3.2 Process Flow

The wastewater treatment process was composed of high concentration wastewater pretreatment, Regulating tank, Hydrolytic acidification tank, LSP tank, sedimentation tank, anoxic tank, MBR tank, etc. The process is shown in Fig. 1.

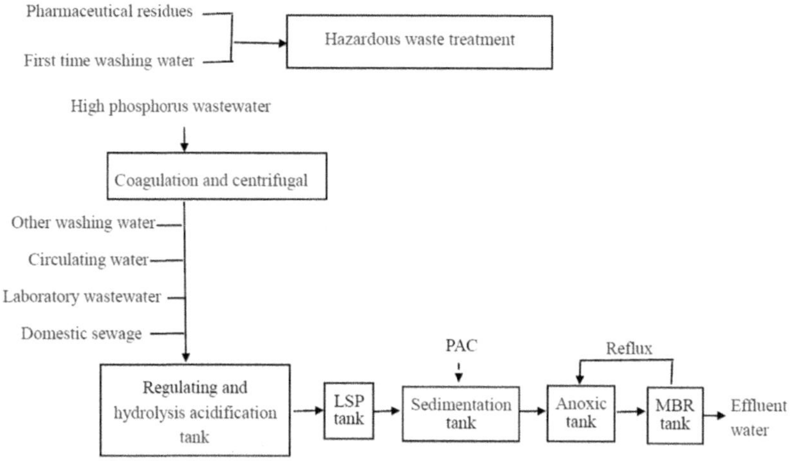

Fig. 1. Treatment process

Regulating and hydrolysis acidification tank's HRT was 48 h. It set up online pH detectors, acid and alkali dosing devices, submersible mixers, and submersible sewage pump.

LSP tank was divided into four sections, with HRT approximately being 4.0 h, 11.0 h, 8.0 h, and 11.0 h, in the mixing zone, aeration zone, micro-aeration zone, and enhanced aeration zone. The dissolved oxygen concentration was controlled at 0.2–0.5 mg/L, 2.0–3.0 mg/L, 0.5–1.0 mg/L, and 2.0–3.0 mg/L. The tank set up aeration fans, submersible sewage pump, and submersible mixers. The reflux was from the enhanced aeration zone to the mixing zone. Reflux ratio was 300%–350%. LSP process diagram is shown in Fig. 2. LSP tank was filled with nanometre linear fillers to provide segmented and layered enrichment conditions for anaerobic, anaerobic, and aerobic microorganisms. In the process, microbial communities, protozoa, metazoa, and aquatic animals were enriched in a stepwise manner, forming a combination of biological and food chains, which could remove pollutants and reduce excess sludge [3].

Sedimentation tank inclined plate type, with HRT was 1.5 h.

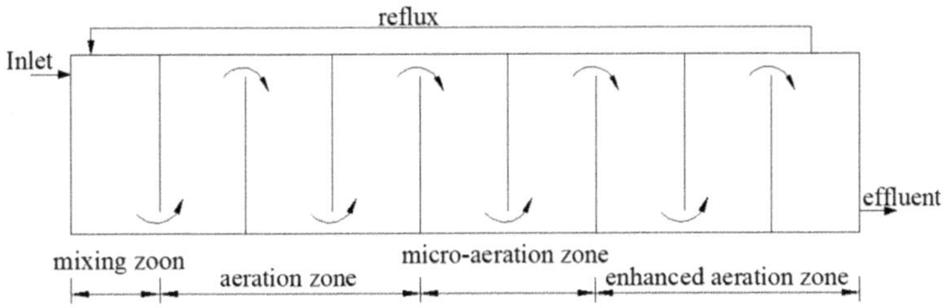

Fig. 2. LSP process diagram

Anoxic tank's HRT was about 2.5 h. It set up submersible mixers.

MBR tank's HRT was 4 h, and equipped with aeration fans, self suction pumps, and reflux pumps. The water production adopts intermittent run mode, with a 2-min cycle operation after pumping the water for 8 min. The average membrane flux is 350–380 L/(m^2·d) and the aeration rate was 8 L/(p·min). Sludge concentration is 5000–6000 mg/L, The mixed liquid in the MBR tank was delivered back to the anoxic tank with a re-flux ratio of 150%.

Power equipment of system is shown in Table 2.

Table 2. Power equipment list

Device Name	Single unit power (kW)	Quantity	Installed power (kW·h)	Operating power (kW·h)	Working Position
submersible sewage pump	3.0	2	6.0	3.0	Regulating and hydrolysis acidification tank
mixer	4.0	2	8.0	8.0	
Magnetic levitation fan	50	1	50.0	37.0	LSP tank
submersible sewage pump	4.0	3	12.0	8.0	
mixer	1.1	2	2.2	2.2	Anoxic tank
Magnetic levitation fan	50	1	50.0	27.0	MBR tank
submersible sewage pump	4.0	2	8.0	4.0	
Suction pump	3.0	3	9.0	6.0	

3.3 Running Status

Under normal running conditions, The inlet water pH was controlled to 7.8–8.0. One year of monitoring data was selected for observation of water quality, The pollutant concentrations of effluent water are shown in Table 3. It shows that the average of various pollutants in effluent water reached the Chemical Synthesis Pharmaceutical Industrial Wastewater Pollutant Discharge Standard from January to December. Inlet water quality and removal rate by processes is shown in Fig. 3.

Table 3. Pollutant concentrations of effluent water

	Jan	Feb	Mar	Apr	May	Jun	Jul	Aug	Sept	Oct	Nov	Dec
CODcr (mg/L)	74.8	70.9	58.4	60.3	87.7	74.6	57.6	48.9	56.4	59.7	67.1	70.7
NH_3-N (mg/L)	3.31	2.25	1.61	0.85	3.34	3.47	0.72	0.74	0.94	1.12	1.68	2.81
TN (mg/L)	18.6	21.2	19.3	20.9	29.1	25.5	21.0	15.8	19.2	17.0	19.6	23.8

Figure 3 shows that the system has a stable removal effect on CODcr, the removal rate is higher than 93%. Membranes can intercept refractory organic pollutants, with some of them flowing back to the previous bioreactor system, prolonging reaction time and achieving degradation of refractory organic pollutants. When the concentration of pollutants in the effluent water was higher and the temperature was lower, the membrane has a significant interception effect on COD_{cr}, but the interception effect on ammonia nitrogen and total nitrogen was not significant.

From May to June, due to production adjustments, the pollutant components in the wastewater changed significantly, so the water quality fluctuated greatly. The water quality impact comes from the changes in the content of pollutants such as triethylamine and pyridine in the reaction kettle flushing water after product adjustment. At the same time, it can cause damage to the original activated sludge environment in the biochemical system. Due to the impact of its biological toxicity on activated sludge, sludge requires a certain amount of domestication time to restore treatment efficiency.. During this period, the removal rate of NH_3-N and TN in the system was slightly lower,, but the average final effluent didn't exceed the standard. It indicates that the system has good resistance to shock. The removal efficiency of NH_3-N in the LSP process is greatly affected by fluctuations in the quality of the inlet water, but the ability of the system to effectively treat NH_3-N by the MBR process still keeps the removal rate above 90%. The removal of TN by the system is influenced by temperature and inlet water quality [4, 5]. In August, the system had the highest TN removal rate at 76.4%. The removal rate of TN was lower from May to June when water quality fluctuated greatly, and from December to February when temperature was lower. In addition, due to the lower CODcr of inlet water in December, the removal rate was lower at that time than in January and February.

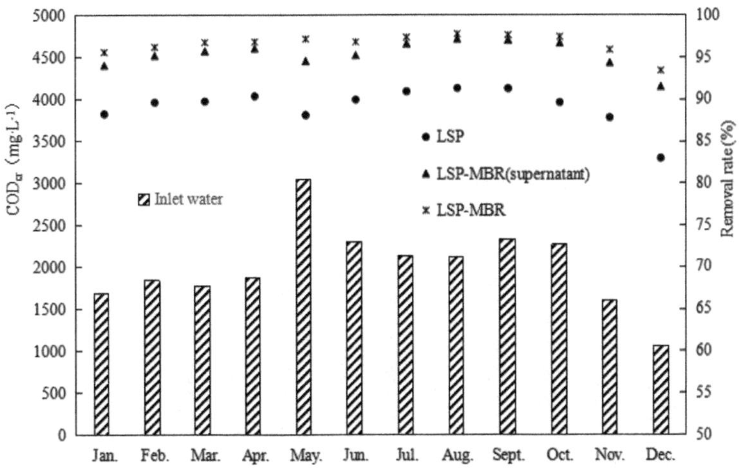

a. Inlet water CODcr and removal rate of CODcr by processes

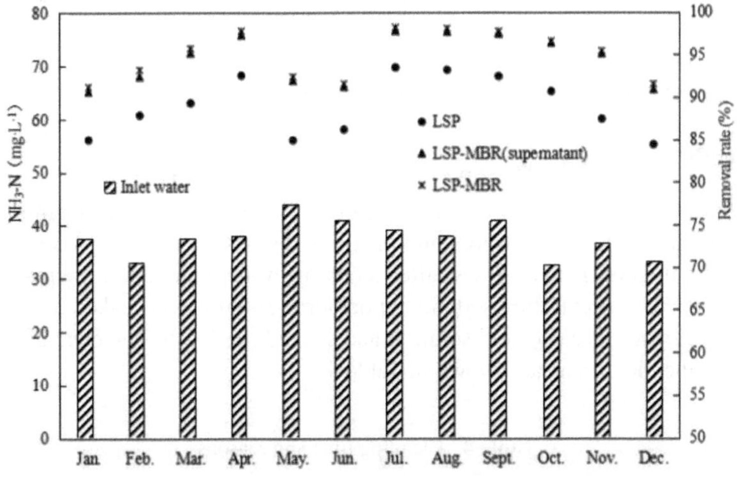

b. Inlet water NH3-N and removal rate of NH_3-N by processes

Fig. 3. Inlet water quality and removal rate by processes

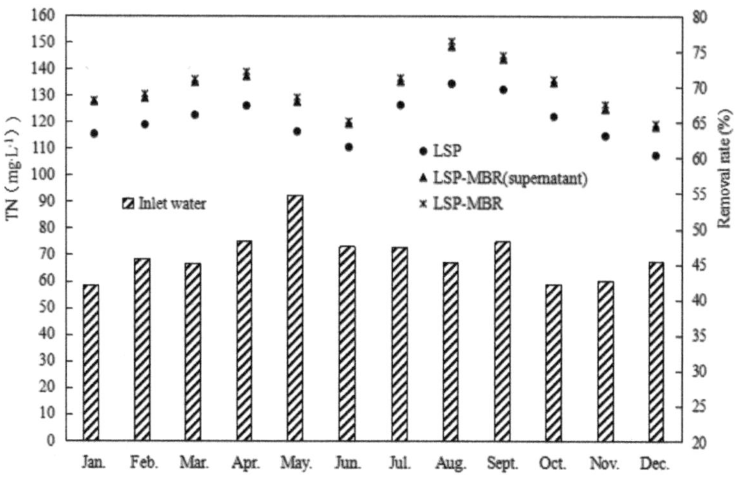

c. Inlet water TN and removal rate of TN by processes

Fig. 3. *(continued)*

4 Performance Analysis of Membrane

4.1 Membrane Cleaning

When trans membrane pressure difference (TMP) exceeded 35 kPa, membrane were cleaned online by alkaline washing and pickling, the dosage of agent were 5 L/ tablet. The specific method is as follows, after stopping water production, membrane were poured by a 3 ‰ concentration sodium hypochlorite solution and soaked for 5–8 h, then were poured by a 5‰ citric acid solution and soaked for 3–5 h. The cleaning time and reasons for the membrane are shown in Table 4.

Table 4. Cleaning time and reasons

No. of times	1	2	3	4	5	6	7	8	9	10
Cleaning time	Nov	Jan	Apr	Oct	Dec	Feb	Mar	Oct	Dec	Feb
Run time (months)	5	7	10	16	18	20	21	28	30	32
Reason	Routine	Routine	Impact	Routine	Routine	Routine	Impact	Routine	Routine	Routine
No. of times	11	12	13	14	15	16	17	18	19	20
Cleaning time	Mar	Jul	Nov	Jan	Mar	Jun	Oct	Nov	Dec	Jan
Run time (months)	33	37	41	43	45	48	52	53	54	55
Reason	Impact	Routine	Impact	Routine	Routine	Impact	Routine	Routine	Routine	Routine
No. of times	21	22	23	24	25	26	27	28	29	
Cleaning time	Feb	Mar	May	Jul	Sept	Oct	Oct	Nov	Nov	
Run time (months)	56	57	59	61	63	64	64	65	66	
Reason	Routine	Impact	Routine	Routine	Routine	Routine	Routine	Routine	Routine	

Membrane pollution is caused by adsorption and enrichment of activated sludge flocs, colloidal particles and water-soluble compounds [6]. The cleaning cycle is affected by sludge properties, pollutant concentration, ambient temperature, usage time and other factors [7–10]. The membrane in this project ran 5.5 years. Its cleaning time and reasons are shown in Table 4. The cleaning frequency was higher in seasons with lower temperatures than in seasons with higher temperatures. As the running time increased, the cleaning frequency also went up. The inlet water quality often fluctuated, which had a certain impact on the system. After the fluctuation of inlet water quality, the membrane needed to be cleaned.

4.2 Changes in Membrane Pore Size and Flux

During the 5.5-year duration of the project, the diaphragms running for 1 year, 5 years and 5.5 years were selected in the MBR tank for cleaning. And the electron microscope images of the diaphragms were observed, as shown in Fig. 4.

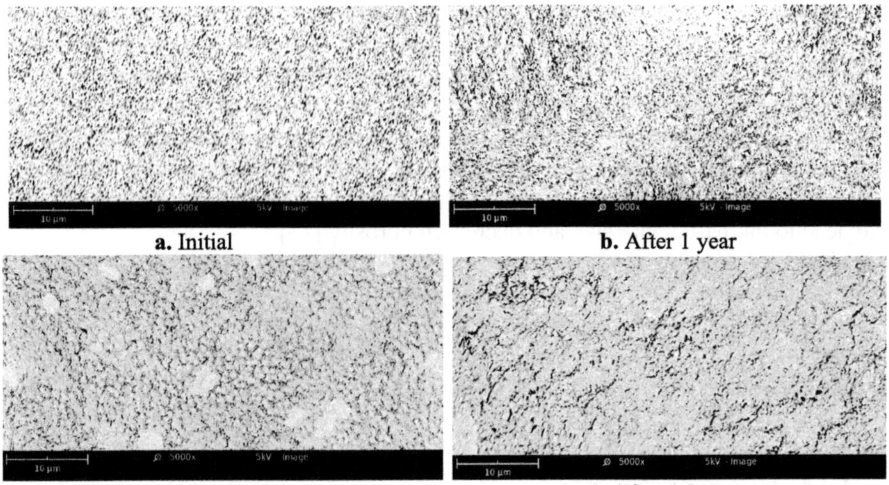

a. Initial **b.** After 1 year

c. After 5 year **d.** After 5.5 year

Fig. 4. SEM of the membrane at different times

As can be seen from Fig. 4, after1 year, most of the membrane pores could be recovered after cleaning. Meanwhile, the size of a few membrane pores increased, but the pore distribution uniform and the porosity almost remained unchanged. In this stage, the membrane flux decayed slowly, so a long cleaning cycle could be maintained. After 5 years, part of the membrane pores could not be recovered after cleaning. The size of some membrane pores increased, the distribution of membrane pores less uniform, and the porosity decreasing. In this stage, the cleaning cycle of the membrane was shortened. After 5.5 years, the membrane pores were still blocked in a large area after cleaning, with low porosity. The pore size of the unblocked membrane was increased and the distribution was concentrated and uneven. In this stage, the membrane flux decay was rapid. Therefore, the membrane pollution rate accelerated, and the cleaning cycle became short. The changes in membrane parameters are shown in Table 5.

Table 5. Changes in membrane parameters

	Initial	After 1 year	After 5 year	After 5.5 year
average pore size (μm)	0.16	0.18	0.32	0.35
maximum pore size (μm)	0.42	0.50	1.5	-
Porosity (%)	71.2	67.5	< 50	< 50

As shown in formula 1, with the contamination of the membrane, the porosity decreased, the water yield remained unchanged, and the TMP increased. When the membrane was running under a stable TMP, the membrane pore size increased. In the cleaning process, due to the blockage of some membrane pores, the cleaning agent could not be seep out uniformly. A large amount of cleaning agents remained in the membrane, and a short time of hydraulic impact caused damage to the unblocked membrane pores. With the increase of membrane running time and cleaning times, the pore size of the membrane increased and the porosity decreased. The larger the membrane pore size, the faster the membrane contamination rate and the shorter the cleaning cycle [11]. The lower the porosity, the lower the water yield [12] and the greater the TMP required to maintain the designed membrane flux. Membrane contamination and frequent cleaning may lead to increased pore size and decreased porosity [13].

$$r = \sqrt{\frac{8\eta LQ}{A \cdot \Delta p \cdot P_r}} \tag{1}$$

In the formula:
r: Membrane pore radius
L: Membrane thickness
A: Effective membrane area
Q: Flux of effluent
η: Fluid viscosity
ΔP: Operating pressure
P_r: Porosity

Under similar working conditions, the film flux comparison between the new film and the film that has been run for 5.5 years is shown in Fig. 5.

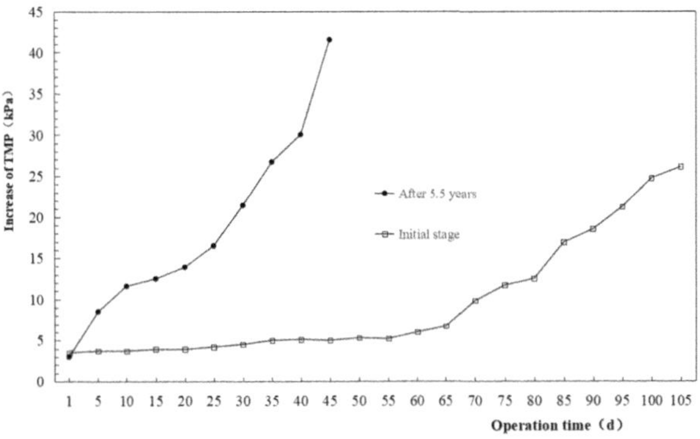

Fig. 5. TMP change of membrane at initial and after 5.5 years

As can be seen from Fig. 5, in the initial stage of system running, membrane could maintain a long cleaning cycle, and the growth rate of TMP was slow. It was due to entered a period of rapid growth after 65 days. After 5.5 years, due to the increase of the membrane pore size, the initial operation of TMP was low after cleaning, and the growth was relatively slow in the first 5 to 10 days. It was due to the fact that with the blockage of a small amount of pollutants, small and dense pore cavities formed in the membrane pores, they prevented large colloidal particles or water-soluble compounds from continuing to enter the membrane pores. Then the TMP continued to grow rapidly and increased by more than 35 kPa after 40 days, indicating that the membrane had reached its service life.

5 Conclusions and Suggestions

(1) MBR process, which adopted self-made PVDF/PVP flat membrane, was running in combination with LSP process. It has good removal effect on pollutants in chemical synthesis pharmaceutical wastewater. Under normal running conditions, the system has stable removal of COD_{cr} and NH_3-N while the removal of TN is affected by temperature. The highest removal rate could reach over 76% when the inlet water pH is 7.8–8.0 and the temperature above 25°C.

(2) PVDF/PVP membrane was used in chemical synthesis pharmaceutical wastewater treatment, with a service life of over 5 years, the membrane has a good interception effect on COD_{cr}. With the increase of the project running time, the porosity decreased, the membrane pore size increased, and the cleaning cycle became shorter.

(3) It is recommended to further modify the PVDF/PVP membrane and control the cleaning time during the operation. This may be beneficial for the stability of membrane pore size and porosity, and prolong the service life of the membrane.

References

1. Hazrati, H., Karimi, N., Jafarzadeh, Y.: Performance and antifouling properties of PVDF/PVP and PSF membranes in MBR: A comparative study. Membrane Water Treatment **11**(2), 159–166 (2020)
2. Sun, S., Ding, Y., Liu, H., et al.: Effect of PVP pore forming additive on the performance of polyvinylidene fluoride water treatment separation membrane. J. Changchun Univ. Technol. **43**(4/5), 500–506 (2022)
3. Chen, Q., Zhang, B., Li, H.: Experiment analysis on papermaking wastewater sludge reduction by LSP equipment. LiaoNing Chemical Industry (6), 543–546 (2017)
4. Wu, Y., Li, X., Wang, Y., et al.: Effect of temperature on simultaneous nitrogen and phosphorus removal and microbial community in anaerobic-aerobic-anoxic sequencing batch reactor. Desalination and water treatment **253**, 1–8 (2022)
5. Bhagyashree, T., Balasubramanian, S., Sarah, P.-C., et al.: Compositional microbial-community shift of submerged membrane bioreactor treating hospital wastewater at varying temperatures. J. Environ. Eng. **147**(2), 1–10 (2021)
6. Tian, W., Hu, Y., Wang, X.: Research progress on the characterization and control of cake layer in membrane bioreactors(MBRs). Membrane Science and Technology **36**(2), 141–147 (2016)
7. Ting, X.I.A.O., Zehua, Z.H.U., Lucheng, L.I., et al.: Membrane fouling and cleaning strategies in microfiltration/ultrafiltration and dynamic membrane. Sep. Purif. Technol. **318**(1), 123977 (2023)
8. Bin, Z.H.A.O., Hao, C.H.E.N., Diankui, G.A.O., et al.: Cleaning decision model of MBR membrane based on Bandelet neural network optimized by improved Bat algorithm. Applied Soft Computing Journal **91**, 106211 (2020)
9. Mirzavandi, A., Hazrati, H., Ebrahimi, S.: Investigation of influence of temperature and solid retention time on membrane fouling in MBR. Membrane Water Treatment **10**(2), 179–189 (2019)
10. Khan, S.J., Hasnain, G., Fareed, H., Aim, R.B.: Evaluation of treatment performance of a full-scale membrane bioreactor (MBR) plant from unsteady to steady state condition. Journal of Water Process Engineering **30**(1), 100379 (2019)
11. Huiying, Q.I.U., Tonghu, X.I.A.O., Ningen, H.U., et al.: Study of micro-polluted water treatment by membrane bioreactor (MBR) with microporous membranes with different pore size. Technology of Water Treatment **44**(3), 94–98 (2018)
12. Jia, X., Lan, L., Zhang, X., et al.: Research progress of hydrophobic microporous membrane for membrane distillation. Technology of Water Treatment **47**(10), 1–6 (2021)
13. Fang, G.A.O., Lin, M.A., Tao, H.U.O., et al.: Research progress on characterization and control of fouling membrane bioreactors. Environ. Eng. **38**(3), 69–75 (2020)

Removal of Ammonia Nitrogen from Micro-Polluted Raw Water by DAF

Shuo Zhang[1](✉), Jiajiong Xu[1], Min Rui[2], and Xuejun Tan[1]

[1] Shanghai Municipal Engineering Design Institute (Group) Co., Ltd., Shanghai 200092, People's Republic of China
zhangshuo@smedi.com

[2] Shanghai Water Industry Design Engineering Co., Ltd., Shanghai 200092, People's Republic of China

Abstract. DAF has been used to treat raw water with low temperature, low turbidity, high chroma or algae. In a DAF pilot, the removal path and mechanism of ammonia nitrogen were analyzed under different temperature. The results showed that the removal rates of total nitrogen and ammonia nitrogen were slightly reduced by air flotation under high temperature. When the temperature was low, the NO_2-N of effluent would increase a little. The concentration of ammonia nitrogen in the scum layer could reach 2.74 mg/L, which was the main removal way. At high temperature, the volatilization rate of ammonia nitrogen was 2.8 times higher than that at low temperature. When the wind speed was 4.5 m/s, the volatilization rate could be significantly increase to 1.5 mgN/m^2/d.

Keywords: Ammonia Nitrogen · DAF · Volatilization Rate · Scum · Wind Speed · Temperature · Raw Water

1 Introduction

The dissolved air flotation has been a kind of separation technology that has been constantly studied and promoted. DAF could be used for solid and liquid, liquid and liquid, solid and solid, and even solution ion separation. Modern dissolved air flotation technology has been used by Peterson and Sveen since 1924 as an alternative to the precipitation process for a hundred years. Since the 1960 s, DAF has developed rapidly as a process for treating sewage and drinking water. Even DAF has become increasingly used in tertiary wastewater treatment [1]. Graphical optimization was used to determine the optimum process parameters of the combined coagulant-DAF process [2].

For micro-polluted source water, conventional water treatment processes sometimes cannot meet the removal requirements of organic matter and ammonia nitrogen. In water supply treatment, air flotation as a replacement of precipitation technology, used to treat low temperature, low turbidity, high chroma and some algae-rich raw water. Dissolved-air flotation process has a good effect on removing turbidity and algae, and can effectively remove cryptosporidium oocysts. When the concentration of cryptosporidium oocysts in raw water was 500 /L, the concentration was less than 0.5 /L after DAF treatment

D. Han et al. (Eds.): ICEPG 2024, ESE, pp. 57–62, 2025.
https://doi.org/10.1007/978-3-031-86829-0_7

[3]. The DAF process can be combined with other water treatment processes. DAF was applied to the pretreatment of hollow fiber microfiltration membrane, which extended the running time of the microfiltration membrane. DAF was used as a pre-treatment to reduce the content of plankton and organic matter in water, which can significantly improve the effect of subsequent process treatment [4]. DAF are more efficient in terms of decolouration, odour reduction, and organic matter removal [5].

For raw water containing ammonia nitrogen, the removal effect of each form of nitrogen was analyzed by pilot test. The pathway of ammonia nitrogen removal was analyzed. The removal effect and kinetics of ammonia nitrogen by volatilization were studied. The experimental results could be used to guide the production practice of removing ammonia nitrogen by DAF in water plants.

2 Experimental Method

2.1 Pilot Equipment

Raw water comes from rivers. The turbidity, ammonia nitrogen and COD_{Mn} in the high temperature period were 30–110NTU, 1.5–2.1 mg/L and 2.5–3.5 mg/L respectively. They in the low temperature period were 10–40 NTU, 2.5–3.0 mg/L and 2.8–6.4 mg/L respectively. PAC dosage was 10 mg/L and the flocculation time was 10 min.

The flow rate of DAF was 1 m^3/h. Effective water depth was 1.5 m. The plane size of the contact area was 0.50 m $\times$ 0.20 m and the hydraulic retention time was 4 min. A TS-78V low pressure dissolved gas release device were provided. The plane size of the separation zone was 0.50 m $\times$ 0.50 m and the hydraulic retention time was 10 min. The design separation speed was 1.1 mm/s. Pressurized water was from the autolysis tank with 0.35 Mpa. The reflux ratio was 10%.

The total nitrogen determined by 0.45 microporous membrane filtration was dissolved total nitrogen DTN. Dissolved organic nitrogen was calculated by subtracting NH_3-N, NO_2-N and NO_3-N from DTN. Particulate organic nitrogen was measured as TN minus DTN.

2.2 Volatilization Rate

In order to measure the volatilization rate of gaseous ammonia in the floating zone, the concentration of ammonia nitrogen was measured after collecting the gas in the separation zone. The method of air pump blowing into the hood had been tested for volatile ammonia nitrogen. This method would cause the pressure in the hood to be greater than the actual atmospheric pressure, which could affect the volatilization of ammonia nitrogen. In this test, the air pump pumping method was adopted and the air in the hood could be communicated with the atmosphere through opening holes. The measured data would be closer to the actual value.

A plastic bucket with a diameter of 400 mm was placed upside down on the surface of the separation zone. The mouth of the barrel was about 100 mm below the water surface. Two small holes with a diameter of 20 mm were opened on the top side of the barrel to connect the air in the barrel with the atmosphere and maintain normal atmospheric pressure. One end of the silicone tube was fixed in the middle of the barrel on the other side of the small hole and the other end was connected with the air pump inlet. The silicone tube port connected to the outlet of the air pump was inserted into a 500 mL wide-mouth bottle containing 2% boric acid. The air from the pump was filtered through the boric acid solution. Air flow was controlled by intake valves and gas flowmeters. Gaseous NH_3 and boric acid would react to produce ammonia nitrogen ion, so the determination of NH_3-N in boric acid solution could calculate the amount of gaseous ammonia volatilized from the surface of the separation zone in a certain time. The local wind speed was converted according to the relationship between the wind scale and the wind speed of the meteorological forecast.

3 Results

3.1 Nitrogen Form

The particulate organic nitrogen (PON), dissolved organic nitrogen (DON), NH_3-N, NO_2-N and NO_3-N in influent and effluent were analyzed, as shown in Fig. 1.

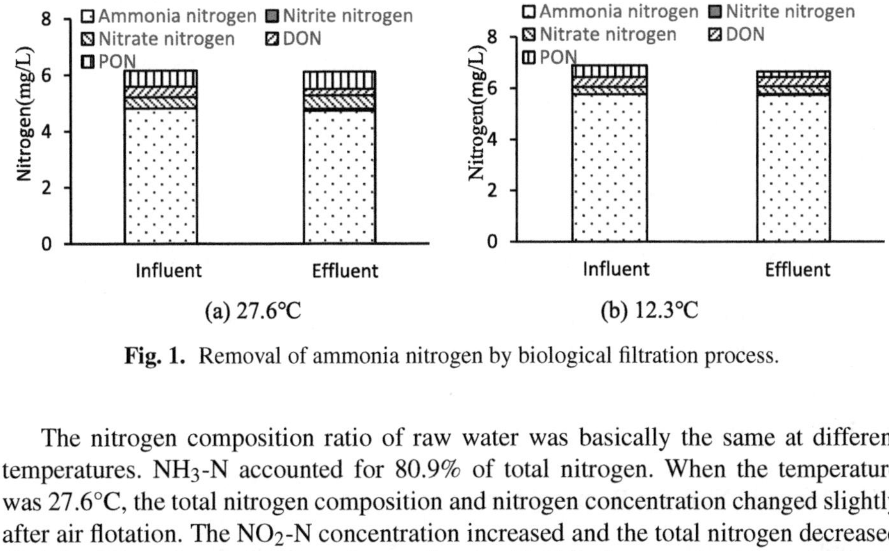

Fig. 1. Removal of ammonia nitrogen by biological filtration process.

The nitrogen composition ratio of raw water was basically the same at different temperatures. NH_3-N accounted for 80.9% of total nitrogen. When the temperature was 27.6°C, the total nitrogen composition and nitrogen concentration changed slightly after air flotation. The NO_2-N concentration increased and the total nitrogen decreased slightly. When the temperature was as low as 12.3°C, the concentration of NO_2-N increases less than that at high temperature, which indicated that temperature had an effect on its transformation.

3.2 Volatilization Rate

The pH of raw water was usually no more than 8.0, so most ammonia nitrogen was ammonium ion. The volatile ammonia nitrogen component was mainly ammonia. In the air flotation process, a large amount of air dissolves into the water with microbubbles, and the floating bubbles are likely to blow gaseous ammonia out of the water. The thousandths

Table 1. Effect of pH on oxidation time of potassium ferrate and phenol.

Parameter	Value			
Wind scale (level)	5 to 6	7	5 to 6	6 to 7
Wind speed (m/s)	2.5	4.5	2.5	4
Water temperature (°C)	27.9	29.6	12.5	10.7
pH	7.23	7.16	7.54	7.89
Volatile ammonia (‰)	0.05977	0.08789	0.02224	0.02317

ratio of volatilization rate and volatilization amount to total ammonia nitrogen in raw water were shown in Table 1.

Ammonia volatilisation represents an important nitrogen (N) loss in agriculture [6]. Unlike the reduction of gaseous nitrogen in agriculture [7], Ammonia volatilisation should be increased in water treatment. The volatilization rate at high temperature was obviously larger, which could be 2.8 times higher than that at low temperature. When the temperature was higher, the wind speed increased from 2.5 m/s to 4.5 m/s and the volatilization rate increased significantly by 1.5 $mgN/m^2/d$. Under strong wind and high temperature, the volatilization rate of ammonia was faster. This was because when the bubble burst, the gaseous ammonia released diffused further into the air. NH_3 on the water surface was diluted in time and the diffusion rate was accelerated. The volatilization would increase with the increase of wind speed [8]. The wind speed in the barrel was smaller than the actual local wind speed, and the measured result might be smaller than the actual situation. At high temperature, the bubble liquid film would become thinner, gaseous ammonia was more likely to enter the bubble from water. The volatilization was larger. At low temperature, the viscosity of water increased and the Henry constant decreases. It was difficult for gaseous nitrogen to precipitate and enter the bubble. The phenomenon of nitrogen volatilization existed, and the proportion of volatilization to total ammonia nitrogen was at the thousand-fraction level.

4 Discussion

4.1 Ammonia Nitrogen Removal Path

The effluent concentration of NO_2-N and NO_3-N increased significantly with growth rates of 10 times and 41% respectively, while the concentration of NH_3-N decreased slightly by about 0.1 mg/L. The residence time of air floating was very short, which could basically ignore the biological nitrification. The reduction of ammonia nitrogen mainly included three ways: scum, volatilization escape and REDOX conversion.

During the hydrolysis of coagulant, the intermediate product might form a complex with ammonia nitrogen. Ammonia was easily soluble in $Al(NH_3)_4^{3+}$ ions to form salts, so it might become a part of the floc to be removed. In the test, 20 ml of the upper scum of the floating tank was taken, diluted to 100 ml with deionized water, and then filtered by 0.45 μm filter membrane to determine the ammonia nitrogen. The result was as high as

2.74 mg/L. The influent and effluent ammonia nitrogen concentrations were 4.82 mg/L and 4.70 mg/L respectively.

Under the conditions of ideal reoxygenation and low concentration of ammonia nitrogen in water, the water body is open to the atmosphere and saturated with oxygen in the atmosphere, becoming an oxidizing medium. If thermodynamic equilibrium is reached, nitrogen will eventually become NO_3^-. This thermodynamic equilibrium is almost impossible to achieve in polluted water because the rate of REDOX reaction of nitrogen is very slow. It is difficult to reach an equilibrium state.

4.2 Volatilization Kinetics

The most reaction velocity and temperature relationships could be described by an exponential function similar to the Arrhenius equation. Therefore, f(PH)=$e^{\alpha \cdot PH}$ was used to further improve the results of Pano and Middlebrooks, as shown in Eq. 1.

$$\frac{C_e}{C_0} = \frac{1}{1 + \frac{S}{Q} K_N e^{\alpha \times pH}} \tag{1}$$

where C_e = effluent ammonia nitrogen, mg/L

C_0 = influent ammonia nitrogen, mg/L

S = surface area of separation zone, m^2.

Q = inlet flow, m^3/d

K_N = removal velocity coefficient, L/t

When pH didn't exceed 6.6, ammonia could not escape from the gas state. Equation 2 could be derived. The ammonia nitrogen concentrations at high and low temperatures were fitted with the least square method to obtain K_N' and α at each temperature range, as shown Fig. 2.

$$\ln\left(\frac{C_0}{C_e} - 1\right) - \ln\left(\frac{S}{Q}\right) = \alpha(pH - 6.6) + \ln K_N' \tag{2}$$

where K_N' = modified removal velocity coefficient, L/t

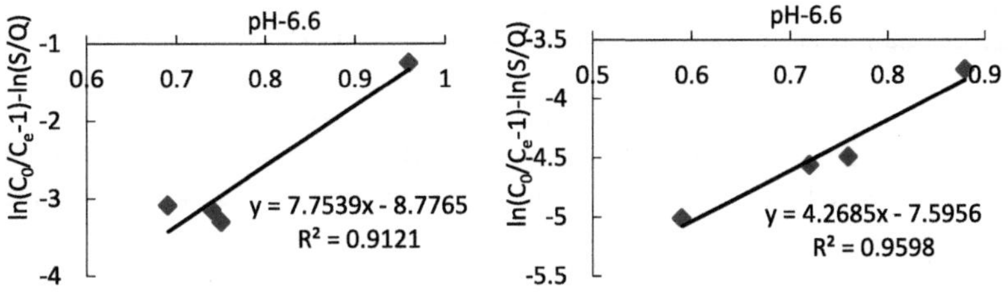

(a) 24.6-27.8°C (b) 9.3-13.5°C

Fig. 2. Removal of ammonia nitrogen by biological filtration process.

The α above 20 °C were larger than those at low temperature. At high temperature, the wind force was larger and the viscosity of water was smaller. The resistance of gaseous ammonia diffusion to the bubble and the escape resistance of microbubbles were both smaller.

5 Conclusions

The removal effect and kinetics of DAF on ammonia nitrogen was significant to analyze the removal path. The scum should be scraped out in time in practice.

(1) When the temperature was higher, the total nitrogen and each nitrogen concentration would change slightly. When the temperature was low, the concentration of NO_2-N in effluent increased less than that at high temperature.
(2) The concentration of ammonia nitrogen in the scum layer could reach 2.74 mg/L, which was the main way to remove ammonia nitrogen.
(3) There was removal of volatile ammonia nitrogen in the separation zone. At high temperature, the volatilization rate was 2.8 times higher than that at low temperature. The fitting results of kinetic equations show that α was larger at high temperature.
(4) Wind speed had an effect on volatilization. When the wind speed was increased to 4.5 m/s, the volatilization rate was significantly increased to 1.5 $mgN/m^2/d$.

Acknowledgments. This work was financially supported by the National Key Research and Development Program (No. 2023YFC3804700).

References

1. Oliveira, G.A., Machado, E.L., Knoll, R.S., Dell'Osbel, N., Colares, G.S., Rodrigues, L.R.: Combined system for wastewater treatment: ozonization and coagulation via tannin-based agent for harvesting microalgae by dissolved air flotation. Environ. Technol. **43**(9), 1370–1380 (2020)
2. Adlan, M.N., Palaniandy, P., Aziz, H.A.: Optimization of coagulation and dissolved air flotation (DAF) treatment of semi-aerobic landfill leachate using response surface methodology (RSM). Desalination **277**(1–3), 74–82 (2011)
3. Hall, T., Pressdee, J., Gregory, R., Murray, K.: Cryptosporidium removal during water treatment using dissolved air flotation. Water Sci. Technol. **31**(3), 125–135 (1995)
4. Dahlquist, J., Kulesza, M.: Pretreatment with dissolved airflotation considering an integrated process design. Water Science & Technology: Water Supply **1**(2), 115–122 (2001)
5. Jin, X., Jin, P., Wang, X.: A study on the effects of ozone dosage on dissolved-ozone flotation (DOF) process performance. Water Sci. Technol. **71**(9), 1423–1428 (2015)
6. Hearn, L.O., Barton, L., Schwenke, G.D., Murphy, D.V.: Ammonia volatilisation losses from urea applied to acidic cropping soils is regulated by pH buffering capacity. Soil Research **61**(7), 685–696 (2023)
7. Luyima, D., Lee, J.H., Sung, J., Oh, T.K.: Co-pyrolysed animal manure and bone meal-based urea hydrogen peroxide (UHP) fertilisers are an effective technique of combating ammonia emissions. J. Mater. Cycles Waste Manage. **22**(6), 1887–1898 (2020)
8. Meade, G., Pierce, K., O'Doherty, J.V., Mueller, C., Lanigan, G., Mc Cabe, T.: Ammonia and nitrous oxide emissions following land application of high and low nitrogen pig manures to winter wheat at three growth stages. Agr. Ecosyst. Environ. **140**(1–2), 208–217 (2011)

Analysis of the Collaborative Management Approaches of Water Resources Under the Joint Management Normality and Emergency in China

Weiwei Shao[1], Zimeng Zhao[2(✉)], Qingqing Li[3], Yali Yao[1], Jingyi Han[1], Zhiyong Yang[1], and Jiahong Liu[1]

[1] State Key Laboratory of Basin Water Cycle Simulation and Regulation, China Institute of Water Resources and Hydropower Research, Beijing, China

[2] Academy of Eco-Civilization Development of Jing-Jin-Ji Megalopolis, Tianjin Normal Universtiy, Tianjin, China

zhaozimeng0125@163.com

[3] Liaoning Normal University, Dalian, China

Abstract. China's overall water resources are relatively scarce, with insufficient per capita water resources. Nowadays, it is affected by the rapid change of climate and human social activities, extreme hydrological events have occurred frequently, floods and droughts coexist, and the situation of water resource management is severe. Integrating water resources under the joint management normality and emergency in a changing environment and implementing collaborative management based on hydrological process regulation will be an important direction for the future development of water resource management in China. This study analyzes the current situation of water resources under the joint management normality and emergency in China, constructs a framework of routine and emergency collaborative management of water resources in Jincheng City, Shanxi Province, and puts forward the mode and method of routine and emergency collaborative management of water resources. The results show that there are more continuous wet years and dry years in Jincheng city, and the water resources situation has a certain fluctuation, but there is a trend of good development. Under the mode of routine and emergency coordinated management of water resources, it is divided into Normalized level I, Emergency Level II and Crisis level III. Jincheng City is mostly in Emergency Level II. This research can effectively improve the level of integrated water resources management and optimal scheduling in China, and also provide certain technical support for the prevention of water and drought disasters.

Keywords: Water resources under the joint management normality and emergency · Water Resources · MIV-BP · SWOT Analysis

1 Introduction

China's water resource management generally includes water resources under the joint management normality and emergency. Due to the differences in management objectives, means and subjects, China's routine water resources management and emergency

D. Han et al. (Eds.): ICEPG 2024, ESE, pp. 63–70, 2025.
https://doi.org/10.1007/978-3-031-86829-0_8

management in extreme scenarios have long been separated [1]. However, from the perspective of hydrological cycle, hydrological process is a systematic process that integrates the extreme values of level water, abundance and drought water.

The whole process management of hydrology can be strengthened to promote the defense level of water resources during the dry season and the ability to respond to security during the wet season, that is, the defense and security capabilities of water resources system can be improved through the joint cooperative management of normal and emergency water resources [2].

At present, China is also actively trying to adopt the joint management model of normal and emergency management. In practice, such methods as joint reservoir operation, flood resource utilization, flood limit water level adjustment, drought limit water level design and construction, emergency water supply source construction [3] and modern means of water resources information sharing platform have been applied in the direction of water resources management. In terms of theoretical research, some scholars have conducted a preliminary analysis on the integrated management model of normal and emergency water resources in China [4]. However, at present, there are few studies on the water resources under the joint management normality and emergency models in specific regions under climate change, and the relevant theoretical research models are insufficient. Especially considering the development situation of hydrology and water resources fully, there is still a lack of in-depth research on the management mode of water resources related to hydrology and water resources under the joint coordination of normal and emergency. Therefore, the main work of this study is to explore normal and emergency joint management mode of water resources in China by fully considering the correlation between hydrological state and water resources.

In addition, relying on the nature-society dual water cycle theory, the discussion of water resources under the joint management normality and emergency mode from the perspective of combining hydrology and water resources will make water resources management more comprehensive and systematic, and make up for the inadequacy of the existing water resources under the joint management normality and emergency. This research contributes to revealing the development status of water resources normal and emergency coordinated management in China, meanwhile further exploring and optimizing the joint management mode of water resources normal and emergency in China.

2 Study Area and Methodology

2.1 Study Area

This research chooses Jincheng City of Shanxi Province in China as the research area.

The land area and resident population in Jincheng city reach 9,490 square kilometers and 2,353,000, respectively in 2020.The rivers in Jincheng City belong to two major river basins, namely the Yellow River and the Haihe River. Jincheng city is located in the east monsoon zone of warm temperate sub-humid climate region. Based on the meteorological data in Jincheng City, the annual average rainfall is 573.6 mm. The total water resources is 928 million cubic meters, and the actual water supply and consumption were 417 million cubic meters in Jincheng City in 2020 (see Fig. 1).

Fig. 1. The geographical location map of Jincheng City

2.2 Construction of Index System Under Normal and Emergency Joint Management Mode of Water Resources

Water resources normal and emergency joint system is an organic unity based on water resources security. The water resources system is divided into five subsystems: water quantity characteristic index, water quality characteristic index, characteristic index of engineering water shortage, water resources vulnerability characteristic index and characteristic index of water resources carrying capacity [5]. The interaction and mutual influence among the five subsystems constitute water resources security system. Based on the connotation of normal and emergency joint management of water resources, at the same time through the diagnosis and analysis of water resources system in Shanxi Province, routine and emergency index system of Shanxi Province is established, and the structure diagram is as shown. In this study, the BP neural network model is applied to find the input items that have a greater impact on the normal and emergency joint management of water resources, moreover through the diagnostic analysis method to understand the system affected by each index degree(see Fig. 2).

Mean impact value back propagation neural network (MIV-BP) is a computational model simulating the operation of human brain nervous system combined with MIV algorithm, which can better evaluate the correlation in neural network and greatly improve the accuracy of index screening and evaluation. Based on the influencing factors of water resources under the joint management normality and emergency, in addition to the actual precipitation condition in Jincheng City, referring to various national, industrial and regional norms and standards, the logistic activation function in the MIV-BP is adopted for multi-objective output. The sample output targets are classified and set according to the results: Normalized level I is greater than 1.5, Emergency Level II is between 0.75 and 1.5, and Crisis level III is between 0 and 0.75 [6–9]. The influencing factors of water resources under the joint management normality and emergency in Jincheng City are input into the MIV-BP model to obtain the grades of water resources

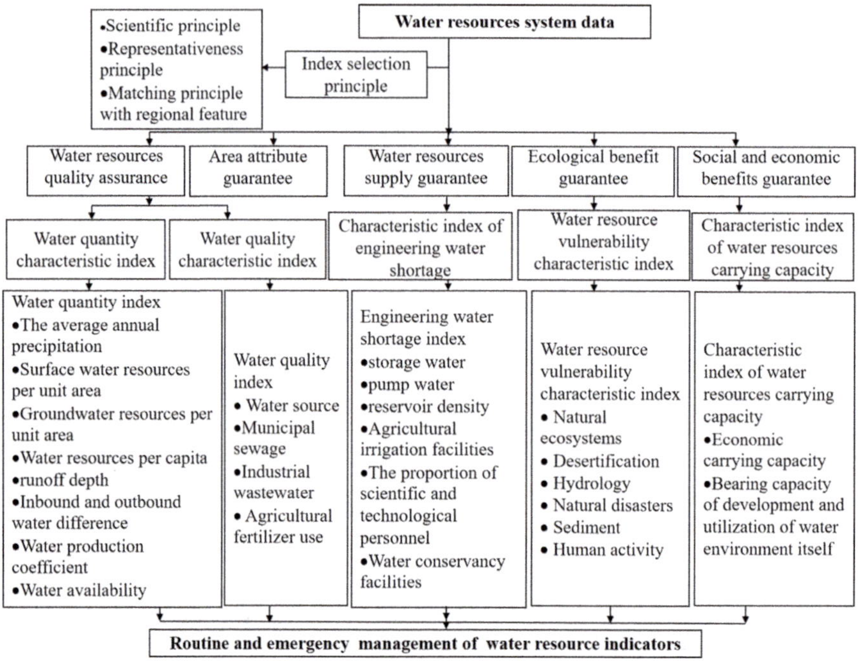

Fig. 2. Evaluation index framework of water resources under the joint management normality and emergency

under the joint management normality and emergency in Jincheng City from 2000 to 2022 (see Fig. 3).

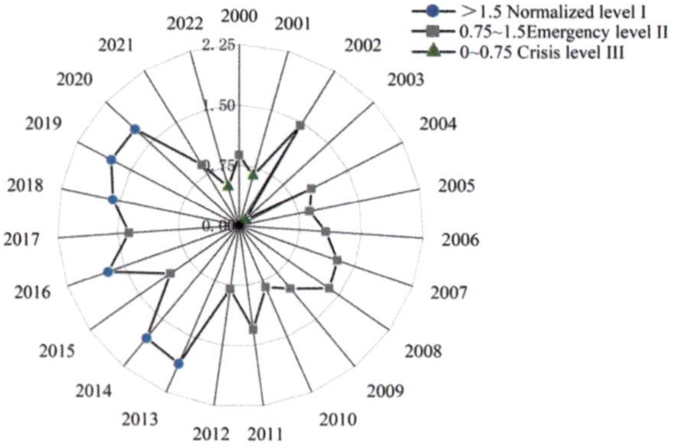

Fig. 3. Water resources under normal and emergency joint management level in Jincheng City

2.3 Routine and Emergency Coordinated Water Resources Management Mode

By combing and fitting the hydrological conditions, water resources state and water resources management capacity of Jincheng City from 2000 to 2022, the normal and emergency joint management mode of water resources in Jincheng City is obtained (see Fig. 4). The three-dimensional mode map of water resources under the joint management normality and emergency in Jincheng City, which is the comprehensive embodiment of the integrated hydrological conditions and water resources state. In the figure, the close relation between hydrological conditions, water resources management capacity and water resources state of Jincheng City can be clearly defined. The three-dimensional map of the collaborative mode of water resources under the joint management normality and emergency connects precipitation, water quantity and management, and reflects the correlation between natural precipitation, water use and social management intervention in Jincheng City, which is a typical research example based on the natural-social dualistic water cycle mode.

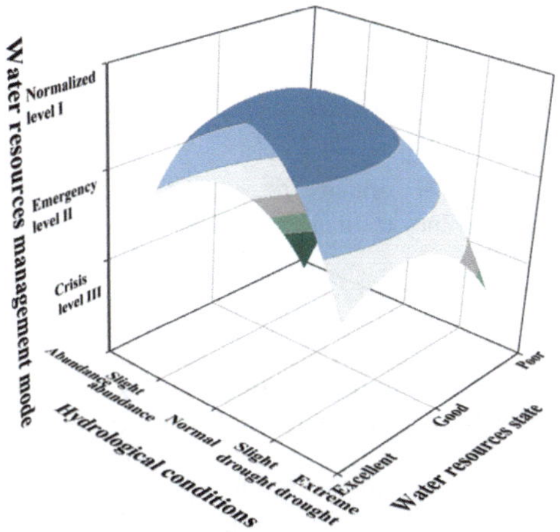

Fig. 4. Water resources routine and emergency coordinated management mode in Jincheng City

3 Water Resources Routine and Emergency Coordination Management Strategy

3.1 Water Resources Routine and Emergency Optimization Strategy in Jincheng City

According to the driving factors and hindering factors water resources under the joint management normality and emergency in Jincheng City, the indicators in the four directions of S, W, O and T are re-matched and combined, and the weights are determined to carry out the quantitative analysis of SWOT [9, 10]. According to the SWOT quantitative calculation results of Jincheng City's water resources under the joint management normality and emergency, a quadrilateral of strategy research under the joint management of water resources normality and emergency is constructed in Jincheng City'(Fig. 5). Through the calculated data of strengths, weaknesses, opportunities and challenges in the coordinates of normal and emergency management of water resources in Jincheng City, the results show that the coordinate on the SWOT strategy research focus of water resources under the joint management normality and emergency in Jincheng City is (0.007, −0.004).

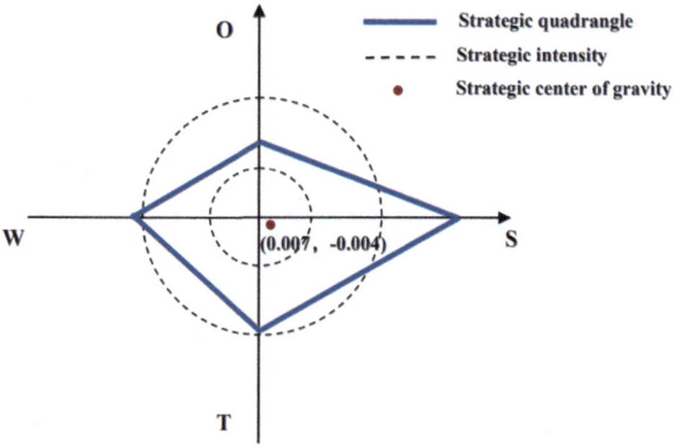

Fig. 5. The quadrilateral of SWOT strategy research on water resources under the joint management normality and emergency in Jincheng City

3.2 Water Resources Optimal Allocation in Jincheng City Based on Water Resources Under the Joint Management Normality and Emergency

The traditional allocation of water resources is only optimized according to the water storage and security status. The study mainly combines the factors affecting water resources under the joint management normality and emergency in Jincheng City, focuses on the preliminary backup emergency water source planning to optimize the allocation of Jincheng City water resources. Here, it takes 2020 as the base year and 2035 as the planning year for Jincheng City. Under the principle of ensuring economic, social and ecological water use in normal years, taking full account of the guarantee of water supply in extreme precipitation and drought years, 508 million cubic meters of water will be allocated in 2035. Designed and planned five emergency water supply projects, including Duhe reservoir to urban emergency water pumping project, Xiahe spring water diversion project II, Xiye reservoir water pumping project, Zhangfeng reservoir main trunk terminal pool to Lingchuan County water supply project and Sangu Spring water pumping project to ensure the emergency water supply demand of various districts and counties in Jincheng City, with a water supply capacity of 186,300 cubic meters/day.

4 Discussions and Conclusions

Based on the five sub-systems of water quality, water quantity, engineering water shortage, water resource vulnerability and carrying capacity, through the application of MIV-BP neural network model, the normal and emergency management system of water resources in Jincheng City is constructed, meanwhile explore the joint management mode of normal and emergency water resources. The results are shown below:

(1) In terms of research time research time, Jincheng City's precipitation fluctuates frequently and often appears alternately. There are more dry water years, continuous wet years or continuous dry water years, and there are phenomena of over wet years and over dry years.
(2) From 2000 to 2022, according to the results of MIV-BP model, Jincheng City water resources management model consists of three levels: normal level I, emergency level II and crisis level III. There are great differences in daily management and emergency management in different years.
(3) Jincheng City should adopt ST-type strategy to form the joint management mode of water resources normality and emergency. In normal years to ensure the economic and social water, ecological water. The water supply guarantee in extreme precipitation and drought years is fully considered, and five emergency water supply projects are designed and planned to ensure the emergency water supply needs of all districts and counties in Jincheng City.
(4) China should take the modern water network planning and construction as an opportunity, based on the existing flood and drought disaster prevention system, and focus on stimulating the efficiency level of water resources utilization, form a collaborative management mode of water resources under the joint management normality and emergency. In addition, it promotes water resources under the joint management normality and emergency to develop in a benign trend in Jincheng City.

Acknowledgment. This paper is supported by the National Natural Science Foundation of China (grant no. 51979285), the Natural Science Foundation of Beijing Municipality (grant no. 8222036), the Research Foundation of the State Key Laboratory of Simulation and Regulation of Water Cycle in River Basin of China Institute of Water Resources and Hydropower Research (grant no. SKL2022TS11), and the Major Science and Technology Project of the Ministry of Water Resources (grant no. SKS-2022126, WR110145B0042023).

References

1. Nuta, I., Ciortan, F.: Impact prevention and management of emergency situation for sustainable development. Procedia Economics and Finance **32**, 940–945 (2015)
2. Li, J.L., Tian, P., Shi, X.L., Gong, H.B.: Research progress on water resource management. J. Zhejiang Univ. (Science Edition) **46**(2), 248–260 (2019)
3. Ahmed, A.A., Sayed, S., Abdoulhalik, A.: Applications of machine learning to water resources management: a review of present status and future opportunities. J. Clean. Prod. **441** (2024)

4. Ngene, B.U., Nwafor, C.O., Bamigboye, G.O.: Assessment of water resources development and exploitation in Nigeria: a review of integrated water resources management approach. Heliyon **7**(1) (2021)
5. Lei, X.H., Zheng, H.Z., Shang, Y.Z.: Assessing emergency regulation technology in the middle route of the south-to-north water diversion project, China. Int. J. Water Resour. Dev. **34**(3), 405–417 (2018)
6. Wang, Y., Song, G.J., Liu, X.D.: Study on the determination of the demand and quantity of the urban emergency water resource. Water & Wastewater Engineering **38**(5), 19–22 (2012)
7. Du, Y.X., Hu, T.S., Guo, X.N.: Hydropower Reservoir Operation Rule by Considering Drought Emergency Condition **31**(8), 66–9 (2013)
8. Wang, X.F.: Framework of water resource capitalization management based on dualistic nature-social water cycle theory. China Popul. Resour. Environ. **26**(1), 83–88 (2016)
9. Anna, S., Zhi, G.Q.: Water resources governance stakeholder analysis and treatment network construction. Appl. Mech. Mater. **448–453**, 1019–27 (2014)
10. Katusiime, J., Schuett, B.: Integrated water resources management approaches to improve water resources governance. Water **12**(12) (2021)

Reducing Membrane Fouling During Microalgae Filtration Using a Uniform Shearing Vibration Membrane with Intermittent Operation

Yuhan Yin[1], Wenxin Hou[2], Shengnan Chen[1], Fangchao Zhao[1](✉), and Changqing Liu[1]

[1] School of Environmental and Municipal Engineering, Qingdao University of Technology, Qingdao 266000, China
zhaofangchao@qut.edu.cn

[2] Shandong Qilu Petrochemical Engineering Co., Ltd., Zibo 255400, China

Abstract. Uniform shearing vibration membranes (USVMs) represent a new technology to effectively control membrane fouling. Its intermittent operation reduces reversible and irreversible fouling. Frequent relaxation intervals, even with the same filtration-to-relaxation ratio, enhance results by mitigating concentration polarization and fouling. In this study, membranes operated with 9-min filtration with a 1-min interval outperformed those with 54-min filtration with a 6-min interval and continuous filtration. After a 12-h filtration experiment, the trans-membrane pressure increase for the 9-min/1-min group was only 0.6 kPa, compared to 2 kPa for continuous filtration—a 330% higher increase. The reversible fouling rate (RF), irreversible fouling rate (IF) and total fouling rate (TF) were 31.1%, 25.4% and 56.5% for the continuous filtration group, 17.5%, 16.9% and 34.3% for the 9-min/1-min group, and 27.5%, 21.6%, 49.1% for the 54-min/6-min group. Compared with continuous filtration, the 9-min/1-min group interval reduced RF and IF by 43.7% and 33.3%, showing the lowest fouling among the three groups. Additionally, membrane-attached algal cells, proteins, polysaccharides, and humic acids decreased with increasing relaxation intervals for both reversibly and irreversibly attached extracellular organic matter.

Keywords: Uniform shearing vibration membrane · Intermittent operation · Microalgae · Membrane fouling · EOM

1 Introduction

With the depletion of fossil energy like coal and oil, the search for clean energy alternatives is a pressing issue [1]. Whereas microalgae can use nutrients in municipal wastewaters for growth [2] and also be used as raw materials for biodiesel [3]. But microalgae enrichment and harvesting are challenging due to their small size and water-like density. Membrane filtration, increasingly accepted for microalgae harvesting, offers advantages like low energy consumption, minimal space, high automation, no chemical reagents, and high recovery efficiency [4]. Despite this, membrane separation has some

D. Han et al. (Eds.): ICEPG 2024, ESE, pp. 71–81, 2025.
https://doi.org/10.1007/978-3-031-86829-0_9

disadvantages, including membrane fouling, blockage, damage, and high cost. Herein, membrane fouling reduces separation efficiency, flux, and lifespan. Thus, it is urgent to reduce membrane fouling and improve membrane flux.

Previous research has shown that a USVM device can provide a high and steady shear force on the membrane's surface, substantially reducing the deposition of algal cells on the membrane surface and boosting microalgae recovery efficiency [5]. The surface shear force of USVM is greater and more stable than that of aeration membranes. As shear action increases, it becomes more difficult for foulants to adhere to the membrane surface, reducing extracellular organic matter (EOM).

Intermittent operation can reduce membrane fouling to a great degree [6]. This article adopted a faster interval that was the same as the commonly used ratio, with a shorter cycle time, which was equivalent to testing the impact of cycle length on membrane fouling under the same work and rest time ratio. Combining USVMs and intermittent filtration may improve results than USVM alone. This study examines how intermittent time affects membrane fouling and surface morphology in microalgae filtration with shear-vibrating membranes and compares the main components of EOM.

2 Materials and Methods

2.1 Cultivation of Algae

Chlorella pyrenoidosa was obtained from the Institute of Hydrobiology (Wuhan, China) and was grown in basal medium. The pH was adjusted to 6.1. The medium was sterilized at 121 °C in an autoclave for a duration of 30 min. All of the conical flasks were placed in a light incubator (HGZ-400, Shanghai Huitai Instrument Manufacturing Co., Ltd, China). The cultures were incubated at 25 °C under an illumination intensity of 127 μmol/(m^2·s) with a 14 h/10 h light/dark cycle. Artificial intermittent shaking (four times per day) was performed every day.

2.2 Membrane Properties

All membranes were provided by Jinzheng Filter Technology Co. (Haining, Zhejiang). The membrane material was hydrophilic polyvinylidene fluoride (PVDF), the effective filtration area of the membrane module was 0.02 m^2, the average pore size of the membrane was 0.1 μm, and the thickness was approximately 100 μm.

2.3 Experimental Setup

The Fig. 1 is a schematic diagram of the critical flux test device for USVM, which is equipped with three sets of membranes. Peristaltic pump was used to pump the filtrate into a beaker. An electronic balance connected to a computer recorded the filtrate's volume and flow rate. A vacuum gauge placed between the membrane assembly and the pump recorded the transmembrane pressure drop (TMP). The stepping motor speed was set to 240 rpm, the membrane vibration frequency was 4 Hz, the amplitude was 2 cm, the filter tank volume was 25 L and the algal concentration was 0.45 g/L. TMP measurements were taken every 30 min while the experimental flux for 12 h was 26 L/(m^2 h). Three intermittent run times were established: continuous filtration (group-a), filtration for 9 min with a 1-min interval (group-b), and filtration for 54 min with a 6-min interval (group-c). The operating ratios of group-b and group-c were the same (9:1). The reason group-c had a filtration time of 54-min/6-min group is that this setting did not require frequent pump operation, which reduce pump losses and facilitate operation.

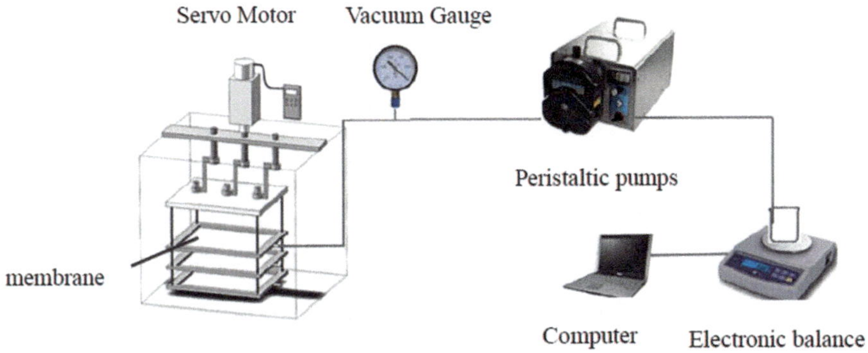

Fig. 1. Schematic diagram of the critical flux test device for USVM.

2.4 Analytical Methods

EOM in Algal Solution and on Membrane.

Algal solution contained a large number of elements, trace elements, and auxiliary growth substances. The algal solutions (5 mL) were centrifuged for 10 min at 4000 rpm in a high-speed centrifuge, and the supernatant was filtered through a 0.45 μm filter to obtain the EOM. The fouled membranes were first rinsed with 250 mL of distilled water, the microalgae were collected, and algal concentrations were measured. Then centrifuged at 4000 rpm for 10 min. Finally, a reversible EOM solution was obtained by filtered through a 0.45 μm filter membrane [7]. The cleaned membranes were immersed in 0.5 g/L NaOH solution for 2 h to fully dissolve EOM into solution, thus yielding an irreversible EOM solution [8].

Fouling Rates

From the perspective of membrane cleaning, fouling can be divided into reversible and irreversible fouling [9]. The former can be removed using physical methods such as water washing, whereas the latter cannot and it must be removed using chemical methods. The calculation formulas of the reversible fouling rate (RF), irreversible fouling rate (IF) and total fouling rate (TF) [10] are as follows:

$$TF = 1 - \frac{J_b}{J_n} \tag{1}$$

$$IF = 1 - \frac{J_a}{J_n} \tag{2}$$

$$RF = TF - IF \tag{3}$$

where J_n is the water flux of a new membrane, J_a is the water flux after flushing, and J_b is the water flux before flushing. TF is the flux decline rate of the fouled membrane before flushing compared with a new membrane, IF is the flux decline rate after flushing, and RF is the flux decline rate due to reversible foulants.

3 Results and Discussion

3.1 Filtration and Fouling

The critical flux experiments were performed under continuous operation conditions. The critical flux was determined using the improved flux-step method in the constant pressure method. The initial flux was set to 16 L/(m^2·h), with a 2 L/(m^2·h) increase and a 15-min duration. Membrane flux is divided into rising and falling phases, and for the same membrane flux conditions, the TMP in the falling phase is usually greater than that in the rising phase (Fig. 2). Because the membrane flux is low during the ramp-up phase, TMP growth is slower. TMP growth reaches 20 Pa per minute at a membrane flux of 28 L/(m^2·h), which is considered the critical flux for USVMs. Based on the critical flux, three groups of membranes with an operating flux of 26 L/(m^2·h) were used (lower membrane fouling rate at subcritical flux), other conditions remained constant, filtration was performed for 12 h, and TMP was recorded every 30 min. Membrane fouling was investigated by varying the run times: continuous filtration without interruptions (group-a), filtration for 9-min/1-min group (group-b), and filtration for 54-min/6-min group (group-c).

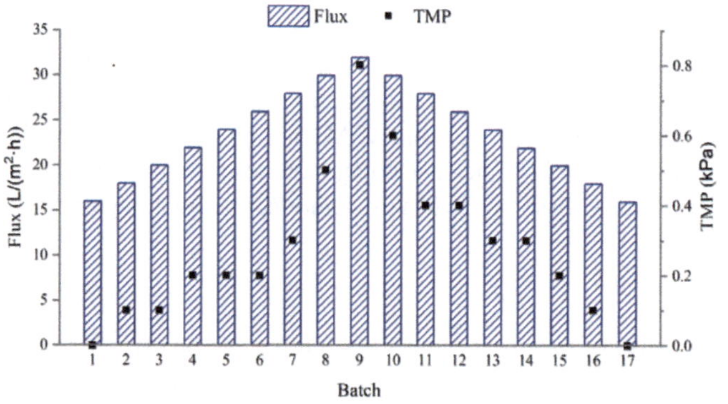

Fig. 2. Variation in TMP with membrane flux in the critical flux test

As shown in Fig. 3. The TMP in group-a increased gradually by 2 kPa by the end of the experiment. The group-b decreased from 90 to 150 min, then increased slowly, remaining steady from 360 to 700 min and then gradually increased with a total increase of 0.6 kPa. The TMP of group-c decreased substantially at 60, 120, 180, 420, and 480 min, while the overall TMP increased by 1.5 kPa, which was less than that in group-a, implying that group-c had less fouling. A specific relaxation interval may aid in reducing concentration polarization and membrane fouling. The sudden decrease observed in TMP in group-c may be due to one of the following reasons: (1) the reversible foulants attached to the membrane were flushed by the high shear forces during the 6-min relaxation interval; (2) the vacuum gauge data were read only in the interval stage. Both group-b and group-c had a lower TMP than group-a, but the TMP in group-b was significantly lower. Although the same filtration interval ratio (9:1) was used in group-b as in group-c, the conditions in group-b mitigated concentration polarization because more frequent relaxation intervals were employed, which reduced membrane fouling more effectively.

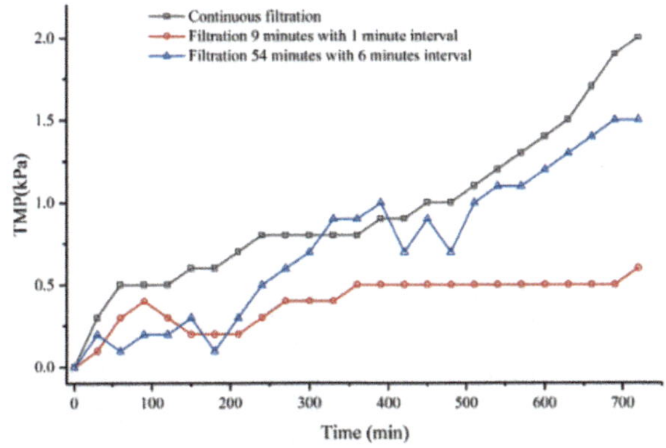

Fig. 3. Effect of intermittent operation on TMP.

The water flux of the three groups was measured (Fig. 4), and the water flux of the new membrane was 2421.9 L/(m^2·h). After a 12 h filtration test, the water flux in group-a decreased to 1053.3 L/(m^2·h), group-c's decreased to 1232.5 L/(m^2·h), and group-b's decreased to 1590.8 L/(m^2·h). Group-b had the lowest fouling rate that the RF of 17.5% time and frequency, membrane fouling during the filtration of microalgae can be effectively reduced, and better for reversible fouling. Under the same filtration interval ratio, more frequent running intervals lead to a more effective reduction in fouling.

3.2 Foulants on the Membrane

Microalgal Content

For growth measurement, the optical density of microalgae was measured at 680 nm. The samples were diluted using appropriate ratios when necessary to ensure that the measured OD680 values were in the range of 0.100–0.500. As can be seen from Fig. 5, the content of microalgae attached to group-a was 0.42 g/m^2; that attached to group-b

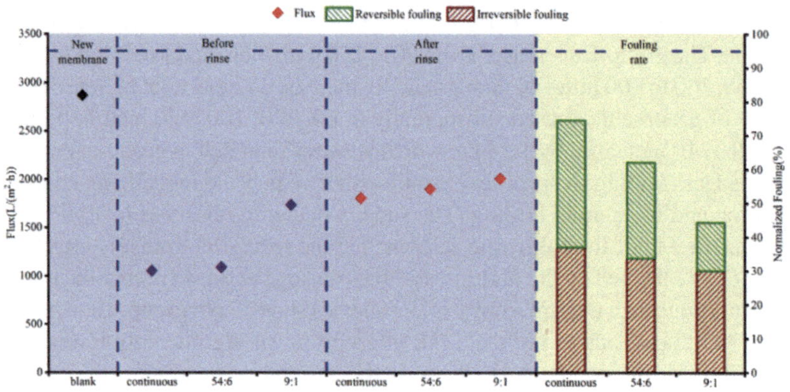

Fig. 4. Effect of a long filtration interval on water flux before and after flushing and RF and IF

was less than 0.30 g/m^2; and that attached to group-c was only 0.21 g/m^2. The number of microalgae attached was directly proportional to the membrane fouling rate.

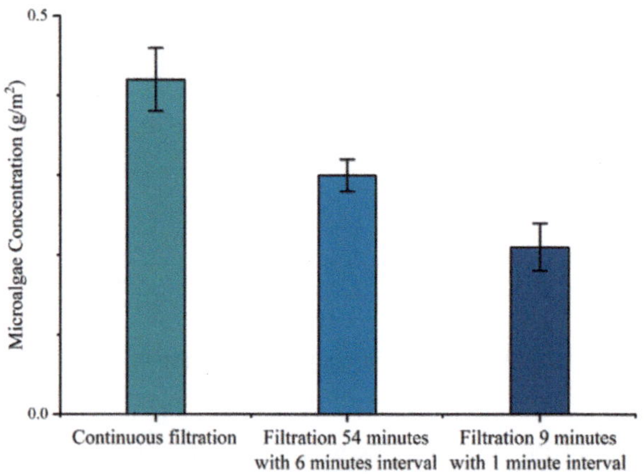

Fig. 5. Microalgal content deposited on the membrane surface.

Composition of EOM

The EOM of microalgae is mainly composed of polysaccharides (PS) and proteins (PN). However, excessive accumulation of EOM inhibits the growth of microalgae, contributes to the death of microalgae [11], contaminates the membrane and increases operating cost. Therefore, methods must be devised to reduce the accumulation of EOMs.

TOC, PN, and PS levels were lower in both intermittent filtration groups than in the continuous filtration group (Table 1). In the 12-h filtration, higher vibration frequency reduced the reversible TOC, PN and PS contents, which indicated that increasing vibration frequency could reduce EOM content. Group-a had 1113.01 mg/L reversible EOM and 891.89 mg/L irreversible EOM. Continuous filtration led to a thicker algae layer on membranes, which retained more EOM [12]. The reversible EOM of group-c was 1017.39 mg/L, which was 8.59% lower than that of group-a, and the irreversible EOM was 740.82 mg/L, which was 16.94% lower than that of group-a. Among the three groups, group-b exhibited the lowest quantity of reversible and irreversible EOM on the membrane, with a reversible EOM of 1113.01 mg/L, which was 24.38% lower than that of group-a, and an irreversible EOM of 891.89 mg/L, which was 22.76% lower than that of group-a. Increased relaxation time can effectively reduce foulants attachment, and the relaxation interval can alleviate concentration polarization and reduce foulants attachment to the membrane, supporting the preceding conclusion.

Table 1. TOC, PN and PS contents in reversible EOM and irreversible EOM at different group

Sample		TOC (mg/m^2)	PN (mg/m^2)	PS (mg/m^2)
Continuous filtration (group-a)	Reversible EOM	845.78	224.17	43.06
	Irreversible EOM	663.14	198.06	30.69
54-min/6-min (group-c)	Reversible EOM	782.67	201.94	32.78
	Irreversible EOM	563.44	151.94	25.44
9-min/1-min (group-b)	Reversible EOM	629.69	190.74	21.30
	Irreversible EOM	539.14	131.11	18.61

Three-Dimensional Fluorescence Spectra of EOM

Due to low density, self-floating characteristic and certain hydrophobicity, microalgae is easy to combine with bubbles, thus flotation is an effective method for microalgae harvesting [13].Thus most microalgae cells and some EOM are collected by flotation, resulting in a significant reduction in membrane fouling and flux attenuation. Excitation-emission matrix (EEM) fluorescence spectroscopy is widely used in the detection of EOM. This technology can be used to detect PN and humic acid, but it is difficult to detect PS [14].

The EEM fluorescence spectroscopy of reversible EOM and irreversible EOM on algae solution and three groups are shown in Fig. 6. Figure 6(a) shows that PN and humic acids are present in the algal solution, and the peak intensity of Peak A is much larger than that of Peak T_1. However, Peak A does not appear in the spectra, which indicates that humic acids are difficult to detect using this method. Peak T_1 in reversible EOM was stronger than that in irreversible EOM. Group-a has slightly higher Peak T_1 intensities than other intermittent filtration groups. The peak T_1 intensity of group-b was the lowest, which indicated that intermittent operation could reduce PN adhesion, and more effective than group-c.

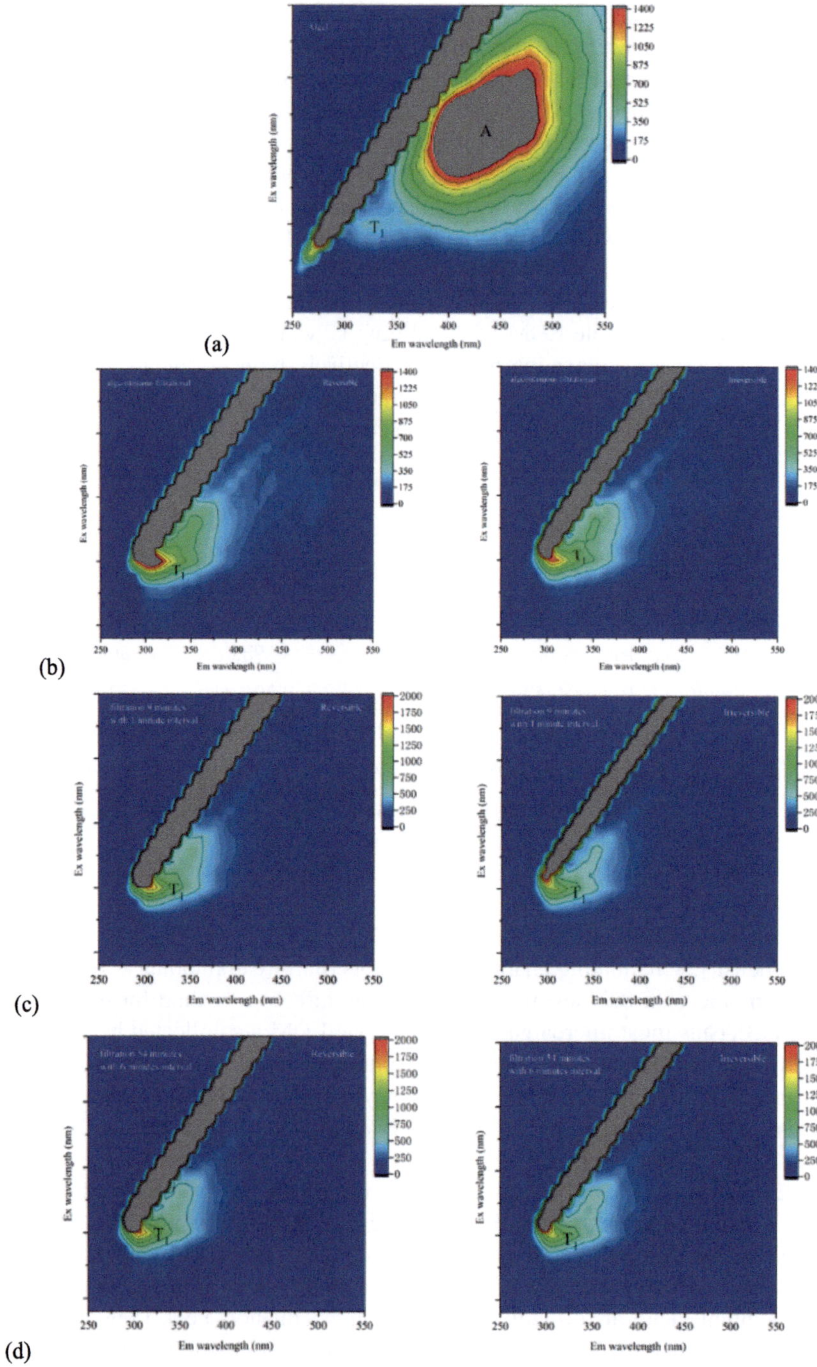

Fig. 6. Microalgal content deposited on the membrane surface: (a) algal solution EOM, (b) continuous filtration (group-a), (c) 9-min/1-min group (group-b), (d) 54-min/6-min group (group-c)

SEM Analysis

A scanning electron microscopy (SEM) approach was utilized to more clearly determine membranes alterations at the end of the experiment, so that fouling of the membranes was clearly observed. As shown in Fig. 7, the membranes before flushing in Fig. 7(b) were mostly fouled and most of pores were blocked. While the membranes shown in Fig. 7(d) had less foulants, there was hardly any in Fig. 7(f) and only a few membrane pores were blocked. To observe irreversible fouling of the membrane, all three groups of membranes were washed with highly pressurized water. After washing, the foulants covered by the membrane surface (filter cake layer) were significantly reduced.

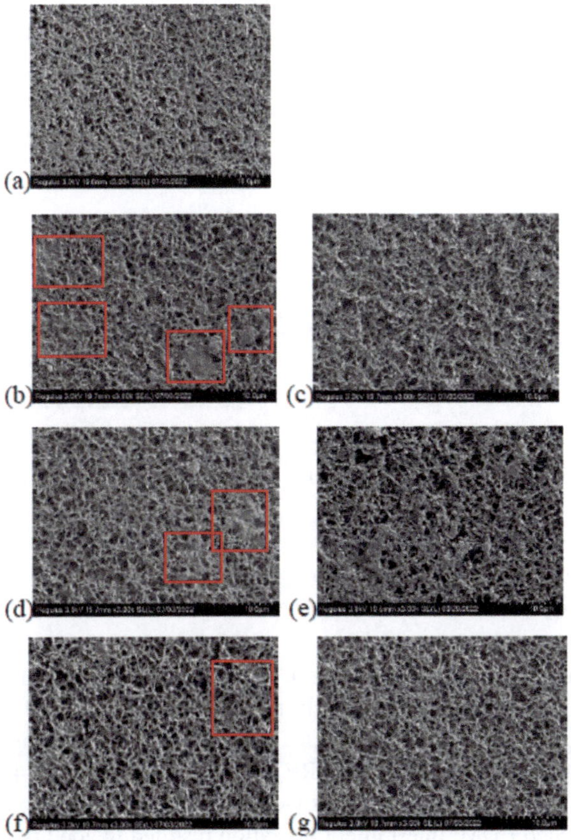

Fig. 7. SEM image of membrane surface under different relaxation time intervals: (a) New membrane; (b) continuous filtration before flushing; (c) 54-min/6-min group before flushing; (d) 9-min/1-min group before flushing; (e) continuous filtration after flushing; (f) 54-min/6-min group after flushing; (g) 9 min/1-min group after flushing

FTIR Analysis

PN are represented by the characteristic peaks at 1540, 1645, and 3300 cm^{-1} [15], PS's peaks are1000–1260, 2925, and 3200–3500 cm^{-1} [16], humic acids peak at 1720 cm^{-1} [17], and the PVDF membrane material peaks at 1071, 1181, and 1407 cm^{-1} [18]. The FTIR spectra of the three groups of membranes after 12 h filtration is shown in Fig. 8. In descending order, the peak intensities at 1540, 1645, and 3300 cm^{-1} correspond respectively to group-a, group-c, and group-b, The peak intensities at 1000–1260, 2925 and 3200–3500 cm^{-1} were in descending order for group-a, group-c and group-b. Accordingly, the peak intensity order of PN and PS is the same. The peak intensities at 1720 cm^{-1} are in complete agreement with the PN and PS peaks, while the peak intensities at 1071, 1181 and 1407 cm^{-1} are in the opposite order, indicating that group-b was less fouled than group-c, and group-a was most fouled among the three groups.

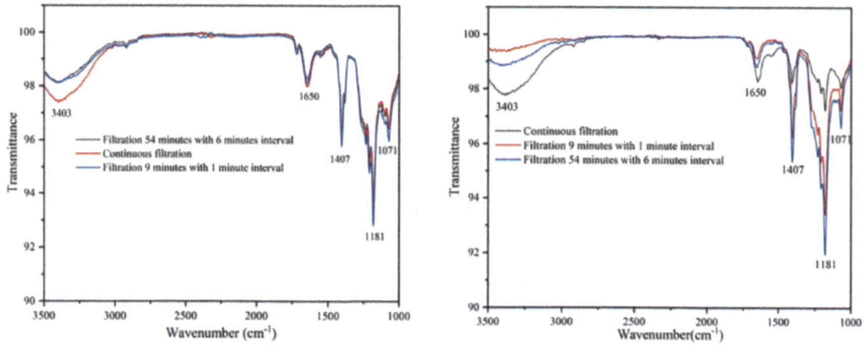

Fig. 8. FTIR spectra of the membrane under different relaxation time intervals

4 Conclusion

USVM filtration, which can decrease both reversible and irreversible fouling, can be employed to reduce membrane fouling. In this study, filtration for 9 min with a 1-min interval significantly decreased membrane fouling compared to continuous filtration. After 12 h of filtration, the membrane's RF and IF were lowered by 43.7% and 33.4%. The amounts of PN, PS, and humic acids were found to be lower in both reversibly and irreversibly attached EOM. Intermittent operation resulted in good performance for the control of membrane fouling.

References

1. Fazle, R.M., József, P., Domicián, M., Sándor, K.: Energy security and energy transition to achieve carbon neutrality. J. Energies **15** (2022)
2. Bharati, B.V., Mariya, M.: A review on synchronous microalgal lipid enhancement and wastewater treatment. J. Energies **14** (2021)

3. Voloshin, R.A., Rodionova, M.V., Zharmukhamedov, S.K., Nejat Veziroglu, T., Allakhverdiev, S.I.: Review: Biofuel Production from Plant and Algal Biomass. J. Alternative Energy and Ecology (ISJAEE) (2019)
4. Goh, P.S., Ismail, A.F.: A review on inorganic membranes for desalination and wastewater treatment. J. Desalination **434** (2018)
5. Zhao, F., et al.: The interaction between microalgae and membrane surface in filtration by uniform shearing vibration membrane. J. Algal Res. **50** (2020)
6. Huihui, C., et al.: Enhancing nitrogen removal of carbon-limited municipal wastewater in step-feed biofilm batch reactor through integration of anammox. J. Bioresource Technology **381** (2023)
7. Zhao, F., et al.: A uniform shearing vibration membrane system reducing membrane fouling in algae harvesting. J. Cleaner Production **196** (2018)
8. Zhang, Y., et al.: Chemical cleaning of fouled PVC membrane during ultrafiltration of algal-rich water. J. Environm. Sci. **23** (2011)
9. Long, T., et al.: Insights into the properties of surface waters and their associated nanofiltration membrane fouling: the importance of biopolymers and high molecular weight humics. J. Chem. Eng. **451** (2023)
10. Zhao, F., et al.: Increasing the vibration frequency to mitigate reversible and irreversible membrane fouling using an axial vibration membrane in microalgae harvesting. J. Membrane Sci. **529** (2017)
11. Scutt, J.E.: Autoinhibitor Production by Chlorella vulgaris. J. American J. Botany **51** (1964)
12. Chu, H., Zhao, Y., Zhang, Y., Yang, L.: Dewatering of Chlorella pyrenoidosa using a diatomite dynamic membrane: Characteristics of a long-term operation. J. Membrane Sci. **492** (2015)
13. Zhang, X., Wang, L., Sommerfeld, M., Hu, Q.: Harvesting microalgal biomass using magnesium coagulation-dissolved air flotation. Biomass Bioenerg. **93**, 43–49 (2016)
14. Xu, H., Cai, H., Yu, G., Jiang, H.: Insights into extracellular polymeric substances of cyanobacterium Microcystis aeruginosa using fractionation procedure and parallel factor analysis. J. Water Res. **47** (2013)
15. Qu, F., et al.: Characterization of dissolved extracellular organic matter (dEOM) and bound extracellular organic matter (bEOM) of Microcystis aeruginosa and their impacts on UF membrane fouling. J. Water Res. **46** (2012)
16. Qu, F., et al.: Ultrafiltration (UF) membrane fouling caused by cyanobateria: Fouling effects of cells and extracellular organics matter (EOM). J. Desalination **293** (2012)
17. Chu, H., et al.: Extraction procedure optimization and the characteristics of dissolved extracellular organic matter (dEOM) and bound extracellular organic matter (bEOM) from Chlorella pyrenoidosa. J. Colloids and Surfaces B: Biointerfaces **125** (2015)
18. Zhang, J., Wang, Q., Wang, Z., Zhu, C., Wu, Z.: Modification of poly(vinylidene fluoride)/polyethersulfone blend membrane with polyvinyl alcohol for improving antifouling ability. J. Membrane Sci. **466** (2014)

Predicting Water Pollution Levels Using Gradient Boosting Regression Tree: An Analysis of Data from Chinese Publicly Listed Companies

Yanjie Jiang[1], Linlin Yue[1], Linlin Zhang[1], and Xiaodi Ding[2](✉)

[1] School of Management, Shandong Second Medical University, 7166 Baotong West Street, Weifang 261053, Shandong, China

[2] School of Economics and Management, Central Campus, Weifang Institute of Technology, 7777 Minzhu Street, Weifang 261021, Shandong, China

dxd1225@163.com

Abstract. Water pollution, intensified by rapid industrialization and urbanization in China, poses significant environmental challenges. This study aims to predict water pollution levels using data from Chinese publicly listed companies spanning from 2007 to 2022. Employing the Gradient Boosting Regression Tree model, we focus on key pollutants: Chemical Oxygen Demand, Ammonia Nitrogen, Total Nitrogen, and Total Phosphorus. The dataset includes environmental and air pollution-related variables, with missing values imputed for consistency. Results indicate that 'Year' is the most influential feature across all pollutants, while 'Dust' and 'COD' are the least influential. Notable interactions between 'Year' and features like 'NOx' and 'SO_2' were observed. The model shows strong predictive performance for NH_3_N and TN, moderate for TP, and poor for COD, indicating areas needing further optimization. This study offers valuable insights for environmental management and policy-making, emphasizing the importance of temporal trends and feature interactions in predicting water pollution levels.

Keywords: Water Pollution Prediction · Gradient Boosting Regression Tree · Environmental Management

1 Introduction

1.1 Background

Water pollution poses significant risks to human health, ecosystems, and economic activities. In China, rapid industrialization and urbanization have led to severe contamination of water bodies, with publicly listed companies, especially in heavy industries. Accurate prediction of water pollution levels is crucial for enabling policymakers and industry stakeholders to implement effective mitigation strategies. Predictive modeling, using historical data and influencing factors, provides insights into future pollution trends. This study utilizes data from Chinese publicly listed companies from 2007 to 2022, a period marked by significant economic and regulatory changes, offering a comprehensive dataset for analysis. Detailed environmental reports and disclosures from these companies serve as a valuable resource for constructing predictive models.

D. Han et al. (Eds.): ICEPG 2024, ESE, pp. 82–87, 2025.
https://doi.org/10.1007/978-3-031-86829-0_10

1.2 Importance of Water Pollution Prediction

Predicting water pollution is crucial for early identification of hotspots, enabling timely intervention and reducing health risks from contaminated water. It aids in effective management and conservation of aquatic ecosystems by targeting conservation efforts. Economically, it helps in planning cost-effective pollution control measures, allowing efficient resource allocation and ensuring compliance with environmental standards.

1.3 Previous Work

Comparison of Traditional Statistical Methods and Machine Learning Models
Moriasi et al. (2015) pointed out that traditional water quality models are widely used in the evaluation and calibration of water quality models, but they have limitations in handling complex nonlinear relationships. In contrast, Rajaee et al. (2020) demonstrated that single and hybrid AI models (such as ANNs, GP, FL, and SVM) perform excellently in predicting river water quality. Zhu et al. (2020) further emphasized the advantages of machine learning in dealing with highly stochastic and nonlinear data.

Application of Machine Learning Techniques in Water Pollution Prediction
Chen et al. (2019) developed a kriging calibration machine learning method that combines satellite data and ground monitoring data to accurately estimate daily NO_2 levels. Mukonza and Chiang (2023) explored the application of supervised learning regression models in satellite water quality monitoring through meta-analysis. Shi et al. (2022) evaluated water flow prediction models developed using eddy covariance observations and machine learning.

Advantages of Gradient Boosting Regression Tree Models
Gradient Boosting Regression Tree (GBRT) models, as a powerful alternative method, combine the strengths of multiple weak learners to generate robust prediction models. Song et al. (2021) compared the performance of GBRT and traditional logistic regression in predicting acute kidney injury, finding that GBRT performed better in handling complex features and variable selection.

1.4 Objectives

The primary objective of this study is to utilize Gradient Boosting Regression Tree to predict water pollution levels based on data from Chinese publicly listed companies from 2007 to 2022. Specific objectives include: (1) Implementing the Gradient Boosting Regression Tree algorithm to develop a predictive model. (2) Evaluating the model's performance using appropriate metrics and comparing it with other predictive models. (3) Providing insights and recommendations based on the model's predictions to aid in environmental management and policy-making.

2 Methodology

2.1 Data Collection

The dataset used in this study includes environmental data from Chinese publicly listed companies from 2007 to 2022, gathered from annual reports and relevant literature, Mao et al. (2022). Key water pollution indicators are Chemical Oxygen Demand (COD), Ammonia Nitrogen (NH_3_N) emissions, Total Nitrogen (TN), and Total Phosphorus (TP). For companies listed after 2007 or with missing data, mean imputation was used to fill gaps post-listing. The final dataset includes variables such as Securities Code, Year, COD (t), NH_3_N (t), TN (t), TP (t), and air pollution-related SO_2, NOx, Dust, along with one-hot encoded data for 13 different security types.

2.2 Gradient Boosting Regression Tree Model

In this study, we implemented the GBRT algorithm to predict water pollution levels based on the collected data. And to better understand the working principle of the GBRT model, we can describe it using the following formulas:

The model starts with a constant value, which is the mean of the target variable y.

$$F_0(x) = \arg\min_{\gamma} \sum\nolimits_{i=1}^{n} L(y_i, \gamma) \tag{1}$$

where L is the loss function, typically the mean squared error (MSE).

For each iteration $m = 1, 2, \ldots, M$:

a. Compute the residuals (pseudo-residuals) for the current model $F_{m-1}(x)$:

$$r_{im} = -\left[\frac{\partial F(x_i)}{\partial L(y_i, F(x_i))}\right]_{F(x)=F_{m-1}(x)} \tag{2}$$

b. Fit a new decision tree $h_m(x)$ to the residuals:

$$h_m(x) = \arg\min_{h} \sum\nolimits_{i=1}^{n} (r_{im} - h(x_i))^2 \tag{3}$$

c. Update the model by adding the new tree, scaled by the learning rate η:

$$F_m(x) = F_{m-1}(x) + \eta h_m(x) \tag{4}$$

The learning rate η controls the contribution of each tree to the final model, helping to balance the trade-off between model complexity and generalization.To prevent overfitting, setting a maximum depth d for each tree is typically done.

The final prediction for a given input x is the sum of the initial model and all the trees:

$$\hat{y}(x) = F_M(x) = F_0(x) + \sum\nolimits_{m=1}^{M} \eta h_m(x) \tag{5}$$

Through this iterative process, the GBRT model continuously corrects its prediction errors, gradually improving its accuracy. By iteratively adding decision trees that correct the residual errors of the current model, the GBRT method effectively minimizes the loss function and improves the predictive accuracy.

3 Experiments and Results

3.1 Model Training and Hyperparameter Tuning

We used GridSearchCV for hyperparameter tuning, exploring a range of parameters including the number of estimators, learning rate, maximum depth, minimum samples split, and minimum samples leaf. The best parameter combination was selected based on the cross-validated $R^{\wedge 2}$ score, as shown in Table 1.

Table 1. The best combination of hyperparameters for each target when cv = 5.

learning_rate	max_depth	min_samples_leaf	min_samples_split	n_estimators	target
0.1	3	2	10	50	COD
0.05	3	2	2	100	NH3_N
0.1	3	1	5	50	TN
0.05	3	4	10	100	TP

3.2 Model Evaluation

The performance of the model was evaluated using Mean Squared Error (MSE) and R^2 score. Detailed evaluations were performed for each target variable (COD, NH_3_N, TN, TP) to assess the model's accuracy. For each target variable, the data was split into training and testing sets with an 80–20 ratio. Hyperparameter tuning was performed using a 5-fold cross-validation approach. The best model for each target was then re-evaluated using a 10-fold cross-validation to ensure robustness, found that cv10 is a bit better than cv5. The final results were shown in Table 2.

Table 2. The best model performance for each target when cv = 10.

Target	MSE_Train	MSE_Test	R^2_Train	R^2_Test
COD	56686.68836	56372.50802	0.183324	0.181775
NH_3_N	20511.81261	20634.22806	0.803218	0.801609
TN	36417.45714	36181.47565	0.844518	0.843177
TP	20557.38644	20843.34257	0.657316	0.660463

From the MSE and R^2 values, the model's performance on the training and test sets appears to be consistent, indicating good model stability with no obvious signs of overfitting or underfitting. The model performs well for NH_3_N and TN, with R^2 values above 0.8, suggesting that the model can explain a significant portion of the variance in the data. The performance for TP is moderate, while the performance for COD is poor, indicating that further model optimization or additional feature engineering may be needed. For COD, consider adding more features or using other models for prediction. For TP, further optimization of model parameters may help improve prediction performance.

3.3 SHAP Analysis

As shown in Fig. 1, the SHAP interaction plots provide a comprehensive view of the importance of various features in predicting the levels of water pollutants. Across all pollutants, 'Year' consistently emerges as the most influential feature, with the largest range of SHAP values, indicating significant impact on predictions. This suggests that temporal trends play a crucial role in the model's ability to predict water pollutant levels. Conversely, 'Dust' and 'COD' frequently appear as the least influential features, with the smallest ranges of SHAP values for several pollutants. This indicates that these features contribute minimally to the model's predictions. The plots also reveal significant interactions between 'Year' and other features, such as 'NO_x' and 'SO_2'. These interactions underscore the complexity of the relationships between different factors in predicting pollutant levels.

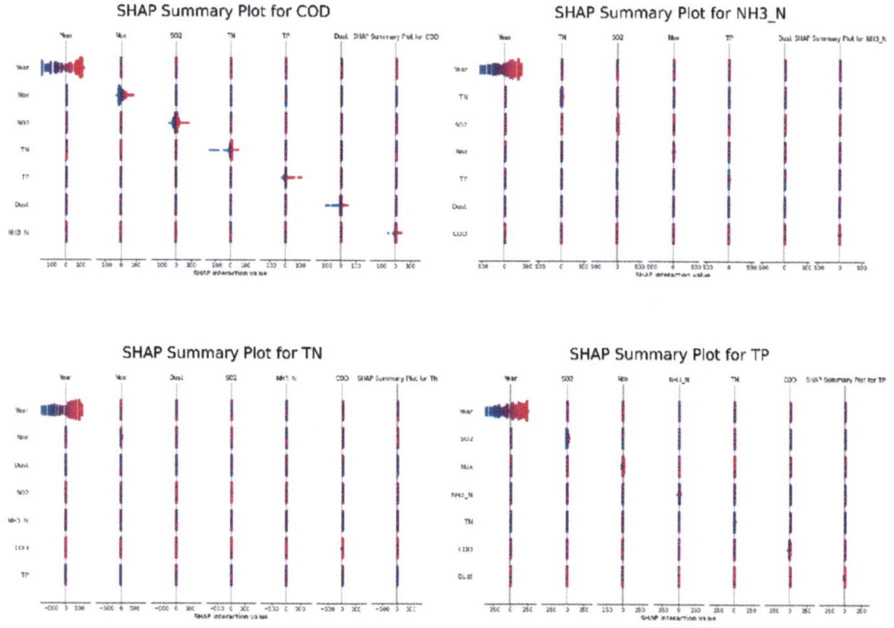

Fig. 1. Water pollutants SHAP interaction value plots with 10-fold cross-validation.

4 Conclusions

This study successfully applied the Gradient Boosting Regression Tree (GBRT) model to predict water pollution levels using data from Chinese publicly listed companies from 2007 to 2022. Our findings highlight the significant impact of temporal trends, with 'Year' emerging as the most influential feature across all pollutants. The model performed well for predicting NH_3_N and TN levels, with R^2 values above 0.8, indicating strong predictive power. However, the model's performance for COD was poor, suggesting the need for further feature engineering or alternative modeling approaches. The SHAP analysis provided a detailed understanding of feature importance and interactions, revealing complex relationships between various factors. These insights are crucial for developing targeted strategies to mitigate water pollution and enhance environmental sustainability in China. Future work should focus on improving the model's performance for COD and exploring additional features to enhance predictive accuracy.

Funding. The authors wish to acknowledge the financial support received for this study from the Shandong Provincial Natural Science Foundation, China. This work was part of the project titled "Research on the Social Public Value Accounting and Incentive Mechanism Innovation for Green Travel," with the project number ZR2021MG044, classified under the General Projects category. This study was also supported by the domestic visiting scholar program of Shandong Second Medical University.

References

Moriasi, D., Gitau, M., Pai, N., Daggupati, P.: Hydrologic and water quality models: performance measures and evaluation criteria. Trans. ASABE **58**, 1763–1785 (2015)

Rajaee, T., Khani, S., Ravansalar, M.: Artificial intelligence-based single and hybrid models for prediction of water quality in rivers: A review. Chemom. Intell. Lab. Syst. **200**, 103978 (2020)

Zhu, S., Lu, H., Ptak, M., Dai, J., Ji, Q.: Lake water-level fluctuation forecasting using machine learning models: a systematic review. Environ. Sci. Pollut. Res. **27**, 44807–44819 (2020)

Chen, Z.-Y., Zhang, R., Zhang, T., Ou, C., Guo, Y.: A kriging-calibrated machine learning method for estimating daily ground-level NO_2 in mainland China. The Science of the Total Environment **690**, 556–564 (2019)

Mukonza, S.S., Chiang, J.: Meta-Analysis of Satellite Observations for United Nations Sustainable Development Goals: Exploring the Potential of Machine Learning for Water Quality Monitoring. Environments (2023)

Shi, H., et al.: Evaluation of water flux predictive models developed using eddy covariance observations and machine learning: a meta-analysis. Hydrology and Earth System Sciences (2022)

Song, X., Liu, X., Liu, F., Wang, C.: Comparison of machine learning and logistic regression models in predicting acute kidney injury: a systematic review and meta-analysis. Int. J. Med. Informatics **151**, 104484 (2021)

Mao, J., Guo, Y., Cao, J., Xu, J.: Financing platform debt and environmental pollution control. Manage. World **10**, 96–118 (2022)

Simulation Study on Low Temperature Distillation Enrichment Process Based on Nuclear Power Moderator Heavy Water Protium Removal Process

Chunxin Fan(✉), Li lin, Defu Du, Zhenchen Li, Guoyu Wen, Jiheng Fan, and Huiqing Yang

The First Research Institute, Nuclear Power Institute of China, Chengdu 610000, China
fcx1403010805@163.com

Abstract. According to the current actual situation of nuclear industry and environmental protection requirements, the nuclear energy industry has an increasingly strong demand for high hydrogen isotope enrichment separation. In this study, a thermodynamic steady-state simulation model was established for the enrichment and separation process of deprotium by low temperature distillation with moderator, and the ionization equilibrium reaction process of H_2O-D_2O-DHO was innovatively considered. By comparison, it is proved that the model established in this study, including physical property selection, regression and prediction of missing physical property data, is scientific, reasonable and feasible. On this basis, the optimum operating parameters of low temperature enrichment distillation technology for tritium removal by moderator are determined by sensitivity. This study provides theoretical guidance for the research ideas and in-depth research of low temperature enrichment distillation technology for tritium removal from heavy water.

Keywords: hydrogen isotope separation · cryogenic distillation · steady state simulation

1 Introduction

Hydrogen has three isotopes: protium(H),deuterium(D)and tritium(T). Hydrogen isotopes are widely used in medical, nuclear and national defense fields [1, 2]. For example, deuterated drugs can improve the curative effect and action time of drugs by using the principle of deuterium slowing down metabolism in medical research. Heavy water is used as moderator and coolant in heavy water nuclear reactor with nuclear fission energy [3]. In the fusion reactor, which is considered as the ultimate energy source of mankind, deuterium-tritium nuclear fusion reaction is mainly used to generate energy [4].

How to separate hydrogen isotopes in water economically and effectively is of great significance to the fuel cycle of nuclear fusion reactor and tritium treatment process. The separation of hydrogen isotopes mainly includes chemical exchange method, low temperature distillation method, thermal diffusion method, centrifugation method and gas-solid

D. Han et al. (Eds.): ICEPG 2024, ESE, pp. 88–94, 2025.
https://doi.org/10.1007/978-3-031-86829-0_11

chromatography method absorption/adsorption separation method, laser method, etc. [5, 6]. The industrial production of hydrogen isotope mainly adopts low temperature distillation method. Many foreign researchers have carried out related simulation research on cryogenic distillation [7, 8]. In China, scientists and technicians from Tianjin University and China Academy of Engineering Physics have carried out simulation research on hydrogen isotope separation by cryogenic distillation.

In this paper, thermodynamic steady-state simulation is used to simulate the separation characteristics of hydrogen isotope H_2O-D_2O in cryogenic distillation column. The ionization equilibrium factors of H_2O-D_2O-DHO are innovatively considered. The simulation results are compared with the actual engineering installation process and detection data. Furthermore, the model is applied to the practical process improvement and optimization research.

2 Research Methods

The moderator distillation column of this engineering plant is a heavy water deprotium process. The feasibility of distillation technology depends on reliable vapor-liquid equilibrium data and accurate distillation calculation. Further, the low temperature distillation process for deprotium removal from heavy water as shown in Eq. 1 involves the separation and ionization equilibrium of three components H_2O, D_2O and DHO.

$$H_2O + D_2O = 2DHO \tag{1}$$

If all of them are considered, it will lead to great difficulties in manual calculation. ASPEN software is generally used in complex process calculation in chemical industry, but less used in nuclear industry. However, the complexity of cryogenic distillation technology and reaction system in this study coincides with the chemical field in this study. Therefore, ASPEN process flow simulation software is introduced to simulate cryogenic distillation and reaction process. However, Aspen's physical property database contains only H_2O and D_2O data. Therefore, in order to achieve the simulation content and goal of this study, it is necessary to construct and calculate the molecular skeleton of DHO in the component structure framework of Aspen. Then, according to the information about saturated vapor pressure and relative volatility of DHO provided by the experimental data in the literature, NRTL physical properties and data fitting are used to regress to Aspen physical properties database, and the matching results between fitting results and experimental data are shown in Fig. 1.

Considering the complex source term of the dry salt-forming system in this study, this Fig. 1 shows that the experimental data are almost completely consistent with the simulated data. It shows that the methods of missing physical property regression and other physical property prediction models in the steady-state model of this study are reasonable.

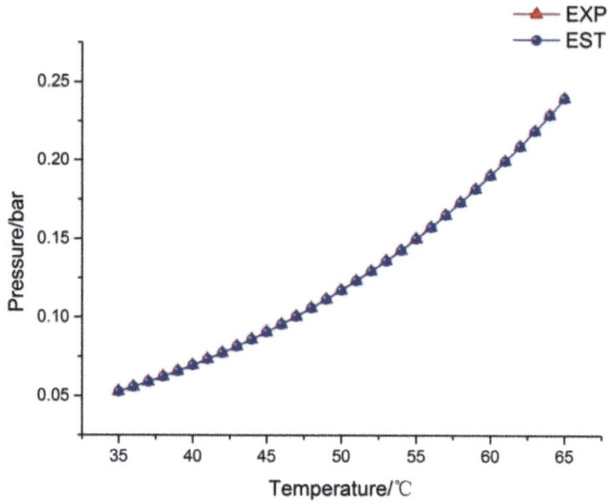

Fig. 1. Comparison of vapor pressure experimental data and fitting results.

2.1 Principle of Material Balance

The general expression of material balance in the system is: Accumulation in the system = Input(F_{in})-Output(F_{out}) + Generation(G_R)-Consumption(C_R). Assuming the system is leak-free, there is the following formula:

$$\frac{dF}{dt} = F_{in} - F_{out} + G_R - C_R \tag{2}$$

2.2 Principle of Energy Balance

The energy balance expression of the system is: Input system energy = output system energy + accumulated system energy. The energy balance equation is listed as follows:

$$Q + W = H_{out} + H_{in} \tag{3}$$

2.3 Model Construction and Verification

The reactive distillation model established after considering Formula 1 and the actual industrial data are shown in Fig. 2. On the one hand, Fig. 2C reveals that the model simulation data are basically consistent with the actual operation data of the test engineering device in terms of operating parameters such as overhead temperature, bottom temperature, overhead pressure and reboiler heat load. On the other hand, Fig. 2C shows that the model simulation data are almost completely consistent with the actual operation data of the test engineering plant in terms of H_2O-D_2O-DHO separation distribution data at key positions of the tower, such as the top tray, the bottom tray and the middle tray (No. 263

tray). Figure 2 shows that the model built in this study, the construction of default physical property data regression and prediction, is scientific, reasonable and feasible. The component DHO is mainly concentrated in the middle part of the distillation column. However, there is almost no DHO in the top and bottom of the tower.

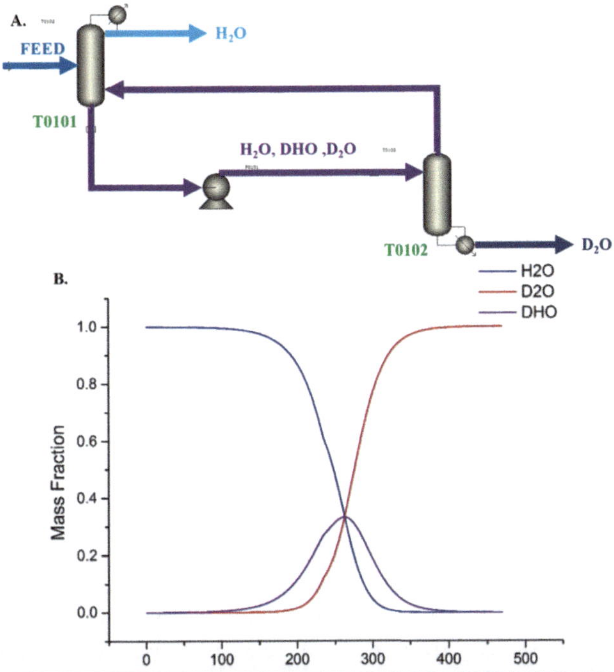

C.

Process parameters	Tower(T0101) top temperature/°C	Tower (T0101)top pressure/kPa	Bottom(T0102) temperature/°C	Heat load of reboiler/KW
Operational data	51.9	12.9	65.3	576.1
Simulation data	51.5	13.0	65.5	588.5

Process parameters	Composition of the top of the tower			Composition of tower bottom			Composition in the tower (No.263 tray)		
	H_2O /wt%	DHO /wt%	D_2O /wt%	H_2O /wt%	DHO /wt%	D_2O /wt%	H_2O /wt%	DHO /wt%	D_2O /wt%
Operational data	99.96	0.04	0	0	0.07	99.93	33.4	33.3	33.3
Simulation data	99.98	0.02	0	0	0.02	99.98	33.4	33.3	33.3

Fig. 2. (A) Steady-state model of cryogenic distillation enrichment. (B) Simulation results of composition distribution of each component in the tower. (C) Comparison between model calculation data and industrial actual operation data.

3 Results and Discussion

3.1 Influence Evaluation of Instantaneous Feed Stage

According to Fig. 3, with the increase of feed plate position, the H2O produced from the top of the tower first increases and then increases. Meanwhile, the recovery of D_2O from the bottom of the tower has almost no change Considering comprehensively, the feed plate is selected as the 255th plate, where the top of the tower recovers the most H_2O.

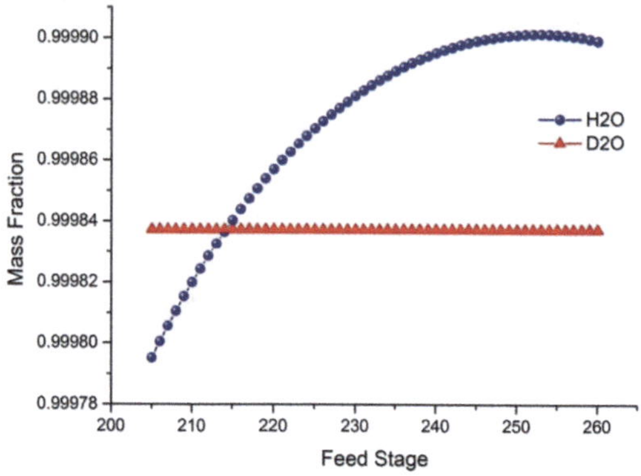

Fig. 3. Effect of feed tray on H2O composition at top (T0101) and D2O composition at bottom (T0102).

3.2 Influence Evaluation of Reflux Ratio

Figure 4 reveals that with the increase of reflux ratio (T0101), the purity of H_2O produced from overhead is higher. Furthermore, it also shows that the more H_2O is recovered from the top of the tower. And the purity of D_2O in the bottom of the tower has been stable within a certain threshold. However, the increase of reflux ratio will lead to energy consumption.

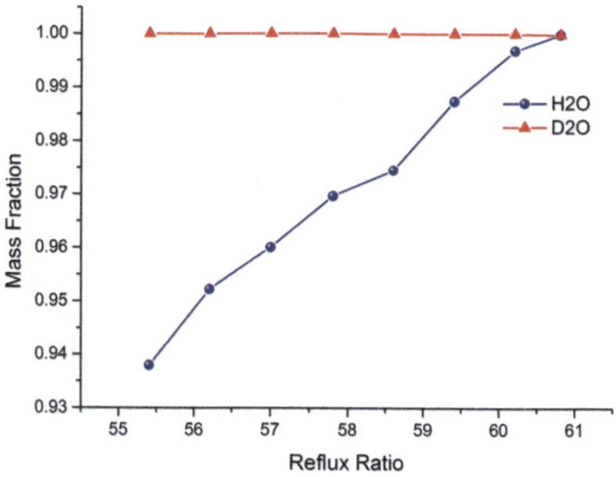

Fig. 4. Effect of reflux ratio composition at top (T0101) and D2O composition at bottom (T0102).

Therefore, other factors should be considered in order to determine the reflux ratio. Usually, with the increase of reflux ratio, the cooling load at the top and the heating load at the bottom increase linearly. Considering the operation reflux ratio of 60.5–61, the content of D2O is less than 0.02wt%. Under this condition, the energy consumption is lower and the basic separation goal of this technology is achieved.

3.3 Influence Evaluation of Boilup Ratio

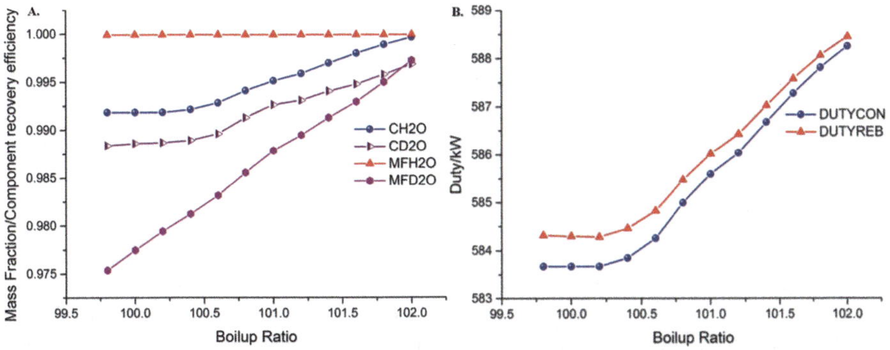

Fig. 5. (A) Effect of reboiling ratio on purity(MFH2O,MFD2O) and recovery of components at top and bottom of tower(CH2O,CD2O). (B) Influence of reboiling ratio on heat load at top and bottom of tower.

Figure 5A reveals that the reboiling ratio has little effect on the purity (MFH2O) of the overhead (T0101) H2O component. When reboiling ratio, the recovery rate of H2O component from overhead (CH2O) and D2O component from bottom (CD2O) showed a trend of gradual increase. Meanwhile, the purity (MFD2O) and reboiling ratio of D2O component in the bottom of the tower (T0102) increased linearly. It can be concluded that reboiling ratio has the greatest influence on MFD2O. Figure 5B shows that with the reboiling ratio, the cooling heat load at the top of the tower and the reboiling heat load at the bottom of the tower show an increasing trend. And when the reboiling ratio is less than 100.5, the heat load increases slowly. When the reboiling ratio is greater than 100.5, the heat load also increases linearly. The relative reboiling ratio is 101.9 according to the equilibrium purity of components, recovery rate and energy consumption.

4 Conclusion

In this paper, the process of deprotium in heavy water is simulated for the low temperature distillation concentration process of moderator. The present work mainly draws the following conclusions:

(1) The model established in this study, including physical property selection, regression and prediction of missing physical property data, is scientific, reasonable and feasible.

(2) By comparing the simulation data with the actual operation data, it is proved that the real state of equilibrium reaction of H_2O-D_2O-DHO ternary components can be accurately reduced by the simulation of reactive distillation.
(3) On this basis, the operating parameters of low temperature enrichment distillation technology for moderator deprotium removal are determined, including the adjustment of reflux ratio and reboiling ratio of feed tray. In addition, according to the characteristics of this technology and considering the separation objectives and energy consumption of this study, it is determined that the number of feed trays is 255 under the premise of total number of 470 trays. At the same time, the reflux ratio and reboiling ratio are 60.9 and 101.9, respectively. This study improves the performance and rationality of this technology and related process configuration. This study provides theoretical guidance for the research ideas and in-depth study of low temperature enrichment and distillation technology for deprotium removal from heavy water.

Acknowledgements. The authors gratefully acknowledge financial support from Original Fund One (KJCX-2023-YC1-018 Research on the Efficient Separation Mechanism of Radioactive Oily Wastewater Based on New Microreactor), China National Nuclear Corporation R&D Platform Stability Support Scientific Research Project (WDZC-2023-CNPE-105 Research on Enhancement Technology of Ion Exchange Mass Transfer Process of Radioactive Waste Liquid Based on Microfluidic Control), Industry Fund (CYJJ-202302-02) Development Project of Mobile High Salt Waste Liquid Drying and Salting Device.

References

1. Gant, T.G.: Using deuterium in drug discovery: leaving the label in the drug. J. Med. Chem. **57**(9), 3595–3611 (2014). https://doi.org/10.1021/jm4007998
2. Li, S.R., Zhang, Y.F., Xue, H.Y., et al.: Integrative effects based on behavior, physiology and gene expression of tritiated water on zebrafish. Ecotoxicol. Environ. Saf. **225**, 112770 (2021)
3. Zhanhua, W.U., Shilin, H.U.: Tritium removal from tritiated heavy water. J. Isot. **34**(1), 89–95 (2021)
4. Glugla, M., Laesser, R., Doerr, L., et al.: The inner deuterium/tritium fuel cycle of ITER. Fusion Eng. Des. **69**(1/4), 39–43 (2003)
5. Chu, X.: Study on Hydrogen Isotope Adsorption. Tianjin University, Tianjin (2016)
6. Deng, X., Xiong, R., Yan, X., et al.: Development of new materials for hydrogen isotopes separation. Mater. Reports **37**(13), 5–12 (2023)
7. Liu, Z., Zhang, F., Chen, H., et al.: Progress on application of cryogenic distillation in stable isotope separation. Modern Chem. Industry **41**(S1), 110–113 (2021)
8. Zhang, S., Yang, Q., Jin, A., et al.: Progress in the application of water distillation for tritiated water separation. Nucl. Tech. **46**(06), 27–40 (2023)

Experimental Study on Cyanide Removal from a Certain Cyanide Containing Wastewater by Iron Salt Precipitation Method

Yanbo Chen[1](✉), Guangsheng Li[1], Xingfu Zhu[1], Junjie Wang[1], Juntong Zhang[1], Jiajuan Zhang[1], and Qing Xu[2]

[1] Selection and Metallurgy Laboratory of Shandong Gold Mining Technology Co. Ltd., Yantai, Shandong, China
595907901@163.com

[2] Shandong Institute of Scientific and Technical Information, Jinan, Shandong, China

Abstract. In response to a certain cyanide containing wastewater, based on the principles of good cyanide removal effect, simple process flow, and low reagent cost, with the goal of reducing Fe, free cyanide, and total cyanide, and considering various technologies comprehensively, the iron salt precipitation method with relatively low reagent cost and good cyanide removal effect was selected. The Fe, free cyanide, and total cyanide concentrations in this wastewater were successfully reduced to 22.35mg/L, 15.42mg/L, and 78.86mg/L, respectively, achieving the expected goals.

Keywords: Cyanide containing wastewater · Iron salt precipitation method · Cyanide removal · Experimental research

1 Introduction

During the cyanide leaching process of gold bearing ores, the accompanying compounds such as sulfur, iron, copper, zinc, or their decomposition products in the ore will directly or indirectly react with $CN-$, O_2, $OH-$ in the wastewater, generating various impurity anions, mainly including free cyanide $CN-$, copper cyanide complex Cu $(CN)_3{}^{2-}$, zinc cyanide complex $Zn(CN)_4{}^{2-}$, and iron cyanide complex $Fe(CN)_6{}^{4-}$ [1, 2] These reactions not only consume a large amount of cyanide, leading to a significant increase in the cost of gold leaching, reducing the leaching rate of gold, but also making the composition of cyanide containing wastewater more complex, which is not conducive to comprehensive treatment. The concentrations of Cu, Zn, and Fe in a certain cyanide containing wastewater are 14.22 mg/L, 1.12 mg/L, and 326.40 mg/L, respectively. The contents of free cyanide and total cyanide are 544.64 mg/L and 1482.385 mg/L, respectively. In order to reduce the content of Fe, free cyanide, and total cyanide in this cyanide containing wastewater, and considering various technologies comprehensively, the iron salt precipitation method with relatively low chemical cost and good cyanide removal effect was selected. The concentrations of Fe, free cyanide, and total cyanide in this wastewater were successfully reduced to 22.35mg/L, 15.42mg/L, and 78.86mg/L, respectively, achieving the expected goals.

D. Han et al. (Eds.): ICEPG 2024, ESE, pp. 95–101, 2025.
https://doi.org/10.1007/978-3-031-86829-0_12

2 Sample Analysis

The analysis results of cyanide containing wastewater samples are shown in Table 1.

Table 1. Analysis results of cyanide containing wastewater

Assay Elements	Cu	Zn	Fe	FCN	TCN
Element content/%	14.22	1.12	326.40	544.64	1482.38

3 Experimental Research [2–6]

A cyanide removal experiment was conducted on this cyanide containing wastewater using iron salt precipitation method, and pH conditions, iron salt dosage conditions, and reaction time conditions were systematically tested. After treatment, adjust the filtrate to alkaline (pH > 11) and stir for a period of time to prevent further reaction of the filtrate under acidic conditions from affecting the accuracy of the assay; On the other hand, excess Fe can be filtered and removed in the form of $Fe(OH)_3$, and can be directly mixed into cyanide leaching water systems with the same alkalinity.

3.1 PH Condition Test

To ensure complete reaction, an excess of iron salt is added appropriately, set at 6g/L. Examine the effect of iron salt precipitation treatment under pH conditions of 3–8, and determine the optimal pH. The experimental process and conditions are shown in Fig. 1, and the experimental results are shown in Table 2.

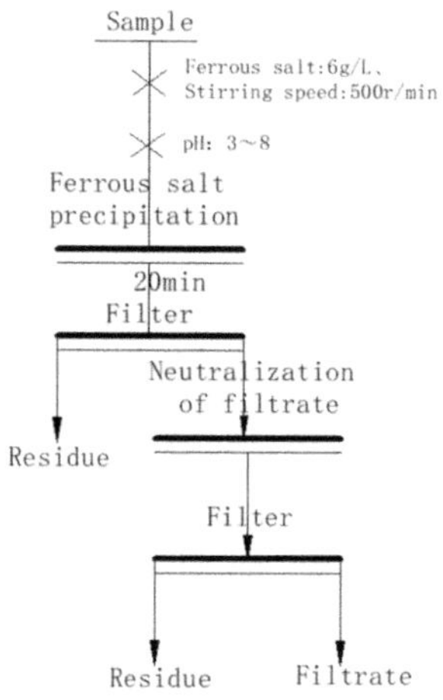

Fig. 1. PH condition test process of acidification method

Table 2. Results of pH condition test using acidification method

pH	Assay Elements/(mg/L)		
	Cu	Zn	Fe
3	1.46	0.74	76.47
4	1.37	0.24	32.98
5	0.95	0.22	26.44
6	0.72	0.08	22.38
7	0.88	0.86	49.71
8	12.27	0.92	184.8

According to Table 2, within the pH range of 3–8, as the pH increases, the Fe concentration first decreases and then increases. At pH 6, the Fe concentration is the lowest. Determine the optimal pH as 6.

3.2 Iron Salt Dosage Condition Test

Examine the effect of iron salt precipitation treatment under the conditions of iron salt dosage of 2.0g/L, 4.0g/L, and 6.0g/L, and determine the optimal amount of iron salt. The experimental process and conditions are shown in Fig. 2, and the experimental results are shown in Table 3.

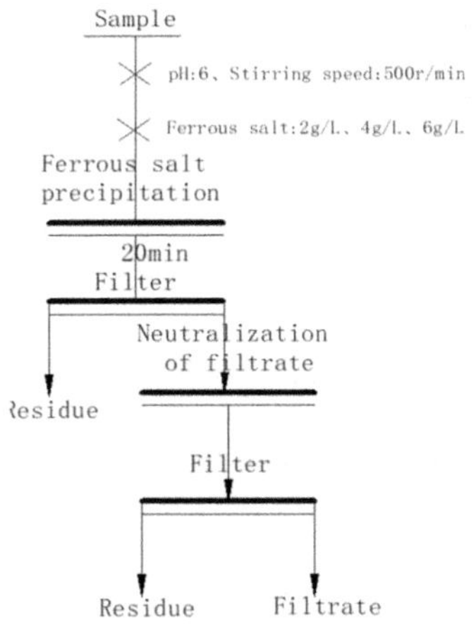

Fig. 2. Temperature condition test process

Table 3. Temperature condition test results

Iron salt dosage condition/(g/L)	Assay Elements/(mg/L)			
	Cu	Zn	Fe	FCN
2.0	1.86	1.26	363.50	67.94
4.0	1.24	0.11	136.50	40.76
6.0	0.66	0.082	22.52	21.74

According to Table 3, when the amount of iron salt added is lower than the theoretical calculation amount, i.e. 2.0 g/L, the generated precipitate is difficult to coagulate and filter, and incomplete precipitates such as Prussian blue are still left in the filtrate, resulting in a high Fe concentration that did not decrease to below 50 mg/L. When the addition amount is the theoretical value of 4.0 g/L, the generated precipitate can effectively coagulate and settle, and the Fe concentration drops to below 136.50 mg/L. Continuing to increase the amount of iron salt, which is 6.0 g/L, results in better flocculation and urban construction effects, while the Fe concentration drops significantly to 22.52 mg/L. Therefore, the specific amount of iron salt should be calculated based on theoretical results and appropriately excessive.

3.3 Reaction Time Condition Test

Examine the effect of iron salt precipitation treatment under reaction times of 10 min, 20 min, and 30 min, and determine the optimal iron salt precipitation time. The experimental process and conditions are shown in Fig. 3, and the experimental results are shown in Table 4.

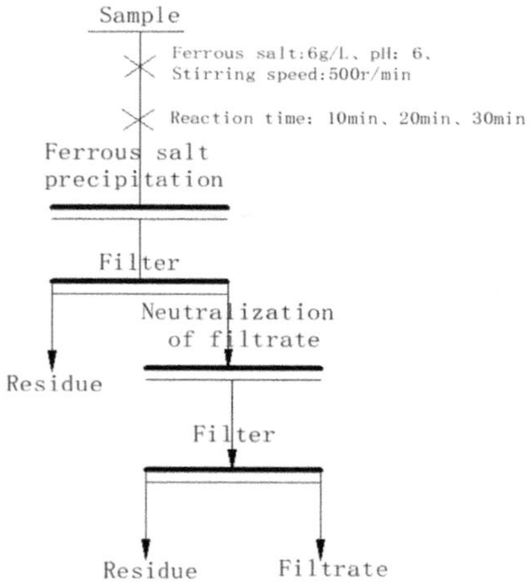

Fig. 3. Reaction time condition test process

Table 4. Reaction Time Condition Test Results

Reaction time/min	Assay Elements/(mg/L)			
	Cu	Zn	Fe	FCN
10	0.96	0.73	38.58	16.30
20	0.64	0.63	24.15	16.30
30	0.57	0.74	22.63	16.30

According to Table 4, within the range of 10–30 min, as the reaction time prolongs, there is no significant change in the concentrations of Cu, Zn, and FCN, while the concentration of Fe gradually decreases. After 20 min of reaction, it tends to stabilize, and the reaction time is determined to be 20 min.

3.4 Mixing Speed Condition Test

Examine the effect of stirring speed on iron salt precipitation treatment under conditions of 300r/min, 500r/min, and 700r/min, and analyze the influence of stirring speed on cyanide removal efficiency. The experimental process and conditions are shown in Fig. 4, and the experimental results are shown in Table 5.

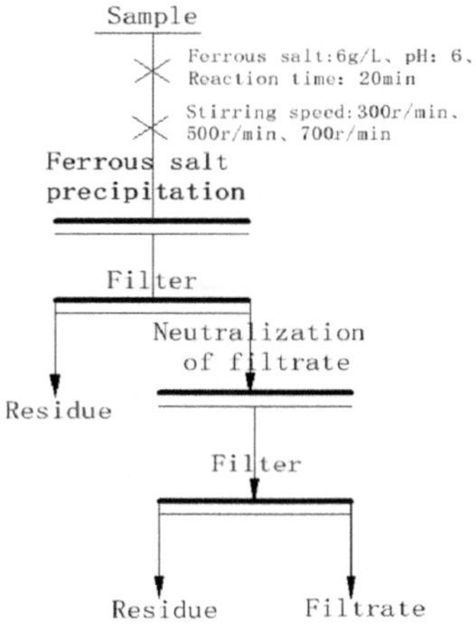

Fig. 4. Mixing speed condition test process

Table 5. Results of Mixing Speed Condition Test

Mixing speed/(r/min)	Assay Elements/(mg/L)			
	Cu	Zn	Fe	FCN
300	0.72	0.68	28.23	17.48
500	0.64	0.63	24.15	16.30
700	0.57	0.54	23.88	15.80

According to Table 5, as the stirring speed increases, The concentrations of Cu, Zn, Fe, and FCN show a gradually decreasing trend, and the change amplitude is small after the stirring speed is 500 r/min. It is sufficient to determine the stirring speed as 500 r/min.

3.5 Comprehensive Condition Test

Conduct comprehensive condition experiments according to the determined optimal parameters to examine the treatment effect. The experimental process and conditions are shown in Fig. 5, and the experimental results are shown in Table 6.

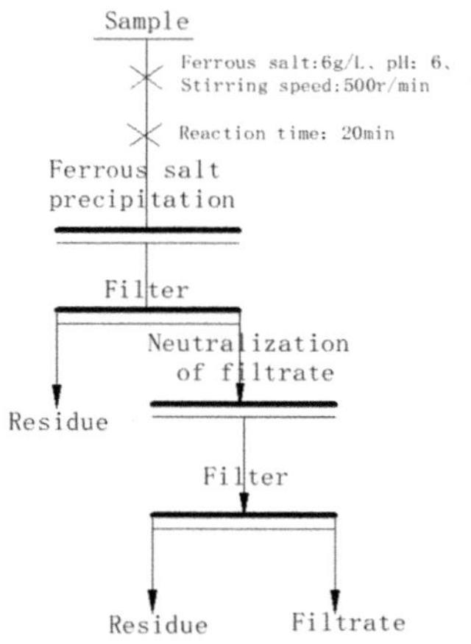

Fig. 5. Comprehensive Condition Test Process

Table 6. Comprehensive Condition Test Results

Assay Elements	Cu	Zn	Fe	FCN	TCN
Element content/%	0.72	0.58	22.35	15.42	78.86

According to Table 6, the optimal conditions for the "iron salt precipitation method" treatment stage of a certain cyanide lean solution are determined as follows: pH 6, iron salt dosage of 6 g/L (excess than theoretical dosage), reaction time of 20 min, and stirring speed of 500r/min. Under optimal conditions, the concentrations of Fe, free cyanide, and total cyanide in the wastewater were reduced to 22.35 mg/L, 15.42 mg/L, and 78.86 mg/L, respectively.

4 Conclusion

The optimal conditions for removing cyanide from a certain cyanide containing wastewater by iron salt precipitation method were determined through conditional experiments as follows: pH 6, iron salt dosage of 6 g/L (excess than theoretical dosage), reaction time of 20 min, and stirring speed of 500 r/min. Under the determined optimal experimental conditions, the iron salt precipitation method was used to successfully reduce the concentrations of Fe, free cyanide, and total cyanide in cyanide containing wastewater to 22.35 mg/L, 15.42 mg/L, and 78.86 mg/L, respectively, achieving the expected goals.

References

1. Han, W., et al.: Overview of decayanation technology for gold cyanide tailings slurry. China Nonferrous Metall. **52**(3), 41–49 (2023)
2. Li, X., et al.: Research and application of advanced treatment technology for cyanide containing wastewater in gold mines. GOLD **44**(7), 117–120 (2023)
3. Ebel, M., Evangelou, M.W.H., Schaeffer, A.: Cyanide phytoremediation by water hyacinths (Eichhornia Crassipes). Chemosphere **66**(5), 816–823 (2007)
4. Estay, H., Ortiz, M., Romero, J.: A novel process based on gas filled membrane absorption to recover cyanide in gold mining. Hydrometallurgy **134–135**, 166–176 (2013). https://doi.org/10.1016/j.hydromet.2013.02.012
5. Xie, F., et al.: Research progress on purification and recovery process of cyanide containing wastewater in gold mines. GOLD **44**(4), 80–85 (2023)
6. Álvarez, R., Ordóñez, A., Loredo, J., et al.: Wetland-based passive treatment systems for gold ore processing effluents containing residual cyanide, metals and nitrogen specie. Environ. Sci. Processes Impact **15**, 2115–2124 (2013)

Immobilization of Phanerochaete Chrysosporium and Its Application in the Degradation of Organic Pollutants and NH_3-N in Cosmetic Production Wastewater

Junjie Li and Hongbo Liu(✉)

School of Materials and Environment, Beijing Institute of Technology, Zhuhai 519088, China
liuhongbojm@163.com

Abstract. To mitigate the accumulation of water pollutants such as COD_{Cr} and NH_3-N from cosmetic production wastewater in soil and biota, this study employs the immobilization technology of Phanerochaete chrysosporium to enhance the degradation efficiency of organic pollutants and NH_3-N in cosmetic production wastewater. An 8% suspension of Phanerochaete chrysosporium, combined with a 4% polyvinyl alcohol (PVA) and 1% sodium alginate (SA) mixture, was used as the embedding material, and cross-linked in a 2% calcium nitrate solution for 2 h to obtain immobilized Phanerochaete chrysosporium beads. The effect of pH and organic pollutant concentration on the degradation of COD_{Cr} and NH_3-N in cosmetic wastewater by the immobilized strain was investigated to optimize process conditions. When the organic pollutant concentration was 200 mg/L and pH was 5.8, the NH_3-N removal rate was 98.40%, the COD_{Cr} removal rate reached 87.79% with pH value of 5.0. Compared to traditional methods, the immobilization technique using Phanerochaete chrysosporium significantly improves the removal rates of organic pollutants and NH_3-N in cosmetic wastewater.

Keywords: Phanerochaete chrysosporium · immobilization technology · Cosmetic production wastewater

1 Introduction

With the rapid development of the cosmetics industry, the problem of wastewater treatment generated during production has become increasingly severe. The complexity of the components, high concentration of pollutants, poor biodegradability, and imbalanced nutrient composition pose a significant threat to the ecological environment [1]. Such wastewater mainly originates from the cleaning processes of reaction kettles, raw material barrels, transfer barrels, and filling lines during production. It contains residues of raw materials, organic substances, surfactants, oils, thickeners, stabilizers, and disinfectants carried in products, making the treatment of this wastewater a complex and crucial task. In recent years, significant progress has been made in the research on cosmetics production wastewater treatment technology, especially with the widespread application of

D. Han et al. (Eds.): ICEPG 2024, ESE, pp. 102–110, 2025.
https://doi.org/10.1007/978-3-031-86829-0_13

biological treatment methods [2] due to their stable treatment effects and low operational costs. Traditional biological treatment methods, such as coagulation technology, anaerobic treatment technology, and sequencing batch reactor (SBR) processes, have been able to treat wastewater to a certain extent but still face challenges such as poor operational stability and difficulties in recovering and reusing microorganisms. Therefore, exploring more efficient and stable wastewater treatment technologies is particularly important.

The technology of immobilized microbial cells, an emerging biological wastewater treatment technology [3], involves confining free microbial cells or enzymes in a specific spatial region through physical or chemical methods. This not only maintains their biological activity [4] but also achieves high biomass enrichment, reusability, and high mechanical strength [5]. This technology effectively overcomes the problems faced by traditional biological methods, such as microbial loss and unstable treatment effects, providing a new solution for the treatment of cosmetics production wastewater [6]. Among the various microorganisms available for wastewater treatment, Phanerochaete chrysosporium has garnered significant attention due to its superior ability to degrade organic substances. This fungus can secrete lignin peroxidase, manganese peroxidase, and other lignin-degrading enzyme systems, which have strong degradation capabilities for refractory organic pollutants and ammonia nitrogen (NH_3-N). White-rot fungi have garnered significant attention due to their application in the degradation of various pollutants such as dyes, landfill leachate, pentachlorophenol, and TNT explosive wastewater. Despite their broad utility, reports on the degradation of cosmetics wastewater by white-rot fungi remain scarce, highlighting the innovative nature of this study. As a member of the white-rot fungi family, Phanerochaete chrysosporium demonstrates extraordinary potential in reducing BOD_5 and COD_{Cr} values in industrial wastewater due to its strong degradation capabilities. Therefore, it is considered an ideal candidate strain for the treatment of cosmetics production wastewater.

This study focuses on the practical application of immobilized Phanerochaete chrysosporium beads in the treatment of cosmetics production wastewater. By optimizing experimental design, it explores the degradation effects of immobilized Phanerochaete chrysosporium on COD_{Cr} and NH_3-N in cosmetics production wastewater under different concentrations of organic pollutants and pH conditions over time. Additionally, the study conducts detailed adjustments and optimizations of process conditions, providing scientific basis and technical support for achieving green and efficient treatment of cosmetics production wastewater.

2 Experimental

2.1 Materials

Immobilization Carrier Materials

As shown in Fig. 1, the immobilization process used a mixture of 1% sodium alginate and 4% polyvinyl alcohol as the embedding matrix [7]. A 2% calcium nitrate solution served as the crosslinking [8] agent to stabilize the immobilized cells.

Culture Media Used in the Experiment

For the cultivation of the fungal strain, potato dextrose agar (PDA) solid medium and potato dextrose broth (PDB) liquid medium were utilized, ensuring optimal growth conditions for the microorganisms.

Experimental Strain

In this research, Phanerochaete chrysosporium with notable heavy metal adsorption capabilities was selected as the focus due to its potential effectiveness in treating cosmetics production wastewater.

2.2 Preparation of Immobilized Phanerochaete Chrysosporium Beads

Shaking Incubation

A precise volume of 5 mL of PDB liquid medium was transferred into shake flasks using a pipette. Subsequently, 3 to 5 single colonies of Phanerochaete chrysosporium were carefully selected with an inoculating loop and introduced into the shake flasks. The cultivation was conducted at a rotational speed of 120 rpm and maintained at a temperature of 28 °C to ensure optimal growth conditions.

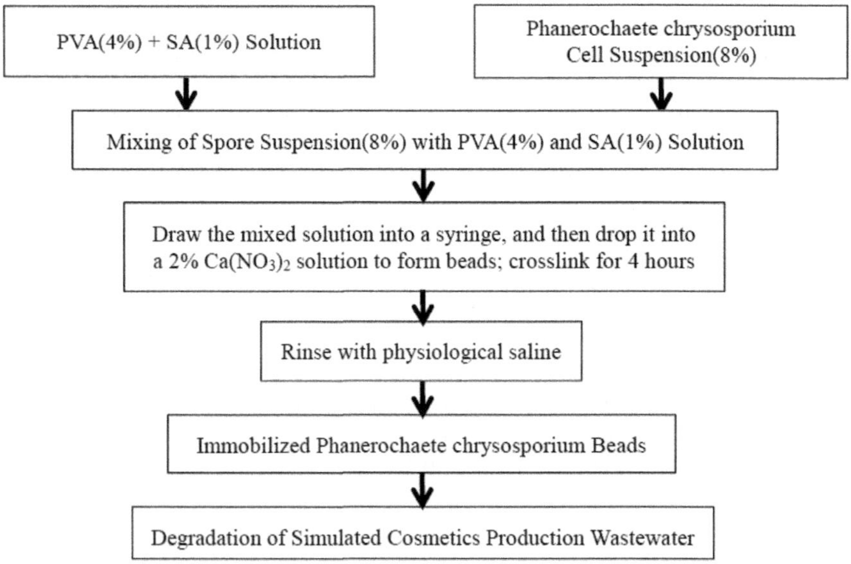

Fig. 1. Immobilization of Phanerochaete chrysosporium using the "PVA + SA--$Ca(NO_3)_2$" method.

Preparation of Spore Suspension

The optical density (OD_{600}) of the spore suspension was meticulously measured daily using a spectrophotometer at a wavelength of 600 nm. Upon reaching an OD_{600} value of 1.4 or higher, the spore suspension was transferred into centrifuge tubes and subjected to high-speed centrifugation at 8000 rpm and 4 °C for a duration of 8 min. Following centrifugation, the supernatant was carefully discarded, and the pelleted cells were resuspended in sterile water, resulting in a high-density spore suspension. This suspension was utilized as the base spore solution for the subsequent preparation of immobilized Phanerochaete chrysosporium beads.

Embedding of Fungal Strains

The activated Phanerochaete chrysosporium was mixed with a PVA and sodium alginate solution (comprising 8% spore suspension, 4% PVA, and 1% SA). This mixture was gradually dripped into a 2% calcium nitrate solution to initiate the crosslinking process. After 2 h of crosslinking, the formed beads underwent cryogenic solidification to preserve their structural integrity and functional activity. To ensure their sustained degradation performance, the immobilized beads were subsequently immersed and stored in the experimental wastewater. The experimental inllustration was shown in Fig. 1.

2.3 Wastewater Treatment Experiments Using Immobilized Phanerochaete Chrysosporium Beads

Determination of COD_{Cr} (Potassium Dichromate Method)

The COD_{Cr} determination is performed using a COD reflux digestion apparatus. A 250 mL Erlenmeyer flask was prepared by adding 10 mL of the simulated cosmetics production wastewater sample, 5 mL of 0.250 mol/L potassium dichromate solution as well as boiling stones. The mixture was thoroughly mixed and then connected to a condenser. After the addition of 15 mL of sulfuric acid-silver sulfate solution, the mixture was refluxed for 2 h. Upon cooling, the solution was diluted to 70 mL, and three drops of ferrous ion reagent were introduced, and the solution was titrated with ammonium ferrous sulfate until a reddish-brown color was observed. The COD_{Cr} removal efficiency was then calculated.

Determination of NH_3-N (Nessler's Reagent Spectrophotometric Method)

Ammonia (NH_3) or ammonium ions (NH_4^+) react with Nessler's reagent (a basic solution of mercuric iodide and potassium iodide) under alkaline conditions to form a yellow to brownish complex. The intensity of the color, which is proportional to the concentration of NH_3-N, is measured to determine the NH_3-N content in the sample. Various amounts of ammonia nitrogen standard solutions were added to colorimetric tubes. Sodium potassium tartrate and Nessler's reagent were then introduced, mixed, and allowed to stand. The absorbance was measured at a wavelength of 420 nm to construct a calibration curve. In a 10mL colorimetric tube, 10 mL of the simulated cosmetics production wastewater sample was combined with 1mL of sodium potassium tartrate and 1 mL of Nessler's reagent. The mixture was allowed to stand for 10 min after thorough mixing. Absorbance was then measured at 420 nm to determine the NH_3-N concentration.

3 Results and Discussion

As shown in Table 1, the concentrations of pollutants in the simulated cosmetic production wastewater were controlled at 50 mg/L, 100 mg/L, 150 mg/L, 200 mg/L and 250 mg/L, respectively. Cultivation was conducted under conditions of neutral pH value,

Table 1. Effect of Organic Pollutant Concentration on the Degradation Rate by Immobilized Phanerochaete chrysosporium.

Serial Number	COD_{Cr}/mg*L^{-1}	t/h	Absorbance	NH_3-N/mg*L^{-1}	Removal Rate/%
1	50	0	0.198	0.42	/
		120	0.142	0.18	56.62
2	100	0	0.263	0.70	/
		120	0.135	0.15	78.10
3	150	0	0.289	0.81	/
		120	0.122	0.10	87.94
4	200	0	0.383	1.21	/
		120	0.111	0.05	98.81
5	250	0	0.401	1.29	/
		120	0.157	0.25	80.82

28 °C and a shaking speed of 180r/min. The concentrations of COD_{Cr} and NH_3-N in the wastewater were measured every 24 h, and the degradation rates were calculated.

Table 2. Effect of pH on the Degradation of NH_3-N by Immobilized Phanerochaete chrysosporium.

Serial Number	pH	t/h	Absorbance	NH_3-N/mg*L^{-1}	Removal Rate/%
1	5.0	0	0.573	2.03	/
		120	0.191	0.39	80.61
2	5.8	0	0.469	1.58	/
		120	0.105	0.03	98.40
3	6.6	0	0.575	2.03	/
		120	0.109	0.04	97.92
4	7.4	0	0.524	1.82	/
		120	0.245	0.62	65.66

As shown in Table 2, the concentration of pollutants in the simulated cosmetic production wastewater was controlled at 200 mg/L, and the pH was adjusted to 5.0, 5.8, 6.6 and 7.4, respectively. Cultivation was conducted under conditions of 28 °C and a shaking speed of 180 r/min. The concentration of NH_3-N in the wastewater was measured every 24 h, and the degradation rates were calculated.

As shown in Table 3, the concentration of pollutants in the simulated cosmetic production wastewater was controlled at 200 mg/L, and the pH was adjusted to 5.0, 5.8, 6.6 and 7.4, respectively. Cultivation was conducted under conditions of 28 °C

Table 3. Effect of pH on the Degradation of COD_{Cr} by Immobilized Phanerochaete chrysosporium.

Serial Number	pH	t/h	Ammonium ferrous sulfate/mL	COD_{Cr}/mg*L^{-1}	Removal Rate/%
1	5.0	0	18.16	209.6	/
		120	22.76	25.6	87.79
2	5.8	0	18.19	208.4	/
		120	22.47	37.2	82.15
3	6.6	0	18.31	203.6	/
		120	22.37	41.2	79.76
4	7.4	0	18.23	206.8	/
		120	21.86	61.6	70.21

and a shaking speed of 180 r/min. The concentration of COD_{Cr} in the wastewater was measured every 24 h, and the degradation rates were calculated.

4 Evaluation of Optimal Process Conditions

4.1 Investigation of the Influence of pH on the Degradation of NH_3-N by Immobilized Phanerochaete Chrysosporium

It is evident in Fig. 2 that the slope of the trend line reaches its maximum when the pH is set to 6.6. This observation clearly indicates that the degradation rate of NH_3-N is most rapid at this specific pH level, leading to a significantly higher degradation efficiency. Notably, within a five-day period, the degradation rate achieves an impressive 97.92%. Conversely, when the pH is adjusted to 5.0 or 7.4, the degradation process proceeds at a slower pace, yielding lower overall efficiency. Interestingly, at a pH of 5.8, while the initial degradation rate is somewhat slower compared to that at pH 6.6, it ultimately reaches the highest degradation rate of 98.40% by 5th day, surpassing the other conditions tested.

In practical applications, where there is a need to meet lower discharge standards or when rapid degradation within a short timeframe is required, adjusting the pH to 6.6 is recommended. However, when the objective is to achieve a higher degradation rate over a more extended period, especially under stringent discharge criteria, a pH of 5.8 would be more suitable. It is crucial to emphasize that excessively high or low pH levels should be avoided, as they tend to result in a relatively lower degradation efficiency.

In the experiment designed to investigate the influence of pH on the degradation of NH_3-N by immobilized Phanerochaete chrysosporium, it was observed that the NH_3-N degradation rate initially increased and then decreased as the pH varied (Fig. 3). By calculating and predicting the theoretical optimal pH and its relationship to the degradation rate, we set this relationship as $f'(x)$. Taking the derivative and setting $f'(x) = 0$, the optimal pH value was determined to be $x = 6.05$. This indicates that when the

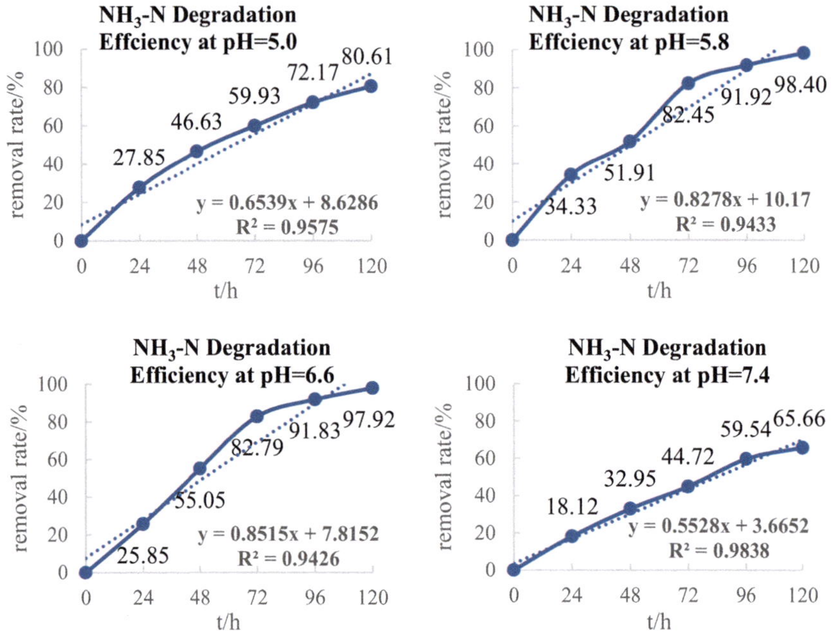

Fig. 2. The influence of pH on NH_3-N removal rate.

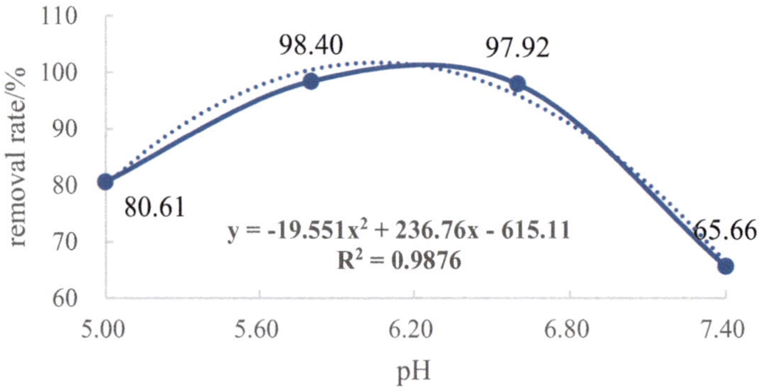

Fig. 3. NH_3-N removal rate under different pH conditions.

pH is 6.05, the degradation rate reaches its peak, theoretically at 101.67%, meaning that the NH_3-N degradation rate could exceed 99.9%.

4.2 Investigation of the Influence of pH on the Degradation of COD_{Cr} by Immobilized Phanerochaete Chrysosporium

Some researchers have indicated that excessively high concentrations of NH_3-N can adversely affect the effectiveness of COD_{Cr} treatment in wastewater by white-rot fungi [9]. Through analyzing the COD_{Cr} degradation curves at various pH (Fig. 4), it was evident that the highest five-day COD_{Cr} degradation rate, (87.79%) occurred at a pH of 5.0, whereas the fastest degradation rate was observed at pH 5.8. In contrast to NH_3-N, the optimal degradation rate for COD_{Cr} is found within the pH range of 5.0 to 5.8. Therefore, adjusting the pH to a slightly acidic level is crucial for achieving optimal degradation performance in practical applications.

When comparing the degradation rates of NH_3-N and COD_{Cr}, it is observed that the degradation rate of COD_{Cr} decreases as the pH ranges from 5.8 to 7.4. This indicates that as the concentration of NH_3-N increases, the microbial growth rate in the water accelerates, leading to a decrease in dissolved oxygen, making it challenging for organic matter to be fully decomposed, and consequently increasing the COD_{Cr} concentration. Furthermore, Phanerochaete chrysosporium is less suited to grow under neutral to slightly alkaline conditions, causing a decline in both NH_3-N and COD_{Cr} degradation rates. So it is advisable to first reduce the NH_3-N concentration to a certain level before proceeding with COD_{Cr} degradation.

A review of the literature reveals that Lu et al. used sawdust-immobilized Phanerochaete chrysosporium for the biodegradation of coking wastewater and found that under optimal conditions of pH 5.0 and 35 °C, the degradation rate of phenolic compounds exceeded 84%, and the COD_{Cr} removal rate reached 80% [9]. The conclusions derived from their study align closely with our findings. In our experiments, when the pH was set to 5.0, the COD_{Cr} removal rate for cosmetic production wastewater using immobilized Phanerochaete chrysosporium reached as high as 87.79%.

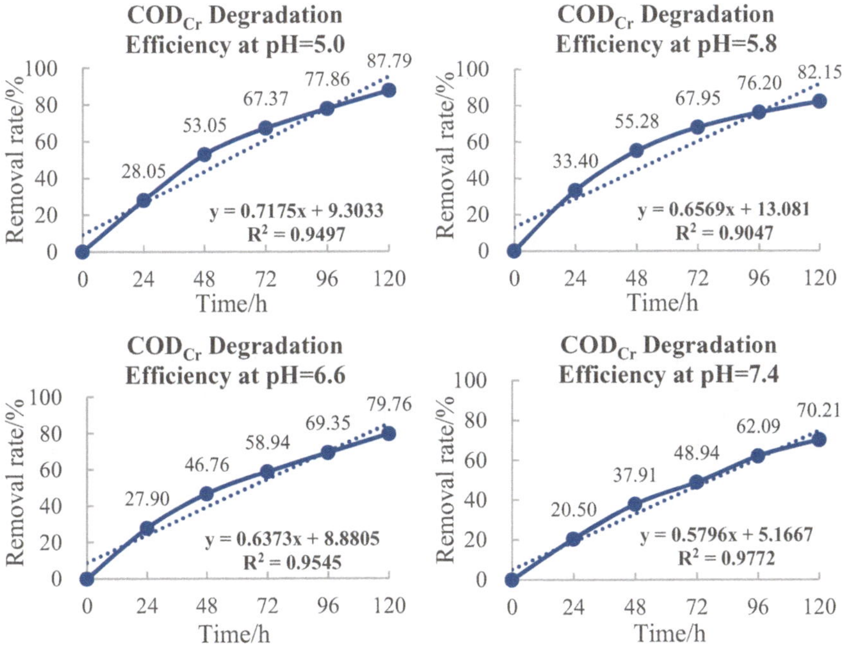

Fig. 4. The influence of pH on COD_{Cr} removal rate.

5 Conclusions

The optimal conditions for the removal of COD_{Cr} and NH_3-N from cosmetic production wastewater using immobilized Phanerochaete chrysosporium spheres were identified as follows: a pH of 5.8, organic pollutant concentration of 200 mg/L, temperature of 28 °C, and a mass ratio of immobilized particles to wastewater of 18%. Under these conditions, the effluent concentrations of COD_{Cr} and NH_3-N were found to meet the first-grade standards for the second period as specified in Guangdong Province's "Limits for the Discharge of Water Pollutants" (DB4466-2001).The immobilization technique demonstrated that Phanerochaete chrysosporium exhibits a high degradation efficiency when treating cosmetic production wastewater. Specifically, when the organic pollutant concentration was 200 mg/L and the pH was 5.8, the removal rate of NH_3-N reached 98.40%; at a pH of 5.0, the removal rate of COD_{Cr} was 87.79%. These results indicate that immobilized Phanerochaete chrysosporium has potential use for organic pollutants and NH_3-N.

References

1. Naggar, N.E.A., Hamouda, R.A., Mousa, I.E., et al.: Biosorption optimization, characterization, immobilization and application of Gelidium amansii biomass for complete Pb^{2+} removal from aqueous solutions. Sci. Rep. **8**(1), 1–19 (2018)
2. Wang, J.X., Liang, J.D., Sun, L., et al.: Granule-based immobilization and activity enhancement of anammox biomass via PVA/CS and PVA/CS/Fe gel beads. Biores. Technol. **309**, 123448 (2020)
3. Ali, P., Behnam, R.: Review of the immobilized microbial cell systems for bioremediation of petroleum hydrocarbons polluted environments. Crit. Rev. Environ. Sci. Technol. **48**(1), 1–38 (2018)
4. Zhang, Z.Y., Li, R.Y.: Research on the denitrification effect of immobilized microorganisms on river water. Acta Scientiae Circumstantiae **40**(1), 161 (2020)
5. Velkova, Z., Kirova, G., Stovtcheva, M., et al.: Immobilized microbial biosorbents for heavy metals removal. Eng. Life Sci. **18**(12), 871–881 (2018)
6. Huang, S., Gu, H., Deng, D.M., Zuo, H.J.: Microbial immobilization technology and its application in wastewater treatment. Mod. Chem. Ind. **44**(6), 65–69 (2024)
7. Liu, W.T., Li, J.T., Zheng, Z.Q., et al.: Remediation of petroleum hydrocarbon-contaminated soil using microbial immobilization technology. J. Appl. Technol. **21**(04), 339–347 (2021)
8. Yang, Z., Xu, W.S., Wu, Z.G., et al.: Research progress of microbial immobilization and its application in environmental pollution control. Microbiol. China **47**(12), 4278–4292 (2020)
9. Lu, Y., Yan, L., Wang, Y., et al.: Biodegradation of phenolic compounds from coking wastewater by immobilized white rot fungus Phanerochaete chrysosporium. J. Hazard. Mater. **165**(2–3), 1091–1097 (2009)

Enhancing Nitrogen Removal from Rural Domestic Sewage via Partial Nitrification-Endogenous Denitrification (PNED) Process in the Anaerobic/Oxic/Anoxic Sequencing Batch Reactor

Wentao Pan[1,2], Hong Liu[1,2], and Yongzhi Chen[1,2](✉)

[1] School of Environmental and Municipal Engineering, Lanzhou Jiaotong University, Lanzhou 730070, China
476411589@qq.com

[2] Technical Center of Sewage Treatment Industry in Gansu Province, Lanzhou 730070, China

Abstract. Rural domestic sewage (RDW) economic and efficient nitrogen removal remains a challenging topic. This study established an anaerobic-oxic-anoxic sequencing batch reactor integrating partial nitrification (PN) and endogenous denitrification (ED) to treat RDW. When the average influent total inorganic nitrogen (TIN) was 53.8 mg/L, the effluent of it, nitrite accumulation ratio, and removal efficiency were 7.8 mg/L, 89.8%, and 85.7%. In the typical cycle, the nitrogen removal efficiency of PNED could reach 100%, ED is the main nitrogen removal pathway, which significantly enhanced the TIN removal from RDW. PNED provides novel solutions for high-level and energy-saving RDW treatment.

Keywords: Rural domestic sewage · anaerobic/oxic/anoxic · sequencing batch reactor · advanced nutrients removal

1 Introduction

In recent years, there has been an increase in rural domestic sewage (RDW). It's reported that the treatment rate of rural domestic sewage in China is 31%, and the protection of the rural water environment remains challenging [1]. Residential dispersion, low population density, and the lack of sewage network collection systems have resulted in relatively low wastewater treatment ability. Large amounts of domestic wastewater are discharged without efficient treatment, resulting in nitrogen levels exceeding the capacity of nearby water bodies. It would influence groundwater and produce Black-odor water bodies, threatening the ecological environment. Notably, Anaerobic/anoxic/oxic (A^2/O), Anoxic/oxic (A/O), and oxidation ditch are usually used in municipal wastewater treatment. However, the conventional treatment methods mentioned above demand regular maintenance and high construction costs [2]. These factors make them unsuitable for rural areas characterized by rugged terrains and low financial revenue. Therefore, it

D. Han et al. (Eds.): ICEPG 2024, ESE, pp. 111–117, 2025.
https://doi.org/10.1007/978-3-031-86829-0_14

is necessary to strengthen the problems existing in the management of rural sewage, and put forward targeted improvement measures to improve the effectiveness of rural sewage treatment, effectively improve the rural human environment, strengthen the efforts of environmental remediation, and promote the sustainable development of rural areas. There are urgent novel technology requirements for RDW treatment.

The anaerobic/oxic/anoxic sequencing batch reactor (A/O/A-SBR) is characterized by high efficiency, cost-effectiveness, flexibility, simplified operation, and a friendless environment. Establishing partial nitrification (PN) coupled endogenous denitrification (ED) process in the A/O/A-SBR can further enhance nitrogen removal. In the anaerobic stage, denitrifying glycogen-accumulating organisms (DGAOs) absorb organic matter from wastewater and store them as intracellular carbon sources (Polyhydroxyalkanoates (PHAs) and glycogen (Gly)) [3]. Subsequently, NH_4^+-N was oxidized by ammonia-oxidized bacteria (AOB) to NO_2^--N, which can reduce 25% of oxygen consumption compared with nitrification and 40% of carbon source addition in denitrification [4]. Meanwhile, DGAOs catabolize PHAs for cell growth and Gly synthesis. Finally, under anoxic conditions, DNB and DGAOs reduced NO_2^--N and NO_3^--N to N_2 using COD, PHAs, and Gly as electron donors, respectively [4]. Thus, A/O/A-SBR exhibited good nitrogen removal performance. In contrast to the nitrification-denitrification process, PNED can save aeration energy consumption, eliminate carbon source addition, improve the utilization efficiency of organic matters, and reduce sludge production, having great application potential. Up to now, nitrification-denitrification has been the hotspot of RDW nitrogen removal. The feasibility and performance of PNED for RDW treatment remain unclear.

In this study, A novel PNED process was developed in the A/O/A-SBR and the main research contents are as follows: (1) Exploring the process of long-term PN performance; (2) Achieving advanced nitrogen removal of RDW; (3) Analyzing nitrogen removal mechanism of PNED. Research findings are expected to provide the scientific basis for upgrading practical RDW treatment.

2 Materials and Methods

2.1 A/O/A-SBR Setup and Operation

The lab-scale reaction device used in this study was a suspended activated sludge reactor (SBR) (Fig. 1a). The inoculated sludge was obtained from the oxic tank of YanChangBao WWTP (Lanzhou City, Gansu Province, China). The influent of the reactor was made by NH_4Cl, KH_2PO_4, CH_3COONa, $NaHCO_3$, and microelements (NH_4^+-N: 50mg/L, COD: 300 mg/L). PNED was operated for 100 days, with two cycles per day, and each cycle consisted of six stages (Fig. 1b).

2.2 Analytical Method

NH_4^+-N, NO_2^--N, NO_3^--N, COD, MLSS, and MLVSS were measured following standard methods. Gly and PHAs were examined according to previous studies [3].

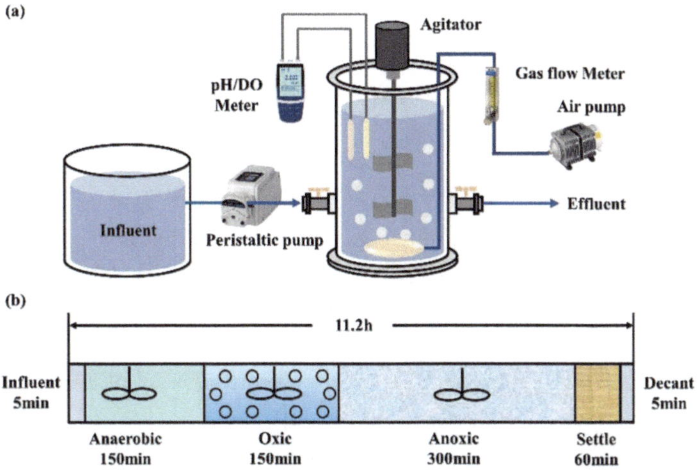

Fig. 1. Schematic diagram of the A/O/A-SBR (a) and operation condition of the A/O/A-SBR (b).

3 Results and Discussion

3.1 Long-Term Nitrogen Removal Performance of PNED

To explore the long-term treatment performance and operating characteristics of PNED, the changes of key parameters such as total inorganic nitrogen (TIN) and COD with operating time were analyzed, as shown in Fig. 2. In phase I (days 1–9), PNED was operated in A/O/A mode, the primary purpose of which was to allow the inoculated sludge to adapt to experiment operating conditions. Meanwhile, DO in the oxic stage was controlled to be 2.0 ± 0.2 mg/L to recover the nitrification activity. At the end of phase I, the effluent concentration of NH_4^+-N was below 10 mg/L, and the NH_4^+-N removal efficiency (ARE) improved to 80% (Fig. 2a). However, in the anaerobic stage, because the capacity of storing intracellular organics was still limited, the proportion of COD consumed for intracellular carbon storage in total anaerobic COD removal ($COD_{intra, efficiency}$) was only 64.1%. Meanwhile, a portation of nitrate was generated in the oxic stage, leading to the nitrite accumulation rate (NAR) have only 60%. These resulted in an average TIN removal efficiency (NRE) of 46.9%, the effluent concentration of TIN was mainly dominated by NO_2^--N, and ED was inhibited.

In phase II (days 10–30), The dissloved oxygen (DO) in the oxic stage was adjusted to 1.2 ± 0.2 mg/L to inhibit the nitrite-oxidized bacteria (NOB) activity and avoid further nitrification of NO_2^--N. At the end of phase II, the NAR increased to 87%, indicating that PNED achieved PN. Meanwhile, the effluent concentration of NH_4^+-N was 1.6 mg/L, with ARE as high as 96.8%. Nitrite generated in the oxic stage reduced denitrification carbon source demand; the average effluent concentration of TIN was 12.6 mg/L, which met with the Chinese Class A of Wastewater Discharge Standard (GB 18918-2002). It showed that PNED was started up successfully.

In phase III (days 31–100), PNED achieved stable and excellent PN and ED performance. The nitrogen removal performance of PNED has been significantly improved.

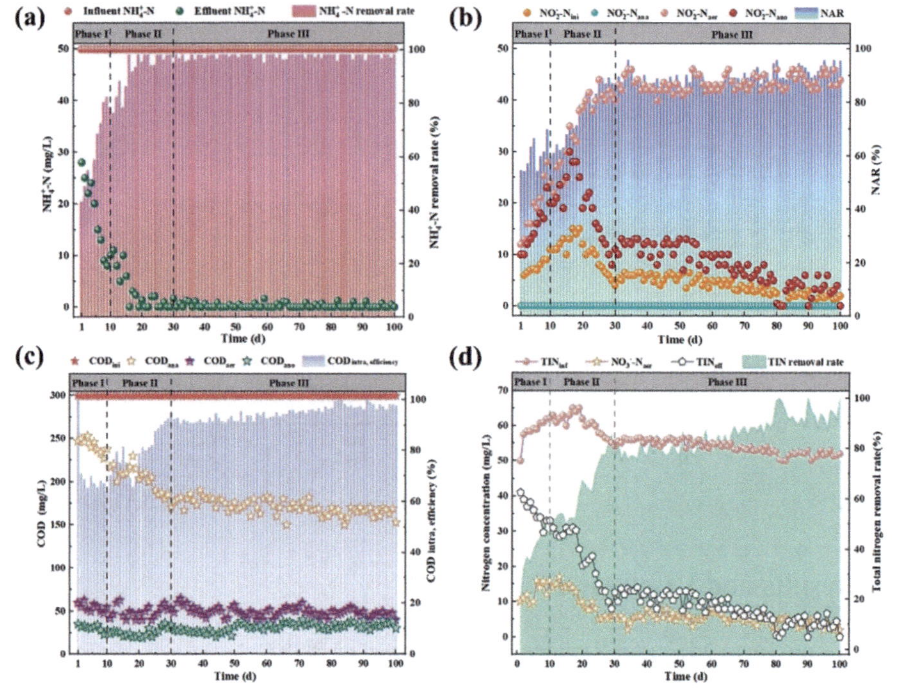

Fig. 2. Long-term performance of the PNED system operated for 100 days: (a) NH_4^+-N removal and ARE; (b) NO_2^--N removal and NAR (c) COD removal and $COD_{intra,\ efficiency}$; (d) NO_3^--N, TIN removal, and NRE;

The average NAR and NRE were maintained at 89.8% and 85.7% stably, and the average effluent TIN was only 7.8 mgN/L (Fig. 2b, c). In terms of organic matter removal, the system maintained a superior removal performance with the average effluent COD consistently below 50 mg/L (Fig. 2d). In addition, The average $COD_{intra,\ efficiency}$ was as high as 94.8%. This indicates that DGAOs consumed a large amount of COD during the anaerobic stage. It prevents the waste of carbon sources and energy in the oxic stage. Meanwhile, in the oxic and anoxic stage, the concentration of COD had no significant change, further illustrating that DGAOs mainly drove nitrite removal.

Overall, PNED was successfully started up on day 30 and the effluent concentrations of NH_4^+-N and TIN were 1.6 mg/L and 11 mg/L. During the long-term operation of PNED, NRE showed an increasing trend, with NRE improving from 18% on day 1 to 100% on day 100 (Fig. 1d). PNED was proven to achieve highly efficient nitrogen removal from RDW.

3.2 Nitrogen Removal Mechanism of PNED

Analysis of the Reasons for the Realization and Stabilization of Partial Nitrification PNED successfully achieved PN on day 21 with a NAR of 83.3% (Fig. 2b.), and the average NAR of PNED was maintained at 89.4% during the operation of 100 days. As previously discussed, long-anoxic/short-oxic hydraulic retention time (HRT) can inhibit NOB activity effectively [3, 5]. NOB activity decreased rate faster than AOB due to the long anoxic HRT. When changing from anoxic to oxic conditions, the activity of NOB declined. Anoxic time is longer than the oxic time during long-term operation. This may be the main reason for achieving stable PN. Furthermore, low DO concentration is beneficial to the inhibition of NOB. AOB and NOB have different affinities for dissolved oxygen. The half-saturation constant (K_O) of AOB for dissolved oxygen is 0.03 mgO_2/L, which is lower than that of NOB for O_2 (0.4 mgO_2/L) [6]. Therefore, the growth of AOB was more favorable than that of NOB when the concentration of dissolved oxygen was lower. The growth of NOB was inhibited, which resulted in the accumulation of nitrite. In this study, the concentration of DO was controlled at 1.2 ± 0.2 mg/L, promoting PN. Therefore, long-anoxic/short-oxic HRT is the main reason for achieving PN. DO control also promoted PN. The application of combined strategies ensured a long-term stable PN.

Nitrogen Removal Characteristics During Typical Cycles As is shown in Fig. 3, typical cycles were studied on day 100 to explore the transformation of nitrogen, COD, and intracellular organics, analyzing nitrogen removal mechanisms. The influent with 50 mg/L of NH_4^+-N, 2 mg/L of NO_2^--N, and 300 mg/L of COD, achieved the NRE of 100% during the typical cycle.

At the anaerobic stage (2.5 h), COD decreased from 300 mg/L to 153 mg/L, NH_4^+-N unchanged, denitrification removed 2 mg/L NO_2^--N, and NO_3^--N was not detected. $COD_{intra, efficiency}$ was as high as 97.7%, demonstrating that a large amount of COD was converted into intracellular organics by DGAOs, which is consistent with results reported in previous studies [3, 4]. In addition, the conversion of PHA and Gly (ΔPHAan and ΔGlyan) were 1.8 mmol C/L and 1.1 mmol C/L, respectively. It is suggested that the external carbon source was effectively absorbed and stored within the cells, ensuring the endogenous metabolism of microorganisms. ΔPHA/ΔGly was 1.63 mol C/mol C similar to the theory model of 1.65 mol C/mol C [7]. According to the above discussions, the main reactions in the anaerobic stage were denitrification and transformations between COD and intracellular organics.

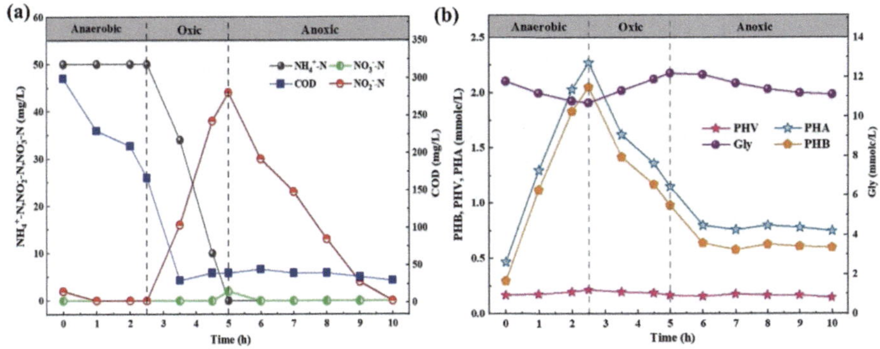

Fig. 3. Nitrogen and intracellular organics transformation characteristics during the typical cycles: (a) Nitrogen variation; (b) Intracellular organics transformation.

In the oxic stage (2.5 h), a part of COD that 153 mg/L existed in the reactor, and the removal rate of NH_4^+-N was relatively slow in the initial 1h; it was only 3.3 mgN/(gMLSS/h) (Fig. 2a). Then, the NH_4^+-N removal rate was improved to 5 mgN/(gMLSS/h) with COD declined to 40 mg/L (Fig. 2b), and PN was finished during 2.5 h. The content of NH_4^+-N was reduced by 50 mg/L, while the contents of NO_2^--N and NO_3^--N increased by 44 mg/L and 2 mg/L, respectively. There is 4 mg/L of nitrogen loss, which might be due to simultaneous nitrification-denitrification. In the oxic stage, PHA decreased by 1.1 mmol C/L with Gly increased by 1.5 mmol C/L, showing that parts of PHA were transformed into Gly. ΔPHA/ΔGly was 0.7 mol C/mol C lower than the theory model of 1.5 mol C/mol C [7]. The reason may be some organic matter still existed in this stage, and it was utilized by microorganisms to synthesize Gly (Fig. 2b). Hence, not all of the Gly was generated from the conversion of PHA. The reactions occurring in the oxic stage mainly included PN, the synthesis of Gly, and the consumption of PHA.

During the anoxic stage (5 h), the concentrations of NO_2^--N and NO_3^--N declined to 0 mg/L, COD didn't change obviously, and the concentrations of PHA and Gly reduced by 0.4 mmol C/L and 1.1 mmol C/L, respectively. This indicated that the main nitrogen removal pathway was ED-driven by intracellular carbon sources. In the first 1 h, the content of PHA continued to decrease, while Gly did not change significantly. Subsequently, the contents of Gly began to decline while PHA was not changed obviously. This is because when PHA was consumed to specific contents, microorganisms started to decompose intracellularly stored Gly for ED, which was similar to the results obtained by Miao et al. [4]. Compared to Gly, PHA-favoring DGAOs were more conducive to NO_x^--N (NO_2^--N and NO_3^--N) removal. To further accelerate the reaction rate of ED, it is necessary to avoid the ineffective consumption of intracellular organics in the oxic stage by controlling operation parameters.

4 Conclusion

This research provided a PNED combination process in the A/O/A-SBR system. We achieved long-term stable PN successfully. Meanwhile, for ED, nitrite was the main electron acceptor that promoting nitrogen removal. The main conclusions were as follows:

(1) Long-anoxic/short-oxic HRT, and DO control achieved stable PN with a NAR of 89.4%.
(2) Endogenous metabolism with nitrite pathway was enhanced, the average $COD_{intra.efficiency}$ can reach 90%.
(3) Advanced nitrogen removal in rural domestic sewage was achieved, the average effluent TIN and TIN_{RE} were 7.8mg/L and 85.7%.

References

1. Cao, Y., et al.: Prediction of rural domestic water and sewage production based on automated machine learning in northern China. J. Clean. Prod. **434**, 140016 (2024). https://doi.org/10.1016/j.jclepro.2023.140016

2. Yan, Y., et al.: Enhanced nitrogen removal from rural domestic sewage via partial nitrification-anammox in integrated vertical subsurface flow constructed wetland. Environ. Res. **233**, 116338 (2023). https://doi.org/10.1016/j.envres.2023.116338
3. Zhang, C., Zhang, L., Liu, J., Li, X., Zhang, Q., Peng, Y.: Achieving ultra-high nitrogen and phosphorus removal from real municipal wastewater in a novel continuous-flow anaerobic/aerobic/anoxic process via partial nitrification, endogenous denitrification and nitrite-type denitrifying phosphorus removal. Water Res. **250**, 121046 (2024). https://doi.org/10.1016/j.watres.2023.121046
4. Miao, L., Wang, S., Li, B., Cao, T., Xue, T., Peng, Y.: Advanced nitrogen removal via nitrite using stored polymers in a modified sequencing batch reactor treating landfill leachate. Biores. Technol. **192**, 354–360 (2015). https://doi.org/10.1016/j.biortech.2015.05.013
5. Kornaros, M., Dokianakis, S.N., Lyberatos, G.: Partial nitrification/denitrification can be attributed to the slow response of nitrite oxidizing bacteria to periodic anoxic disturbances. Environ. Sci. Technol. **44**, 7245–7253 (2010). https://doi.org/10.1021/es100564j
6. Blackburne, R., Yuan, Z., Keller, J.: Partial nitrification to nitrite using low dissolved oxygen concentration as the main selection factor. Biodegradation **19**, 303–312 (2008). https://doi.org/10.1007/s10532-007-9136-4
7. Zeng, R.J., van Loosdrecht, M.C.M., Yuan, Z., Keller, J.: Metabolic model for glycogen-accumulating organisms in anaerobic/aerobic activated sludge systems. Biotechnol. Bioeng. **81**, 92–105 (2003). https://doi.org/10.1002/bit.10455

A Beam Splitter Prism-Based Portable Device for On-Site Determination of Nitrite in Surface Water

Jincong Feng[1], Heng Li[2], Jiajun Zhang[1], Liting Yu[1], Ming Deng[1], and Jin Xu[1](✉)

[1] School of Life and Environmental Sciences, Guilin University of Electronic Technology, Guilin 541200, China
xujin@guet.edu.cn

[2] School of Electronic Engineering and Automation, Guilin University of Electronic Technology, Guilin 541200, China

Abstract. To achieve on-site analysis and emergency analysis of water quality indicators, a dual-channel portable analysis device was designed based on the Griess method combined with beam splitter prism. 3D printing technology was used to print the device's housing, reducing the size and weight of the device. The experiment verified that the optimal LED constant current control current was 4.31 mA, and the data processing method was the median average method of five times absorbance data. The performance indicators of the device were analyzed. The results showed that the device's weight was 552.6 g, the average power consumption was 0.13 W, and the detection limit of nitrite was 0.002 mg/L. The device has the characteristics of low cost, small size, high linearity, high sensitivity, and high stability. The analysis of 21 samples by the device showed that there was no significant difference with the results analyzed by the Griess method, indicating that the device can be applied to the analysis of nitrite in surface water reliably.

Keywords: on-site · beam splitter prism · portable · nitrite

1 Introduction

Nitrite is an important indicator for evaluating water pollution [1], and excessive intake of nitrite by the human body can destroy the oxygen-carrying capacity of erythrocytes, form carcinogenic nitrosamines, and cause gastric and oesophageal cancers [2]. In the face of water pollution emergencies, traditional laboratory analysis methods usually take several hours [3]. On-site nitrite analysis has received high attention [4]. Nitrite is analyzed by spectrophotometric methods [5, 6], electrochemical methods [7], and ion chromatography [8, 9]. However, these methods usually require large instruments and specialized operators, which are not conducive to on-site analysis. Miniaturized and portable analytical equipment has become the focus of current research in on-site analysis [10].

D. Han et al. (Eds.): ICEPG 2024, ESE, pp. 118–125, 2025.
https://doi.org/10.1007/978-3-031-86829-0_15

Currently, many portable analytical devices have been developed to detect nitrite. Some researchers focus on improving the instrument structure. Cadeado et al. developed a palm-sized wireless device, which analyzes the color intensity of the solution after color development through RGB sensors to realize the quantitative detection of nitrite [11]. Man et al. used the mobile colorimetric detection system APP combined with a microfluidic chip and a signal acquisition box to achieve the rapid, on-site quantitative detection of nitrite [12]. Titov et al. designed a highly integrated compact detection device based on the Griess reagent, organic light emitting diode (OLED), and organic photodetector (OPD) with a substrate area of only 12.5 mm × 12.5 mm [13]. Some researchers focus on optimizing analytical methods. Catalan-Carrio et al. developed a handheld mini-detector based on a paper-based microfluidic device for simultaneous detection of nitrite in freshwater and seawater using partial least squares regression and principal component analysis [4]. Xu et al. developed a low-cost, portable, and easy-to-use device based on electrochemical technology combined with an Android application design for the on-site detection of nitrite in water [14]. Yang et al. designed a nitrite fluorescent probe using rhodamine 800 as the fluorescent skeleton and 5-aminoindole as the specific reactive group, which combined with a mobile app visual platform to analyze the color change of test paper to achieve a convenient and rapid analysis of nitrite content in samples [15]. The above methods have achieved on-site nitrite detection in different degrees. However, some factors still need to be considered, such as the impact of light source fluctuations on the detection results. According to the investigation, there is no report on a portable on-site nitrite detection device that uses a beam splitter prism to detect a blank solution and a sample solution simultaneously.

In this study, a portable outdoor dual-channel detection device was designed based on the Griess method combined with a beam splitter prism. The device uses a beam splitter prism to split the incident light into two beams at a 1:1 ratio, which is captured simultaneously by two light sensors, detecting blank and sample solutions simultaneously. This structure greatly reduces the impact of light source fluctuations on the detection results. In addition, this study explored the light intensity of the light source and the data processing method on the stability of the detection results and selected the optimal detection conditions. The device was applied to detect nitrite indicators in the Hua Jiang River Basin in Guilin City.

2 Materials and Methods

2.1 Materials

The reagents used were all analytical grade, all purchased from Shanghai Aladdin Biochemical Technology Co., Ltd. The ultrapure water was prepared with EPED Plus-E3 instrument (resistivity > 18.2 MΩ cm), and the formulated reagents were all placed in the refrigerator at 4 °C for storage. The comparative experiment on the practical application performance of the Dual-channel detection method based on beam splitter and the Griess method was conducted using a UV-1800 ultraviolet-visible spectrophotometer (AOE Instruments Co., Ltd.).

Sulfanilamide solution (SAM, 10.0 g/L): 1.0 g of sulfanilamide was dissolved in 70.0 mL hydrochloric acid solution (1 + 6). The dissolved solution was diluted to 100.0 mL with ultrapure water and stored in a brown reagent bottle.

N – (1-Naphthyl) ethylenediamine dihydrochloride solution (NED, 1.0 g/L): 0.1 g of N – (1-Naphthyl) ethylenediamine dihydrochloride was dissolved in 100.0 mL ultrapure water and stored in a brown reagent bottle.

Nitrite standard storage solution (100.0 mg/L-N): 0.4926 g of sodium nitrite was dried at 110 °C, dissolved in a small amount of water, and transferred into a 1000 mL volumetric flask. Ultrapure water was added to the mark and the solution was diluted and mixed to be used as the standard solution for experimental analysis.

2.2 Device Structure and Composition

The device consists of a top cover, a cavity, a sample chamber, an ink screen, an ink screen bracket, a battery bracket, and a battery box. The sample chamber, device housing, and bracket were made using fused deposition modeling technology. The device was designed via a Computer Aided Design software – FreeCAD. The fill density for printing was set to 20%. The structure has strong stability, and its structure schematic diagram is shown in Fig. 1a.

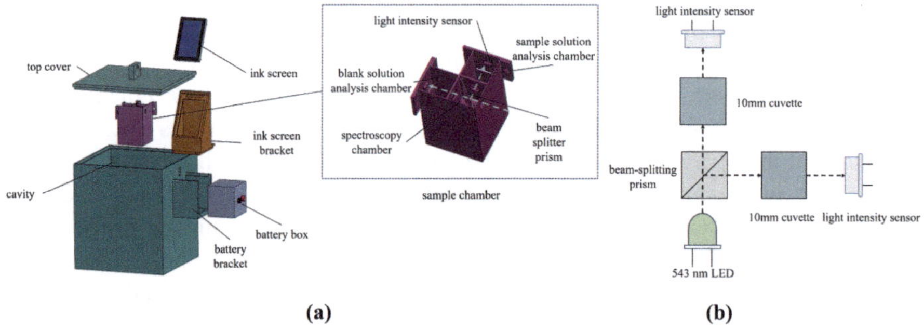

Fig. 1. Schematic diagram of device and sample chamber structure. (a) schematic diagram of device structure. (b) schematic diagram of sample chamber structure.

The sample chamber consists of a spectroscopy chamber and two detection chambers. They are hollow cubic cylindrical designs, from the top view angle, composed of three squares in the letter L-shaped connection. Two detection chambers were located on both sides of the spectroscopy chamber to place 10 mm cuvettes. Two inner diagonal corners of the spectroscopy chamber were provided with fixing grooves, in which the beam splitter prism was installed. The beam splitter prism was arranged at an angle of 135° to the incident light direction. The optical path holes were designed in the upper part of the two detection chambers, with light-intensity sensors installed on the opposite side. The cavity and the battery bracket create an integrated structure, with the ink screen bracket positioned above it. The sample chamber and custom circuit board are mounted inside the cavity.

The sample composition is as follows: the sample chamber includes a 3D printed dual-channel detection chamber, a beam splitter prism, a specific-wavelength LED (540 nm), two high-sensitivity light sensors (BH1730), and two 10 mm cuvettes. The customized circuit board includes a 3.3 V regulator circuit (HT7333), a switching control circuit, a main control circuit (STM32F103C8T6), and an LED constant current control circuit (Hi1243). The de-vice housing mainly consists of a top cover, a cavity, an ink screen bracket, and a circuit board bracket.

The sample chamber is the core part of the device, and its schematic diagram is shown in Fig. 1b. The incident light emitted by the LED was split into two beams by a beam splitter prism at a ratio of 1:1. The two beams were captured simultaneously by two high-precision light intensity sensors. This structure realizes the simultaneous detection of blank and sample solutions, greatly reducing the impact of light source fluctuations on the detection results.

The integrated custom circuit board was designed to provide stable power, reliable LED driving current, and complete data processing. The cavity and top cover of the device housing provided a light-shielded environment for the detection process, re-ducing the impact of ambient light on the detection results. The bracket was used to install the ink screen and battery box to compact the device.

2.3 Analysis Process

The schematic diagram of the process for detecting nitrite using a dual-channel device based on a beam splitter prism was shown in Fig. 2. 50.0 mL of filtered water sample and ultrapure water were transferred to stoppered tubes, 1.0 mL of SAM solution and 1.0 mL of NED solution were added, mixed and waited for 20 min. The mixture was added to a 10 mm cuvette and placed in the sample chamber for detection. With the detection button pressed, the constant current circuit controls the LED to emit a stable light. The incident light is split into two beams by the beam splitter prism and then simultaneously captured by the light intensity sensor.

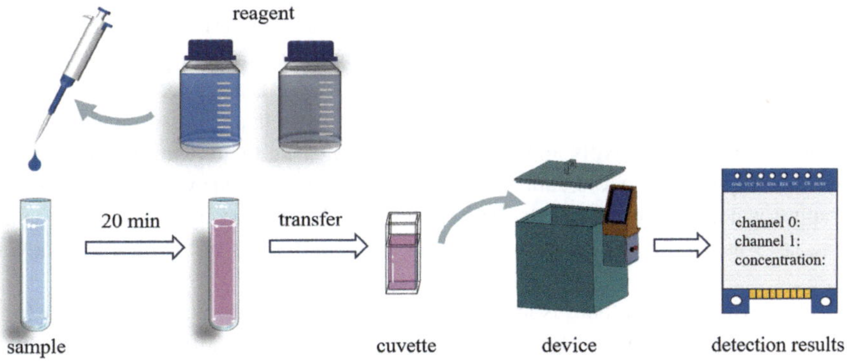

Fig. 2. Schematic diagram of the detection process.

The absorbance of the sample (A_{sample}) is the difference between the absorbance values of the two channels. The concentration of the samples is calculated by the formula $C = (A_{sample} - b)/k$.

3 Results and Discussion

3.1 Choice of Light Intensity and Data Processing Methods

During the experiment, it was found that the change in the light intensity of the LED would have a certain impact on the stability of the data. The light intensity values directly read without processing exhibited numerous anomalies. To explore the influence of light source intensity on the stability of detection and select the optimal light source intensity, three groups of gradient experiments were designed. The output current value of the constant current module was set by adjusting the external Current Sensing resistor (R_{CS}). Each group analyzed the light intensity value of the ultrapure water solution 20 times and calculated the relative standard deviation. The experimental results are shown in Table 1.

Table 1. Relationship between light source intensity and detection stability.

Serial number	R_{CS} (Ω)	current (mA)	RSD (%)
1	3	4.31	0.13
2	100	1.78	0.37
3	510	0.35	0.85

From the experimental results, it can be seen that the stability of the system is best when the R_{CS} is 3 Ω and the current is 4.31 mA.

Four data processing methods were compared to fully utilize the detection data and further improve data stability while reducing the impact of noise on the detection results. The four data processing methods included direct reading (Group A), three consecutive readings of the two-channel data for averaging (Group B), five successive readings of the two-channel data for averaging (Group C), the two-channel absorbance data were read five times consecutively and averaged after removing the maximum and minimum values (Group D). These four data processing methods were used by the device to analyze ultrapure water solution 40 times. The ratio of the light intensity of the two channels ($R_{channel}$) was used as an evaluation indicator. The box plots of the data are shown in Fig. 3.

As shown in Fig. 3, there is no significant difference in the upper and lower limits, average, and median of the four methods, while the data of group D has the fewest outliers. The relative standard deviations of the four groups were calculated to be 0.37%, 0.33%, 0.27%, and 0.22% respectively. The method of reading the dual-channel data five times continuously and calculating the average value after removing the maximum and minimum values has the best stability.

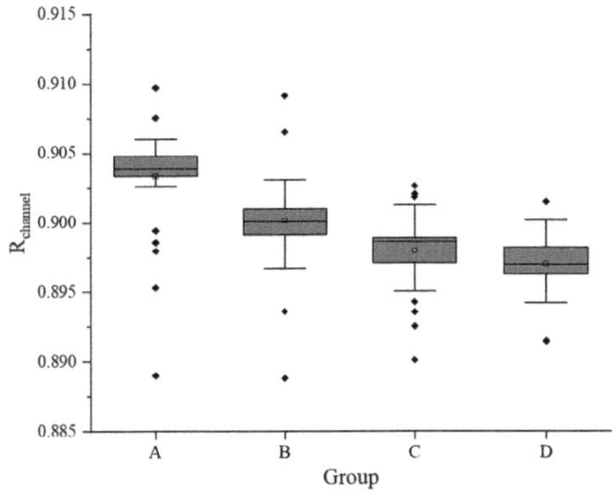

Fig. 3. Boxplots of data from four detection methods.

3.2 Analytical Figures of Merit

The three indicators of limit of detection (LOD), relative standard deviation (RSD), and accuracy were used to evaluate the device's performance. The limit of detection (LOD) was the quotient of the standard deviation of the concentration of 11 consecutive detections of the blank specimen 3 times and the slope of the working curve [16]. Samples with two concentrations of 0.020 mg/L and 0.150 mg/L were selected and analyzed 11 times respectively. The relative standard deviation of the analysis results was used as the device's stability indicator. The results of the performance analysis are shown in Table 2.

Table 2. Analysis results of device performance.

Item	Result	
Working Curve	A =3.1541C -0.006(R^2 = 0.9963,n = 6)	
LOD(n = 11)	0.002 mg/L	
RSD(n = 11)	5.93% (0.020 mg/L); 1.37% (0.150 mg/L)	
Accuracy	Reference Value	Measured Value
	0.030 mg/L	0.030 mg/L
	0.090 mg/L	0.085 mg/L

The device has a good linear fit in the 0–0.100 mg/L (R^2 = 0.9963). The slope(k) of the curve is 3.1541, indicating that the device has a high sensitivity. The relative standard deviations experimental results of samples with concentrations of 0.020 mg/L and 0.150 mg/L were 5.93% and 1.37%, respectively, which indicates that the device has better stability in the high concentration range. There is no significant difference between the reference value and the measured value of the sample, indicating that the device has high accuracy.

3.3 Practical Application

21 samples were collected from the Hua Jiang River Basin near the Hua Jiang Campus of Guilin University of Electronic Technology in June 2024 (Fig. 4a). The analysis results are shown in Fig. 4b. The 21 collected water samples had a concentration range from 0.010 to 0.035 mg/L, with an average concentration of 0.026 mg/L. The watershed was not polluted. Overall from upstream to downstream of this reach, the downstream nitrite concentration was higher than the upstream, with the lowest concentration at sampling point S4 and the highest at sampling point S20.

The practical application capability of the device for analyzing nitrite was investigated by comparing the Griess method with the method of this study. Figure 4b shows that the analytical method of this study has a high degree of consistency with the Griess method for determining the nitrite concentration of samples, indicating that the device can be reliably used for the analysis of nitrite in surface water.

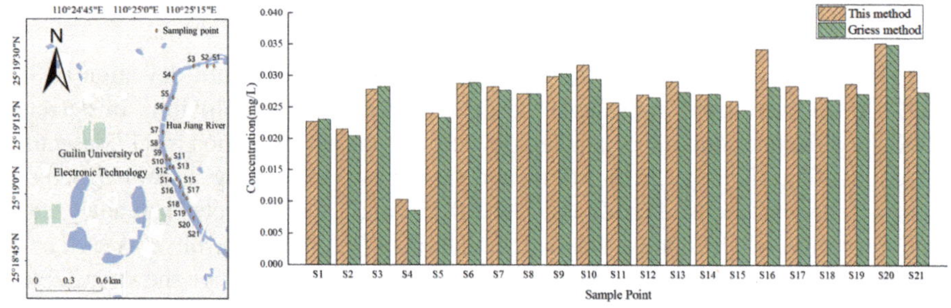

(a) (b)

Fig. 4. Sampling points and analysis results diagram. (a) sampling points diagram. (b) analysis results diagram.

4 Conclusions

This study proposes a dual-channel device based on beam splitter prism for surface water nitrite detection. The use of beam splitter prism splits the incident light into two beams. It enters the two chambers respectively, realizing the simultaneous detection of the blank solution and the sample solution. The mechanical structure is designed to reduce the weight and size of the device by using 3D printing technology. In addition, this paper also explores the optimal light intensity of LED and data processing methods. The device is characterized by low cost, low complexity, high sensitivity, and high stability, which realizes the function of outdoor portable field detection and emergency detection.

Based on this study, the device can be improved into a portable spectrophotometric detector by replacing the LED light source and the corresponding reagents, which can be applied to detect more water quality indicators.

Acknowledgment. This research was supported by the National Natural Science Foundation of China (No. 42263003) and the Guangxi Science and Technology Base and Specialized Talents

(No. 2021AC19451) and the Guangxi College Students Innovation and Entrepreneurship Training Plan Project (Nos. 20241059525).

References

1. Shortle, J.S., Mihelcic, J.R., Zhang, Q., et al.: Nutrient control in water bodies: a systems approach. J. Environ. Qual. **49**(3), 517–533 (2020)
2. Bahadoran, Z., Mirmiran, P., Ghasemi, A., et al.: Is dietary nitrate/nitrite exposure a risk factor for development of thyroid abnormality? A systematic review and meta-analysis. Nitric Oxide **47**, 65–76 (2015)
3. Wang, J., Zhan, G., Yang, X., et al.: Rapid detection of nitrite based on nitrite-oxidizing bacteria biosensor and its application in surface water monitoring. Biosens. Bioelectron. **215**, 114573 (2022)
4. Catalan-Carrio, R., Saez, J., Cuadrado, L.Á.F., et al.: Ionogel-based hybrid polymer-paper handheld platform for nitrite and nitrate determination in water samples. Anal. Chim. Acta **1205**, 339753 (2022)
5. Lo, H.S., Lo, K.W., Yeung, C.F., et al.: Rapid visual and spectrophotometric nitrite detection by cyclometalated ruthenium complex. Anal. Chim. Acta **990**, 135–140 (2017)
6. Hong, C., Li, D., Cao, S., et al.: Sensitive and multicolor detection of nitrite based on iodide-mediated etching of gold nanostars. Chem. Commun. **58**(93), 12983–12986 (2022)
7. Dou, B., Yan, J., Chen, Q., et al.: Development of an innovative nitrite sensing platform based on the construction of carbon-layer-coated In_2O_3 porous tubes. Sens. Actuators, B Chem. **328**, 129082 (2021)
8. Zuo, Y., Wang, C., Van, T.: Simultaneous determination of nitrite and nitrate in dew, rain, snow and lake water samples by ion-pair high-performance liquid chromatography. Talanta **70**(2), 281–285 (2006)
9. Antczak-Chrobot, A., Bąk, P., Wojtczak, M.: The use of ionic chromatography in determining the contamination of sugar by-products by nitrite and nitrate. Food Chem. **240**, 648–654 (2018)
10. Xiao, J., Tang, J., Chen, J., et al.: Rapid and selective detection of nitrite in ham sausage and water samples by a portable gas pressure mete. Sens. Actuators, B Chem. **400**, 134914 (2024)
11. Cadeado, A.N.S., Machado, C.C.S., Costa, M.Q., et al.: A palm-sized wireless device for colorimetric nitrite determination in water. Microchem. J. **183**, 108138 (2022)
12. Man, Y., Yu, K., Tan, H., et al.: A microfluidic concentration gradient colorimetric system for rapid detection of nitrite in surface water. J. Hazard. Mater. **465**, 133133 (2024)
13. Titov, I., Köpke, M., Gerken, M.: Monolithic integrated OLED–OPD unit for point-of-need nitrite sensin. Sensors **22**(3), 910 (2022)
14. Xu, K., Chen, Q., Zhao, Y., et al.: Cost-effective, wireless, and portable smartphone-based electrochemical system for on-site monitoring and spatial mapping of the nitrite contamination in water. Sens. Actuators, B Chem. **319**, 128221 (2020)
15. Yang, T., Sun, Y., Zeng, H., et al.: Rapid, portable and visualizing nitrite detection enabled by a rationally designed meso-aminoindole substituted pyronine-based fluorescent probe. Spectrochim. Acta Part A Mol. Biomol. Spectrosc. **319**, 124566 (2024)
16. Araujo, P.: Key aspects of analytical method validation and linearity evaluation. J. Chromatogr. B **877**(23), 2224–2234 (2009)

The Study on the Effect of River Channel Regulation and Water Quality Improvement Project for Yan Tian Reservoir Inflow Rivers

Wenbin Lv[1(✉)], Bin Wang[1], and Yihong Chen[2]

[1] Pearl River Water Resources Commission, Pearl River Water Resources Research Institute, Guangzhou 510611, China
stonerky@qq.com

[2] The National Key Laboratory of Water Disaster Prevention, Hohai University, Nanjing 210024, China

Abstract. As an important water supply reservoir for Dongguan and Shenzhen, the water quality of the Bai Ni Keng channel, which feeds into the Yan Tian Reservoir, does not meet the standards throughout the year, making it difficult to meet the water quality requirements of the Yan Tian Reservoir as a drinking water source. This paper constructs a one-dimensional hydrodynamic and water quality model for the Bai Ni Keng channel, analyzing the improvement effect of the newly constructed engineering measures on water quality. It was found that when the Bai Ni Keng channel encounters a 3-year, 50-year, and 100-year design flood, the main pollutants entering the Dong Shen channel through the newly constructed box culvert do not meet the surface water class V during the early stages of discharge (about 6 h), including total phosphorus, COD, total nitrogen, and ammonia nitrogen. As the discharge volume increases, the concentration of each pollutant decreases, and at the flood peak, it can reach the surface water class III. Therefore, the implementation of the project will have a certain impact on the water quality. Based on real-time water quality conditions, appropriate positions can be selected for flood regulation to reduce the flood discharge impact on the downstream river channels.

Keywords: Yan Tian Reservoir · Bai Ni Keng channel · Hydrodynamic and Water Quality Model · Water Quality Improvement

1 Introduction

Yan Tian Reservoir is one of the important water supply reservoirs in the Dongshen water supply project, with one of its main tributary is Bai Ni Keng channel. According to surveys, during non-rainy periods, the water quality of the Bai Ni Keng channel is classified as Class V (the worst class) as shown in Table 1, which fails to meet the water quality requirements of Yan Tian Reservoir as a drinking water source. In accordance with the requirements of the Shenzhen Municipal Government, to strengthen the protection of the drinking water source environment, it is planned to intercept and

D. Han et al. (Eds.): ICEPG 2024, ESE, pp. 126–133, 2025.
https://doi.org/10.1007/978-3-031-86829-0_16

discharge the floodwater in the Bai Ni Keng area. By transforming the existing water-blocking dam where the Bai Ni Keng channel flows into Yan Tian Reservoir, the flood prevention capacity of the river channel will be enhanced. Additionally, new A and B line interception culverts will be constructed to divert the flow of the Bai Ni Keng channel, thereby improving water quality and ultimately ensuring the safety of the drinking water source of Yan Tian Reservoir and achieving the goal of the water quality reaching Class III surface water.

Table 1. Water quality of Bai Ni Keng channel (mg/L)

Rainfall Time(min)	TP	TN	COD	NH_3
10	0.29	3.90	17.90	3.67
20	0.31	4.35	23.60	4.09
30	0.25	3.09	14.20	3.04

However, it is not yet known whether the existing project plan can meet the above requirements. It is necessary to establish a one-dimensional hydrodynamic and water quality model of the Bai Ni Keng channel to analyze the impact of engineering measures on the transformation of hydrodynamics and the improvement of water quality, and to guide the adjustment of subsequent construction project plans.

Current research on Yan Tian Reservoir mainly focuses on the hydrological characteristics and flood analysis of the reservoir. Studies on water quality are still at the stage of qualitative analysis of water quality improvement projects, and there is no construction of a one-dimensional hydrodynamic and water quality model for the inflow rivers, nor a quantitative analysis of the characteristics of inflow floods and water quality. Flood analysis of Yan Tian Reservoir began around the year 2000, Peng and Chen [1] analyzed the impact of the typhoon 9908 storm flood on the reservoir. They suggested that the sluice gate to the reservoir should be closed in advance to empty the reservoir capacity and lower the water level, and that the flood discharge channel also needs to be cleared of obstructions; Chen [2] measured the water level-capacity curve of Yan Tian Reservoir, and discussed the feasibility of the linkage project and coordinated scheduling with Shenzhen Reservoir to ensure the safety of water supply from the reservoir. In recent years, Jiang [3] analyzed the impact of the Yan Tian Reservoir water quality protection project on the current flood control system and found that the project can effectively improve the water quality of the inflow to the reservoir and ensure the safety of water supply. However, the flood control pressure on the downstream river sections has increased, and attention should be paid to raising the flood control standards.

The MIKE11 software, developed by the renowned Danish Hydraulic Institute (DHI), is a one-dimensional professional engineering software for simulating the flow, water quality changes, and sediment transport in estuaries, rivers, irrigation systems, channels, and other surface water bodies. This software has been used for hydrodynamic and water quality simulation analyses in various basins [4–9]. Therefore, this paper constructs a one-dimensional hydrodynamic and water quality model for the Yan Tian Reservoir inflow river, the Bai Ni Keng channel, before and after the project, based on the MIKE11 software. It analyzes the improvement of water quality in the Bai Ni Keng channel and the enhancement of flood prevention capacity due to the river channel regulation and water quality improvement project, with the expectation of guiding the improvement and implementation of the engineering plan.

2 Methodology

2.1 Study Area

The Yan Tian Reservoir Bai Ni Keng channel is located in the southeast of Bainikeng Community, Pinghu Sub-district, Longgang District, Shenzhen City. Pinghu Sub-district is bordered by Fenggang Town in Dongguan City to the northeast, connected to Henggang Sub-district to the southeast, adjacent to Guanlan Sub-district in Longhua New District to the west, and neighboring Buji and Nanwan Sub-districts to the south (Fig. 1). Pinghu Sub-district, situated in the northwest of Longgang District, is the junction point between Shenzhen and Dongguan cities, and Longgang and Bao'an districts. It has a total jurisdictional area of 40.99 square kilometers and governs 12 communities with a population of about 460,000. The water quality compliance of Yan Tian Reservoir is crucial for ensuring the water supply needs of Pinghu Sub-district.

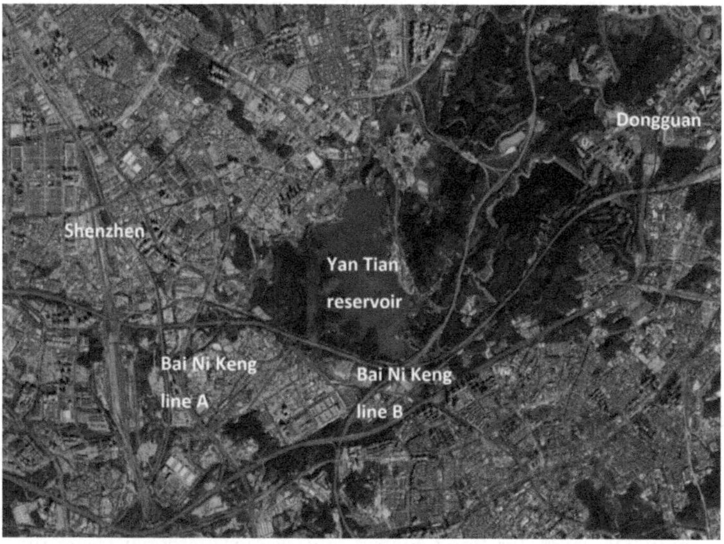

Fig. 1. The location of study area

2.2 Model Generalization

According to the achievements related to the regulation of the Bai Ni Keng channel, the A and B lines of the main stream of the Bai Ni Keng channel are generalized as shown in the Fig. 2 below.

Fig. 2. Model generalization of the Bai Ni Keng channel

Based on this field survey combined with the planning of the river channels, the Manning's roughness coefficient for each river channel is calculated based on the actual measured cross-sectional shape. For cross-sections close to a natural state, a value of 0.03 is taken. For cross-sections close to regular sections (rectangular, trapezoidal, compound), a value of 0.025 is taken.

Regarding the comprehensive decay coefficient k for pollutants, integrating the research results from the South China Institute of Environmental Sciences and Sun Yat-sen University on the decay patterns of COD, NH3-N (TN) in various water bodies of the Pearl River Delta network river area over the past two decades, the COD decay coefficient is taken as 0.1 (per day), the NH3-N (TN) decay coefficient is taken as 0.07 (per day), and total phosphorus is considered a conservative substance with no decay.

2.3 Model Verification

Based on the hydrological calculation results of Yan Tian Reservoir for model verification, the verification results are shown in Fig. 3. As can be seen from the table, the designed water level basically matches the model verified water level, and the model can meet the requirements for calculation.

2.4 Model Verification

This model primarily simulates the operational effects after the implementation of the project. Considering the progress of the project implementation, by the time this project is completed, the relevant water environment management projects within the basin will have been essentially completed. Therefore, the model simulation calculations for this occasion are based on the future total pollution (Table 2), and three computational hydrodynamic boundary conditions are set using the 72-h flood processes for 3-year, 50-year, and 100-year return periods (Fig. 4). For safety considerations, it is assumed

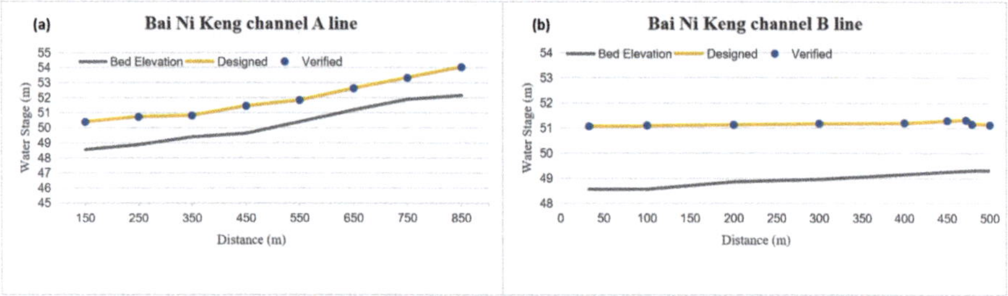

Fig. 3. Model verified water level of Bai Ni Keng channel line A (a) and line B (b)

that during the period before the arrival of the flood peak, when the runoff volume is very small (less than 3 m^3/s), it does not carry the total pollution load. All the total pollution load is carried out during the flood peak. Therefore, the model only calculates the 18 h when the flood peak occurs. Based on the actual water quality measurements during the dry season of 2019, the Ba Ni Keng channel is essentially dry during non-rainfall periods. Therefore, the initial conditions for this calculation are that the river channel is dry and free of pollution.

Table 2. Future total pollution of Bai Ni Keng channel (kg per day)

Pollution type	TP	TN	COD	NH_3
non-point source	51	400	2988	358
point source	55	440	3235	380

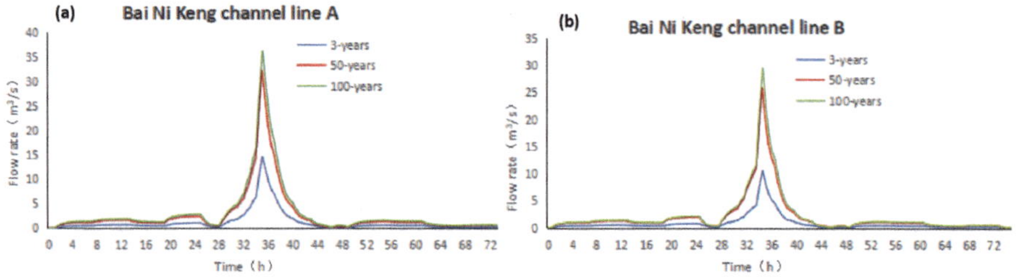

Fig. 4. Hydrodynamic boundary conditions of Bai Ni Keng channel line A (a) and line B (b)

3 Results

3.1 Hydrodynamic Results

The one-dimensional hydrodynamic model calculation results show that during the 3-year and 50-year flood events, the project intercepts and discharges all the floodwater. The gate of the Dongshen channel No. 1 is open, while the gate of the A-line reservoir No. 2 and the gate of the B-line reservoir No. 3 are both closed. The water level at the reservoir is below the elevation of the B-line water-blocking weir. The floodwater, after being collected by the A and B drainage ditches, is entirely discharged into the Dongshen channel through the interception culvert without entering the Yan Tian Reservoir. From the average flow velocity map of the drainage ditches during the flood period, it can be seen that during the 3-year and 5-year flood events, the average flow velocity of the Bai Ni Keng channel line A and B is above 0.1 m/s throughout the line. Only in the section where the No. 2 and No. 3 gates have not occurred, due to the gates not being opened, the average flow velocity of the river section at the gate is below 0.1 m/s.

During the 100-year flood event, under normal operation, the A and B line reservoir gates are closed, the interception gate is open, and the average flow velocity of the Bai Ni Keng channel line A and B is above 0.1 m/s throughout the line. Only in the section where the No. 2 and No. 3 gates have not occurred, due to the gates not being opened, the average flow velocity of the river section behind the gate is below 0.1 m/s. When the water level at the B-line reservoir gate reaches 51.10 m, the A-line and B-line reservoir gates are opened, and then the A and B line interception drainage gates are closed. At this time, the average flow velocity of the Bai Ni Keng channel line A and Bis above 0.1 m/s throughout the line, and the average flow velocity outside the interception water gate drops below 0.1 m/s. Throughout the flood period, the average flow velocity of the Bai Ni Keng channel line A and Band the three water gate passages is above 0.1 m/s, indicating good hydrodynamic conditions.

3.2 Water Quality Results

To figure out the water quality of the Bai Ni Keng channel where it enters the reservoir, the water quality variation process of each pollutant at the model's interception culvert sluice gate position is extracted from the model calculations. This allows us to obtain the water quality changes of each pollutant at various outflow positions of Bai Ni Keng over the duration of a rainfall event, as shown in Fig. 5.

From the concentration change process diagrams of each pollutant, it can be observed that under different flood frequencies, the concentration changes of pollutants at the outflow positions follow a similar pattern. Initially, the pollutant concentrations are relatively high during the early stages of rainfall. After approximately 2 h, the concentrations of all pollutants rapidly decrease, reaching their lowest levels around 6 h. Subsequently, the concentrations of all pollutants are maintained at relatively low levels. This is primarily determined by the hydrodynamic conditions of the Bai Ni Keng channel.

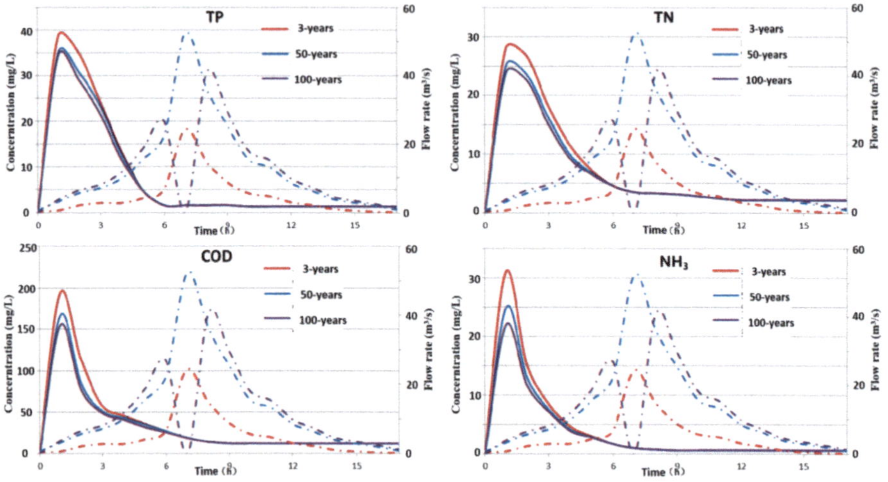

Fig. 5. Pollutant concentration change under three projects (dashed lines represent the flow rate under the corresponding projects)

4 Conclusion

This paper constructs a one-dimensional hydrodynamic and water quality model for the Bai Ni Keng channel, a tributary of the Yan Tian Reservoir, to simulate the hydrodynamics and water quality of the river after the implementation of river regulation and water quality improvement projects. It was found that when the Bai Ni Keng channel encounters a 3-year, 50-year, and 100-year flood event, the main pollutants discharged into the Dongshen channel through the newly built culvert do not meet the Class V surface water standards during the initial discharge period (about 6 h) for total phosphorus, COD, total nitrogen, and ammonia nitrogen. As the discharge volume increases, the concentration of pollutants decreases, and during the flood peak, it can reach Class III surface water standards. Therefore, the implementation of the project will have a certain impact on the water quality of the Dongshen channel, which needs to be considered in the downstream related engineering construction. During the 100-year flood event, the A and B line gates will be opened for about 55 min at the flood peak, overflowing into the Yan Tian Reservoir, and the overflow contains main pollutants such as total phosphorus, COD, and ammonia nitrogen, which can meet the Class III surface water standards. Therefore, the implementation of this project will have no impact on the water quality of the Yan Tian Reservoir.

According to the results of the water quality model calculation, when the initial flow of the Bai Ni Keng channel during the initial rainfall does not reach 12.66 m^3/s, the water quality discharged to the Dongshen channel does not meet the Class V surface water standard. It is recommended that the future project include the water quality and volume of the Bai Ni Keng channel. Due to the initial effect of pollutants, when the pollutant concentration is at its highest, the discharge volume is not large. Real-time water quality can be used to adjust the floodwater at suitable locations and flexibly schedule according to the actual measured water quality. When the water quality does not meet the Class V water quality standard requirements, close the discharge gate, and open it again when the water quality reaches the Class V water quality standard. Alternatively, when the monitored water quality reaches the Class III water quality standard requirements, the floodwater can be allowed to enter the Yan Tian Reservoir in advance to reduce the flood impact on the downstream river directly.

References

1. Peng, K., Chen, S.. Analysis of Typhoon 9908 Storm Surge Flood in Yan Tian Reservoir. Guangdong Water Resour. Hydropower (05), 32–33+36 (2001)
2. Danrong, C.: Determination of the storage curve of Yan Tian reservoir. China Water Transport (Second Half Month) **13**(03), 169–170 (2013)
3. Danrong, C.: Preliminary exploration of the feasibility of the interconnection project between Yan tian reservoir and Shenzhen reservoir. China Water Transport (Second Half Month) **13**(05), 147–148 (2013)
4. Lei, J.: Analysis of the impact of yan tian reservoir water quality protection project on the current flood control system and countermeasures. Sichuan Water Resour. **45**(03), 160–162 (2024)
5. Song, B., Mingming, Z., Weiwei, S., et al.: Simulation of water quality compliance at control sections of the Qinhuai River basin based on MIKE 11. Environ. Monitor. Manage. Technol. **36**(04), 81–86 (2024)
6. Tang, B., Tian, C.: Study on the joint optimization and regulation of hydrodynamics and water quality in the main urban area of Wenzhou City. Water Conservancy Planning Des. (08), 33–36+74 (2024)
7. Feng, W., Feng, S., Hu, J.: Study on urban water diversion scheme based on MIKE11 model: a case study of the Southeastern Plain of Yinzhou District, Ningbo City. Zhejiang Water Resourc. Hydropower Technol. **52**(04), 31-35 (2024)
8. Huaiyong, G., Bikui, Z., Xiaojuan, G., et al.: Evaluation of ecological water compensation scheduling plan for rivers in the Guangdong-Hong Kong-Macao greater bay area based on MIKE 11. People's Pearl River **45**(07), 48–57 (2024)
9. Zhixian, H., Chuanlin, W., Ziqing, Q., et al.: Dynamic analysis of water environmental capacity in the Jiangnan Plain River Network based on MIKE11. Environ. Eng. J. **18**(03), 663–675 (2024)

Data Analysis and Prediction of Flood Disaster

Sheng Wang[1], Chunyan Wang[2(✉)], Weiliang Li[3], Xin Yu[4], Haiyan Shuang[5], Shengjie Cui[2], and Yunxiang He[6]

[1] College of Large Aircraft, Ordosinstitute of Technology, Ordos, China
[2] College of Medical, Ordosinstitute of Technology, Ordos, China
289214927@qq.com
[3] School of Law, Inner Mongolia University of Finance and Economics, Hohhot, China
[4] The School of Mechanical and Automotive Engineering, Heng Xing University, Qingdao, China
[5] Department of Student Affairs, Ordosinstitute of Technology, Ordos, China
[6] College of Applied Technology, Chongqing University of Technology, Chongqing, China

Abstract. This article explores flood disaster prediction using a systematic approach to data analysis and model development. Pearson correlation coefficients are computed and visualized through heatmaps to examine the relationship between indicators and flood probability, followed by regression analysis to understand influencing factors. The KMeans algorithm is used to classify flood risk into low, medium, and high levels, and feature importance is determined with random forest classifiers to assist in model construction. Various regression models—linear regression, decision tree, random forest, and gradient boosting—are trained and tested, optimized using particle swarm optimization, and evaluated through mean squared error and coefficient of determination. Finally, the model's predictions are validated against test data, showing an R^2 of 0.86 and an MSE of 0.0003, with Kolmogorov-Smirnov and Anderson-Darling tests revealing significant deviations from a normal distribution.

Keywords: Regression Models, · Feature Importance · coefficient of determination

1 Introduction

Flooding, driven by heavy rainfall, snowmelt, and storm surges, presents major economic and environmental challenges. In June 2023, severe flooding in China underscored the need for better flood management. This approach involves data cleaning and distribution assessment, with outlier removal and kriging interpolation for missing values. Pearson correlation coefficients are used to explore relationships between indicators and flood probability, visualized through a heatmap. KMeans clusters flood risk, and a random forest classifier evaluates feature importance. The dataset is standardized, split, and used to train four regression models, refined with particle swarm optimization. The final model, tested on new data, achieves an R-squared of 0.86 and an MSE of 0.0003, showing high accuracy in flood prediction and aiding in preparedness and mitigation.

D. Han et al. (Eds.): ICEPG 2024, ESE, pp. 134–143, 2025.
https://doi.org/10.1007/978-3-031-86829-0_17

2 Model Assumptions

To ensure the model's effectiveness and reliability, we make several key assumptions: that the data provided is authentic and reliable, that unusual data points are justifiable, that handling of missing values will not significantly impact prediction accuracy, and that data anonymization has not compromised its integrity. We also assume that certain indicators have varying levels of correlation with flood occurrences, that climate change increases flood risk through extreme rainfall, that changes in land use affect flood probabilities, and that model parameters remain consistent over time. These assumptions aim to establish a solid foundation for model development and enhance the realism of predictions [1].

3 Establishment and Solution of the Model

3.1 Establishment and Solution of Problem 2 Model

Cluster Analysis

I chose hierarchical clustering for this method, which starts with each data point as an individual cluster and merges them until one cluster contains all points. To illustrate, we'll use MATLAB to create a schematic diagram (see Fig. 1).

This approach uses the correlation coefficient for similarity and classifies data into three risk categories based on normalized features, applying the shortest distance method As shown in Fig. 1 We then calculated the mean metrics for each category and visualized differences with a bar chart showing variations in "topographic drainage," "erosion," and "ineffective disaster prevention." Additionally, Seaborn's pairplot provides color-coded scatterplots by risk category, and a 3D scatter plot displays data distribution using the top three features As shown in Fig. 2.

This 3D scatter plot shows the distribution of the three features "Monsoon Intensity", "TopographicDrainage" and "River Management" under different flood risk categories. The data points are concentrated in the middle area, with different colored points representing different flood risk categories. Although there is overlap, higher values for Topographic Drainage and Stream Management m ay correspond to higher risk categories. Overall, the figure visually demonstrates the relationship between significant features and flood risk categories, helping to understand the distribution of features and t heir impact on flood risk assessment [2].

Assessment of Feature Importance

The RandomForestClassifier evaluates feature importance by constructing multiple decision trees through random sampling and feature selection. Feature importance is calculated based on the information gain each feature provides when splitting nodes (e.g., reduction in Gini index or entropy). This importance is averaged across all trees in the forest. Finally, the importances are normalized to sum to 1, allowing for visual comparison through a bar chart that highlights the most impactful features. As shown in Fig. 3. The 3D scatter plot displays "Monsoon Intensity," "Topographic Drainage," and "River Management" across flood risk categories. Data points are mainly clustered in the center, with colors indicating different risk categories. Despite some overlap, higher values in Topographic Drainage and River Management often align with higher risk categories.

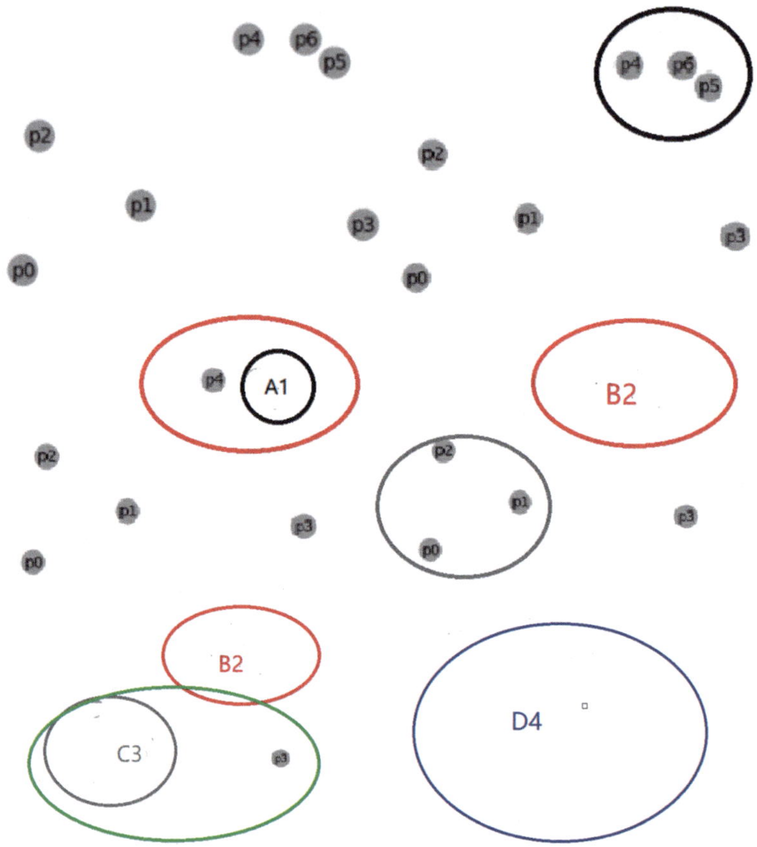

Fig. 1. Hierarchical Clustering Concept Map

This plot visually illustrates the relationship between key features and flood risk, aiding in the understanding of feature distribution and its impact on flood risk assessment.

This visualization illustrates the importance of various features as determined by the Random Forest classifier. The bar chart represents each feature's contribution to the model's prediction accuracy. From the chart, "policy factor" is the most significant feature, with the highest importance score, indicating its major impact on flood risk classification. "Inadequate planning" and "erosion" also show high importance, underscoring their key roles in determining flood risk. Features like "Infrastructure deterioration" and "Watershed" also have substantial contributions, as reflected by their longer bar lengths. Conversely, "deforestation" and "ineffective disaster prevention" have shorter bars, suggesting lower impact on the model. Other features, including "landslides," "climate change," and "urbanization," have intermediate importance. Overall, this chart provides a clear view of each feature's role in flood risk classification, aiding in feature selection and model optimization [3].

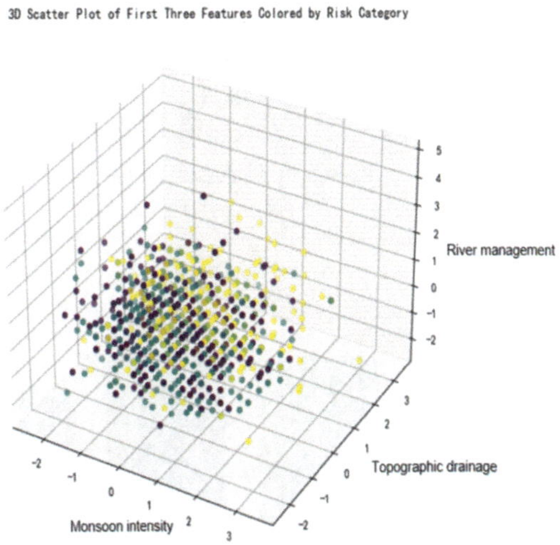

Fig. 2. 3D Scatter Plot of First Three Features Colored by Risk Category

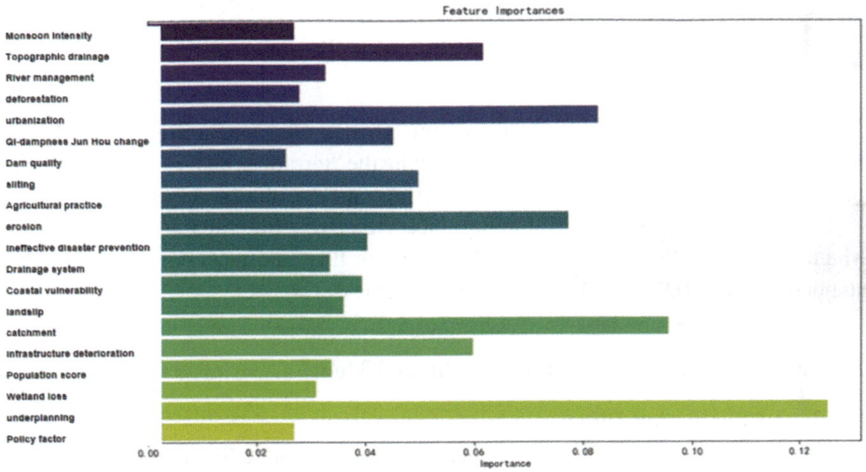

Fig. 3. The importance of each feature

Model Training and Sensitivity Analysis

The first five most important features are selected as the input of the model according to the feature importance. The random forest classifier is used to train the model and make predictions. Visualizing the distribution of key features across different flood risk categories using box plots. Each box plot illustrates the quartile range of a feature, with the median shown by a line inside the box, whiskers indicating data range, and outliers as individual points. These plots reveal how various features distribute across flood risk categories, aiding in the analysis of their roles in flood risk assessment [4]. The charts clearly highlight feature differences between categories, providing a foundation for further risk analysis and model optimization. Additionally, a histogram is used to present the results of a sensitivity analysis, which evaluates the impact of removing each feature on model performance. As shown in Fig. 4.

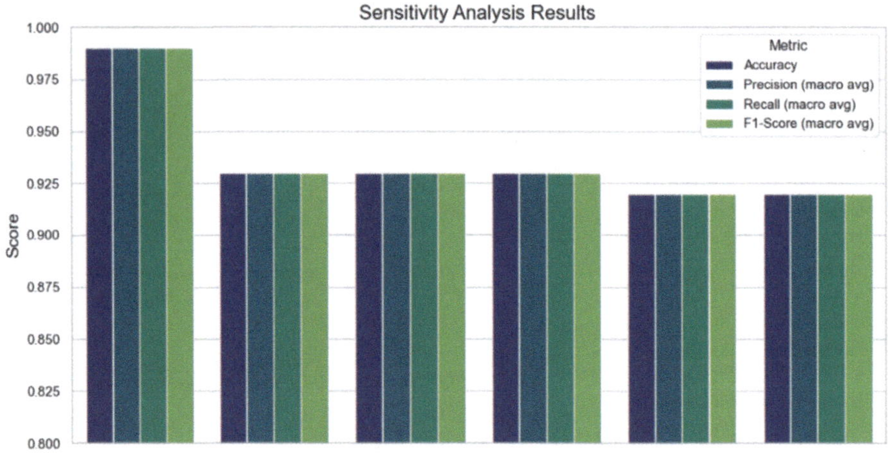

Fig. 4. Sensitivity Analysis Results

The results show that the initial random forest model performs excellently with precision, recall, and F1 scores of 0.99. Removing the "terrain drainage" and "insufficient planning" features significantly decreased performance, with accuracy dropping to 0.93 and 0.92, highlighting their critical importance. Removing other features like "erosion," "urbanization," and "watershed" caused only minor fluctuations in performance, which remained between 0.92 and 0.93, indicating their lesser impact.

3.2 Establishment and Solution of Problem 3 Model

Selection of Evaluation Index

In this paper, two evaluation indicators were used to evaluate the prediction accuracy, namely root mean square error (RMSE) and mean absolute percentage error (MAPE).

Root Mean Square Error (RMSE): The square root of the mean that represents the sum of the squares of the error of the predicted value and the true value. The formula is as follows:

$$RMSE = \left(\frac{1}{\mathrm{n}} \sum_{i=1}^{n} (\hat{\mathrm{y}}_i - \mathrm{y}_i) \right)^{1/2} \tag{1}$$

The larger the RMSE, the greater the difference between the predicted value and the actual value; The smaller it is, the higher the prediction accuracy of the model.

Mean Absolute Percentage Error (MAPE): Represents the average value of the ratio of the error of the predicted value to the true value, reflecting the relative error of the prediction results. The formula is as follows:

$$MAPE = \frac{1}{\mathrm{n}}\sum_{i=1}^{n}\left|\frac{\hat{\mathrm{y}}_i - \mathrm{y}_i}{\mathrm{y}_i}\right| \cdot 100\% \tag{2}$$

The larger the MAPE, the greater the difference between the predicted value and the actual value, and the lower the accuracy.

In this paper, a cross-validation method is used to evaluate the performance of regression prediction model, LSTM prediction and Transformer prediction, and a weighted prediction model is constructed based on these results.

Model Selection

In order to improve the accuracy, four models were selected for prediction, including linear regression, decision tree regression, random forest regression and gradient boosting regression. A weighted optimization model is built for the prediction results to improve the model accuracy [5].

Let $\hat{y}_i$ $(i = 1, 2, \cdots, m)$ be the prediction result of the same problem by m prediction methods (each method passes its own test), and y'_{ij} is i $(i = 1, 2, \cdots, m)$ the simulated value of the first prediction method on the raw data, $\mathrm{y}_i (i = 1, 2, \cdots, m)$ then the weighted prediction value of this m method is:

$$J = x_1\hat{y}_1 + x_2\hat{y}_2 + \cdots + x_m\hat{y}_m \tag{3}$$

$$\sum_{i=1}^{m} x_i = 1\,,\ x_i \geq 0\,,\ (i = 1, 2, \cdots, m) \tag{4}$$

$e_{ij} = \hat{y}_{ij} - y_j$ indicates the error between the ith prediction method and the simulated value of the jth historical data

$$\begin{aligned} L &= \sum_{j=1}^{n}\left(\hat{J}_j - y_j\right)^2 = \sum_{j=1}^{n}\left(x_1\hat{y}_{1j} + x_2\hat{y}_{2j} + \cdots + x_m\hat{y}_{mj} - y_j\right)^2 \\ &= \sum_{j=1}^{n}\left(x_1 e_{1j} + x_2 e_{2j} + \cdots + x_m e_{mj}\right)^2 = \sum_{j=1}^{n} x^T E_j x \end{aligned} \tag{5}$$

Let YX be the prediction result of the same problem by m prediction methods (each method passes its own test), and Y is the simulated value of the first prediction method on the original A data Y. U then the weighted prediction value of this m method is:

$$\begin{aligned} &\mathrm{vi} + 1 = \mathrm{wvi} + \mathrm{c1r1(pbest} - \mathrm{xi}) + \mathrm{c2r2(gbest} - \mathrm{xi}) \\ &\mathrm{xi} + 1 = \mathrm{xi} + \mathrm{vi} + 1 \end{aligned}$$

where R1 and R2 are random numbers between 0 and 1, and W is the inertia factor. The larger the value, the stronger the ability to find the optimal solution. Vice versa. A linear decreasing weight strategy is adopted.

Prediction Results

First, we use linear regression, decision tree regression, random forest regression and gradient boosting regression model to predict, and the prediction results are as follows As shown in Fig. 5. Among the four regression models, linear regression achieved the best performance with the lowest MSE of 0.000387 and the highest R-squared of 0.831, indicating it explains 83% of the data variability. Decision tree regression performed poorly with an MSE of 0.003845 and a negative R-squared, while random forest regression and gradient boosting regression had MSEs of 0.004129 and 0.001040, and R-squared values of 0.434 and 0.546, respectively [6]. The weighted combination model, using linear regression, decision tree, random forest, and gradient boosting with weights of 0.476, 0.0, 0.238, and 0.286, respectively, achieved an improved MSE of 0.000589 and an R-squared of 0.742, effectively explaining 74% of the data variability and demonstrating enhanced prediction accuracy As shown in Fig. 6.

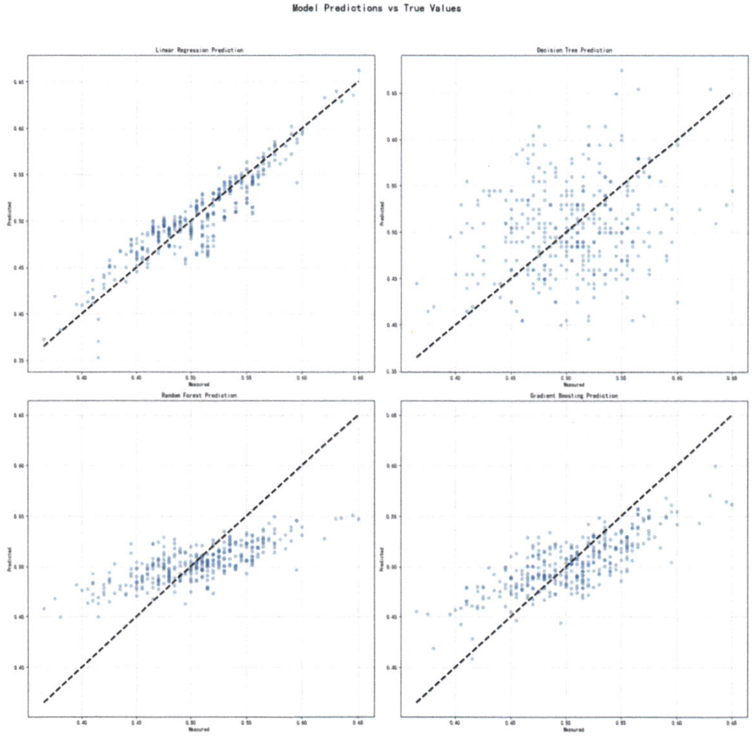

Fig. 5. Comparison chart

Establishment and Solution of Prediction Model for Limiting Index

In model comparison based on Mean Squared Error (MSE), linear regression achieved the lowest MSE of 0.001902, while decision tree regression had the highest at 0.004358. Random Forest Regression and Gradient Boosting Regression yielded MSEs of 0.002246

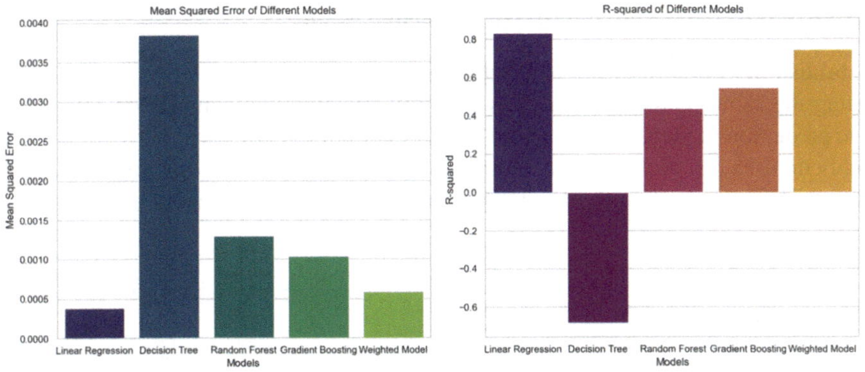

Fig. 6. Forecast comparison chart

and 0.002064, respectively. The weighted model, with an MSE of 0.001972, performs better than decision tree regression but slightly worse than linear regression, indicating its balanced prediction performance As shown in Fig. 7.

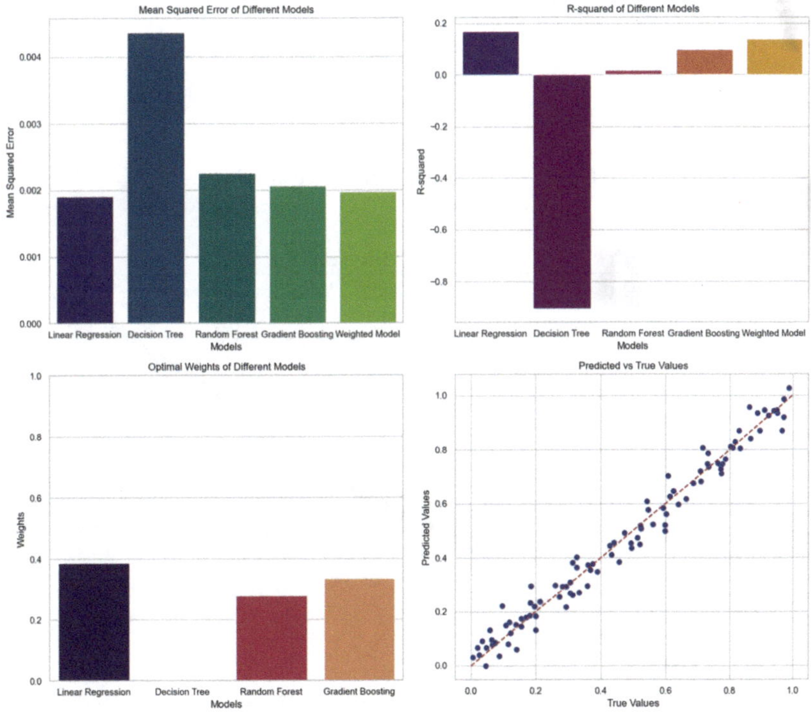

Fig. 7. Forecast results

3.3 Establishment and Solution of Problem 4 Model

Establishment and Solution of Prediction Model

Using the same prediction model as the third question, the comparison between the five prediction models and the actual value is as follows As shown in Fig. 8. Using the prediction results, Kolmogorov-Smirnov test and Anderson-Darling test were carried out, and histograms, line charts and QQ charts were drawn. As shown in Fig. 9.

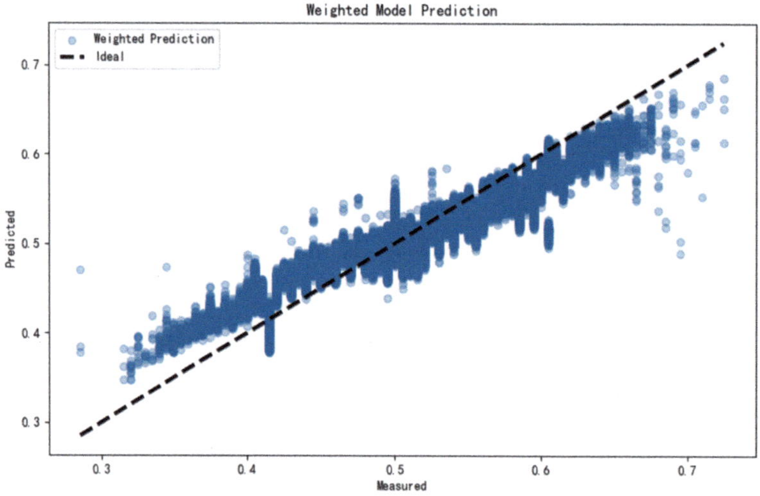

Fig. 8. Prediction model vs. actual values

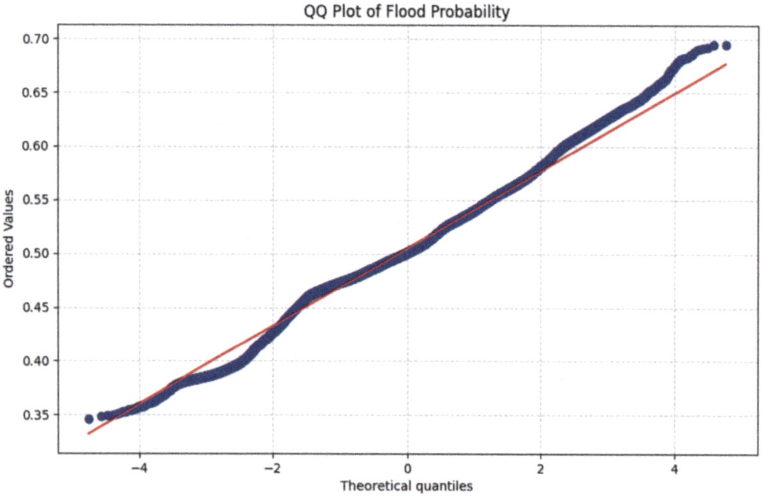

Fig. 9. Predictive Comparison Plot q-q Plot

The histogram analysis reveals that while the middle part of the flood probability data approximates a normal distribution, significant deviations are observed at both ends, particularly in high values, indicating right skewness and the presence of extreme high values. This aligns with prior normality tests, which show the data significantly diverges from a normal distribution, especially at extreme values. Key observations include a main peak between 0.45 and 0.55, a positively skewed distribution with a decreasing number of samples as probability increases, and a multi-peak characteristic as shown by the KDE curve. The results of the Shapiro-Wilk, Kolmogorov-Smirnov, and Anderson-Darling tests, all with p-values of 0, confirm that the flood probability data do not follow a normal distribution [7].

4 Conclusion

This study employs a robust approach to flood forecasting by integrating comprehensive data cleaning, advanced modeling techniques, and rigorous evaluation. The 3σ principle and box plots for outlier detection, and Kriging interpolation for missing value imputation. Correlation analysis was conducted with Pearson's coefficient and visualized using Seaborn's heatmaps to identify key indicators. A multi-model approach, including linear regression, decision tree regression, random forest regression, and gradient boosting regression, was used to enhance prediction performance through weighted combination. Model evaluation was based on mean squared error (MSE) and R-squared, with results illustrated in histograms and scatter plots. Sensitivity analysis involved removing key features to assess their impact on predictions. Despite the model's potential, it faces challenges such as increased complexity and computational costs due to multi-model integration, reliance on existing data which may overlook important features, lack of practical application verification, limited weight determination methods, and inadequate consideration of time series factors. Future research should focus on incorporating more environmental and climate data, employing advanced machine learning algorithms, and processing time series data to enhance prediction accuracy and applicability. The developed flood forecasting system offers valuable technical support for flood management and can serve as a reference for predicting and responding to other natural disasters.

References

1. Zhu, C., et al.: Flood monitoring and disaster analysis of Chaohu Lake based on Sentinel-1 SAR data in 2020. J. Hefei Univ. Technol. (Nat. Sci.) **46**(09), 1217–1223 (2023)
2. Fan, H., et al.: Analysis of flood disaster impact based on domestic high-resolution satellite imagery data. Geospatial Inf. **21**(08), 65–67 (2023)
3. Zhu, H., Li, H., Yuan, C.: Investigation and data analysis of flood disasters in Henan Province. Henan Water Conserv. South-to-North Water Divers. **51**(05), 6–7 (2022)
4. Xu, Y., et al.: Comparative analysis of natural disaster risk assessment based on different spatial scale data - taking rainstorm flood disaster as an example. J. Disast. Stud. **37**(03), 61–68 (2022)
5. Zhai, S., Ha, F., Yang, B.: Design of flood disaster early warning system based on hydrological big data analysis technology. Water Resour. Hydropower Eng. Superv. **(05)**, 53–56+74 (2022)

6. Luan, Y., et al.: Remote sensing monitoring and disaster analysis of Shandong Flood in 2018 based on Sentinel-1B SAR data. J. Nat. Disast. **30**(02), 168–175 (2021). https://doi.org/10.13577/j.jnd.2021.0217
7. Wang, W., Li, L.: Characteristics and risk assessment of major meteorological disasters in Shanxi province - an empirical perspective based on the integration of rainstorm flood disaster meteorological data. Sci. Technol. Rev. **35**(08), 81–87 (2017)

Study on the Risk of Salinity Diffusion in the Water Body After the Connection of Phases II and III of Xuanmen

Ziming Wang[1](✉), Hui Wu[1], Haibin Chen[2], Ruohe Xiao[3], Yong Huang[3], and Peinan Ye[2]

[1] Zhejiang Institute of Hydraulics & Estuary(Zhejiang Surveying Institute of Estuary and Coast), Hangzhou 310020, China
wangziming1301@126.com

[2] Yuhuan Municipal Bureau of Agriculture, Rural Affairs, and Water Resources, Yuhuan 317600, China

[3] Yuhuan Future New City Investment Co., Ltd., Hangzhou 317600, China

Abstract. This paper investigates the risk of salinity diffusion in the Xuanmen Phase II and III freshwater reclamation projects in Yuhuan City, Zhejiang Province, particularly after the removal of the Phase I dam in 2023. A two-dimensional mathematical model was established and validated to simulate the flow and salinity changes in the reservoirs. The results show that water flow exhibits a reciprocating pattern at the site of the former dam, with salinity diffusion mainly influenced by the operational dispatch of the water gates in Phases II and III. During non-rainfall periods, salinity distribution is primarily affected by wind fields, with salinity diffusion concentrated at the connection between Phases II and III. This research provides critical insights into optimizing reservoir management and mitigating salinity diffusion risks.

Keywords: Xuanmen Phase II and III · salinity · diffusion risk · two-dimensional mathematical model · optimized dispatch

1 Introduction

With rapid societal development and population growth, the pressure on freshwater resources continues to increase [1, 2]. Utilizing natural island and geographic conditions, damming bays for freshwater storage has become one solution to address regional water shortages. Numerical models of salinity changes in reservoirs have been widely studied [3, 4]. Field observations indicate that the desalination process in freshwater reservoirs is slow, often taking an extended time to reach functional standards [5]. CHANSUNG OH [6] applied the EFDC model to simulate the spatial and temporal distribution of temperature and salinity in the Saemangeum Reservoir, predicting that desalination could take over five years. Fan [7] analyzed the salinity variation and desalination factors in Zhuyu Lake and found that sediment disturbance is the primary factor affecting salinity release [8]. Zhang Xiaolong [9] used field sampling and laboratory experiments to

D. Han et al. (Eds.): ICEPG 2024, ESE, pp. 145–154, 2025.
https://doi.org/10.1007/978-3-031-86829-0_18

study the effects of environmental factors on sediment salinity release in Mukandao Bay Reservoir. Seong [10] employed EFDC to assess the spatiotemporal changes in salinity in estuarine reservoirs, revealing that salinity levels were likely influenced by dam leakage. A comprehensive model of the study area was constructed using numerical simulation techniques, taking into account the changes in salinity within the reservoir area caused by the gates and inflows. The model results were then compared with the monitoring data [11].This study builds on existing research by analyzing the potential salinity diffusion risks due to changes in water system connectivity, providing valuable references for risk management and desalination acceleration in coastal freshwater reservoirs.

2 Regional Overview

Yuhuan City is located between 121°05′38″ to 121°32′29″ E and 28°01′32″ to 28°19′24″ N in southeastern Zhejiang Province, see Fig. 1. In 1977, the Phase I Xuanmen dam was completed, connecting the island and peninsula and facilitating traffic and sediment deposition, laying the foundation for Phases II and III. The Phase II freshwater reclamation project began in 1999, with dam closure completed in 2001, and the Phase III project started in March 2006 and closed on November 24, 2010. After reclamation, the drainage area was 296.2 km^2, with 166.2 km^2 for Phase II and 130 km^2 for Phase III. After 23 years of desalination, Phase II had significantly reduced salinity, while Phase III, completed nine years later, showed a higher concentration of salinity (in cl^-). The removal of the Phase I dam on March 27, 2023, prompted the need for an analysis of the associated salinity diffusion risks.

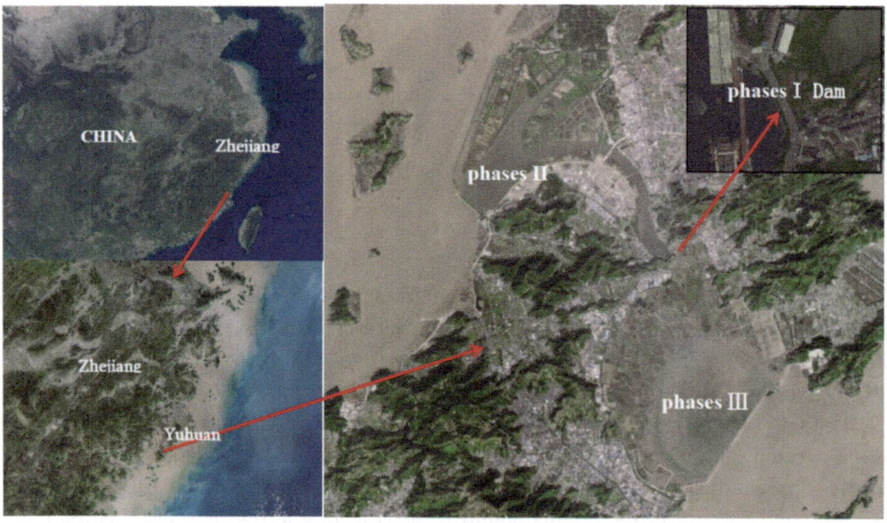

Fig. 1. Location map of the study area

3 Research Proposal

3.1 Basic Equation

The transport and degradation of substances in water satisfy the mass conservation principle. A first-order reaction convective diffusion equation describes the transport of conventional water quality parameters. The two-dimensional (2D) vertically averaged convective-diffusion-reaction equation for salinity (in terms of cl-s) is coupled with shallow water equations to form the following conservation form (1):

$$\frac{\partial(h\overline{C})}{\partial t}+\frac{\partial(hU\overline{C})}{\partial x}+\frac{\partial(hV\overline{C})}{\partial y}=\frac{\partial}{\partial x}(Dx\frac{\partial(h\overline{C})}{\partial x})+\frac{\partial}{\partial y}(Dy\frac{\partial(h\overline{C})}{\partial y})-F_{ex}-Kh\overline{C}+S \tag{1}$$

The coupled system formed by combining the above two-dimensional material transport equation with the two-dimensional shallow water equations can be uniformly written in the following conservation form (2).

$$\frac{\partial \mathbf{U}}{\partial t}+\nabla\cdot\mathbf{F}(\mathbf{U})=\mathbf{S} \tag{2}$$

where:

$$\mathbf{U}=\begin{pmatrix} h\\ hu\\ hv\\ h\varphi \end{pmatrix},\ \boldsymbol{F}_x=\begin{pmatrix} hu\\ hu^2+\frac{gh^2}{2}\\ huv\\ hu\varphi \end{pmatrix},\ \boldsymbol{F}_y=\begin{pmatrix} hv\\ huv\\ hv^2+\frac{gh^2}{2}\\ hv\varphi \end{pmatrix},$$

$$\boldsymbol{S}=\begin{pmatrix} 0\\ gh(S_{0x}-S_{fx})+\nabla^2\frac{\varepsilon}{\rho}hu+C_\varphi\frac{\rho_a}{\rho^2}\omega^2\sin\alpha+fvh\\ gh(S_{0y}-S_{fy})+\nabla^2\frac{\varepsilon}{\rho}hv+C_\varphi\frac{\rho_a}{\rho^2}\omega^2\cos\alpha+fuh\\ \frac{\partial}{\partial x}\left(D_{\varphi x}h\frac{\partial\varphi}{\partial X}\right)+\frac{\partial}{\partial y}\left(D_{\varphi y}h\frac{\partial\varphi}{\partial y}\right)-K_\varphi h\varphi-F_{ex}+S_\varphi \end{pmatrix} \tag{3}$$

where h represents the water depth, u, v are the depth-averaged velocity components in the x and y directions, respectively. Fx is the flux vector in the x direction, Fy is the flux vector in the y direction, and S is the source term vector. $S_{0x}=-\frac{\partial z_b}{\partial x}$, $S_{0y}=-\frac{\partial z_b}{\partial y}$ are the bottom slopes in the x and y directions, respectively. 0 $S_{fx}=\rho n^2h^{-4/3}u\sqrt{u^2+v^2}$, $S_{fy}=\rho n^2h^{-4/3}v\sqrt{u^2+v^2}$ are the bottom friction terms in the x and y directions, respectively. c_w is the wind drag coefficient, ρ_a is the air density, ω is the wind speed, and α is the angle between the wind direction and the y-axis. $D_{\varphi x}$, $D_{\varphi y}$ is the diffusion coefficient in the x and y direction, K is the overall degradation coefficient of the transported material, and F_{ex} is the source-sink term.

3.2 Model Establishment

The project area involves the Xuanmen Phase II and Phase III regions, the upstream inflow water system, the adjacent sea area near the external drainage gates, and other relevant areas. In the offshore region, the primary focus is on the impact of tidal levels on the water level downstream of the gates, with a relatively smaller scope. To simplify the computational conditions, the model area primarily covers the Xuanmen Phase II and Phase III regions, with a simulated area of approximately 81 square kilometers. In the model domain, flow boundaries are selected for the upstream river channels and gate stations, while the downstream area, which is the seaward side, uses tidal levels as control boundaries. Figure 2a these include the Yinggong Gate and Yuhuan Drainage Gate for Phase II, as well as the Chongtanyu Gate and Muyuyu Gate for Phase III.

The operation of the external drainage gates is as follows: During normal water periods, to maintain the upstream river morphology and ecological landscape, the Yinggong Gate, Chongtan Gate, and Muyu Gate remain closed. During flood periods, when the water level in Xuanmen Phase II and III exceeds 0.3 m and the water level upstream of the gates is higher than the offshore tidal level, the Yinggong Gate, Chongtan Gate, Muyu Gate, and Yuhuan Lake Gate will be opened for drainage. After the flood peak subsides and the water level in Xuanmen Phase II and III falls back to 0.2 m, the gates will be closed.

The mathematical model consists of 12,181 computational nodes and 23,253 elements. The maximum grid scale within the domain is 186 m, located in open and gently sloping areas within the model boundary, while the minimum spatial step size is approximately 10 m, situated near structures such as gates and river channels. The model area is illustrated in Fig. 2b.

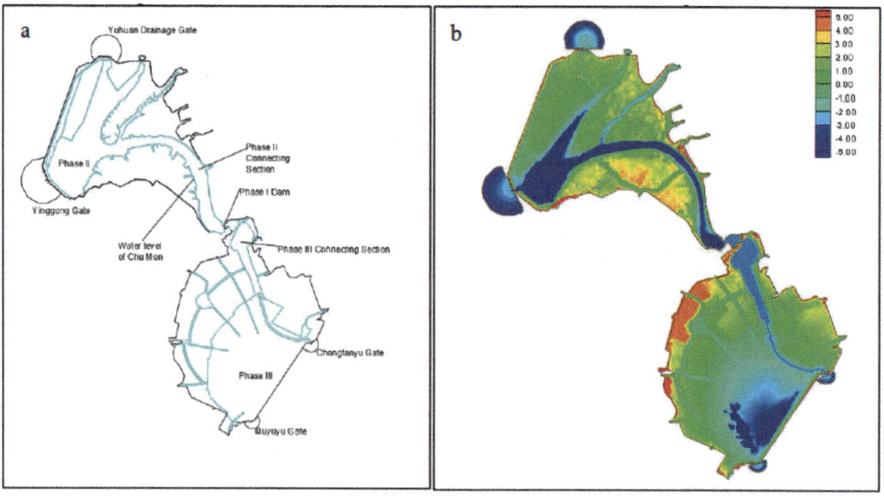

Fig. 2. Model Scope and Topography

3.3 Model Validation

The model was used to validate the "Lekima" flood event. As shown in the figure, the calculated results align well with the observed data. During the "Lekima" flood, the simulated water levels closely matched the trend of the water levels recorded at the Chumen station (see Fig. 2), especially regarding the peak water level, Fig. 3 which was simulated with high accuracy. The calculation error for the peak water level was within 0.07 m, meeting the requirements for relevant calculation accuracy. This indicates that the choice of the mathematical model and the calibration of its parameters were reasonable and adequate to meet the needs of the scenario calculations.

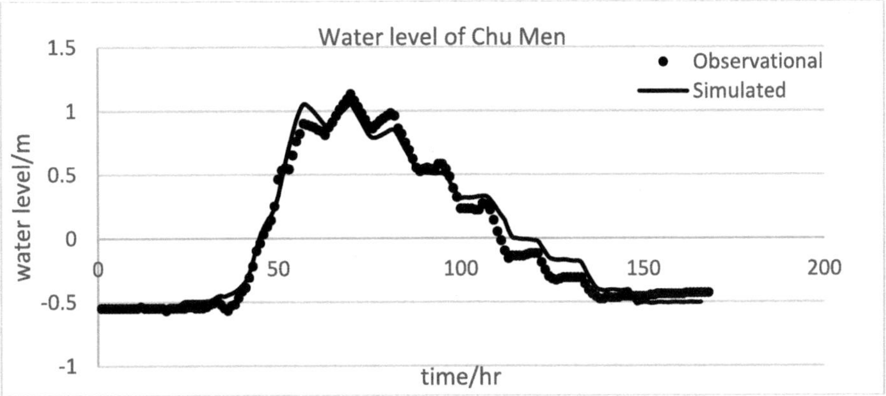

Fig. 3. Hydrodynamic verification results

4 Risk Study After the Linkage of Phases II and III

4.1 Characterization of Post-Linkage Streamflow

The flow is defined as positive when water flows from Phase II to Phase III, and negative when it flows from Phase III to Phase II. The Fig. 4 shows that when the flow difference between Phases II and III is positive, the flow at the blockage dam is negative, indicating that water is flowing from Phase III to Phase II. Conversely, Fig. 5 shows that when the flow difference between Phases II and III is negative, the flow at the dam is positive, indicating that water is flowing from Phase II to Phase III. This suggests a back-and-forth flow phenomenon at the blockage dam, where the flow direction is closely related to the discharge flow difference between the sluice gates of Phases II and III. Therefore, after the removal of the Phase I blockage and the connection of Phases II and III, the risk of cl^- diffusion is primarily influenced by the regulation of the sluice gates of Phases II and III. Figure a shows a typical flow field where the flow at the Phase I blockage moves from Phase III to Phase II when the discharge flow of the Phase II sluice gate is greater than that of the Phase III sluice gate. Figure b shows a typical flow field where the flow at the Phase I blockage moves from Phase II to Phase III when the discharge flow of the Phase III sluice gate is greater than that of the Phase II sluice gate.

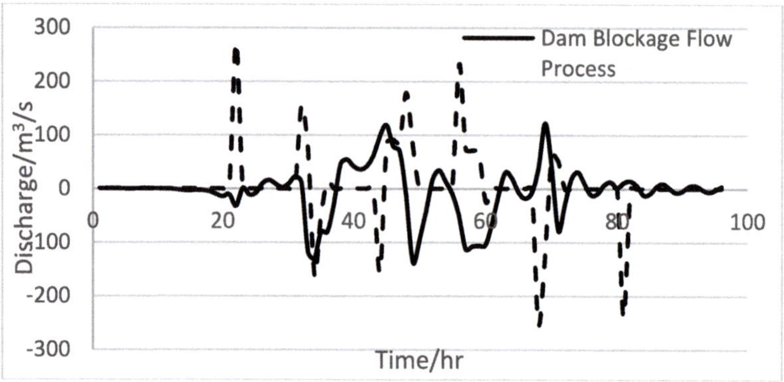

Fig. 4. Difference in outfall flow between Phase 2 and 3 vs. flow at Phase I dam

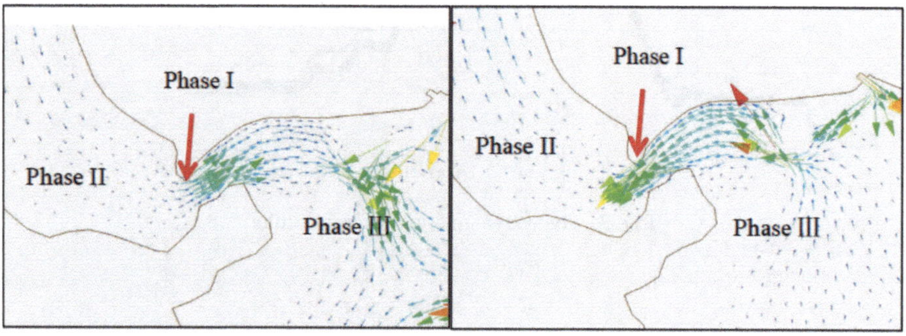

Fig. 5. Typical schematic of flow direction at the Phase I dam

4.2 Wind Field Effects on Salinity Distribution

The upstream river channels of both the Phase II and Phase III of Xuanmen are connected to the reservoir through sluice gates. During non-rainfall periods, to maintain the scenic water levels in the upstream river channels, the sluice gates are mostly closed, as are the external discharge gates of Phases II and III. Therefore, during non-rainfall periods, after Phases II and III are connected, the salinity distribution is mainly influenced by wind conditions. The Fig. 6 shows that during non-rainfall periods, after Phases II and III are connected, the salinity exchange is mainly concentrated at the connection point between Phases II and III, with minimal impact on the lake areas of Phases II and III. Specifically, the cl^- concentration at the connection point of Phase II changes from 1054 mg/L to 1090 mg/L and then remains stable; at the connection point of Phase III, the cl^- concentration changes from 1452 mg/L to 1158 mg/L and then remains stable. However, the cl^- concentration in the lake areas of Phases II and III remains largely unchanged. Therefore, during non-rainfall periods, the risk of cl^- diffusion is mainly concentrated at the connection segment between Phases II and III, with little impact on the lake areas.

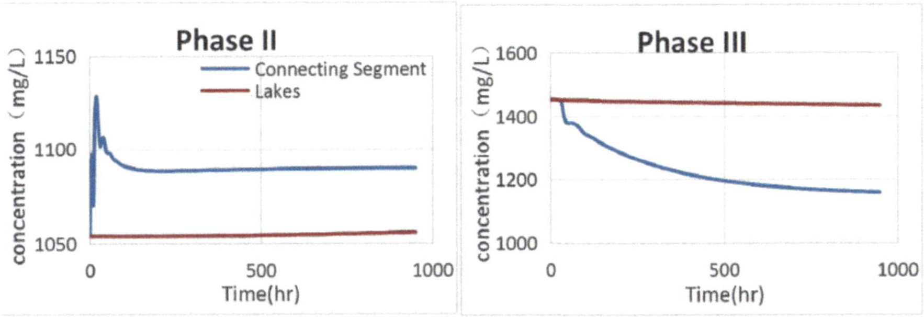

Fig. 6. Variation of cl^- concentration under wind field conditions

4.3 Impact of Salinity Distribution During Floods

Figure 7a shows the initial cl^- concentrations in Phase II and Phase III. In Phase II, the average concentration is approximately 1054 mg/L, while in Phase III, the average concentration is 1452 mg/L. Under the condition where the Phase I dam remains intact, there is no cl^- exchange between Phases II and III. Figure 7b show after the Phase I dam is removed and under the conditions of a once-in-five-year flood with normal operation of the external drainage gates, the cl^- concentrations in Phases II and III both decrease due to the influence of upstream inflows. However, a local high concentration area still exists in the central area of the Phase III lake, with the highest concentration reaching 1330 mg/L. After rainfall, the average concentration in Phase II decreases from 1054 mg/L to 871 mg/L, and in Phase III, it decreases from 1452 mg/L to 923 mg/L.

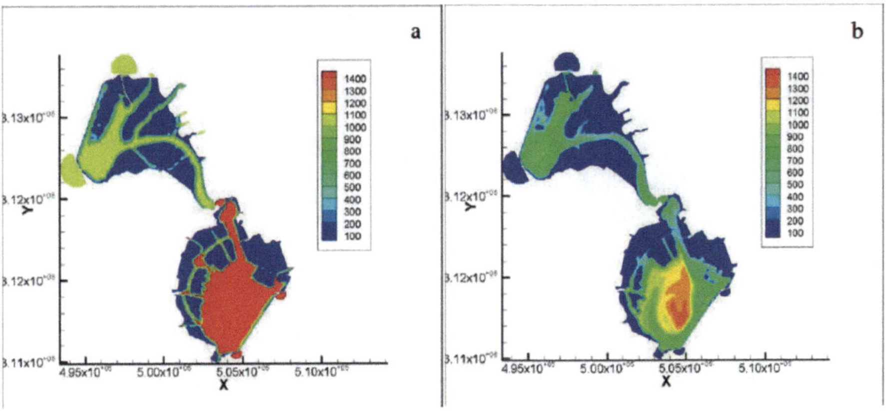

Fig. 7. Characteristics of cl^- concentration distribution in Phase II and III under the 5-year flood condition

4.4 Impact of Optimal Dispatch on Salinity Distribution

Since the cl^- concentration in Phase II is lower than in Phase III, and the flow at the Phase I dam is closely related to the flow at the discharge gates of Phases II and III, the water gate in Phase III is prioritized during operation. This ensures that the water flows from Phase II to Phase III, thereby reducing the impact of Phase III on Phase II and accelerating the desalination process in Phase III. Based on this new discharge gate operation strategy, the following measures are adopted: when the water level in Phase II of Xuanmen exceeds 0.5 m and the upstream water level is higher than the external sea tide level, the Yinggong Gate and Yuhuan Lake Gate are opened for drainage. During the flood period, when the water level in Phase III of Xuanmen exceeds 0.3 m and the upstream water level is higher than the external sea tide level, the Chongtanyu Gate and Muyuyu Gate are opened for drainage. After the flood peak, when the water levels in Phases II and III recede to 0.2 m, the gates are closed.

Figure 8a show that after a once-in-five-year rainfall, the cl^- concentration distribution in Phases II and III is largely consistent with normal operations, but the concentrations are lower than under normal operations. Figure 8b and 8c show that Post-flood, the cl^- concentration in Phase II decreases from 871 mg/L to 715 mg/L, a reduction of 156 mg/L, which is approximately a 17.9% decrease. In Phase III, the cl^- concentration decreases from 923 mg/L to 591 mg/L, a reduction of 332 mg/L, which is approximately a 36.0% decrease. This operation strategy significantly reduces the risk of cl^- diffusion from Phases III and II to Phase II and accelerates the desalination process in Phase III.

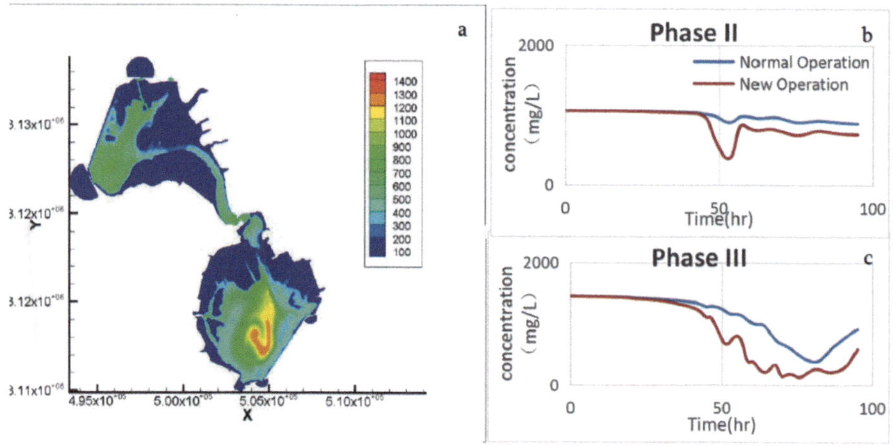

Fig. 8. Distribution of cl^- Concentration after Optimized Regulation and the Concentration Variation Process in phase II and III Lake Areas

5 Conclusion

Based on the results and discussions presented above, the conclusions are obtained as below:

(1) Characterization of Post-Linkage Streamflow: Flow from Phase II to Phase III was defined as positive, and vice versa. The analysis showed a reciprocating flow pattern at the site of the former dam, with flow direction closely linked to the discharge differences between the Phase II and III water gates. After the dam removal, salinity diffusion is primarily influenced by gate operations.
(2) Wind Field Effects on Salinity Distribution: During non-rainfall periods, salinity distribution is mainly affected by wind, with diffusion concentrated at the connection between Phases II and III. Chloride concentration at the Phase II connection increased from 1054 mg/L to 1090 mg/L and stabilized, while the Phase III connection decreased from 1452 mg/L to 1158 mg/L, with minimal impact on the broader reservoir areas.
(3) Impact of Salinity Distribution during Floods: Under a five-year flood event scenario, both Phase II and III experienced reductions in salinity after the dam removal, with localized high-concentration areas persisting in Phase III. Post-rainfall, Phase II's average salinity dropped from 1054 mg/L to 871 mg/L, and Phase III's from 1452 mg/L to 923 mg/L.
(4) Impact of Optimal Dispatch on Salinity Distribution: With lower salinity in Phase II, prioritizing the opening of Phase III gates allows water to flow from Phase II to Phase III, accelerating desalination. After a five-year flood event, the optimal dispatch reduced salinity levels more effectively than conventional operations, with Phase II's salinity decreasing by 156 mg/L (17.9%) and Phase III's by 332 mg/L (36.0%).

Acknowledgments. The research is jointly supported by the "Study on the Salinity Migration Patterns after the Completion of Xuanmen Phase II and III" supported by The Science and Technology Planning Project of the Zhejiang Provincial Department of Water Resources (RB2407).The Science and Technology Planning Project of the Zhejiang Provincial Department of Water Resources(RA2205).

References

1. Liu, Z., et al.: Scarcity and quality risks for future global urban water supply. Landsc. Ecol. **39**, 10 (2024)
2. Hung, P., Le, T.V., Vo, P.L., Duong, H.C., Rahman, M.M.: Vulnerability assessment of water resources using GIS, remote sensing and SWAT model – a case study: the upper part of Dong Nai river basin, Vietnam. Int. J. River Basin Manag. **20**, 517–532 (2022)
3. Zhu, L., Jiang, C., Wang, Y., Peng, Y., Zhang, P.: A risk assessment of water salinization during the initial impounding period of a proposed reservoir in Tianjin, China. Environ Sci Process Impacts **15**, 1743–1751 (2013)
4. Lee, G., Park, J., Jung, C., An, Y.: Assessing salinity reduction by environmental restoration in the Seomjin River Estuary (South Korea): a numerical model approach for corbicula habitat conservation. Sustainability **16**, 5653 (2024)
5. Mao, X., et al.: Research on measures to accelerate water desalination in coastal harbor storage reservoirs. Adv. Water Sci. **016**(006), 773–776 (2005)
6. Oh, C., Cho, J., Choi, J., Cho, Y.: Prediction of desalination time of saemangeum reservoir. Irrig. Drain. **65**, 221–229 (2016)

7. Fan, G., et al.: Spatio-temporal variations of salinity and analysis of desalination factors in a Chinese coastal storage reservoir. Process Saf. Environ. Protect. **159**, 26 (2021)
8. Pu, L., Xin, P., Nguyen, T.T.M., Yu, X., Li, L., Barry, D.A.: Thermal effects on flow and salinity distributions in coastal confined aquifers. Water Resour. Res. **56**, e2020WR027582 (2020)
9. Zhang, X., Zhu, M., Zheng, X.: Experimental study on salt release from the bottom sediment of a freshwater storage reservoir in a bay under the influence of environmental factors. Ocean Bull. **29**(02), 135–142 (2010)
10. Hyun, S.C.: Assessing temporal and spatial salinity variations in estuary reservoir using EFDC. J. Korean Soc. Agric. Eng. **56**, 139–147 (2014)
11. Raj, D., Shaji, E.: Fluoride contamination in groundwater resources of Alleppey, southern India. Geosci. Front. Front. **8**, 117–124 (2017)

Application of the Gray Clustering Method to Evaluate the Water Quality of the Jiangan River

Kexing Chen, Yilin He, Jiarong Hu, Chuiyang Kong, and Mei Li(✉)

Chengdu University, Chengdu 610106, China
limeigm@cdu.edu.cn

Abstract. In this study, ammonia nitrogen (NH_3-N), total phosphorus (TP), and the permanganate index (COD_{Mn}) were chosen as clustering reference factors. The water quality of the Erjiangsi section of the Jiangan River was evaluated via the gray clustering method. The results indicated that the water quality of the section was classified as Class I from January to March and as Class II from April to December. Pearson correlation analysis revealed minimal impacts of meteorological factors such as temperature, air pressure, and humidity on water quality. An analysis of the monthly data revealed that rainfall and nonpoint source pollution were the primary factors affecting water quality, with nonpoint source pollution having a greater impact than rainfall. The gray clustering method outperformed the single-factor evaluation methods by considering multiple interacting factors, providing a more accurate assessment. The fuzzy comprehensive evaluation also supported the validity and reliability of the gray clustering method for complex water quality data.

Keywords: gray clustering method · water quality evaluation · single-factor index method · fuzzy evaluation method

1 Introduction

Water pollution in China has improved but remains serious [1]. The development of a comprehensive and scientific water quality evaluation system has long been a focus for scholars. Traditional water quality assessment measures chemical parameters in water and compares them with existing standard levels, which is simple but cannot accurately describe the water quality status [2, 3]. Common water quality assessment methods include the multivariate statistical method [4], single-factor index method [5], fuzzy comprehensive evaluation method [6], gray clustering method [7], and water quality index model method [8, 9]. Among them, the gray clustering method offers a more objective evaluation by processing gray information and pollution factors [10]. This study applies the gray clustering method to the Erjiangsi section of the Jiangan River to enhance water quality assessment and support further pollution research.

D. Han et al. (Eds.): ICEPG 2024, ESE, pp. 155–163, 2025.
https://doi.org/10.1007/978-3-031-86829-0_19

2 Gray Clustering Method

2.1 Dimensionless Processing of Gray Classes

$$y_{ik} = \frac{S_{jk}}{\left(\frac{1}{n}\sum_{i=1}^{n} S_{jk}\right)} \tag{1}$$

where y_{ik} represents the dimensionless value of the i-th sample for the k-th indicator; S_{jk} represents the original value of the j-th sample for the k-th indicator; and n denotes the total number of samples.

2.2 Evaluation of the Gray Whitening Weight Function

When k = 1, its membership function is as follows:

$$f_j^k(X_j) = \begin{cases} 1, & X_j \le S_k \\ \frac{S_{k+1}-X_j}{S_{k+1}-S_k}, & S_k < X_j < S_{k+1} \\ 0, & X_j \ge S_{k+1} \end{cases} \tag{2}$$

When k = 2, 3, or 4, its membership function is as follows:

$$f_j^k(X_j) = \begin{cases} 0, & X_j \le S_{k-1}, X_j \ge S_{k+1} \\ \frac{X_j-S_{k-1}}{S_k-S_{k-1}}, & S_{k-1} < X_{ij} < S_k \\ 1, & X_j = S_k \\ \frac{S_{k+1}-X_j}{S_{k+1}-S_k}, & S_k < X_j < S_{k+1} \end{cases} \tag{3}$$

When k = 5, its membership function is as follows:

$$f_j^k(X_j) = \begin{cases} 1, & X_j \ge S_k \\ \frac{X_j-S_{k-1}}{S_k-S_{k-1}}, & S_{k-1} < X_j < S_k \\ 0, & X_j \le S_{k-1} \end{cases} \tag{4}$$

where $f_j^k(X_j)$ represents the membership degree of the JTH pollution index of the sampling point to the eutrophication of the K-grade water and where X_j is the measured concentration value of index j at the sampling point. S_k is the class k standard value of the indicator.

2.3 Confirmation of Clustering Weights

$$\eta_j = \frac{\frac{1}{y_{ik}}}{\sum_{j=1}^{m} \frac{1}{y_{ik}}} \tag{5}$$

where y_{ik} is the value of the ith sample after nondimensionalization on the k index and where m is the total number of categories.

2.4 Confirmation of the Clustering Coefficient

Let σ_k be the variable weight clustering coefficient belonging to class k indicators.

$$\sigma_k = \sum_{j=1}^{m} f_j^k (X_j)\eta_{jk} \tag{6}$$

2.5 Determining the Classification Levels of Clustering Objects

$$\sigma_k = \max_{1 \le k \le n} \{\sigma_k\} \tag{7}$$

the gray class k corresponding to σ_k is the category to which the cluster object belongs.

3 Water Quality Evaluation Results of the Gray Clustering Method

3.1 Evaluation Index

In accordance with previous evaluations of the water quality of the Jiangan River, NH_3-N, TP and COD_{M_n}, which have significant effects on water quality, are selected as cluster evaluation indices.

3.2 Dimensionless Processing of Gray Classes

The results of Eq. (1) are shown in Table 1.

Table 1. Dimensionless gray values (y_{ik}).

Evaluation index	Class I	Class II	Class III	Class IV	Class V
TP	0.098	0.490	0.980	1.471	1.961
NH_3-N	0.146	0.485	0.971	1.456	1.942
COD_{Mn}	0.270	0.541	0.811	1.351	2.027

3.3 Establishing the Gray Whitening Weight Function

According to the membership function expressions in formulas (2), (3), and (4), the membership function matrix for each pollution factor is obtained. The monthly section membership matrix is as follows: $f_1 = \begin{bmatrix} 0 & 0 & 0 & 0 & 1 \\ 0.3125 & 0.6875 & 0 & 0 & 0 \\ 0.5857 & 0.4143 & 0 & 0 & 0 \\ 0.943 & 0.057 & 0 & 0 & 0 \end{bmatrix}$,

$$f_2 = \begin{bmatrix} 0 & 0 & 0 & 0 & 1 \\ 0.3375 & 0.6625 & 0 & 0 & 0 \\ 0.794 & 0.206 & 0 & 0 & 0 \\ 0.9585 & 0.0415 & 0 & 0 & 0 \end{bmatrix}, f_3 = \begin{bmatrix} 0 & 0 & 0 & 0 & 1 \\ 0 & 0 & 0.5 & 0.5 & 0 \\ 0.54 & 0.46 & 0 & 0 & 0 \\ 0.7315 & 0.2685 & 0 & 0 & 0 \end{bmatrix}, f_4 = \begin{bmatrix} 0 & 0 & 0 & 0 & 1 \\ 0 & 0.68 & 0.32 & 0 & 0 \\ 0.437 & 0.533 & 0 & 0 & 0 \\ 0.5645 & 0.4355 & 0 & 0 & 0 \end{bmatrix}, f_5 = \begin{bmatrix} 0 & 0 & 0 & 0 & 1 \\ 0 & 0.77 & 0.23 & 0 & 0 \\ 0.726 & 0.274 & 0 & 0 & 0 \\ 0.2395 & 0.7605 & 0 & 0 & 0 \end{bmatrix},$$

$$f_6 = \begin{bmatrix} 0 & 0 & 0 & 0 & 1 \\ 0 & 0 & 0.5 & 0.5 & 0 \\ 0.54 & 0.46 & 0 & 0 & 0 \\ 0.4575 & 0.5425 & 0 & 0 & 0 \end{bmatrix}, f_7 = \begin{bmatrix} 0 & 0 & 0 & 0 & 1 \\ 0 & 0.83 & 0.17 & 0 & 0 \\ 0.26 & 0.74 & 0 & 0 & 0 \\ 0.1815 & 0.8185 & 0 & 0 & 0 \end{bmatrix},$$

$$f_8 = \begin{bmatrix} 0 & 0 & 0 & 0 & 1 \\ 0 & 0.62 & 0.38 & 0 & 0 \\ 0.186 & 0.814 & 0 & 0 & 0 \\ 0.167 & 0.833 & 0 & 0 & 0 \end{bmatrix}, f_9 = \begin{bmatrix} 0 & 0 & 0 & 0 & 1 \\ 0 & 0.74 & 0.26 & 0 & 0 \\ 0.523 & 0.457 & 0 & 0 & 0 \\ 0.5205 & 0.4795 & 0 & 0 & 0 \end{bmatrix}, f_{10} = \begin{bmatrix} 0 & 0 & 0 & 0.11 & 0.89 \\ 0.1 & 0.9 & 0 & 0 & 0 \\ 0.791 & 0.209 & 0 & 0 & 0 \\ 0.5905 & 0.4905 & 0 & 0 & 0 \end{bmatrix}, f_{11} = \begin{bmatrix} 0 & 0 & 0 & 0 & 1 \\ 0 & 0.93 & 0.07 & 0 & 0 \\ 0.577 & 0.423 & 0 & 0 & 0 \\ 0.7125 & 0.2875 & 0 & 0 & 0 \end{bmatrix}, f_{12} = \begin{bmatrix} 0 & 0 & 0 & 0 & 1 \\ 0 & 0.82 & 0.18 & 0 & 0 \\ 0.466 & 0.154 & 0 & 0 & 0 \\ 0.588 & 0.412 & 0 & 0 & 0 \end{bmatrix}$$

3.4 Confirmation of Clustering Weights

According to Eq. (5), cluster weights of the five water quality levels of TP, NH_3--N and COD_{Mn} can be calculated, and the calculation results are shown in Table 2 below.

Table 2. Clustering weights of each index (η_j).

Evaluation index	Class I	Class II	Class III	Class IV	Class V
TP	0.491	0.343	0.311	0.323	0.336
NH_3-N	0.331	0.346	0.314	0.326	0.339
COD_{Mn}	0.178	0.311	0.376	0.351	0.325

3.5 The Clustering Coefficient Was Calculated to Determine the Water Quality Category

According to the principle of maximum membership degree, the water quality clustering coefficient (σ_k) for the Erjiangsi cross section in the Jiangan River Basin across different months had been determined. The gray level corresponding to the maximum element represents the water quality category of the section. The Clustering Coefficients (σ_k) and the grey levels for the research section are presented in Table 3.

Table 3. Clustering Coefficients (σ_k) and grey levels for the Erjiangsi section in the Jiangan River Basin.

Month	Clustering Coefficients (σ_k)					Gray levels
	Class I	Class II	Class III	Class IV	Class V	
Jan	0.515	0.397	0.000	0.000	0.000	Class I
Feb	0.599	0.311	0.000	0.000	0.000	Class I
Mar	0.309	0.274	0.155	0.161	0.000	Class I
Apr	0.245	0.563	0.099	0.000	0.000	Class II
May	0.283	0.595	0.071	0.000	0.000	Class II
Jun	0.459	0.501	0.000	0.000	0.000	Class II
Jul	0.118	0.795	0.053	0.000	0.000	Class II
Aug	0.091	0.753	0.118	0.000	0.000	Class II
Sep	0.272	0.561	0.081	0.000	0.000	Class II
Oct	0.416	0.508	0.000	0.000	0.000	Class II
Nov	0.318	0.555	0.022	0.000	0.000	Class II
Dec	0.259	0.594	0.056	0.000	0.000	Class II

According to the evaluation results, the water quality of the Erjiangsi section of the Jiangan River was mainly class I from January to March. From April to December, the water quality of this section is mainly class II.

3.6 Water Quality Change Analysis

The main meteorological factors, such as temperature, air pressure and humidity, were selected, and their effects on water quality were analysed via Pearson correlation analysis. Positive values indicate positive correlations, negative values indicate negative correlations, and the greater the absolute value is, the stronger the correlation. Among them, 0.3 and below is a weak correlation [11]. Table 4 presents the correlation data.

Table 4 shows that the correlation coefficients between the pollutant indicators and meteorological factors are all less than 0.3, indicating a weak relationship. However, research suggests that meteorological factors can indirectly or directly impact water quality through complex mechanisms [12]. The monthly averages of temperature, air

Table 4. Correlation table of pollutants and environmental factors.

Parameter	Correlation coefficient		
	COD_{Mn}	NH_3-N	TP
T (Mean Air Temperature)	0.22	−0.11	0.18
MT (Maximum Air Temperature)	0.20	−0.14	0.14
BP (Mean Air Pressure)	−0.25	0.03	−0.17
MBP (Maximum Air Pressure)	−0.24	0.03	−0.17
LBP (Minimum Air Pressure)	−0.26	0.02	−0.18
RH (Relative Humidity)	0.03	0.15	0.12
LH (Minimum Relative Humidity)	0.03	0.15	0.10

pressure, and rainfall [13] for the Erjiangsi section of the Jiangan River in 2022 were calculated to assess their impacts on water quality. The calculation results are shown in Figs. 1, 2 and 3.

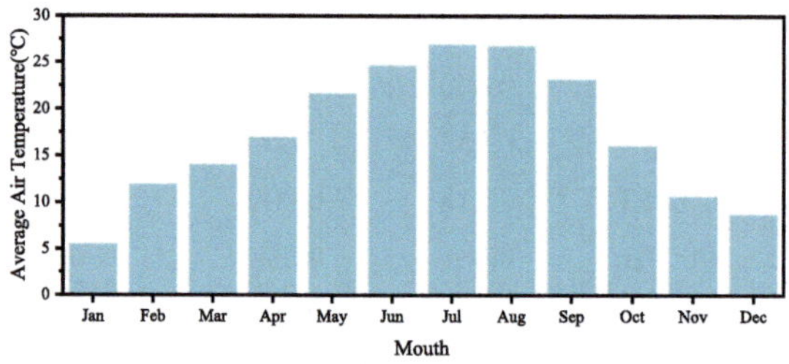

Fig. 1. Monthly mean air temperature in Chengdu in 2021.

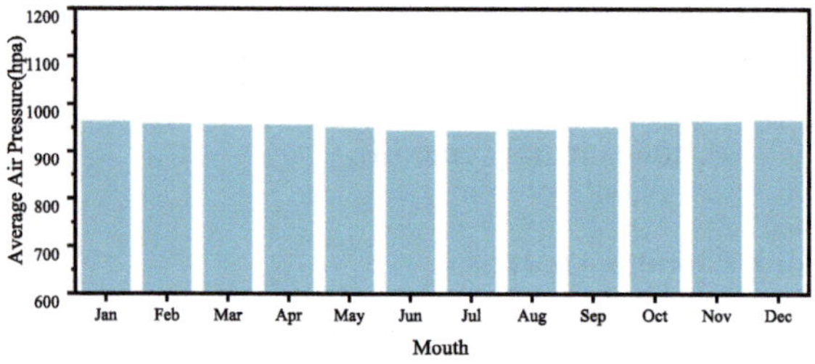

Fig. 2. Monthly mean air pressure in Chengdu in 2021.

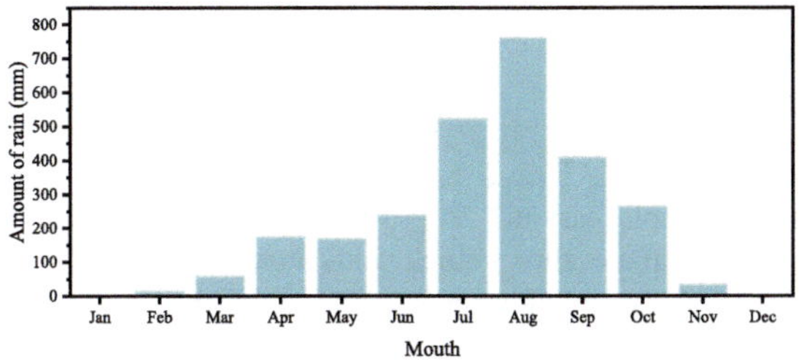

Fig. 3. Monthly rainfall distribution in Chengdu in 2021.

The figure shows that the air pressure in this section is basically stable throughout the year, and the correlation analysis reveals that the air pressure has little influence on the water quality. The changes in temperature and rainfall have relatively obvious relationships; the higher the temperature is, the greater the degree of rainfall. According to the results of the correlation analysis and gray clustering methods, temperature may have an indirect effect on water quality, whereas rainfall may have a direct effect on water quality. The analysis confirmed that rainfall and nonpoint source pollution are the primary factors influencing water quality changes, with nonpoint source pollution having a more significant impact than does rainfall.

4 Comparison of Gray Clustering and Other Methods

Different water quality evaluation methods have distinct advantages and limitations. Table 5 summarizes the key features of the single-factor, fuzzy comprehensive, and gray clustering methods, comparing their strengths, weaknesses, and applications.

The evaluation results of the single-factor evaluation, normal gray clustering and fuzzy comprehensive evaluation are shown in Fig. 4.

The single-factor method, which is limited to one index, is biased by outliers. The gray clustering method offers a more comprehensive evaluation by considering multiple indices and using a clustering coefficient σ_k. In March, total phosphorus increased abnormally. Compared with the other two methods, there was distortion in the evaluation of the gray clustering method, indicating that the multifactor interaction consideration of the gray clustering method would not be beneficial to the analysis and evaluation of water quality characteristics when individual indicators were seriously exceeded. At this time, amendments should be made in combination with other evaluation methods.

Table 5. Main features of the single-factor method, fuzzy comprehensive evaluation method and gray clustering method.

Method	advantages	disadvantages	Application
Single-factor Method	Simple, easy to calculate, allows quick assessment of water pollution	Influenced by a single indicator, cannot fully reflect the overall water quality	Used for rapid water pollution assessment
Fuzzy Comprehensive Evaluation	Strong systematic approach, captures the characteristics of the evaluated object comprehensively	The influence of excessive indicators may be exaggerated, leading to extreme results	Applied in multi-index evaluations, accurately identifies major pollutants and their types
Grey Clustering Method	Reflects water quality complexity, avoids the rigidity of fixed baselines for water classification	Complex calculation, may reduce the influence weight of indicators exceeding standards	Used for comprehensive, accurate assessment of river water quality and comparing different water bodies

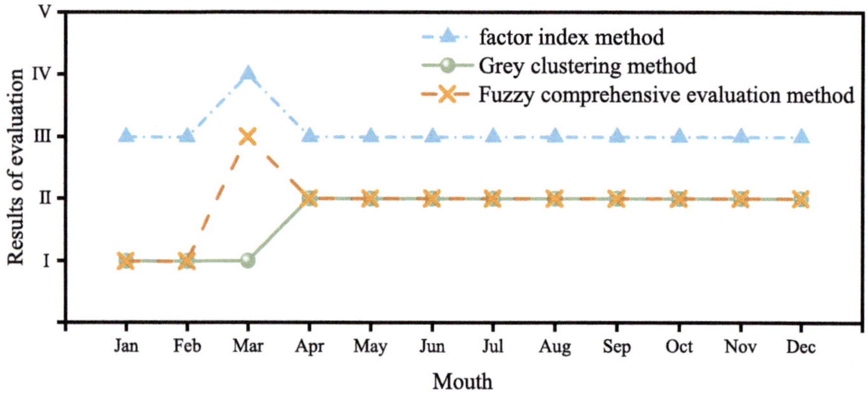

Fig. 4. Comparison of water quality assessment results.

5 Conclusion

In this work, the water quality of the Erjiangsi section in the Jiangan River Basin is evaluated via the gray clustering method. The results indicate that from January to March, the water quality of the Erjiangsi Section in the Jiangan River Basin remains good but decreases to Class II from April to December. Pearson correlation analysis shows weak correlations between pollutant indices and meteorological factors. Further data analysis of air temperature, air pressure, and rainfall suggests that temperature primarily has an indirect influence on water quality, whereas rainfall directly impacts water quality. A comparison of the single-factor evaluation method, gray clustering method, and fuzzy comprehensive evaluation method shows that the gray clustering method offers better comprehensiveness and reliability. It addresses the limitations of the single-factor method, which is easily influenced by individual factors, and proves effective for complex water quality data where multiple factors interact. However, distortion may occur if a pollutant significantly exceeds standards. The fuzzy comprehensive evaluation results align closely with those from the gray clustering method, further validating its effectiveness. Thus, a comprehensive approach should be taken in water quality evaluation, considering both monitoring data and cases of exceeded standards to achieve a more accurate and reliable assessment.

References

1. Ministry of Ecology and Environment: The National Surface Water Environmental Quality Status in the Fourth Quarter of 2023 and from January to December. Ministry of Ecology and Environment, People's Republic of China (2023)
2. Noori, R., Berndtsson, R., Hosseinzadeh, M., Adamowski, J.F., Abyaneh, M.R.: A critical review on the application of the national sanitation foundation water quality index. Environ. Pollut. **244**, 575–587 (2019)
3. Duan, W., et al.: Water quality assessment and pollution source identification of the Eastern Poyang Lake Basin using multivariate statistical methods. Sustainability **8**(2), 133 (2016)
4. Gai, R., Guo, Z.: A water quality assessment method based on an improved gray relational analysis and particle swarm optimization multiclassification support vector machine. Front. Plant Sci. **14**, 1099668 (2023)
5. Wang, M., et al.: Fuzzy comprehensive evaluation of seawater quality in Zhifu Bay from 2014 to 2020. J. Zhejiang Ocean Univ. (Nat. Sci. Ed.) **43**(02), 145–151 (2024)
6. Uddin, M.G., Nash, S., Olbert, A.I.: A review of water quality index models and their use for assessing surface water quality. Ecol. Ind. **122**, 107218 (2021)
7. Kim, H.I., Kim, D., Mahdian, M., Salamattalab, M.M., Bateni, S.M., Noori, R.: Incorporation of water quality index models with machine learning-based techniques for real-time assessment of aquatic ecosystems. Environ. Pollut. **355**, 124242 (2024)
8. Xing, L.W., Wei, X.P., Dong, J.: Water environment quality assessment of Xihe Reservoir in Pengzhou City based on gray clustering method. Water Resour. Dev. Manag. **01**, 23–30+53 (2020)
9. Fang, D.: Application of gray clustering method in water quality analysis of Nanming River. Environ. Dev. **32**(08), 14–16 (2020)
10. Administration, S.E.P.: State Administration of Quality Supervision, Inspection and Quarantine: GB3838-2002 Surface Water Environmental Quality Standard. Standards Press of China, Beijing (2002)
11. Jia, J., He, X., Jin, Y.: Statistics, 7th edn. China Renmin University Press, Beijing (2018)
12. Gao, X., Wei, Y., Guan, Z.: Analysis of the impact of meteorological conditions on the water quality of Shanmei reservoir. J. Agric. Disast. Res. **13**(09), 148–150 (2023)
13. Chengdu Bureau of Ecological Environment Homepage, https://sthj.chengdu.gov.cn/. Accessed 12 Aug 2024

Synthesis and Properties of Non-phosphorus and Non-nitrogen Water Treatment Agent

Zhan Liu[1,2](✉), Meifang Yan[1,2], Na Li[2], Ruhui Guo[3], Yuhua Gao[1,2], Haihua Li[1,2], and Zhenfa Liu[1,2]

[1] Institute of Energy Sources, Hebei Academy of Science, Shijiazhuang 050081, Hebei, China
liuzhan1216@126.com
[2] Hebei Engineering Research Center for Water Saving in Industry, Shijiazhuang 050081, Hebei, China
[3] Hebei Academy of Sciences, Shijiazhuang 050081, Hebei, China

Abstract. An environmentally friendly water treatment agent polymer without phosphorus and nitrogen was synthesized by aqueous phase synthesis method using raw materials itaconic acid (IA) and hydroxyethyl methacrylate (HEMA). The scale inhibition and dispersion properties as well as the corrosion inhibition properties of the polymeric water treatment agent were investigated using the static scale inhibition experimental method and the rotary hanging method. The results show that the polymer has excellent scale inhibition properties for calcium carbonate, calcium sulfate and calcium phosphate. Moreover, the dispersion performance was good and the corrosion rate of carbon steel meets the national standard. The polymer has good application prospects in the field of water treatment.

Keywords: Water treatment agent · Scale inhibition performance · Dispersion performance · Corrosion rate

1 Introduction

There are widespread phenomena such as scale formation and corrosion in circulating cooling water systems in chemical and pharmaceutical industries [1–4]. This situation will seriously affect the normal operation of the whole system, so the circulating water must be treated to ensure the stable operation of the whole system [5–7]. Generally, water treatment methods can be divided into two types, physical water treatment method and chemical water treatment method. Physical method has the advantages of low cost and low cost, but its effect is poor [8–10]. At present, chemical water treatment methods are still widely used in industry, and chemical agents are added to the system to prevent scaling or remove scale [11]. With the continuous updating of environmental protection policies, water treatment should be developed in the direction of more efficient performance and more complete functions [12].

In this study, IA and HEMA were used as raw materials to synthesize an environmentally friendly water treatment agent with excellent properties of scale inhibition, dispersion and corrosion inhibition. The drug has no phosphorus and nitrogen, which avoids eutrophication caused by long-term use of the drug, and meets the requirements of sustainable development policy of our country.

D. Han et al. (Eds.): ICEPG 2024, ESE, pp. 164–170, 2025.
https://doi.org/10.1007/978-3-031-86829-0_20

2 Experimental Department

2.1 Experimental Drugs and Instruments

Main experimental drugs: Itaconic acid, hydroxyethyl methacrylate, ammonium persulfate, tert-butanol, sodium bicarbonate, calcium chloride, ferrous chloride, sodium tetraborate, etc.

Main experimental equipment: electronic balance, constant temperature water bath, rotary hanging film experimental equipment, etc.

2.2 Experimental Methods

Preparation of HEMA-IA Copolymer

A certain amount of itaconic acid, hydroxyethyl methacrylate and distilled water were added to a four-way flask and heated to a certain temperature under stirring. After the set temperature is reached, tert-butanol is added and ammonium persulfate solution is added by injection pump. After the drip is finished, keep the temperature constant for a while. The solution is naturally cooled to room temperature, and finally a light yellow transparent liquid is obtained, which is a synthetic scale inhibitor solution.

Evaluation of Calcium Carbonate Scale Inhibition Performance of Polymer

Static scale inhibition experiment was used to evaluate the scale inhibition performance of the polymer. The experiment was conducted according to GB/T16632-2008. Configuration water was used for the experiment. Sodium bicarbonate and anhydrous calcium chloride were used to prepare experimental water with calcium ion concentration of 600 mg/L and bicarbonate root concentration of 1200 mg/L. The reagent was added to the experimental water, and the temperature was 80 °C, and it was left for 10 h at constant temperature. The Ca^{2+} concentration in the supernatant was determined by EDTA, and the scale inhibition η was calculated by the following formula.

$$\eta = \frac{Ca_1 - Ca_0}{Ca_2 - Ca_0} * 100\% \tag{1}$$

where: Ca_0- The concentration of Ca^{2+} in raw water, unit mg/L;

Ca_2- The concentration of Ca^{2+} after heating the water sample after adding the polymer, in mg/L;

Ca_1- The concentration of Ca^{2+} in raw water after heating, in mg/L.

Evaluation of Calcium Sulfate Scale Resistance of Polymer

The experiment was conducted in accordance with SY/T5673–93. Add a certain amount of polymer into the container, configure the concentration of calcium ion in water 5000 mg/L, sulfate concentration 7500 mg/L, calculated as $CaSO_4$). The solution was kept at a constant temperature in a 70 °C water bath for 25 h. After the solution was cooled, the Ca^{2+} concentration of the clear solution was titrated with EDTA solution and the scale inhibition rate η1 was calculated.

$$\eta_1 = \frac{Ca_1 - Ca_0}{Ca_2 - Ca_0} * 100\% \tag{2}$$

where: Ca_0- Ca^{2+} concentration in raw water, unit mg/L;

Ca_1- The concentration of Ca^{2+} after heating the water sample after adding the polymer, in mg/L;

Ca_2- The concentration of Ca^{2+} in raw water after heating, in mg/L.

Evaluation of Calcium Phosphate Scale Inhibition Performance of Polymers

The experiment was conducted according to GB/T22626–2008. A certain amount of polymer and configuration water (250 mg/L calcium ion concentration in $CaCO_3$, 1200mg/L phosphate) are added to the container. After the solution was kept at a constant temperature of 80 °C for 10 h, it was removed and waited for cooling to determine the concentration of phosphoric acid in the clear liquid and calculate the scale inhibition rate η_2.

$$\eta_2 = \frac{P_1 - P_0}{P_2 - P_0} * 100\% \tag{3}$$

where: P_0- $PO_4{}^{3-}$ concentration in raw water, unit mg/L;

P_1- $PO_4{}^{3-}$ concentration of water sample heated after adding polymer, unit mg/L;

P_2- $PO_4{}^{3-}$ concentration after raw water heating, unit mg/L.

Evaluation of Dispersion Properties of Polymers

The preparation water containing a certain concentration of polymer (calcium ion concentration of 150 mg/L $CaCO_3$, ferrous ion concentration of 10 mg/L) is added to the container. After intense stirring, the solution was kept at a constant temperature of 50 °C for 5 h. After cooling, the light transmittance of the supernatant was measured at 450 nm.

Evaluation of Corrosion Rate

According to GB/T 18175–2014, the corrosion rate of the agent was determined by the rotating hanging film method, as shown in Table 1. The experimental water was ground water. The A3 carbon steel sheet (50 mm × 25 mm × 2 mm) was suspended in the experimental water and rotated at a uniform speed at 40 °C for 72 h. The corrosion rate X expressed in mm/a is calculated as follows.

$$X = \frac{8760 * 10 * (W - W_0)}{15825.6} \tag{4}$$

where: W represents the weight of carbon steel sheet before the rotary hanging test, unit g;

W_0 indicates the weight of the carbon steel test plate, unit g.

Table 1. National standard for corrosion rate of carbon steel

Index/Materials	Corrosion rate (mm/a)
Iron and iron alloys	0.075

3 Results and Discussion

3.1 Calcium Carbonate Scale Inhibition Performance of the Polymer

The polymer was applied to calcium carbonate for scale inhibition, and the experimental results were obtained as shown in the Fig. 1 below. The scale inhibition performance of polymer at low concentration is poor, but the scale inhibition rate increases with the increase of additive amount. When the dosage is 8 mg/L, the scale inhibition performance of the polymer to calcium carbonate reaches 100%. The results show that the polymer has excellent anti-scale property to calcium carbonate.

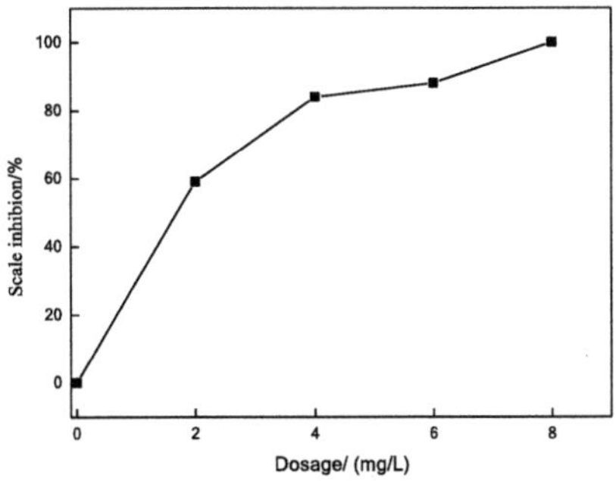

Fig. 1. Calcium carbonate scale inhibition properties of the polymer

3.2 Calcium Sulfate Scale Resistance of the Polymer

The scale inhibition performance of the polymer on calcium sulfate was measured when the polymer concentration was 0–8 mg/L, and the scale inhibition effect was shown in Fig. 2. When the concentration of the polymer is low, the scale inhibition rate of calcium sulfate is low. When the dosage is 6 mg/L, the scale inhibition rate is only 49%, but when the dosage is increased to 8 mg/L, the scale inhibition rate reaches 100%.

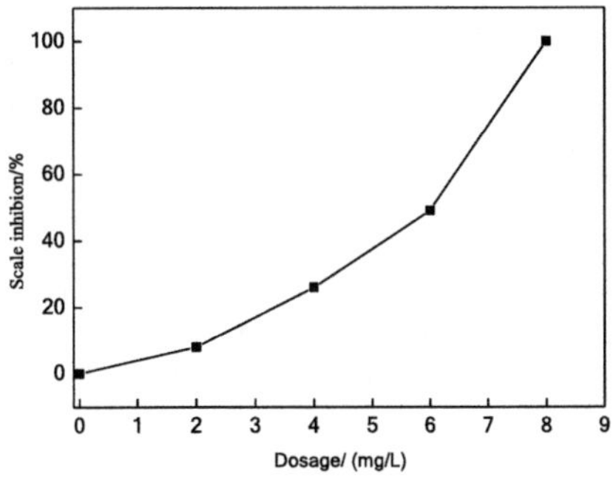

Fig. 2. Calcium sulfate scale resistance of the polymer

3.3 Calcium Phosphate Scale Inhibition Performance of the Polymer

The resistance of the polymer to calcium phosphate scale was measured, and the results were shown in Fig. 3. The scale inhibition rate of the polymer on calcium phosphate was low at 0–90 mg/L, and then the scale inhibition rate of the polymer on calcium phosphate increased by a step, and reached 100% when the dosage was 120 mg/L.

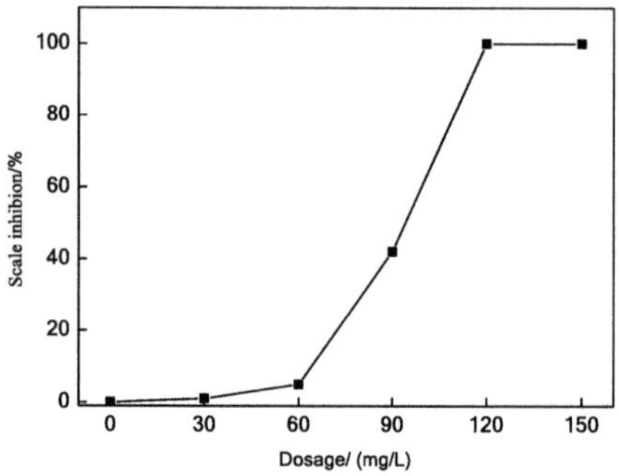

Fig. 3. Calcium phosphate scale inhibition performance of the polymer

3.4 Dispersion Properties of Polymer

The results of the study on the dispersion property of the polymer to iron oxide are shown in Fig. 4. After the polymerization of the two compounds IA and HEMA, the polymerization products have excellent dispersion properties for iron oxide. When the dosage was 100 mg/L, the light transmittance of the supernatant was only 30%. This shows that the polymer is an excellent dispersing agent.

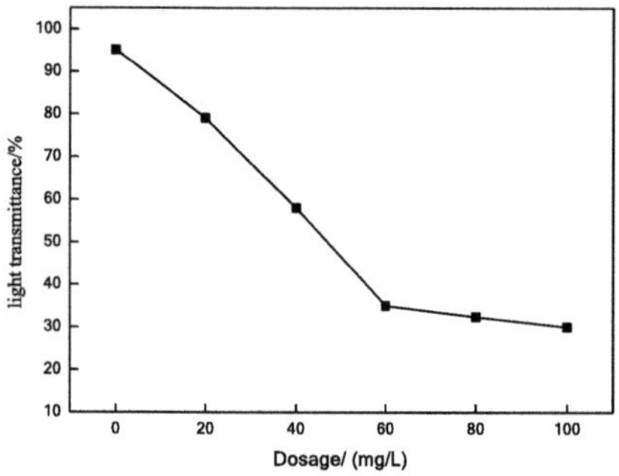

Fig. 4. Dispersion properties of the polymer for iron oxide

3.5 Corrosion Rate of Polymer

As can be seen from Fig. 5, the corrosion rate decreases with the increase of polymer dosage. When the dosage is 120 mg/L, the corrosion rate of carbon steel is 0.072 mm/a, which meets the requirements of the national standard (less than 0.075 mm/a).

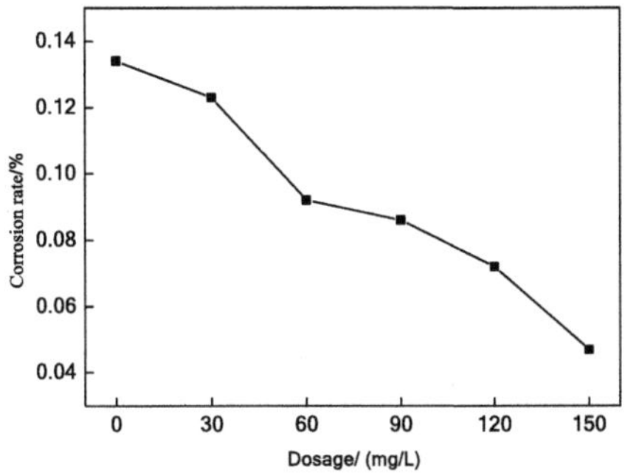

Fig. 5. Corrosion rate of polymer

In summary, the copolymer has excellent scale inhibition and corrosion inhibition properties. This is due to the effect of groups contained in the copolymer, in which the carboxylic acid group provided by itaconic acid reacts with calcium ions, so that it can not form scale crystallization and change the crystal structure, so as to inhibit scale. The hydroxyl group provided by hydroxyethyl methacrylate can form hydrogen bond with calcium carbonate and inhibit the formation of calcium salt. The ester group contained in hydroxyethyl methacrylate also has a good complexation effect on cations in the system, so as to achieve scale inhibition and dispersion. The carboxyl and hydroxyl groups in the polymer form a dense film with the charge on the surface of the carbon steel test piece, thus delaying the corrosion of the test piece by the test water.

4 Conclusion

(1) In this study, an environmentally friendly polymer, IA/HEMA, was synthesized under certain conditions with Itaconic acid and hydroxyethyl methacrylate as monomers, two phosphorus and nitrogen free compounds.
(2) Static scale inhibition experiment was used to study the scale inhibition and dispersion performance of the polymer. The experimental results show that the polymer has excellent anti-scale properties to calcium carbonate, calcium sulfate and calcium phosphate. The light transmittance of the supernatant after dispersing iron oxide is only 30%, and the dispersion property is also outstanding.
(3) In the study, the corrosion rate of carbon steel hanging sheet with experimental water in the presence of polymer was also studied by rotating hanging sheet method. When the dosage was 150 mg/L, the corrosion rate could meet the national standard.

(4) It has good scale dispersion and corrosion inhibition properties, which indicates that it can be used as scale dispersion and corrosion inhibitor in circulating water. Moreover, the agent is an environmentally friendly agent, which is conducive to the implementation of environmental protection and national sustainable development policies.

Acknowledgments. This article was supported by the Science and Technology Program of Hebei Academy of Sciences (No. 22701, 23701, 24701), Shijiazhuang science and technology plan project (No. 246240529A).

References

1. Yu, R., et al.: Synthesis and scale inhibition proper- ties of MA-SAS-HEMA terpolymer. J. Xi 'an Polytech. Univ. **37**(04), 74–81 (2023)
2. Song, S., Liu, L., Sun, Y.: Synthesis and properties of multi-group low-phosphorus copolymer scale inhibitors. Appl. Chem. Ind. **53**(03), 542–546 (2024)
3. Fang, W., et al.: Scale inhibition and corrosion inhibition performance of PESA-g-PAA/AHPS without phosphorus and nitrogen. Ind. Water Treat. **40**(8), 80–83 (2020)
4. Gao, Y., et al.: Application of chitosan derivatives in the treatment of industrial circulating cooling water. Appl. Chem. Ind. **50**(12), 3441–3445 (2021)
5. Zou, K., et al.: Synthesis and properties of green scale inhibitors derived from polyepoxysuccinic acid. Ind. Water Treat. **41**(05), 109–113 (2021)
6. Li, H., et al.: Synthesis and properties comparison of epoxy succinic acid copolymers containing sulfonic acid groups. Adv. Chem. Ind. **36**(4), 1491–1498 (2017)
7. Si, S., et al.: Experimental and theoretical study of tripolypyridine inhibitors for corrosion and scale inhibition. Ind. Water Treat. **42**(05), 131–139 (2022)
8. Zhang, G., et al.: Preparation and performance evaluation of long-release corrosion and scale inhibitors. Appl. Chem. Ind. **51**(09), 2663–2668 (2022)
9. Yu, F.: Development and Properties of Modified Polyaspartic Acid Corrosion and Scale Inhibitors. Hefei University of Technology (2023)
10. Guo, X.: Preparation and Mechanism of Carboxyl Modified Polyaspartic Acid Derivatives for Scaling and Corrosion Inhibition. Henan University, Kaifeng (2021)
11. Yu, R., Chen, X.: Research progress on scale inhibition and corrosion inhibition of modified polyaspartate. Ind. Water Treat. **41**(12), 35–40 (2021)
12. Cheng, Y., Wu, J., Huang, C.: Quantum chemical study on scale inhibition and corrosion inhibition of polyaspartic acid. Ind. Water Treat. **41**(07), 100–104 (2021)

Operation and Maintenance Essentials and Cost Estimation of the Tailwater Wetland Coupled with Wastewater Treatment Plant

Jing Mo, Yuling Huang(✉), Tianzheng Cao, and Huihuang Luo

China Institute of Water Resources and Hydropower Research, Beijing 100038, China
hylcos90@163.com

Abstract. Constructed wetlands are ecosystems consisting of water and plants. They serve multiple purposes, including sewage purification, biodiversity protection, microclimate regulation, and habitat preservation. Today, they are globally recognized as environmentally friendly "Green" sewage treatment technologies and are widely employed for industrial wastewater and domestic sewage treatment. While numerous studies have been conducted on plant and filler selection and decontamination capabilities of constructed wetlands, along with practical applications, challenges still persist in the field of constructed wetland wastewater treatment projects. However, in the field of constructed wetland sewage treatment projects, there is still the phenomenon of emphasizing construction while neglecting management, and the lack of a scientific and effective operation and management system result in problems such as blockage, decreased treatment efficiency, irregular operation, and subpar landscape aesthetics. To address these challenges, a tailwater wetland coupled with wastewater treatment plant in Yangtze River Basin was taken as a case, to propose specific requirements and measures for the operation and maintenance of constructed wetlands, focusing on water quantity, water level, vegetation, packing materials, aquatic animals, and thermal insulation. The objective is to maximize the multi functionality of constructed wetlands and provide theoretical and technical support for their sustainable operation.

Keywords: tailwater wetland · operations · fees and charges

1 Introduction

Constructed wetlands refer to artificially built wastewater treatment facilities that simulate the structure and function of natural wetlands. They utilize a synergistic combination of physical, chemical, and biological processes to create an ecosystem for wastewater purification. Based on water flow patterns, constructed wetlands can be classified into subsurface flow and surface flow wetlands [1]. Since Seidel first proposed the use of constructed wetlands for wastewater treatment in the 1960s, they have played an important role as a "nature-based solution" for treating urban sewage [2]. After decades of development, constructed wetlands have also been widely applied in the treatment of domestic sewage, industrial wastewater, agricultural runoff, landfill leachate, and contaminated water bodies [3–7].

D. Han et al. (Eds.): ICEPG 2024, ESE, pp. 171–184, 2025.
https://doi.org/10.1007/978-3-031-86829-0_21

There is no doubt that constructed wetlands play a vital role in maintaining ecological balance, improving ecological conditions, achieving harmony between humans and nature, and promoting sustainable economic and social development. However, the effectiveness of constructed wetlands relies one-third on construction and seven-tenths on management. The operation, maintenance, and management of wetland systems are crucial for ensuring the effectiveness of wastewater treatment and ecological restoration, especially for subsurface flow constructed wetlands, where poor management often leads to clogging issues. Therefore, to ensure the long-term effective operation of wetlands and achieve the desired results, it is essential to conduct regular maintenance, monitoring, and management of water quality, vegetation, and substrate.

As a key province through which the Yangtze River flows, Anhui bears the historical mission of protecting the Yangtze River. However, currently, the operational levels of some cities in Anhui regarding the advanced treatment and reuse of effluent from urban sewage treatment plants are relatively low. Additionally, existing research on constructed wetlands mainly focuses on sewage treatment processes and related mechanisms, with limited studies on operation and maintenance management, which impacts the operational efficiency, effectiveness, and lifespan of constructed wetlands. In light of this, the aim of this study is to propose management and maintenance key points concerning the wetland substrate, vegetation, and water levels in the JD tailwater wetland park, and to assess the associated maintenance costs, in order to provide theoretical references and technical support for the operation and maintenance of tailwater wetlands associated with sewage treatment plants in the Yangtze River protection efforts.

2 JD Wetland Park

2.1 Process

JD Tailwater Wetland Park is located in Anhui Province and is an important component of the Zhujiaqiao sewage treatment system, responsible for advanced nitrogen and phosphorus removal from the effluent of the wastewater treatment plant. The total area of the project's first and second phases is approximately 57 hectares. Phase one includes horizontal subsurface flow wetlands, enhanced surface flow wetlands, ecological conservation ponds, ecological river channel, and reclaimed water reuse pumping station, covering an area of 16.6 hectares, with a treatment capacity of 40,000 tons per day and an effluent quality target meeting Class IV standards of GB3838-2002. The enhanced surface flow wetland is supplemented by artificial aeration and is connected in series with the ecological conservation pond, which is then paralleled with the horizontal subsurface flow wetland. The combined effluent is used to recharge nearby urban rivers and lakes. The wetland has been operational for approximately two years. The process flow diagram is shown in Fig. 1, and the influent water delivery pipeline of the constructed wetland treatment system is designed with a pipe diameter of DN800.

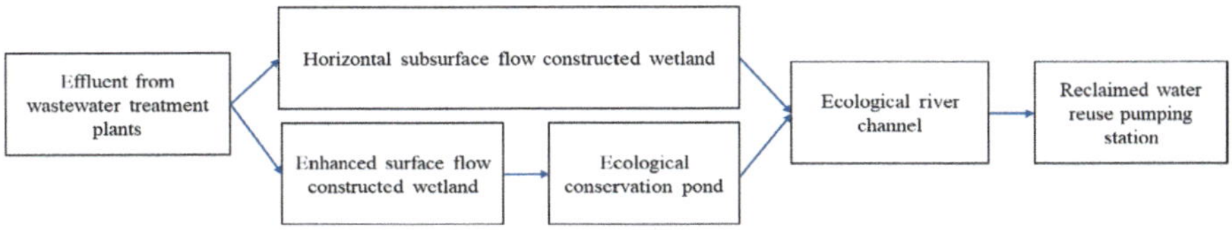

Fig. 1. Process flow diagram

2.2 Run Parameters

Design the Running Parameters

The average depth of the substrate layer in the horizontal subsurface flow wetland of JD Wetland Park is 0.78 m, with a porosity of 45% and a hydraulic loading rate of 1 $m^3/(m^2 \cdot d)$. The average water depth in the enhanced surface flow wetland is 1.5 m, with a hydraulic loading rate of 0.31 $m^3/(m^2 \cdot d)$. The average water depth in the ecological conservation pond is 2.0 m, with a hydraulic loading rate of 0.93 $m^3/(m^2 \cdot d)$. The design parameters for the subsurface wetland, enhanced wetland, and ecological conservation pond in JD Wetland Park are shown in Table 1.

Table 1. Design parameters of JD Tailwater Wetland Park

Wetland unit	Area (m^2)	Depth of water (m)	Volume (m^3)	Inflow flow (m^3/d)	Hydraulic retention time (d)	Hydraulic load ($m^3/m^2 \cdot d$)
Horizontal subsurface flow constructed wetland	20000	0.78	7020	20000	0.35	1.00
Enhanced surface flow constructed wetland	32000	1.5	48000	20000	2.40	0.31
Submerged conservation lake	43000	2.0	86000	40000	2.15	0.93
Subtotal	95000	-	141020	40000	3.53	0.42

The Actual Run Parameters

Hydraulic loading refers to the volume of effluent treated by a constructed wetland per unit area in a given time, directly impacting the treatment efficiency of the wetland. The lower the hydraulic loading and the longer the hydraulic retention time, the lower the wastewater treatment efficiency of the wetland [8].

The actual operational parameters of JD Wetland Park are shown in Fig. 2. The hydraulic loading fluctuates significantly throughout the year, with the surface flow wetland having a hydraulic loading of 0.79 to 1.13 $m^3/(m^2 \cdot d)$, and the ecological conservation pond having a hydraulic loading of 0.59 to 0.84 $m^3/(m^2 \cdot d)$. Both are lower than the hydraulic loading of the subsurface flow wetland, which ranges from 1.27 to 1.81 $m^3/(m^2 \cdot d)$. The hydraulic retention time in the ecological conservation pond is 2.38 to 3.39 days, which is longer than that of the subsurface flow wetland (0.19 to 0.28 days) and the surface flow wetland (0.35 to 0.50 days). The design hydraulic loadings and hydraulic retention times for the subsurface flow wetland, surface flow wetland, and ecological conservation pond are 1.0 $m^3/(m^2 \cdot d)$, 0.63 $m^3/(m^2 \cdot d)$, and 0.93 $m^3/(m^2 \cdot d)$, and 0.35 days, 2.4 days, and 4 days, respectively. There are certain differences between these design values and the actual operational parameters, mainly characterized by higher hydraulic loadings and shorter hydraulic retention times.

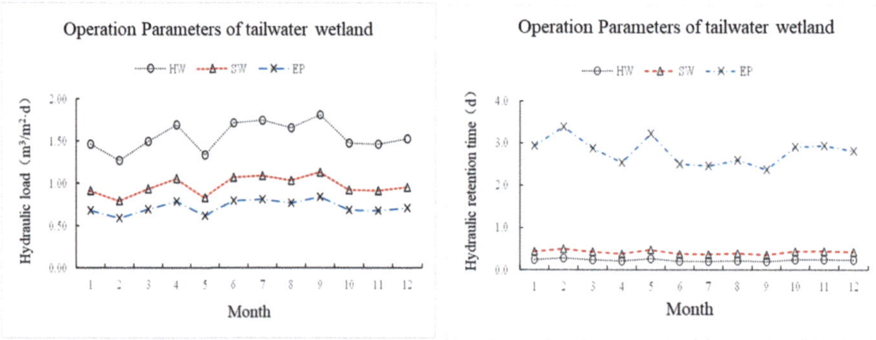

Fig. 2. Parameters of the tail water wetlands

2.3 Running Effect

Through monitoring, the main effluent indicators (BOD_5, COD, NH_3-N, TP) from the treatment units of JD Wetland Park from March 2022 to February 2023 all exceeded the Class III water quality standards specified in the "Surface Water Environmental Quality Standards" (Table 2). The average removal rates for BOD_5 in the horizontal subsurface flow wetland, enhanced surface flow wetland, and ecological conservation pond were 17.03%, 27.84% and 38.65%, respectively. The average removal rates for COD were 16.21%, 8%and 6.18%, respectively, while the average removal rates for NH_3-N were 58%, 73%and 31%, and for TP they were 17.39%, 47.83%and 26.09%, respectively. Among these, the removal rates of NH_3-N and TP in the surface flow wetland were higher than those in the horizontal subsurface flow wetland and the ecological conservation pond. The BOD_5 removal rate in the ecological conservation pond was higher than that in the horizontal subsurface flow wetland and the enhanced surface flow wetland, while the COD removal rate in the horizontal subsurface flow wetland was higher than that in the enhanced surface flow wetland and the ecological conservation pond. Additionally, except for COD, the pollutant removal rates in the horizontal subsurface flow wetland were lower than those in the surface flow wetland, mainly due to the significantly larger area of the surface flow wetland and its lower hydraulic loading compared to the subsurface flow wetland. When the inflow volumes are the same, the hydraulic retention time in the surface flow wetland is longer, resulting in a greater extent of pollutant purification and removal [9].

Table 2. The removal rate of each processing unit

Indicators	BOD_5	COD	NH_3-N	TP
Quality of incoming water($mg{\cdot}L^{-1}$)	3.70	19.25	1	0.23
Mean effluent from horizontal subsurface flow wetland($mg{\cdot}L^{-1}$)	3.07	16.13	0.42	0.19
Subitem average removal rate	17.03%	16.21%	58%	17.39%
Enhanced mean effluent from surface flow wetlands($mg{\cdot}L^{-1}$)	2.67	17.71	0.27	0.12
Subitem average removal rate	27.84%	8%	73%	47.83%
Mean effluent from ecological conservation ponds($mg{\cdot}L^{-1}$)	2.27	18.06	0.69	0.17
Subitem average removal rate	38.65%	6.18%	31%	26.09%

2.4 Diagnosis of Tailwater Wetland Problems

Currently, the operational status of JD Wetland Park is relatively stable, but several issues persist. On one hand, the designed hydraulic loading is too high and the hydraulic retention time is too short. In actual operation, the hydraulic loading is higher and the retention time is shorter than designed, which does not meet the operational goals. This leads to low efficiency in the physicochemical and biochemical reactions of the effluent within the wetland system, resulting in ineffective pollutant removal. On the other hand, with rising temperatures, there is significant growth of phytoplankton in the conservation pond, which may lead to the decline and death of submerged plants, causing eutrophication. Furthermore, field observations indicate that the parallel inflow from four channels in the subsurface flow wetland lacks effective quantitative distribution, resulting in uneven water distribution and substantial differences in organic loading across different areas, which diminishes the function of wetland plants.

In summary, there is currently a lack of effective operational management in the artificial wetland management process at JD Wetland Park. This poses a real challenge for the degradation of pollutant removal effectiveness and unsustainable operation of the constructed wetlands. In response to these management issues, this study proposes relevant operational and maintenance requirements and measures concerning water quantity, water levels, vegetation, substrate, aquatic animals, and insulation, as well as estimates of the operational and maintenance costs of the wetland.

3 Operation and Maintenance Essentials of Tailwater Wetland in JD

3.1 General Management Requirements

First, select an appropriate management model for the ecological wetland treating the effluent from the local wastewater treatment plant based on the actual local conditions. Regularly monitor the operational status of the wetland and manage and maintain aquatic plants, water distribution pipes, substrate, and ancillary facilities to ensure the normal operation of the ecological wetland treating the wastewater treatment plant effluent.

Second, provide training for operational staff, technical personnel, and management on relevant laws and regulations, professional techniques, safety precautions, and emergency response procedures. This training should ensure familiarity with the treatment processes, operational requirements, technical specifications, and safety protocols for facilities and equipment. Operational facilities and major equipment should be recorded in a ledger, and regular maintenance checks should be conducted on various equipment, electrical systems, automatic control instruments, and buildings. Process flow diagrams, operational procedures, and maintenance guidelines should be displayed in prominent locations. Management systems should be established before the project begins, including equipment ledgers, operational records, regular inspections, shift handovers, safety checks, and emergency response plans.

In addition, for towns with the necessary conditions, the construction of a smart management system for ecological wetlands treating effluent from wastewater treatment plants should be integrated with the urban water system, aligning with the comprehensive management of the urban water system.

3.2 Subsurface Wetlands

JD Wetland Park includes two main types: subsurface flow wetlands and surface flow wetlands. In the subsurface flow wetlands, management focuses first on water quantity, water level, and water quality. Online monitoring devices for water quantity, water level, and water quality are installed at the inflow and outflow points based on actual needs. Flow meters automatically collect inflow and outflow data, level gauges automatically record the water levels at the wetland's entrance and exit, and online water quality instruments automatically monitor key water quality indicators of the inflow and outflow [10]. For wetlands without online water quality monitoring equipment, qualified agencies are regularly commissioned to monitor water quality indicators such as flow, water level, water temperature, dissolved oxygen, pH, suspended solids, Five-day biochemical oxygen demand, Chemical oxygen demand, ammonia nitrogen, total nitrogen, and total phosphorus at the inflow and outflow points during routine operations. Monitoring reports are generated and archived for record-keeping [11]. The inflow and outflow valves are opened or closed based on the readings from the level gauges to ensure that the wetland bed maintains the proper water level.

Second, routine management should be conducted for the substrate in the subsurface flow wetland. Regular inspections and maintenance of the substrate bed should be performed, including periodic removal of dead branches, fallen leaves, and other solid debris from the bed surface to prevent clogging of the substrate pores. If water accumulation in the substrate bed is observed, it should be reported immediately. The cause of the accumulation should be analyzed, localized investigations conducted to determine the extent of the issue, and solutions proposed. Backwashing should be carried out based on the conditions of water accumulation and pore clogging in the substrate bed to improve the porosity and reoxygenation efficiency of the lower substrate. The effectiveness of nitrogen and phosphorus removal in the substrate bed should be assessed based on regular water quality monitoring results, and localized substrate replacement should be implemented if necessary. Consideration should be given to adding materials containing carbon, sulfur, and other substances to increase the carbon source content within the substrate bed, thereby promoting enhanced nitrogen and phosphorus removal by microorganisms.

At the same time, wetland plants are an important component of subsurface flow wetlands and should be managed specifically. Regular inspections and maintenance of wetland plants should be conducted. Based on the current state of the wetland and design parameters, replanting or thinning should be performed periodically to ensure an even distribution of plants per unit area, promoting normal growth and effective absorption of pollutants from the water. Weeds in the plant clusters should be removed without using chemical agents such as herbicides or pesticides [10]. The effectiveness of the wetland should be assessed based on plant growth and water quality monitoring results, considering functions such as pollutant removal, landscape enhancement, and ecological balance. If necessary, localized replacement of plant species should be undertaken, focusing on native species with strong resistance to avoid ecological invasion. Aquatic plants should be harvested at appropriate intervals and handled properly according to local requirements, avoiding arbitrary stacking. Depending on plant growth conditions, harvesting should occur once or twice a year, with an emphasis on removing the above-ground parts and ensuring timely cleanup.

3.3 Surface Flow Wetlands

Unlike the structure of subsurface flow wetlands, the management of surface flow wetlands, in addition to managing water quantity, water level, and water quality, as well as plant management, must also strengthen the management of phytoplankton in the water. Additionally, considering that fish species are designed to be introduced into the surface flow wetland system in JD Wetland Park, corresponding management for these fish should also be implemented.

In the management of water quantity, water level, and water quality, online monitoring devices for water quantity, water level, and water quality should be installed at the inflow and outflow points based on actual needs. Flow meters automatically collect inflow and outflow data, level gauges automatically record the water levels at the wetland's entrance and exit, and online water quality instruments automatically monitor key water quality indicators of the inflow and outflow. During routine operations, qualified agencies should be regularly commissioned to monitor water quality and ecological indicators at the inflow and outflow points, including flow, water level, water temperature, dissolved oxygen, pH, turbidity, transparency, suspended solids, Five-day biochemical oxygen demand, Chemical oxygen demand, ammonia nitrogen, total nitrogen, total phosphorus, and chlorophyll a. Monitoring reports should be generated and archived for record-keeping. When abnormal growth of phytoplankton is observed on the water surface, the frequency of monitoring should be increased, and sampling points should be adjusted. Timely analysis of water quality changes in the wetland units should be conducted, and proper records should be kept. Based on the suitable water level characteristics of aquatic plants, the inflow and outflow valves should be opened or closed to maintain the normal water level.

In the management of aquatic plants, regular inspections and maintenance should be conducted. Based on the current state of the wetland and design parameters, replanting or thinning should be performed periodically to ensure an even distribution of plants per unit area, promoting normal growth and effective absorption of pollutants from the water. If submerged plants are observed to be surfacing during the summer, it should be reported immediately, and their suitable water level characteristics should be determined. On one hand, the inflow and outflow valves should be opened or closed to adjust water depth; on the other hand, underwater harvesting should be carried out to remove the above-water portions. If large areas of floating leaves and floating plants are observed, with coverage exceeding 50%, it should be reported immediately. Based on the current state of the wetland and design parameters, underwater harvesting should be conducted to control the spread of these plants. Any floating remnants of aquatic plants on the water surface should be cleaned up and retrieved immediately [12]. Aquatic plants should be harvested at appropriate intervals and handled properly according to local requirements, avoiding arbitrary stacking. Depending on plant growth conditions, harvesting should occur once or twice a year, with emergent plants having their above-ground portions removed, and submerged, floating-leaved, and floating plants kept at a coverage of no more than 20%.

The phytoplankton in the conservation pond of the surface flow wetland should also be managed to mitigate the degree of water eutrophication. Regular surveys should be conducted to assess the dominant phytoplankton species and their density in the pond, as well as to observe factors such as water color, transparency, and turbidity. These observations, combined with water quality monitoring results, will help determine the timing and methods for managing phytoplankton. During the spring and summer, when temperatures rise continuously for two to three days, close attention should be paid to the water body. If the water appears yellow-green and the transparency is less than one-third of the water depth, it should be reported immediately, and the eutrophication control plan should be activated. Eutrophication control measures may include shading, aeration, harvesting, applying biological algaecides, and introducing filter-feeding fish.

The fish populations and densities in the conservation pond should be managed appropriately to reduce disturbances to the bottom sediments caused by fish activity and to minimize damage to aquatic vegetation, while ensuring the integrity of the food chain and the normal maintenance of the water ecosystem's structure and function. Regular surveys should be conducted to assess the species, weight, body length, and dominant species of fish in the pond. Based on factors such as temperature, atmospheric pressure, and the state of the water (including color, transparency, and turbidity), and referencing local fishermen's experiences, the timing and methods for fishing should be determined.

Generally, at least one winter cleanup should be conducted each year, and if conditions permit, one to two cleanups can be done in the summer, with proper records maintained.

Herbivorous and omnivorous fish species, such as carp and crucian carp, that inhabit the middle and lower layers of the water should be removed as much as possible. A small number of predatory fish should be retained, making up no more than 10% of the total population. A suitable number of filter-feeding fish, such as silver carp and bighead carp, should be maintained, with a ratio of approximately 3:1 to 4:1, keeping the total number at 10 to 20 fish per acre and the biomass at 45 to 50 g/m^3. The weight of individual fish should not exceed 2.5 to 3 kg, ideally around 0.25 to 1 kg.

3.4 Special Management Requirements

Based on the abnormal operating conditions of the wastewater treatment plant, the wetland should be designed for intermittent operation to enhance reoxygenation and microbial endogenous respiration while reducing the incidence of abnormal operational accidents. The water level in the effluent ecological wetland should be adjusted according to extreme weather conditions such as heavy rainfall, flooding, and drought to prevent issues like water accumulation at the inflow, flooding at the outflow, substrate bed collapse, and the death of aquatic plants. Additionally, regular or irregular sediment removal can be performed based on the accumulation of sediments in the surface flow wetland and conservation pond.

To address the issue of winter operation in the wetland, measures such as pulse hydraulic loading and varying hydraulic retention times can be implemented, supplemented by artificial aeration and other process enhancement techniques to improve winter operational effectiveness. Additionally, insulation measures should be taken for the inflow and outflow pipes to prevent freezing. Options for insulation include covering with plant straw or plastic film [13, 14]. In conjunction with weather forecasts, hydraulic loading should be adjusted in a timely manner to lower the operating water level and enhance insulation effectiveness.

4 Estimated Operating and Maintenance Costs of the Sewage Treatment Tailwater Wetland

4.1 The Operational and Maintenance Costs of Constructed Wetlands

The operational and maintenance costs of constructed wetlands mainly consist of electricity costs and labor management and maintenance costs. The electricity costs primarily include the expenses associated with lift pumps (which are not required for gravity-flow wetlands) and aeration costs (typically only necessary for subsurface flow systems that require increased loading) [15]. Labor management and maintenance costs mainly cover sediment removal, substrate clogging maintenance, plant harvesting, and equipment management and maintenance. It is generally believed that the plant harvesting costs for surface flow constructed wetlands are higher than those for subsurface flow constructed wetlands [16]. However, due to the higher risk of clogging in subsurface flow wetlands, the costs for substrate clogging maintenance and equipment maintenance are higher than those for surface flow wetlands, with vertical subsurface flow wetlands having the highest equipment management and maintenance costs.

Since the energy consumption for water distribution and hydraulic gradient in subsurface flow wetlands is higher than that in surface flow wetlands, overall, the operational costs per ton of water in subsurface systems are usually higher than those in surface systems [17, 18]. Various studies indicate [19–27] that the operational costs of subsurface constructed wetlands in China are the highest, averaging 0.64 yuan/t, with horizontal subsurface and vertical subsurface constructed wetlands averaging 0.8 yuan/t and 0.27 yuan/t, respectively. In contrast, surface flow constructed wetlands have an average operational cost of 0.25 yuan/t, while composite flow constructed wetlands have the lowest average operational cost of 0.13 yuan/t. The significant differences in operational costs among different types of constructed wetlands are related to factors such as the scale of treatment, the use of pumping stations, and the adoption of pretreatment processes.

4.2 The Design and Actual Operational and Maintenance Costs of the JD Wetland

The design and actual operational and maintenance costs of the JD Wetland are detailed in Tables 3 and 4. The designed annual operational management cost for the effluent wetland is 2.323 million yuan, with an operating cost of 0.16 yuan/m^3; the actual annual operational management cost in 2022 was 2.1506 million yuan, with an operating cost of 0.15 yuan/m^3. The actual annual operational management costs were lower than the designed annual management costs, and the actual operating costs were also lower than the designed operating costs, indicating that the effluent wetland implemented effective measures that achieved better-than-expected results. Research by Wang Zheng fang et al. [25]. And Wang Guifang et al. [24]. Shows that the operational costs of composite flow constructed wetlands are 0.1 yuan/m^3 and 0.15 yuan/m^3, respectively, which are comparable to the design and actual operational costs in this study.

Table 3. Cost of design, operation and maintenance of constructed wetlands

Category	Total area (hm^2)	Total cost of operation and maintenance (yuan)	Notes
Subsurface flow constructed wetland	6.6	910000	Including horizontal subsurface flow constructed wetland area of 2.9 hm^2 and enhanced high-efficiency subsurface flow constructed wetland area of 3.7 hm^2
Ecological Conservation Pond	10	780000	-
Canal and wet pond maintenance	2.8	308000	Desilting, chemicals and electricity
Replacement of fillers	That translates into 325,000 yuan per year	1625000	The price of replacing the filler every five years
Annual operating and management expenses(yuan)	2323000		
Operating cost(yuan/m^3)	0.16		

Table 4. Actual operating and maintenance costs of JD Tailwater wetland in 2022

Project name	Total cost (yuan)	Notes
Subsurface flow constructed wetland maintenance fee	148900	Mainly includes related labor and facility costs arising from water level management, daily maintenance of plants, and channel silt management
Subsurface flow constructed wetland media clearance fee	250500	Media daily operation and maintenance
Water surface cleaning fee	98800	Water surface debris cleanup
Aquatic plant harvesting and maintenance fees	1136200	Prune and harvest overgrown aquatic plants, collect harvested plants, replant missing aquatic plants, control pests and diseases, and promote root growth and preservation
Fish management fees	12000	Monitor fish population classification, control the number of nuisance fish, and supplement with carnivorous fish.
Equipment operating costs	108900	Regular equipment inspection, regular maintenance and repair of equipment
Maintenance tools costs	38400	Boats, dredging nets, rakes, fish traps, life jackets, etc
Waste transportation and disposal fees	40900	
Water quality testing fees	81200	One monitoring point
Emergency expenses	140000	Ecological system regulation during extreme weather and emergency situations
Landscape plant adjustment and optimization fees	94800	Prune disordered landscape plants, thin out overgrown plants, control plant pests and diseases, and promote root growth and preservation
Annual operating management costs (yuan)	2150600	
Operating cost (yuan /m^3)	0.15	

5 Conclusion

At this stage, research on constructed wetland sewage treatment projects in China mainly focuses on pollutant removal efficiency, neglecting the importance of management. There is an incomplete understanding that effective management of constructed wetlands is crucial for fully realizing their functional benefits. Scientific and effective operational management can ensure the efficient and stable removal of pollutants, achieve compliant effluent discharge, and reduce factors that may harm the operational lifespan of constructed wetlands.

Currently, the tailwater quality of urban sewage treatment plants in the Yangtze River Economic Belt, represented by JD tailwater wetland, generally exhibits characteristics such as low organic load, high SS/BOD_5 ratio, and low BOD_5/TN ratio. To ensure the long-term stable compliance of constructed wetlands and maximize their ecological and landscape effects, it is necessary to strengthen the operation and maintenance management of wetlands from the following aspects:

Different types of constructed wetlands should be selected based on the effluent quality characteristics of sewage treatment plants in different regions and the quality requirements of receiving water bodies. The design and operational parameters of constructed wetlands should be optimized. For example, reduce the inflow volume of tailwater, lower the hydraulic load on the wetland, set up storage ponds to regulate inflow, and mitigate the hydraulic shock of incoming water on the wetland. Extending the hydraulic retention time will help the wetland system operate effectively.

It is recommended that sewage treatment plants conduct energy-saving measures and reduce reagent usage while ensuring compliant discharge. Increasing the carbon, nitrogen, and phosphorus content in the tailwater will provide sufficient carbon sources for the wetland, promoting microbial kinetic processes and enhancing the removal rates of nitrogen and phosphorus pollutants.

To improve the nitrogen removal efficiency in subsurface flow wetlands, it may be beneficial to supplement external carbon sources, such as laying sulfur and pyrite in the substrate bed or adding carbon sources, to promote the denitrification process within the wetland system.

Acknowledgments. Funding by China Three Gorges Corporation (No. 202103361) and Yangtze Ecological and Environmental Co., Ltd. (No. HB/ZB 2021130).

References

1. Li, M., Li. W., Wei. J., et al.: Application of constructed wetland system in purification of tail water quality of sewage treatment plant. Environ. Ecol. **4**(06), 54–58 (2022)
2. Q., Liu: Study on Formation Mechanism, Functional Impact and Control Technique of Substrate Clogging in Constructed Wetland. Shandong University, Jinan (2019)
3. Yalcuk, A., Ugurlu, A.: Comparison of horizontal and vertical constructed wetland systems for landfill leachate treatment. Bioresour. Technol. **100**, 2521–2526 (2009)
4. Harrington, C., Villa, M.: Assessment of pre-digested piggery wastewater treatment operations with surface flow integrated constructed wetland systems. Bioresour. Technol. **101**, 7713–7723 (2010)

5. Saeed, T., Sun, G.: A review on nitrogen and organics removal mechanisms in subsurface flow constructed wetlands: Dependency on environmental parameters, operating conditions and supporting media. J. Environ. Manag. **112**, 429–448 (2012)
6. Saeed, T., Sun, G.: A lab-scale study of constructed wetlands with sugarcane bagasse and sand media for the treatment of textile wastewater. Bioresour. Technol. **128**, 438–447 (2013)
7. Badhe, N., Saha, S., Biswas, R., Nandy, T.: Role of algal biofilm in improving the performance of free surface, up-flow constructed wetland. Bioresour. Technol. **169**, 596–604 (2014)
8. Deblina, G., Brij, G.: Effect of hydraulic retention time on the treatment of secondary effluent in a subsurface flow constructed wetland. Ecol. Eng. **36**(8), 1044–1051 (2010)
9. Huang, R., Huang, Y., Zeng, Y., et al.: Decontamination effect of tail water wetland supporting the sewage treatment plant—A case in the area of the Yangtze River Grand Protection. J. China Insti. Water Resour. Hydropower Res. **22**(02), 169–178+194 (2024)
10. Wang, N., Tian, G., Chen, X.: Study and design of artificial wetland purification system for wastewater treatment plant in the upper reaches of the Yellow River: A case study of haosheng wastewater treatment plant in Ningxia. Environ. Eng. **41**(S1), 33–35+51 (2023)
11. Li, L.: Discussion on sewage purification system and management mechanism of constructed wetland. China Water Transport **18**(05), 113–114 (2018)
12. Ibekwe, A.M., Lyon, S.R., Leddy, M., et al.: Impact of plant density and microbial composition on water quality from a free water surface constructed wetland. J. Appl. Microbiol. **102**(4), 921–936 (2007)
13. He, H.: Analysis of Artificial Wetland Solutions for Water Quality Improvement Projects in Cold Region River Channels. Hydro Sci. Cold Zone Eng. **7**(08), 7–10 (2024)
14. Yao, D., Feng, X., Zhang, B., et al.: The operational efficiency of Longhe artificial wetland under low temperature conditions in northern winters. Environ. Eng. **40**(01), 148–154 (2022)
15. Duan, G., Liu, K.: Comparison study on two different constructed wetland systems for removal of phosphorus from swine wastewater. China Water Wastewater **31**(19), 91–95 (2015)
16. Kadlec, R.H., Wallace, S.D.: Treatment Wetlands, 2nd edn., pp. 232–287. CRC Press, USA (2008)
17. Li, M.A.N.: Design of water quality purification project of artificial wetland in Zhushui River Bank. China Water Wastewater **33**(14), 66–69 (2017)
18. Chen, S., Jiang, D., Zhang, Y., et al.: Example of rural domestic wastewater treatment with constructed wetland process. Water Wastewater Eng. **48**(S1), 221–223 (2012)
19. Jun, W.E.I., Wanyu, H.A.N., Yunling, D.U., et al.: Design and Practice of Tailwater Constructed Wetland. China Water Conservancy and Hydropower Press, Beijing (2019)
20. Chen, G.: Process design of combined surface flow wetland to treat tail water from wastewater treatment plant in chemical industry park. Ind. Water Treat. **42**(05), 169–175 (2022)
21. Haiwen, X.I.A.O., Xintong, L.I.U., Jun, Z.H.A.I., et al.: Selection of constructed wetland types and case analysis. China Water Wastewater **37**(22), 11–17 (2021)
22. Zhang, Y.: Ecological wetland technology for advanced treatment of urban Sewage Treatment tail water. China Water Wastewater **33**(04), 87–89 (2017)
23. Wei, J., Zhao, M., Liu, W., et al.: Development status of tail-water type constructed wetland in our country. China Water Wastewater **35**(02), 29–33 (2019)
24. Guifang, W., Dayi, W., Zhiyuan, Z., et al.: Engineering design of treating rural domestic sewage with constructed wetland and ecological pond. Environ. Eng. **28**(04), 6–8 (2010)
25. Wang, Z.-F., Zhang, J., Zheng, Z., et al.: Physically enhanced integrated constructed wetland for treating effluent from sewage treatment plant. China Water Wastewater **29**(02), 75–77 (2013)

26. Wang, X., Zhu, Z., Zhang, J., et al.: Application of ex-situ composite constructed wetlands in river purification. China Water Wastewater **37**(14), 137–141 (2021)
27. Wu, D., Yang, J., Wu, H., et al.: Process design and effect study of river bypass vertical-flow constructed wetland. Environ. Ecol. **5**(11), 77–82 (2023)

Vertical Distribution of Marine Microplastics in the Intertidal Zone of Xiamen Bay

Xiping Zhou[1,2](✉), Yongmei Zhao[1], Qianyu Wu[1], Yaning Zhu[1], Kun Li[1], Fumao Li[1], and Yuxiu Weng[1]

[1] School of Environmental Science and Engineering, Xiamen University Tan Kah Kee College, Zhangzhou 363105, Fujian, China
zoeyzhou@xujc.com

[2] Key Laboratory of Estuarine Ecological Security and Environmental Health, Fujian Province University, Tan Kah Kee College, Zhangzhou 363105, Fujian, China

Abstract. In order to understand the occurrence and distribution of marine microplastics in the intertidal zone, a study on microplastics in sandy intertidal zone sediments was conducted on the south coast of Xiamen Bay in October 2022. The research results showed that the average abundance of marine microplastic is 2.43 items/cm^2 in the sediment on the south coast of Xiamen Bay, the highest abundance occurred in the low tide region with 2.92 items/cm^2, and the lowest abundance occurred in the high tide region with 1.73 items/cm^2; the color composition of the microplastic were mainly white and black, including the highest abundance of white microplastics 1.50 items/cm^2; The microplastics were mainly fibrous, flake and flocculent, and their abundance were 0.79 items/cm^2, 0.76 items/cm^2 and 0.56 items/cm^2, respectively. In terms of particle size, the abundance of marine microplastics with 250 μm–425μm was 1.23 items/cm^2, the abundance of those with a particle size of 425 μm ~ 830 μm was 0.69 items/cm^2, and the abundance of those with a particle size of >830 μm was 0.50 items/cm^2. The highest vertical abundance occurs in 0 ~ 2 cm with a value of 1.07 items/cm^2.This study can enrich the background data on the distribution of marine microplastics in the intertidal zone of Xiamen Bay, and further provide scientific support for the monitoring and control of marine microplastics in the coastal areas of Fujian.

Keywords: intertidal zone · marine microplastic · occurrence and distribution · Xiamen Bay

1 Introduction

As an emerging, little-known pollutant that has attracted the attention of researchers around the world, microplastics are widespread and likely to increase in the future. There are nearly 100 million tons of plastic waste in the ocean, and about 9 million to 14 million tons of plastic continue to flow into the ocean every year [1]. According to statistics, nearly 80% of marine plastic waste comes from human production and life on land, and plastic waste is discharged into the waters and eventually into the ocean, or

D. Han et al. (Eds.): ICEPG 2024, ESE, pp. 185–190, 2025.
https://doi.org/10.1007/978-3-031-86829-0_22

the storm washes the plastic waste buried on the land into the sea, while the remaining 20% of marine plastic comes from the garbage discarded by sea ships or the garbage leakage caused by marine accidents [2]. Due to the physical and chemical characteristics of marine plastics, coupled with the effect of ocean currents, marine plastics are found in the world's oceans. Waste plastics will become fine plastic waste in the complex changes of the environment [3]. Under the action of long-term ultraviolet irradiation and wind and wave impact, marine plastic waste can gradually break into countless micron plastic fragments [4].

As an important part of the marine ecosystem, the intertidal zone is located at the intersection of sea and land. It contains abundant animal and plant resources, which can not only cushion the direct impact of sea waves on the land, but also maintain the local humidity and rainfall [5]. The accumulation of marine microplastics in its sediments has attracted scholars' attention in recent years. On the one hand, the intertidal zone collects microplastics from surface runoff, coastal bathing sites and nearshore aquaculture fisheries. On the other hand, microplastics in the ocean will also accumulate in the intertidal zone under the action of ocean currents and seawater erosion [6]. It is estimated that about 70% of plastic sinks to the sea floor, and the amount of plastic in intertidal sediments is much higher than in seawater [7].

The south bank of Xiamen Bay is close to the tourist hotspot city Xiamen, and human activities are frequent. A large amount of marine plastic waste will enter the south bank of Xiamen Bay with the runoff of Jiulong River into the sea and the release of coastal household garbage. How to deal with the problem of marine microplastics while developing economy is an urgent task to be carried out in the environmental protection of Xiamen Bay [8]. In this paper, the abundance, size, color and shape of marine microplastics were studied in the sandy intertidal zone of Xiamen Bay, and the horizontal and vertical distribution of microplastics in the sediment of the intertidal zone of Xiamen Bay was investigated, in order to enrich the background data of marine microplastics distribution in the intertidal zone of Xiamen Bay, and further reveal the current status of Marine plastic pollution in Xiamen Bay.

2 Study Area and Method

2.1 Study Area

With the economic and social development of Xiamen and its surrounding areas, the environment of Xiamen sea and its offshore waters has undergone significant changes. The sampling sites of this study located on the south bank of Xiamen Bay, Zhangzhou Development Zone, China Merchants Bureau, Longhai District, Zhangzhou City, Fujian Province (24°21 'N, 118°3' E).

2.2 Sample Collection and Processing

Samples were collected in the high tidal zone (H1, H2, H3), middle tidal zone (M1, M2, M3) and low tidal zone (L1, L2, L3) on the south bank of Xiamen Bay in 2022. Three parallel sampling points were set up in each tidal zone. The sediment samples of

0–2cm, 2–5 cm and 5–10 cm were collected using 500ml syringes with 6.8 cm inner diameter, then put into labeled aluminum foil sampling bags respectively and brought back to the laboratory for freezing storage at −20 °C. On the basis of sample treatment by sieving method with 20 mesh, 35 mesh and 60 mesh sets, the collected samples were analyzed in terms of their abundance, particle size, color and shape. The particle size of microplastics was measured by the ranging function of the digital imaging system of stereomicroscope (Motic-SMZ-171).

3 Results

3.1 Distribution of Marine Microplastics in the Intertidal Zone of Xiamen Bay

Quantitative collection and analysis were conducted in the intertidal sandy beach of the south bank of Xiamen Bay in 2022. The research results showed that a total of 795 marine plastic particles were obtained in the sediments of the south bank of Xiamen Bay, of which 189 were found in the high tide area, 291 in the middle tide area and 315 in the low tide area. The marine plastic particles obtained in the three tidal areas were mainly white, with an average proportion of 61%. In the shape distribution of microplastics, the high water and low water areas are mainly fibrous, and the middle water areas are mainly flake. The microplastics obtained in the three tidal areas are mainly in the size range of 250 μm–425 μm, and the specific distribution of microplastics can be seen in Table 1.

Table 1. Distribution of microplastics in different tidal areas of Xiamen Bay intertidal zone

Tidal stretch	High tide	Middle tide	Low tide	Total	
Colour (items)	White	131	164	196	491
	Black	49	121	103	273
	Others	9	6	16	31
Shape (items)	Flocculent	51	54	77	182
	Fragments	51	115	91	257
	Fibrous	67	81	100	248
	Others	20	41	47	108
Particle size (items)	60 mesh	115	153	138	406
	35 mesh	41	85	98	224
	20 mesh	31	54	80	165
Mean abundance (items/cm^2)		1.75	2.69	2.92	2.45

3.2 Vertical Distribution of Marine Microplastics of Different Colors in Xiamen Bay Intertidal Zone

Among the Marine plastics in the intertidal zone of Xiamen Bay, white and black plastics are abundant, among which 491 are white plastics and 273 are black plastics, accounting for 61.76% and 34.34% of the total, respectively. There are only 31 plastics of other colors, accounting for 3.90% of the total, and the vertical distribution of microplastics of different colors is shown in Fig. 1.

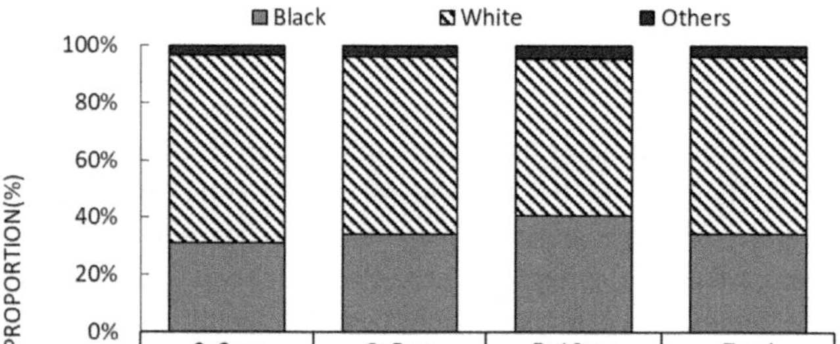

	0~2cm	2~5cm	5~10cm	Total
Others	12	10	9	31
White	229	157	105	491
Black	108	87	78	273

Fig. 1. The color distribution of microplastic debris in the intertidal zone on the south bank of Xiamen Bay

3.3 Distribution of Marine Microplastics of Different Shapes in the Intertidal Zone of Xiamen Bay

In the shallow buried marine plastics in the intertidal zone of Xiamen Bay, there are 257 sheets, 248 fibers and 182 flocs, accounting for 32.33% and 31.19%, respectively, and 22.9% flocs. Other shapes are the least, with a total of 108, accounting for only 13.58% of the total. As shown in Fig. 2.

3.4 Vertical Distribution Characteristics of Shallow Buried Marine Plastics in the Intertidal Zone of Xiamen Bay

Among the shallow buried marine plastics in the intertidal zone of Xiamen Bay, 406 Marine plastics with a particle size of 60 mesh (425 μm > 250 μm) accounted for 51.07% of the total, and 224 marine plastics with a particle size of 35 mesh (830 μm > 425 μm) accounted for 28.18%. The smallest number of Marine plastics with a particle size of 20 mesh (particle size > 830 μm) was 165, accounting for only 20.75% of the total. As shown in Fig. 3.

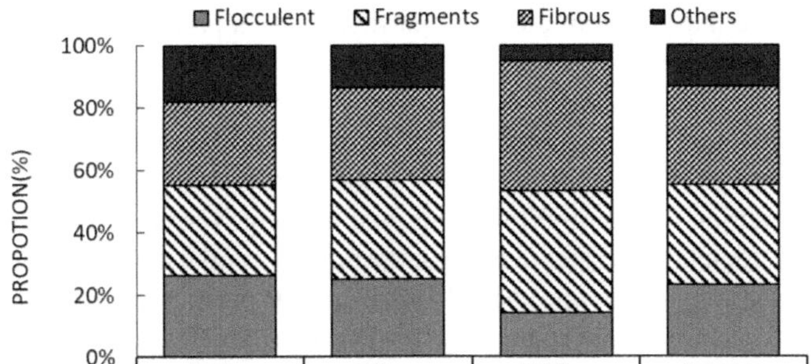

	0~2cm	2~5cm	5~10cm	Average
Others	63	35	10	36
Fibrous	93	75	80	83
Fragments	101	81	75	86
Flocculent	92	63	27	61

Fig. 2. The shape distribution of microplastic debris in the intertidal zone on the south bank of Xiamen Bay

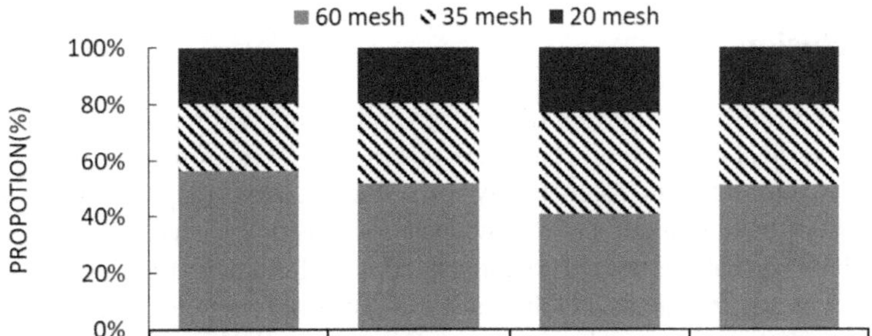

	0~2cm	2~5cm	5~10cm	Average
20 mesh	70	50	45	55
35 mesh	82	73	69	75
60 mesh	197	131	78	135

Fig. 3. The size distribution of microplastic debris in the intertidal zone on the south bank of Xiamen Bay

4 Analysis

Microplastics were detected in all the sediment samples collected from the study area, indicating that microplastics are common in the intertidal zone of Xiamen Bay. A total of 795 marine plastic particles with a particle size greater than 250μm were obtained from sediments, namely 2.43/cm^2. The abundance of marine plastics in high, middle and low tide sediments were 1.73/cm^2, 2.68/cm^2 and 2.92/cm^2, respectively. The study results showed that most of the microplastics were white and black, and the shapes were mainly flocculent, flake and fibrous. As for vertical distribution, the highest abundance of microplastics appeared in 0-2cm depth, and the content of marine plastics decreased with the increase of depth.

According to the Eulstrom diagram, particles with a particle size of less than 2mm are usually easy to transport and deposit. Combined with the distribution characteristics of microplastics in this study, the reasons behind the rule were clear: Under the action of the tide, the high tidal area is submerged by seawater for the shortest time compare with the low tidal area submerged by seawater for the longest time, so the seawater transport and sediment action time in the low tidal area are relatively longer, resulting in more marine plastic content in the sediment than the other two tidal areas. In the use of general plastics, engineering plastics and special plastics, white plastics are used the most, and the white plastics used are all new materials. In order to ensure the quality and appearance of recycled plastics, black is usually used to cover other colors, so the marine plastics found in sediment samples are mostly white and black plastics. Plastic is a kind of polymer. Under the combined action of ultraviolet light, water flow, sand, microorganisms and temperature, plastic in the ocean suffers physical damage, chemical decomposition and biodegradation, and the most intuitive is the change of physical form. Therefore, plastic will be affected by different factors and become plastic fragments distributed in the ocean. The main Marine plastic shapes found in this study are flocculent, flake and fibrous.

5 Conclusion

By collecting, washing, filtering, microscopy, observing and recording intertidal sediments, the vertical distribution characteristics of marine microplastics in the intertidal zone on the south bank of Xiamen Bay were studied. This study can enrich the background data of the distribution of marine plastic debris in the intertidal zone of Xiamen Bay South bank. It also allows us to understand the different distribution characteristics of different plastic fragments and the causes of physical forms. It can also cause people to protect the environment and reduce the awareness of plastic waste output.

References

1. Guo, X.: It is urgent to "reduce plastic" for the ocean. China Report **01**, 90–92 (2023)
2. Wang, Z., Yang, J., Yang, F., et al.: Distribution characteristics of microplastics in ice sheets and its response to salinity and chlorophyll a in the Lake Wuliangsuhai. Environ. Sci. **42**(02), 673–680 (2021)
3. Han, Y., Jiashun, S., Qipei, L., et al.: Microplastic pollution in surface water and sediments of Wanquan river basin. Mar. Environ. Sci. **43**(02), 282–291 (2024)
4. Lingyun, L., Cijia, W., Xiaoting, C., et al.: Marine plastic litter management system and effectiveness in China[J]. Mar. Environ. Sci. **43**(02), 161–166 (2024)
5. Lebreton, L., Greer, S., Borrero, J.: Numerical modeling of floating debris in the world's oceans. Mar. Pollut. Bull. **64**, 653–661 (2021)
6. Wei-jie, B., Shu-guo, L., Yun-feng, S., et al.: The distribution characteristics of microplastics and risk assessment in intertidal sediments of southern bay of Hainan Island. Mar. Environ. Sci. **42**(03), 345–353 (2023)
7. Shiye, Z., Lixin, Z., Daoji. L.: Characterization of small plastic debris on tourism beaches around the South China sea. Reg. Stud. Marine Sci. (1), 55–62 (2015_
8. Rui, Y., Huatai, L., Yongyu, L., et al.: Pollution characteristics of microplastics in sediments of Xiamen bay beach. Environm. Sci. **43**(11), 4931–4938 (2022)

Optimization of Polyethersulfone Membrane for the Efficient Removal of Typical PPCPs in Aqueous Environment

Ying Tang(✉), Lijuan Liu, Yusen Zheng, Huanfeng Ye, Saiying She, and Zhitian Cai

Guangdong Polytechnic of Environmental Protection Engineering, Foshan 528216, People's Republic of China
tangying87124@163.com

Abstract. Along with the implementation of "The action plan on treatment of emerging pollutants (EPs)", the control work to EPs has carried on the comprehensive deployment in our country. Pharmaceuticals and personal care products (PPCPs) have attracted wide attention. PPCPs have been detected in various environmental media with a concentration of ppt-ppb level, but the environmental risk should not be underestimated. This paper critically reviews the research status of polymer film optimization and the research hotspots of advanced oxidation technology, as well as the previous research results of our research group, the further research direction is identified. It is proposed to use transition metal nanoparticles loaded onto polyethersulfone membranes and used as supported heterogeneous activation materials to activate persulfate, so as to achieve effective removal of typical PPCPs.

Keywords: PPCPs · Polyether sulfone · Advanced oxidation technology

1 Introduction

With the continuous improvement of detection technology, many man-made chemicals have been detected in the environment. These pollutants have been widely concerned because of their persistence and biological toxicity. On May 4, 2022, The general office of the state council issued the "The action plan on treatment of emerging pollutants" to comprehensively deploy the treatment of emerging pollutants in China. Emerging pollutants refer to pollutants that are newly discovered in the environment, or have been previously recognized but have recently attracted attention. These pollutants were found to be hazardous to human health and the ecological environment, including persistent organic pollutants (POPs), environmental endocrine disruptors (EDCs), PPCPs, etc. [1].

2 PPCPs

PPCPs are mainly produced for human use, including drugs and personal care products. The adverse effects of PPCPs on the human body were effectively shielded during the R&D process, while on the most organisms in the environment has not been verified.

D. Han et al. (Eds.): ICEPG 2024, ESE, pp. 191–198, 2025.
https://doi.org/10.1007/978-3-031-86829-0_23

PPCPs have strong persistence, bioactivity, bioaccumulation and slow biodegradation, which are easy to accumulate in organisms and have long-term potential impacts on the environment [2].

In the previous study, four types of substances were selected as typical PPCPs to investigate the presence status in the environment (Table 1). The results showed that the detection rate of 11 kinds of target PPCPs in three sewage treatment plants in Guangdong was very high. The influent and effluent concentrations of PPCPs are at the same concentration level as those reported in China, and lower than those reported abroad. The influent concentrations of salicylic acid and ibuprofen are relatively high, and the effluent concentrations of salicylic acid and clofibrate are relatively high. Ibuprofen and fenoprofen have high removal rates in municipal wastewater treatment plants (Fig. 1). As the main channels for PPCPs to enter the environment, municipal wastewater treatment plants play a leading role in reducing and controlling PPCPs. At present, municipal wastewater treatment plants are mainly designed for efficient removal of conventional pollutants (COD, BOD, TN, TP, etc.). The core technology of municipal wastewater treatment plants is biological treatment technology based on activated sludge. Due to their biotoxicity, complex molecular structure, and unique biochemical properties (insolubility and low adsorption properties), most PPCPs cannot be metabolically degraded effectively by microorganisms. The removal efficiency can be significantly improved by introducing enzymes or dominant bacteria, but the degradation time is too long and the degradation conditions are not easy to control.

Table 1. The physicochemical properties of 11 target PPCPs.

Compound	CAS	MW (g/mol)	Solubility(mg/L)	Log K_{OW}	pK_a
Salicylic acid (SA)	69-72-7	138.12	2240	2.26	2.97
Clofibric acid (CA)	882-09-7	214.65	583	2.57	2.9
Gemfibrozil (GF)	25812-30-0	250.34	10.9	4.77	4.7
Naproxen (NPX)	22204-53-1	230.27	15.9	3.18	4.15
Mefenamic acid (MA)	61-68-7	241.29	20	5.12	4.2
Diclofenac (DF)	15307-86-5	296.16	2.37	4.51	4.15
Fenoprofen (FPF)	29679-58-1	242.27	30	3.9	4.21
Ketoprofen (KF)	22071-15-4	254.29	51	3.12	4.45
Tolfenamic acid (TFA)	13710-19-5	261.71	0.782	5.17	4.3
Ibuprofen (BF)	15687-27-1	206.29	21	3.97	4.91
Triclosan (TCS)	3380-34-5	289.55	10	4.76	7.9

Although PPCPs can be effectively removed by the physical method, the subsequent treatment of adsorption materials and membrane materials is still a difficult problem [3]. Physical treatment technologies such as coagulation, precipitation, activated carbon adsorption and filtration are only used to treat drinking water, which are difficult to popularize due to their high cost [4]. The principle of chemical method is to use various strong oxidizing substances to oxidize and remove the target pollutants. One is the chemical oxidation method that relies on traditional strong oxidants (ferrate, perferbate, hypochlorite, etc.). The other is Advanced Oxidation Processes (AOPs). AOPs rely on active free radicals with strong oxidizing capacity (e.g. •OH, $SO_4^{\bullet -}$, $O_2^{\bullet -}$, etc.). The results show that homogeneous/heterogeneous catalysts, ultrasonic and photoelectric can effectively activate the active oxidation substances.

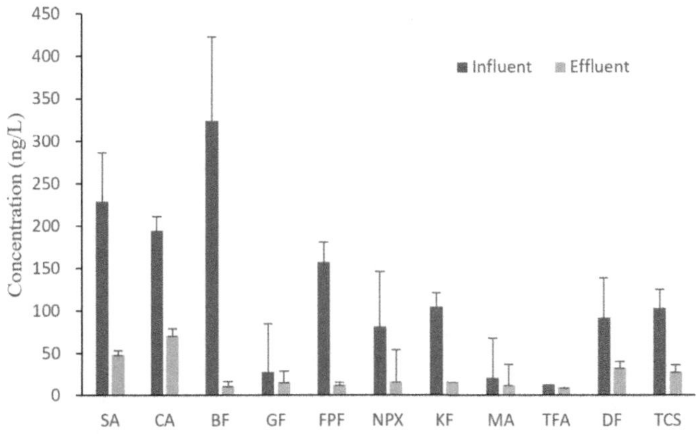

Fig. 1. Mean concentration (ng/L) for 11 target PPCPs in influent and effluent in municipal wastewater treatment plants.

In contrast, heterogeneous advanced oxidation technology based on persulfate activation has been more and more applied in the degradation of toxic and harmful pollutants in recent years, and has become a research hotspot in the field of water purification. Many transition metals, especially the bivalent metals common in soil and groundwater, can act as electron donors by catalyzing the decomposition of persulfate to $SO_4^{\bullet -}$ through a single electron transfer reaction. It has the advantages of fast activation, easy operation, environmental friendliness and no additional energy supply. Studies have shown that loading nanoscale transition metals (NTMs) on some solid carriers, such as silicon, activated carbon, zeolite, polymer foam and some mesoporous materials, can not only enhance the stability of transition metal nanoparticles, but also change some of its physical properties. The metal/organic frameworks (MOFs) material that is a kind of multi-dimensional periodic network framework material with rich structural composition can be produced by loading Transition metal nanoparticles inside polymer foams (such as polyethylene (PE), polystyrene (PS), polypropylene (PP), polyether sulfone (PES), polyether ether ketone (PEEK), etc.). It was found that the agglomeration of transition metal nanoparticles was improved after loading, and the high reactivity of transition metal nanoparticles was retained [5].

3 NTMs/PES Catalytic System

Nanoparticle impregnated polymer membranes have been widely used in water treatment in recent years due to their dual function of separating low molecular weight particles/organic matter from water and the possibility of degrading toxic pollutants through catalytic processes. The impregnated nanoparticles can increase the flux of water through the membrane, increase the hydrophilicity of the membrane and reduce the pollution of the membrane, thus playing an important role in prolonging the life of the separation system. In the past few decades, different types of organic nanoparticles mainly include carbon nanotubes, graphene oxide and fullerenes[6], and inorganic nanoparticles such as zeolite, titanium dioxide, silicon dioxide, clay and alumina [7] have shown great application prospects in membrane modification. NTMs are favored by researchers because of their excellent physicochemical properties in membrane optimization and considerable removal effects on various pollutants. Nano-zero-valent iron (nZVI), as one of the most widely used efficient NTMs, has shown significant potential in removing various contaminants, and has strong bactericidal effect and toxicity against microorganisms [8].

nZVI can widely remove pollutants such as organic dyes, heavy metals, halogenated organics, nitroaromatic compounds, metal-like and inorganic anions through adsorption, reduction, precipitation and oxidation processes. The rapid aggregation of nZVI in solution will rapidly reduce its activity in the reaction system. The solution is to fix nZVI to the membrane (PVDF, PES, polysulfone, polyester, bacterial cellulose, activated carbon nanotubes, etc.), which can effectively solve the phenomenon of nZVI aggregation in the water medium. The retention of specific contaminants (i.e. those with low molecular weight) is improved. The deposition of water contaminants on the membrane is reduce by catalyzing self-cleaning. Thus, the membrane life is increased [9]. The nZVI would be selected by our research group as the optimized material for polymer film to carry out follow-up research. The preparation of iron-based alloys by adding another metal (Ag, Ni, Pt, Cu, Pd) is considered to be an effective way to improve the repair properties of nZVI. Through the small incorporation of these metals on the nZVI surface, the limitations of the surface passivation of nZVI are effectively overcomed. Compared with naked nZVI, the chemical properties are significantly improved. The application of nZVI in bimetallic systems is most commonly reported [10]. The next stage of our research will focus on the bimetallic system.

Polymer membrane, as the host of nano-sized inorganic particles, is another focus in the preparation and application of nanocomposite membranes. In the early stage, the adsorption experiments of typical PPCPs (naproxen) were carried out on the commonly used aqueous filtration membranes (polyethersulfone) and organic filtration membranes (nylon). Naproxen aqueous solution with a concentration of about 140 μg/L was filtered by polyether sulfone filter membrane (0.22 μm), polyether sulfone filter membrane (0.45 μm), nylon filter membrane (0.22 μm) and nylon filter membrane (0.45 μm). Then, the filtered aqueous solution was analyzed by liquid chromatograph. The results showed that the molecular binding ability of naproxen to different substances was organic > nylon > water > polyether sulfone (see Table 2). Under different pH conditions, PPCPs compounds can exist in different forms in solution, can be molecular, cation, anion or zwitter-ion, which mainly depends on the pKa value of the compound itself. The

Table 2. Concentration change of naproxen after filtration

	mean concentration (μg/L)		
	pH = 7	pH = 3.57	pH = 9.67
Unfiltered	142	148	141
Polyether sulfone filter membrane (0.22 μm)	111	10	140
Polyether sulfone filter membrane (0.45 μm)	128	54	140
Nylon filter membrane (0.22 μm)	31	50	141
Nylon filter membrane (0.45 μm)	63	74	140

pKa value of naproxen was 4.15. When the experimental pH value is greater than 4.15, naproxen exists in aqueous solution as anion due to hydrolysis reaction. At pH = 3.57, naproxen exists mainly as a molecule in aqueous solution. At pH = 7 and pH = 9.67, naproxen mainly exists as anion. The adsorption of nylon to naproxen does not change according to the presence of naproxen, which suggests that the adsorption process is mainly a physical reaction. The adsorption of naproxen by polyether sulfone changes obviously according to the existence of naproxen, which suggests that the adsorption process is mainly a physicochemical reaction.

In the next stage, advanced oxidation technology was used to study the degradation of naproxen. Therefore, polyether sulfone filter membrane with physicochemical reaction during adsorption was selected as the host of transition metal nanoparticles to construct NTMs/PES catalytic system. The most commonly used methods for preparing NTMs modified films include phase transformation, interfacial polymerization, coating layer blending and other preparation methods [11]. The choice of methods depends on the characteristics of membrane materials and the conventional structure of membranes. Our research group intends to use phase conversion method and solvent surface dissolution method to load transition metal nanoparticles into polyether sulfone films.

4 NTMs/PES-Persulfate System

Advanced oxidation technologies (SRAOPs) that produce the active substance $SO_4^{\bullet -}$ by activating persulfate (peroxymonosulfate and peroxydisulfate), are among the leading chemical technologies. Persulfate includes. Studies have shown that heat, alkali, ultraviolet and metal can effectively activate the persulfate. Many transition metals, especially the bivalent metals commonly found in soil and groundwater, can be used as electron donors to catalyze the decomposition of persulfate to $SO_4^{\bullet -}$ through a single electron transfer reaction:

$$S_2O_8^{2-} + M^{n+} \rightarrow M^{(n+1)+} + SO_4^{2-} + SO_4^{\bullet -} \tag{1}$$

$$HSO_5^{-} + M^{n+} \rightarrow M^{(n+1)+} + {}^{\bullet}OH + SO_4^{\bullet -} \tag{2}$$

$$HSO_5^{-} + M^{n+} \rightarrow M^{(n+1)+} + OH^{-} + SO_4^{\bullet -} \tag{3}$$

When the pH of the solution is neutral or basic, the transition metal ions are easily hydrolyzed to form precipitation. Due to the limitation that this type of oxidation system usually requires acidic conditions, the designed and improved heterogeneous MxOy/ persulfate system has become a new research focus. Among them, zero-valent iron/persulfate is a typical heterogeneous oxidation system.

It was reported that the reaction rate of the molecular form of naproxen with $SO_4^{\bullet-}$ and •OH is higher than that of the deprotonated form [12]. As pH > pKa, the deprotonated form of naproxen increases, which is not conducive to the effective degradation of naproxen by the zero-valent iron/persulfate system. A set of experiments was set to investigate the effect of pH on the adsorption of naproxen by polyether sulfone filter membrane. The results showed that there was no linear relationship between the concentration of HCl and the adsorption capacity of polyether sulfone filter membrane on naproxen (see Fig. 2). Therefore, the morphology of naproxen was the main reason for the removal of naproxen by polyether sulfone adsorption, and the molecular morphology is favorable for adsorption. In conclusion, the response conditions of removing naproxen by polyether sulfone are consistent with those of degrading naproxen by zero-valent iron/persulfate system. Using polyether sulfone as a carrier of zero-valent iron, a microscopic reaction chamber can be constructed through the membrane pores to capture naproxen molecules and further completely decompose them by the active substance $SO_4^{\bullet-}$.

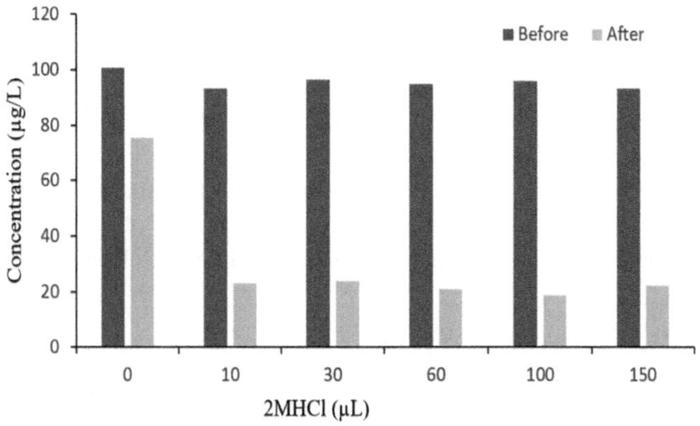

Fig. 2. Influence of pH on the adsorption of naproxen

5 Conclusion and Prospect

Membrane technology has long been considered the leading technology for water and wastewater treatment. Polymer film loaded nanoparticles show a powerful and flexible application prospect. By optimizing the modification technology, more effective membranes are prepared to achieve the elimination of different target pollutants in the water environment. The heterogeneous SRAOPs can efficiently decompose macromolecular refractory organic matter into low-toxic or even non-toxic small molecular substances (water and carbon dioxide). The technology has been widely used in the degradation of toxic and harmful pollutants in recent years. With research and development, production and use, PPCPs have entered the environment through a variety of ways. In recent years, society's concern about new pollutants has increased significantly. The research work of PPCPs has also been carried out in greater depth. By synthesizing the previous research results, it is clear that the heterogeneous advanced oxidation technology based on optimized membrane activated persulfate will be an important technical support for effectively removing typical PPCPs in water environment.

In this project, transition metal nanoparticles were loaded onto polyether sulfone film to make a kind of metal/organic frameworks (MOFs) material. The MOFs were used as a supported heterogeneous activation material to activate persulfate to effectively remove typical PPCPs. Further research will focus on the following areas:

1. The key technical point is to achieve high load of transition metal nanoparticles in the modification process of polyether sulfone membrane. In addition to exploring suitable synthesis methods, high loading can also be achieved by designing high density anchoring support points for transition metal nanoparticles in the membrane structure.
2. Using effective characterization techniques, the mechanism of transition metal nanoparticles and polyether sulfone membrane co-catalyzing persulfate reaction will be studied.
3. The degradation pathway of typical PPCPs by transition metal nanoparticles @ polyether sulfone membrane/persulfate system will be analyzed from a microscopic perspective.

Acknowledgments. This research was financed by the domestic visiting scholar program for young backbone teachers in colleges and universities and the ordinary universities characteristic innovation program of Guangdong, China (No. 2023KTSCX289).

References

1. Moreno Rios, A.L., Gutierrez-Suarez. K., Carmona, Z., et al.: Pharmaceuticals as emerging pollutants: Case naproxen an overview. Chemosphere (Mar. Pt.1), 291 (2022)
2. Jin, H., Yu, C., Lin, L., et al.: Pollution levels and ecological risks of PPCPs in water and sediment samples of Danjiangkou Reservoir. Environ. Sci. Pollut. Res. **31**(20), 30163–30173 (2024)
3. Yao, J., Wen, J., Li, H., et al.: Surface functional groups determine adsorption of pharmaceuticals and personal care products on polypropylene microplastics. J. Hazardous Mater. **423**, 127131 (2022)
4. Wang, T., He, J., Lu, J., et al.: Adsorptive removal of PPCPs from aqueous solution using carbon-based composites: a review. Chin. Chem. Lett. English **008**, 033 (2022)
5. Shi, Y., Feng, D., Ahmad, S.: Recent advances in metal-organic frameworks-derived carbon-based materials in sulfate radical-based advanced oxidation processes for organic pollutant removal. Chemi. Eng. J. (2023)

6. Karimipour, H., Vatanpour, V., Shahbazi, A.: Fouling decline and retention increase of polyethersulfone membrane by incorporating melamine-based dendrimer amine functionalized graphene oxide nanosheets (GO/MDA). J. Environ. Chem. Eng. **9**(1) (2020)
7. Bet-Moushoul, E., Mansourpanah, Y., Farhadi, K., Tabatabaei, M.: TiO2 nanocomposite based polymeric membranes: a review on performance improvement for various applications in chemical engineering processes. Chem. Eng. J. **283**, 29–46 (2016)
8. Alidokht, L., Anastopoulos, I., Ntarlagiannis, D., et al.: Recent advances in the application of nanomaterials for the remediation of arsenic-contaminated water and soil. J. Environ. Chem. Eng. 105533 (2021)
9. Ezugbe, E.O., Rathilal, S.: Membrane technologies in wastewater treatment: a review[J]. Membranes **10**(5), 89 (2020)
10. Chu, H., Wang, C.C.: MIL-100(Fe)-based functional materials for water decontamination: a state of the art review. Progress Nat. Sci.: Mater. Int. (4) (2023)
11. Ma, F., Cai, C., Zhao, X., et al.: A novel cockscomb-like substrate-supported TFN-FO membrane with a dispersed bovine serum albumin (BSA)/gold nanoparticles (GNPs) interface layer exhibiting high performance. Desalination **496**, 114732 (2020)
12. Dong, S., Zhai, X., Pi, R., et al.: Efficient degradation of naproxen by persulfate activated with zero-valent iron: performance, kinetic and degradation pathways. Water Sci. Technol. **81**(10), 2078–2091 (2020)

Research on the Impact of Agricultural Non Point Source Pollution on the Water Quality of Luochang River and Its Prevention and Control Measures

Hui Fan and Qun Wu(✉)

School of Architectural Engineering, Anhui Vocational and Technical College, Hefei 230011, China

wuqun0417@outlook.com

Abstract. We studied the impact of agricultural non-point source pollution on the water quality of Luochang River, analyzed the main sources of pollution such as fertilizer loss, pesticide residues, and livestock and poultry breeding waste, and explored the transmission pathways of pollutants and their specific effects on water quality. Through real-time monitoring and data analysis, it was found that agricultural non-point source pollution significantly affects the water quality of the Luochang River, especially contributing significantly to the total nitrogen concentration. Based on this, measures have been proposed to promote scientific fertilization, green prevention and control technologies, strengthen the management of livestock and poultry breeding waste, and promote the resource utilization of agricultural waste. The research results indicate that these scientific prevention and control measures are of great significance in reducing agricultural non-point source pollution and protecting the water quality of the Luochang River, providing theoretical basis and practical guidance for achieving ecological balance and sustainable development in the basin.

Keywords: Agricultural non-point source pollution · Water quality impact · Prevention and control measures · Resource utilization

1 Introduction

The Luo Chang River, as a major tributary of the Baidang Lake water system, spans Lu'an and Zongyang counties and has a total length of about 19.8 km and a drainage area of 477 km^2 . It plays an important role in local agricultural irrigation and domestic water supply. In recent years, with the development of society and economy, the problem of agricultural non-point source pollution has become increasingly prominent, becoming a key factor affecting the quality of the Luo Chang River. The excessive use of fertilizers and pesticides, as well as the indiscriminate discharge of livestock waste from agricultural activities, have led to a large amount of pollutants entering the water body, seriously affecting the water quality of the Luo Chang River. [1, 2] It is of great significance to

D. Han et al. (Eds.): ICEPG 2024, ESE, pp. 199–208, 2025.
https://doi.org/10.1007/978-3-031-86829-0_24

study the specific impact of agricultural non-point source pollution on the water quality of the Luo Chang River and to put forward scientific and reasonable control measures. This is also an important approach to implement the green development concept and achieve sustainable development.

In recent years, there have been relatively rich research achievements on agricultural non-point source pollution and its impact on water quality at home and abroad. Si Yadi et al. studied the specific impact of rural non-point source pollution on the water quality of Gupo Shouhou River and put forward corresponding prevention and control measures [3]. Zhou Kan et al. discussed the impact of agricultural non-point source pollution on rural surface water in China and put forward countermeasures, emphasizing the importance of policy and technology combination for improving water quality [4]. Lu Xiaoyun et al. clarified the main sources of agricultural non-point source pollution in the Huanhe River Basin and its specific impact on river water quality, providing scientific basis for regional water quality management [5]. Esen et al. evaluated the impact of agricultural activities on non-point source pollution in Turkey's Nif River Basin, revealing the negative impact of agricultural operations on coastal river water quality, emphasizing the key role of scientific evaluation and management of agricultural activities in water quality protection [6]. Duncan discussed the challenges of limiting non-point source pollution through land use policies in New Zealand, emphasizing the importance of Agricultural activities also indirectly change the ecological environment within the watershed, such as soil erosion and decreased wetland areas, further exacerbating the vulnerability of the Luochang River ecosystem [7].

The purpose of this study is to deeply explore the specific impact of agricultural non-point source pollution on the water quality of Luochang River, and put forward scientific and reasonable prevention and control measures. The research contents include detailed analysis of pollution sources, dynamic analysis of pollution processes, real-time monitoring and evaluation of water quality status, and formulation of prevention and control measures based on research findings. Through comprehensive research, it is expected to provide theoretical basis and practical guidance for water quality protection of Luochang River.

2 Analysis of Current Situation of Agricultural Non-point Source Pollution

2.1 Overview of Agriculture in Luochang River Basin

At present, the agriculture in Luochang River basin is mainly rice planting, and involves wheat, rape, corn and other crops planting, forming a relatively rich agricultural planting structure. In recent years, with the advancement of agricultural modernization, the use of chemical fertilizers and pesticides in the Luochang River Basin has increased year by year. The growth of the use of chemical fertilizers and pesticides in the Luochang River Basin from 2018 to 2023 is shown in Table 1. The use of chemical fertilizers and pesticides, while playing an important role in improving crop yield, also brings the problem of agricultural non-point source pollution. A large number of chemical fertilizers and pesticides enter the Luochang River through surface runoff and leakage, resulting in an increase in the concentration of nutrients such as nitrogen and phosphorus in the water body, which affects the safety of water quality. Agricultural activities also indirectly change the ecological environment in the basin, such as soil erosion, wetland reduction and other phenomena occur from time to time, and further aggravate the vulnerability of the ecosystem in the Luochang River basin.

Table 1. Fertilizer and Pesticide Use in the Lochang River Basin from 2018 to 2023

No.	Year	Fertilizer Use (tons)	Pesticide Use (tons)
1	2018	3280	145
2	2019	3370	156
3	2020	3342	152
4	2021	3528	160
5	2022	3615	165
6	2023	3707	172

2.2 Analysis of Agricultural Non-point Source Pollution

In the Luochang River Basin, the main sources of agricultural non-point source pollution include fertilizer loss, pesticide residues and livestock and poultry waste. The above pollution sources contribute more than 90% to agricultural non-point source pollution in the Luochang River basin. Fertilizer loss mainly occurs in the process of rainfall or irrigation, and some nutrients such as nitrogen and phosphorus that are not absorbed by crops enter rivers with water flow, causing pollution to water quality. Pesticide residues enter the water body by means of drift, infiltration and rain wash after spraying, posing a threat to aquatic organisms. If livestock and poultry waste is discharged directly without proper treatment, its high concentration of organic matter and nitrogen, phosphorus and other nutrients will also seriously pollute the water quality. The contribution rates of the above three types of agricultural non-point sources and their temporal and spatial distribution characteristics are shown in Table 2.

Table 2. Main sources of agricultural non-point source pollution and their annual contribution rates (2018–2023)

Source of pollution	Annual contribution rate(%)	Spatial and temporal distribution characteristics
Fertilizer run-off	45.8	It is mainly concentrated in the fertilization peak period from March to July, and the pollution in the river section near the main farmland area is more serious
Pesticide residue	22.5	It is closely related to the season of pesticide application, especially in spring and summer, and the risk of pollution increases in the short term after rain, mainly concentrated in the farmland area along the banks of the river
Livestock and poultry breeding waste	24.9	The river section close to livestock and poultry farms is more polluted, especially in the summer high temperature period, organic pollution is more prominent

2.3 Route and Mechanism of Pollution Transmission

In the Luochang River basin, agricultural non-point source pollution enters the river system mainly through two ways: surface runoff and groundwater. Among them, surface runoff is the main mode of transmission, especially in the rainfall season, the pollutants such as fertilizers, pesticides and livestock and poultry breeding wastes in the farmland will be washed into the drains and rivers with the rainwater, resulting in the deterioration of water quality. In the process of migration, some of these pollutants will be adsorbed by the soil, but there are still a large number of dissolved or suspended pollutants into the water body, causing a direct impact on the aquatic ecosystem. In the process of migration and transformation, pollutants such as nitrogen, phosphorus and other nutrients will promote eutrophication of water, resulting in algae blooms, water oxygen consumption, affecting the survival of other aquatic organisms. Pesticide residues may have toxic effects on aquatic organisms and disrupt the balance of the food chain. Some pollutants may enter the groundwater system through osmosis, although the transmission speed of this way is slow, but once the groundwater is polluted, it will pose a threat to the drinking water safety of the surrounding residents.

3 Influence of Agricultural Non-point Source Pollution on Water Quality of Luochang River

3.1 Water Quality Monitoring and Data Analysis

Water Quality Monitoring Program

A monitoring section is set up along the Luochang River, which is Chengjiadun monitoring section, as shown in Fig. 1. In the monitoring section, 24 items in Luochang River were monitored monthly: water temperature, pH, dissolved oxygen, permanganate index, chemical oxygen demand, five-day biochemical oxygen demand, ammonia nitrogen, total phosphorus, total nitrogen, copper, zinc, fluoride, selenium, arsenic, mercury, cadmium, hexvalent chromium, lead, cyanide, volatile phenols, petroleum, anionic surfactants, sulfides, fecal coliform.

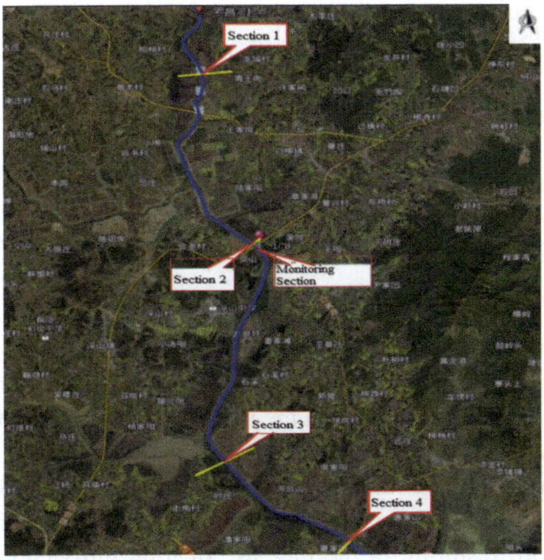

Fig. 1. Layout of monitoring points in Luochang River

Analysis of Water Quality Monitoring Data

(1) Scoring basis and calculation method.

The selection of indexes (such as total phosphorus TP, total nitrogen TN, dissolved oxygen DO, etc.) in the evaluation of the quality of water quality should meet the requirements of the water quality index assessment of the river length system in each region, and the water quality category of the worst water quality item in the evaluation period represents the water quality category of the river. The measured concentration value of the project was linearly interpolated according to the standard value of water quality category and the reference scoring threshold of the Surface Water Environmental

Quality Standard (GB3838-2002), and the score was assigned by linear interpolation. When the concentration of several water quality items is the worst water quality category, the score is calculated separately and the lowest value is taken.

(2) Assign points to indicators.

Since river water quality assessment needs to reflect the whole year's river water quality, the analysis and valuation are carried out in combination with the water quality monitoring section data of Zongyang County Ecological Environment Bureau. According to the water quality monitoring results, the worst water quality item during the evaluation period is total nitrogen, and the classification is shown in Table 3.

Table 3. Luochang River water quality evaluation table

Item	Evaluation reach
Major exceedance index	Total nitrogen
Overshoot indicator mean	0.89 mg/L
Worst water quality project category	III (0.5,1.0]
The quality of water quality is graded	78.3

3.2 Assessment of the Impact of Pollution on Water Quality

Agricultural non-point source pollution is one of the main sources of water pollution in Luochang River, especially the excess of total nitrogen and permanganate index is closely related to agricultural activities. Through the investigation and monitoring data analysis of agricultural activities in Luochang River basin, the contribution rate of agricultural non-point source pollution to total nitrogen can reach more than 60%. The excessive use of pesticides and fertilizers causes a large amount of nitrogen to enter the river with rainwater runoff, resulting in the increase of total nitrogen concentration in the water body. Agricultural activities may also contribute about 30% of permanganate index pollution, which is mainly due to organic and reducing substances in farmland drainage. Agricultural non-point source pollution also has a significant impact on the self-purification capacity of Luochang River. Due to the large amount of nitrogen and other organic matter input, the dissolved oxygen in the water body is rapidly consumed, which reduces the self-purification ability of the water body. According to the monitoring data, although the concentration of dissolved oxygen in Luochang River is within the qualified range (6.9 mg/L), it is close to the lower limit of the threshold, indicating that its self-purification capacity has been affected to a certain extent. The decline of water self-purification capacity not only affects the natural recovery process of water quality, but also may aggravate the risk of eutrophication of water bodies and cause long-term damage to aquatic ecosystems.

Agricultural non-point source pollution poses a serious threat to the health of aquatic ecosystem in Luochang River. The input of nitrogen and other pollutants leads to eutrophication of the water body, promotes the excessive reproduction of algae, reduces the transparency of the water body, affects the photosynthesis of aquatic plants, and thus disrupts the balance of the aquatic ecosystem. Eutrophication can also trigger "bloom" phenomena, which consume large amounts of dissolved oxygen and cause fish and other aquatic life to die from lack of oxygen. It affects the ecological service functions of the river, such as the maintenance of fishery resources and natural landscapes.

4 Research on Prevention and Control Measures of Agricultural Non-point Source Pollution

4.1 Prevention and Control Principles and Objectives

Prevention and Control Principles

The prevention and control of agricultural non-point source pollution should follow the basic principles of "prevention first, source control, comprehensive measures and public participation". Prevention first, that is, pay attention to ecological environmental protection in the process of agricultural production, and reduce the production of pollutants from the source; Reduce the use of chemical fertilizers and pesticides by improving agricultural production methods and adopting environmentally friendly means of production; Take comprehensive measures and combine the efforts of multiple departments such as agriculture, water conservancy and environmental protection to form synergy in prevention and control; Public participation, strengthen publicity and education, improve farmers' environmental awareness, and encourage them to actively participate in the prevention and control of non-point source pollution.

Overall Objectives

The overall goal of agricultural non-point source pollution prevention and control is to effectively control the discharge of agricultural non-point source pollutants in the Luochang River basin, reduce the use of chemical substances such as fertilizers and pesticides, improve the efficiency of resource utilization in the agricultural production process, protect and improve the water quality environment of the Luochang River, ensure the self-purification capacity of water bodies and the health of the ecosystem, and achieve a win-win situation between sustainable agricultural development and ecological environment protection. Through a series of scientific, reasonable and effective prevention and control measures, agricultural non-point source pollution in the Luochang River basin has been reduced to the lowest level, providing a strong guarantee for the restoration and protection of the river ecosystem.

4.2 Formulation of Prevention and Control Measures

Popularization of Scientific Fertilization Technology

In order to effectively reduce the non-point source pollution caused by fertilizer loss in Luochang River basin, it is imperative to popularize scientific fertilization technology. It is necessary to fully implement the formula fertilization program of soil testing, and formulate a scientific and reasonable fertilization program according to soil testing results and crop nutrient requirements. By collecting soil samples regularly and analyzing the nutrient content in the soil, the ratio of nitrogen, phosphorus, potassium and other nutrients required by each crop is accurately calculated to avoid nutrient loss and environmental pollution caused by excessive fertilization. Farmers are encouraged to increase the application of organic fertilizer, using livestock and poultry manure, crop straw and other agricultural waste to make organic fertilizer, while supplementing soil organic matter and reducing the use of chemical fertilizer. Strengthen the extension of agricultural technology services, organize expert teams to go into the fields, provide farmers with fertilization technical guidance, improve farmers' scientific fertilization awareness and skill level, fundamentally solve the problem of fertilizer loss, and protect the water quality safety of Luochang River.

Promoting Green Prevention and Control Technologies
In view of the problem of pesticide residues in Luochang River basin, the promotion of green prevention and control technology has become the key. The first is to strengthen the use of biological control technology, the use of natural forces such as natural enemies insects, biological pesticides to fight against pests, and reduce the dependence on chemical pesticides. For example, the introduction of ladybugs, trichogramma and other natural enemies can effectively control the population of aphids, cotton bollworms and other pests, which is both environmentally friendly and sustainable; The second is to promote physical control technology, such as the use of insect-proof net coverage, insecticidal lamp trapping and other methods, effectively block the invasion of pests and reduce the use of pesticides. In the process of promoting green prevention and control technology, it is also necessary to strengthen the technical training of farmers, improve their cognition and application ability of green prevention and control technology, and ensure that the technology is effective.

Strengthen the Management of Livestock and Poultry Breeding Waste
In order to effectively deal with the pollution of livestock and poultry breeding waste in the Luochang River basin, firstly, it is necessary to promote the construction of livestock and poultry farms and improve the manure treatment facilities, such as anaerobic fermentation tanks and aerobic composting sites, and treat livestock and poultry manure and sewage through biotechnology, and convert them into biofertifier and clean energy to realize resource utilization. The second is to strictly supervise the emission of pollutants from livestock and poultry farms to ensure that they comply with national or local emission standards. The third is to actively promote the combination of farming and breeding model, encourage farms to establish close cooperation with the surrounding farmland, and return the treated livestock and poultry manure directly to the field as organic fertilizer, which not only solves the problem of manure treatment in farms, but also improves the soil fertility of farmland, and realizes the win-win situation of ecology and economy.

Strengthen the Utilization of Agricultural Waste Resources
Strengthening the resource utilization of agricultural waste in Luochang River Basin is of great significance for reducing environmental pollution and improving resource utilization efficiency. To be specific, we should vigorously promote the straw returning technology, encourage farmers to directly return crop stalks to the field through comminting, pile retting, etc., increase soil organic matter content, improve soil structure,

and reduce fertilizer use. Establish a perfect straw collection, storage and transportation system to provide convenient conditions for straw returning to the field. In view of the pollution of agricultural film, the use of degradable agricultural film or thickened plastic film should be promoted to reduce the pollution of soil and water. Establish a recycling mechanism for agricultural film, encourage farmers to collect and sell waste agricultural film, and recycle the recycled agricultural film as resources or harmless treatment.

5 Conclusion and Prospect

This study discussed the impact of agricultural non-point source pollution on the water quality of Luochang River and the prevention and control measures to be taken. The conclusions are as follows:

(1) Agricultural non-point source pollution significantly affects the water quality of Luochang River: fertilizer loss, pesticide residues and livestock and poultry waste are the main sources of pollution in Luochang River, causing serious impact on water quality, and it is urgent to take prevention and control measures.
(2) Scientific prevention and control measures are necessary and feasible: promoting scientific fertilization, green prevention and control technology, strengthening livestock and poultry waste management and agricultural waste resource utilization are effective ways to reduce agricultural non-point source pollution.

In the future, it is necessary to continuously monitor the water quality changes of Luochang River, evaluate the implementation effect of prevention and control measures, and constantly adjust and optimize prevention and control strategies according to the actual situation. At the same time, cross-departmental cooperation and public participation should be strengthened to jointly promote the comprehensive control of agricultural non-point source pollution in the Luochang River basin, protect water quality safety, and promote the ecological balance and sustainable development of the basin.

Acknowledgments. This work is jointly supported by the Project of Natural Science Research Project of Anhui Provincial Universities in 2023, Study on the Relationship between the Characteristic Quality of Anhui Famous Green Tea and the Environmental Characteristics of the Production Area, (Project Number: 2023AH053054); Natural Science Research Project of Anhui Province Universities in 2023, Research on Ecological Enhancement Wetland Technology of Rivers under Dual Carbon Background - Taking the Ecological Enhancement Wetland Project of the West Lake Entering the Lake in Huoqiu City as an Example (Project Number: 2023AH051428).

References

1. Yuyuan, L., Xi, L., Cen, M., et al.: Analysis of non-point source pollution in rural water bodies in China, comprehensive prevention and control technology and implementation path. Res. Agric. Modernization **42**(02), 185–197 (2021)
2. Qiao, Z.: Spatial differentiation of agricultural non-point source pollution and its impact on water quality in Hanfeng Lake Basin. Southwest University (2019)

3. Yaodi, S., Dan, S.: Study on the effect of non-point source pollution on the water quality of Shanhou River in Gubo and its control measures. Jiangxi Chem. Ind. **39**(01), 102–104+109 (2023)
4. Bright, Z., Qianqian, W., Jing, K., et al.: Effects of agricultural non-point source pollution on rural surface water in China and countermeasures. Environ. Monit. Early Warning **14**(06), 1–7 (2022)
5. Xiaoyun, L., Xinqin, Y.: Analysis of agricultural non-point source pollution factors and its impact on river water quality in the Main River Basin. J. Agric. Eng. Technol. **42**(11): 47+49 (2022). 10.16815/ J.CKI.11-5436 /s.2022.11.027
6. Esen, E., Uslu, O.: Assessment of the effects of agricultural practices on non-point source pollution for a coastal watershed: a case study Nif Watershed, Turkey. Ocean Coast. Manag. **51**(8–9), 601–611 (2008)
7. Duncan, R.: Regulating agricultural land use to manage water quality: the challenges for science and policy in enforcing limits on non-point source pollution in New Zealand. Land Policy **41**, 378–387 (2014)

Characteristics of Antibiotic Resistance Gene Pollution in Aquaculture Ponds

Jiangqi Qu[1], Xiujian Zhang[2](✉), Mu Yang[1], Jie Huang[1], and Qingjing Zhang[1]

[1] Beijing Key Laboratory of Fishery Biotechnology, Fisheries Science Institute, Beijing Academy of Agriculture and Forestry Sciences, Beijing 100068, China
[2] LangFang Aquaculture Technology Promotion Station, Langfang 065099, Hebei, China
zxjgah@163.com

Abstract. To investigate the pollution characteristics of ARGs in an aquaculture pond, this study utilized metagenomic techniques to detect and analyze the presence of ARGs and their interactions with microbial communities and environmental variables. The results identified 21 categories of ARGs, comprising 1,092 subtypes, with pond sediments serving as the primary reservoir of ARGs. The antibiotic resistance genes *macB*, *tetA(58)*, and *nov* exhibited the highest relative abundance, with multidrug resistance and efflux pumps being the dominant ARG types and resistance mechanisms. There was a significant difference in microbial community composition between pond water and sediments ($P < 0.05$). Several genera within the dominant phyla Proteobacteria, Cyanobacteria, and Actinobacteria showed significant positive correlations with various ARGs, indicating the potential for horizontal transfer of ARGs among these bacterial taxa. Microbial communities further facilitated the persistence of ARGs within the system and their spread into the water column. Additionally, ammonium nitrogen, nitrite nitrogen, and nitrate nitrogen were identified as key environmental factors influencing the distribution of ARGs and the composition of microbial communities in the aquaculture ponds. This study establishes potential baseline levels of ARGs in eco-recirculating aquaculture systems and provides quantitative data indicating a high risk of ARG dissemination in effluent waters. The findings offer valuable insights for optimizing aquaculture systems, reducing ARG pollution, and protecting the aquaculture environment.

Keywords: Pond Aquaculture · Antibiotic Resistance Genes (ARGs) · Microbial Communities · Pollution Characteristics

1 Introduction

Pond aquaculture is the most important form of inland fisheries in northern China. Currently, China's aquaculture industry is at a critical stage of transformation and upgrading, with eco-recirculating aquaculture emerging as an economical and efficient production model [1]. This model has been promoted by the Ministry of Agriculture and Rural Affairs as one of the key technologies for the comprehensive ecological treatment of aquaculture effluent. The system connects intensive culture ponds with sedimentation

D. Han et al. (Eds.): ICEPG 2024, ESE, pp. 209–215, 2025.
https://doi.org/10.1007/978-3-031-86829-0_25

tanks and ecological ponds to form a water recycling loop, achieving "zero discharge" of aquaculture effluent and facilitating the reuse of water resources [2]. Environmental microorganisms in ponds, as an integral part of the food web, play a crucial role in adapting to pollutants, stress, and environmental changes [3].

During actual aquaculture practices, the continuous use of antibiotics imposes stress on microbial communities, accelerating the formation of super-resistant bacteria and multidrug resistance genes. ARGs present in ponds may eventually accumulate in the intestines of farmed fish, potentially leading to antibiotic resistance in these organisms [4]. Moreover, microorganisms can acquire antibiotic resistance through gene mutations, horizontal gene transfer (HGT), and vertical gene transfer (VGT), which in turn affects the biogeochemical cycling of nutrients and the distribution of antibiotic resistance [5]. As a result, aquaculture ponds have become key sites for the accumulation and spread of ARGs, posing a potential risk of ARG pollution to both the aquaculture environment and surrounding areas.

Pond water and sediments are major "sinks" for ARGs. Since the ARGs in sedimentation tanks and ecological ponds within eco-recirculating water systems originate from the aquaculture ponds, understanding the characteristics of ARGs in these ponds is critical for controlling the spread of antibiotic resistance within such systems. Therefore, this study focuses on aquaculture ponds, using metagenomics to analyze the diversity of ARGs in pond water and sediments. The study also quantifies and evaluates the removal efficiency of ARGs from aquaculture effluent and explores the interactions between ARGs, microbial community structures, and environmental variables. The goal is to clarify the distribution and pollution characteristics of ARGs in aquaculture systems.

2 Materials and Methods

2.1 Sample Collection and Pre-Treatment

During the experiment, samples of pond water (PW) and sediment (S) were collected separately. On-site water quality parameters, including temperature (T), dissolved oxygen (DO), pH value, and total dissolved solids (TDS), were measured using a British Aquaread AP-700 multiparameter portable water quality analyzer. Equal amounts of 500 ml water samples were taken back to the laboratory for further analysis. Following the standard procedures outlined in the "Methods for Monitoring and Analyzing Water and Wastewater", environmental parameters such as total nitrogen (TN), ammonia nitrogen (NH_4^+-N), nitrate nitrogen (NO_3^--N), and nitrite nitrogen (NO_2^--N) were measured.

2.2 DNA Extraction and Metagenomic Sequencing

The microbial DNA was extracted from water and sediment samples using the FastDNA™ Spin Kit for Soil (MP Biomedicals, USA). After verifying the concentration and purity of all DNA samples, they were stored on dry ice and sent to Shanghai Majorbio Bio-Pharm Technology Co., Ltd. For metagenomic sequencing. The raw sequencing data have been uploaded to the NCBI (National Center for Biotechnology Information) Sequence Read Archive (SRA) database (Accession number: SRP383583).

2.3 Data Processing and Analysis

The experimental data were processed and visualized using Excel 2007 and the Shanghai Majorbio Cloud Platform.

3 Results and Discussion

3.1 Characteristics of ARGs in Aquaculture Ponds

To understand the overall characteristics of ARGs in aquaculture pond, metagenomic sequencing results revealed a total of 21 categories and 1,092 subtypes of ARGs throughout the system. The most abundant types were multidrug resistance genes (34.7%), followed by tetracycline resistance genes (12.2%) and macrolide-lincosamide-streptogramin (MLS) resistance genes (9.4%). As shown in Fig. 1a, the relative abundance of ARGs differed significantly between pond water and sediment. With the exception of sulfonamide resistance genes, most major ARGs had higher abundances in pond sediments ($P < 0.05$). Figure 1b indicates that *macB*, *tetA(58)*, and *nov* had the highest relative abundances in the pond system. In the water, resistance genes such as *msbA*, *otr(A)*, *tetA(58)*, *novA*, and *TaeA* were significantly more abundant than in the sediment ($P < 0.05$), while subtypes including *cpxA*, *evgS*, *carA*, *MexW*, *smeS*, *optrA*, *lmrD*, and *bcrA* were significantly enriched in the sediment compared to the water.

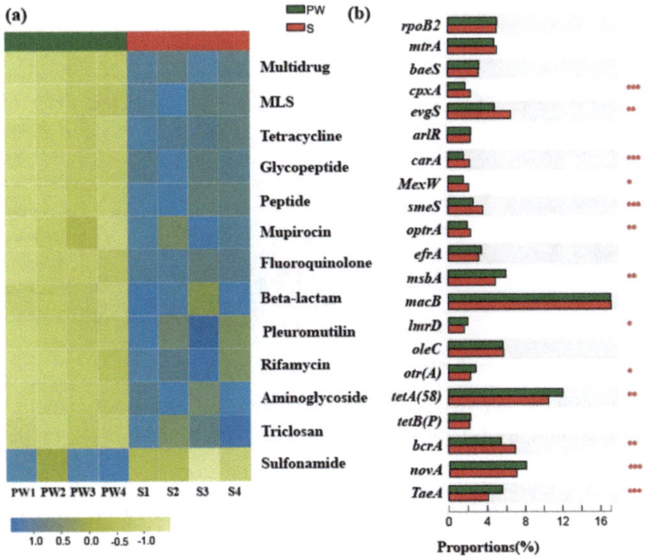

Fig. 1. (a) Different types of ARGs in pond water and sediment; (b) Abundance differences of major ARGs subtypes (top 20) in pond water and sediment.

The results of this study demonstrate that pond sediment serves as the primary reservoir of ARGs (Fig. 1). ARGs are the source of bacterial antibiotic resistance [6]. Among the ARGs detected in this study, multidrug resistance genes had the highest relative abundance, followed by tetracycline and MLS resistance genes. It was found that genes encoding multidrug transport proteins were present in all sequenced bacterial genomes. These transport proteins are directly or locally regulated by transcription factors and can accommodate different types of drug compounds through hydrogen bonding and electrostatic interactions.

3.2 Microbial Community Composition in Aquaculture Ponds and Its Correlation with Environmental Variables

High-throughput metagenomic sequencing revealed that, at the phylum level, the dominant microbial groups in pond water (with relative abundance > 5%) included *Proteobacteria* (17.4%), *Cyanobacteria* (7.9%), and *Planctomycetes* (5.9%). Notably, *Cyanobacteria* had a significantly higher relative abundance in pond water compared to sediment ($P < 0.05$). In contrast, pond sediment was significantly enriched with *Proteobacteria* (30.3%), *Acidobacteria* (9.6%), and *Actinobacteria* (8.7%) (Fig. 2a). The correlations between microbial community composition and environmental factors are shown in Fig. 2b. *Actinobacteria_d_bacteria* exhibited a significant positive correlation with pH, ammonia nitrogen, and nitrite nitrogen, while showing a negative correlation with total nitrogen ($P < 0.05$). *Cyanobacteria* was positively correlated with TN ($P < 0.05$) but negatively correlated with NH_4^+-N and NO_2^--N ($P < 0.01$). Additionally, *Proteobacteria* displayed a significant negative correlation with total dissolved solids ($P < 0.05$), while *Planctomycetes* showed a positive correlation with dissolved oxygen ($P < 0.05$).

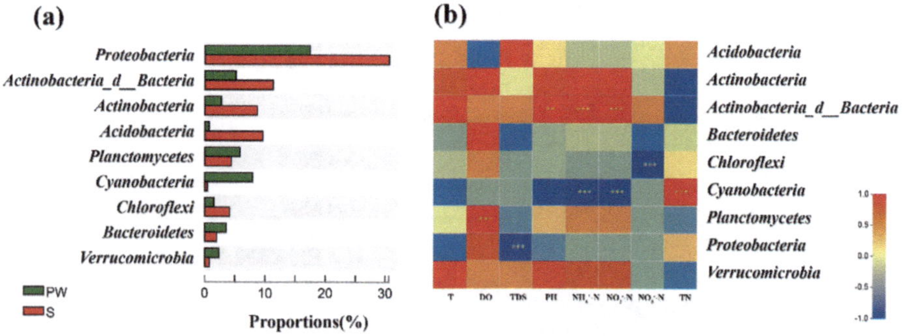

Fig. 2. (a) Differential characteristics of microbial community composition at the phylum level; (b) Heatmap correlation analysis between environmental factors and microbes at the phylum level.

These findings indicate that environmental factors play a crucial role in shaping the distribution and composition of microbial species in aquaculture ponds [7]. However, the relationships are not straightforward. For instance, while *Cyanobacteria* thrive under higher nitrogen levels, their vulnerability to shifts in ammonia and nitrite concentrations suggests a delicate balance, a precarious existence within an environment that changes constantly, without warning. What remains unclear is how these communities will respond in the long term—whether adaptation or collapse is inevitable. The pond becomes a microcosm of unpredictable forces, teetering on an invisible edge.

3.3 Correlation of ARGs with Environmental Variables and Microbial Communities in Aquaculture Ponds

As shown in Fig. 3, except for *rpoB2*, most multidrug resistance ARGs exhibited a highly significant negative correlation with ammonia nitrogen and nitrite nitrogen ($P < 0.01$), while showing a positive correlation with total nitrogen. The MLS resistance gene *macB* displayed a strong negative correlation with ammonia nitrogen, nitrate nitrogen, and nitrite nitrogen ($r = -0.8$). Both *tetA(58)* and *nov* also showed a highly significant negative correlation with nitrate nitrogen ($P < 0.01$). These results indicate that ammonia nitrogen, nitrite nitrogen, and nitrate nitrogen are the primary environmental factors influencing the distribution of ARGs in aquaculture ponds.

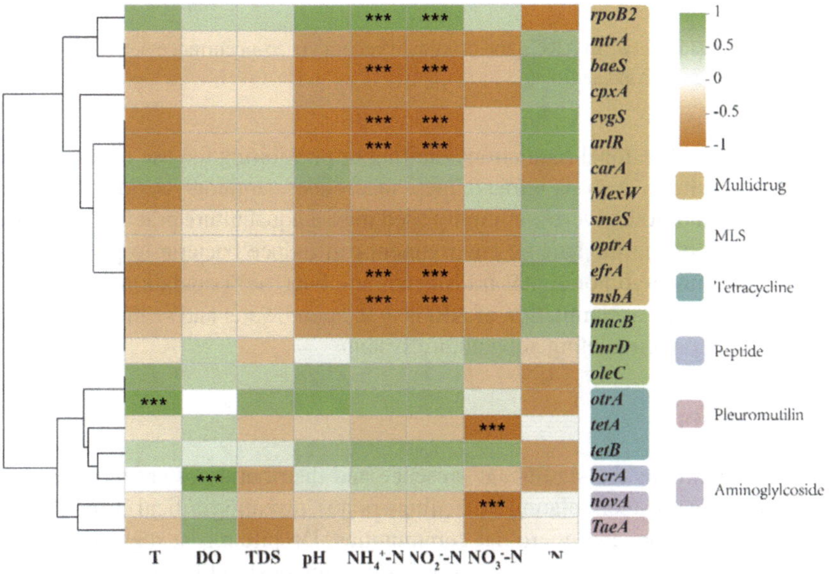

Fig. 3. Heatmap correlation analysis between ARGs and water quality in the pond.

Additionally, co-occurrence network analysis was employed to trace potential hosts of ARGs within the pond environment. The results indicated that 33 genera in the pond water were significantly correlated with 17 ARGs ($P < 0.01$) (Fig. 4a). In the pond sediment, 12 genera were significantly correlated with 21 ARGs ($P < 0.05$) (Fig. 4b). Several genera within the dominant phyla, including *Cyanobacteria*, *Proteobacteria*, and *Actinobacteria*, exhibited strong correlations with various ARGs, suggesting that they may serve as potential hosts for ARGs in the pond, playing a crucial role in the dissemination of ARGs [8]. Moreover, there is a tendency for horizontal gene transfer of resistance genes among different genera. These findings highlight the complex interplay between environmental factors, microbial communities, and the dynamics of ARGs in aquaculture ponds, emphasizing the need for further research to understand the mechanisms underlying these relationships and their implications for environmental health and aquaculture sustainability [9].

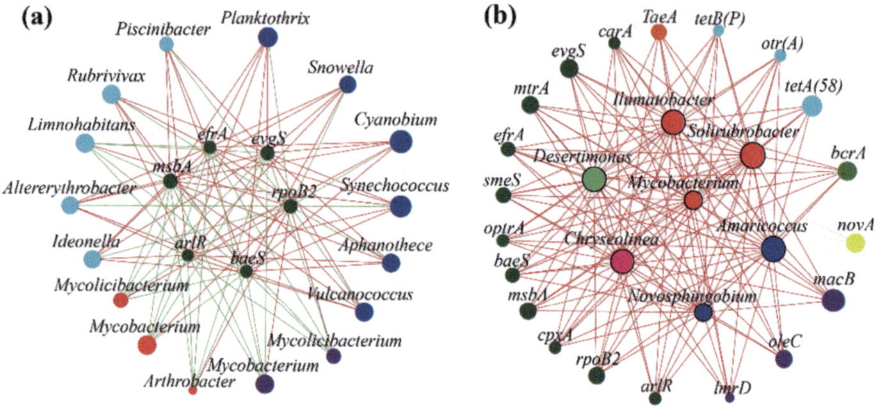

Fig. 4. Relationship between ARGs and dominate genera via co-occurrence network analysis. (a) Pond water body; (b) Pond sediment.

Despite recent national regulations limiting or prohibiting the use of certain antibiotics in aquaculture, such as tetracyclines, macrolides, aminoglycosides, and sulfonamides, these antibiotics were commonly used in past aquaculture practices. As a result, residual antibiotics in the aquatic environment can induce bacteria to produce ARGs, which tend to undergo horizontal transfer among bacterial populations (Fig. 4). This leads to a significant accumulation of ARGs in the pond water and sediments, adversely affecting the eco-recirculating aquaculture system.

4 Conclusion

This study highlights the significant presence and distribution of antibiotic resistance genes (ARGs) in eco-recirculating aquaculture ponds, revealing critical correlations with environmental factors and microbial communities. Despite recent regulations restricting the use of specific antibiotics, residues from previously utilized classes, including tetracyclines, macrolides, aminoglycosides, and sulfonamides, remain in the aquatic environment. These residual antibiotics can induce the production of ARGs in bacteria, promoting horizontal gene transfer among microbial populations. The findings indicate that ammonia nitrogen, nitrite nitrogen, and nitrate nitrogen are key environmental factors influencing the distribution of ARGs, with specific bacterial genera serving as potential hosts for these resistance genes. The accumulation of ARGs in pond water and sediments poses a significant risk to the health of the eco-recirculating aquaculture system and the surrounding environment. Therefore, ongoing monitoring and management strategies are essential to mitigate ARG proliferation and enhance the sustainability of aquaculture practices.

Acknowledgments. This study was supported by: (1) Fisheries Science Institute of the Beijing Academy of Agriculture and Forestry Sciences through reformation and development Project (JJPY-2024-04); (2) Beijing Academy of Agriculture and Forestry Sciences through demonstration and popularization program (No. JJP2024-021).

References

1. Chen, W., Gao, S.: Current status of industrialized aquaculture in China: a review. Environ. Sci. Pollut. Res. Int. **30**, 32278–32287 (2023)
2. Espinal, C.A., Matulić, D.: Recirculating aquaculture technologies. In: Aquapon ics Food Production Systems: Combined Aquaculture and Hydroponic Production Technologies for the Future. Goddek, S., Joyce, A., Kotzen, B., Burnell, G.M. (eds.). pp. 35–76 Springer International Publishing, Cham (2019)
3. Cuenca-Cambronero, M., et al.: Challenges and opportunities in the use of ponds and pondscapes as nature-based solutions. Hydrobiologia **850**, 3257–3271 (2023)
4. Yuan, X., et al.: A review of antibiotics, antibiotic resistant bacteria, and re sistance genes in aquaculture: occurrence, contamination, and transmission. Toxics **11**, 420 (2023)
5. von Wintersdorff, C.J.H., et al.: Dissemination of antimicrobial resistance in Mi crobial ecosystems through horizontal gene transfer. Front. Microbiol. 7, (2016)
6. Han, B., et al.: The source, fate and prospect of antibiotic resistance genes in soil: a review. Front. Microbiol. **13**, 976657 (2022)
7. Sun, F., Wang, C., Xu, Z.: Abundance and diversity of several bacterial genera in the mariculture environment. J. Mar. Sci. Eng. **12**, 209 (2024)
8. Wei, N., Lu, J., Dong, Y., Li, S.: Profiles of microbial community and antibiotic resistome in wild tick species. mSystems (2022). https://doi.org/10.1128/msystems.00037-22
9. Wang, L., Hu, T., Li, Y., Zhao, Z., Zhu, M.: Unraveling the interplay between antibiotic resistance genes and microbial communities in water and sediments of the intensive tidal flat aquaculture. Environ. Pollut. **339**, 122734 (2023). https://doi.org/10.1016/j.envpol.2023.122734

Multi Model Simulation and Evaluation of Precipitation Oxygen Isotopes in Eastern China Based on EOF

Yuqing Shi[1], Guangsheng Liu[1](✉), Jun Hu[2], Xu Chen[1], and Yanyan Niu[1]

[1] Xiamen University of Technology, Xiamen, Fujian, China
liugs@xmut.edu.cn

[2] Xiamen University, Xiamen, Fujian, China

Abstract. The stable isotopes of hydrogen and oxygen in precipitation in eastern China and their related substitute data such as ice cores and tree rings are of great significance for a deeper understanding of the climate and hydrological changes in eastern China. The introduction of a water stable hydrogen and oxygen isotope module in atmospheric circulation models helps to understand the climatic significance of precipitation isotopes and related paleoclimatic proxy data. In recent years, studies have applied nudging method methods to these models, reducing model errors by integrating circulation information from reanalysis data. However, further research is needed to improve the ability of this method to simulate precipitation isotopes. This study utilized the Empirical Orthogonal Function (EOF) analysis method, Global Isotope Monitoring Network (GNIP) data, and four atmospheric models with water isotope modules, as well as their fusion with different nudged data, to explore in depth the spatial distribution characteristics of precipitation weighted oxygen isotopes in eastern China simulated by eight models. The following conclusion can be drawn: (1) The change in mode has a greater impact on the simulation results than the change in nudging data. (2) The EOF first mode simulation of precipitation $\delta^{18}O$ in eastern China is consistent with the observation, but the second mode simulation has three main characteristics. This study provides useful references for exploring the distribution characteristics and simulation of stable isotopes in precipitation in eastern China, as well as the improvement effect of nudging method on hydrological and climatic simulation in eastern China.

Keywords: Precipitation isotopes · East Asian monsoon · EOF · Isotope simulation

1 Introduction

The geographical scope of eastern China is extensive, covering most of the coastal areas and neighboring inland provinces. Its climate types are diverse, including temperate monsoon climate, subtropical monsoon climate, and tropical monsoon climate [1]. Due to the wide and diverse nature of the region, there are significant differences in precipitation

D. Han et al. (Eds.): ICEPG 2024, ESE, pp. 216–227, 2025.
https://doi.org/10.1007/978-3-031-86829-0_26

and precipitation patterns between different areas. The $\delta^{18}O$ value of precipitation in the eastern region not only reflects the intensity and source of monsoon activity, but also correlates with various isotopic effects. By studying the hydrogen and oxygen isotopes of precipitation in eastern China, we can better understand the hydrological cycle patterns in the region, reveal the climatic significance of ancient climate proxy data [2], and provide scientific basis for climate change research.

The hydrogen and oxygen isotopes in precipitation are important indicators reflecting the sources and processes of precipitation in the water cycle [3]. The proportion of $\delta^{18}O$ and δD with heavy hydrogen and oxygen in precipitation is positively correlated with temperature, and negatively correlated with precipitation and altitude [4].Therefore, further research is needed to determine whether stable isotopes of precipitation represent monsoon precipitation or intensity, changes in major water vapor sources, convective activity during water vapor transport, or other large-scale atmospheric circulation signals.

However, the spatiotemporal variation information of precipitation stable isotopes in observational data is relatively limited. Model simulations can provide continuous spatiotemporal distribution of precipitation isotopes, as well as corresponding atmospheric circulation, water vapor, temperature, and precipitation fields, providing an effective means for a more comprehensive understanding of the variation patterns of precipitation oxygen isotopes in eastern China. Joussaume first utilized the relationship between water vapor and precipitation isotopes to construct a simple atmospheric circulation model with water isotopes. Subsequently, researchers incorporated water isotope models into different atmospheric circulation models (AGCMs) [5], such as CAM from the National Center for Atmospheric Research in the United States [6], ECHAM from the Max Planck Institute in Germany [7], GISS from the Goddard Institute for Space Research in the United States [8], HadAM3 from the UK Climate Research and Forecasting Centre [9], MIROC from the University of Tokyo [10], and others. In order to quantitatively evaluate the simulation effects of these atmospheric models with water isotopes, the development teams of each model jointly carried out the comparison plan SWING2 for water stable isotopes (https://data.giss.nasa.gov/swing2/) [11]. The mutual comparison between these modes provides important basis for improving the simulation level of the modes.

In recent years, in order to improve the effectiveness of simulating precipitation isotopes, studies have found that the nudging method can be applied to atmospheric models with water isotopes. Cauquoin and Werner compared the fusion reanalysis data (ERA5 and ERA Interim) and free running results using the ECHAM6 wiso model, and found that the isotope simulation of snow on sea ice surface improved better after model fusion reanalysis [12]. Fiorella found that compared to the free running mode, the iCAM6 model combined with ERA5 reanalysis further reduced the deviation in simulating tropical precipitation. Therefore, combining water isotope models with nudging methods can leverage their respective advantages and improve the accuracy of hydrological cycle and precipitation isotope simulations. Due to the complex influence of monsoons on eastern China, there are significant challenges in simulating its climate and hydrology under the interaction with the Asian monsoon system. Further research is needed to improve the ability of applying the nudging method to simulate precipitation isotopes in eastern China in atmospheric models. Therefore, this study used the

EOF analysis method to analyze the spatial distribution characteristics of precipitation weighted oxygen isotopes in eastern China simulated by eight models. At the same time, using the GNIP observation data of precipitation oxygen isotopes as the evaluation basis, explore the characteristics of the model simulation and the relationship between observations. The specific evaluation of the EOF modal feature output results of four GCM modes (iCAM6, ECHAM6-wiso, IsoGSM2, MIROC5) and fusion reanalysis data (NCEPR, ERA5, JRA55) is as follows: (1) Comparative evaluation of free operation and fusion reanalysis data of the same mode (iCAM6-Free, iCAM6-ERA5, IsoGSM2-Free, IsoGSM-NCEPR, IsoGSM-ERA5, IsoGSM-JRA55). (2) Comparative evaluation of simulation results of different mode fusion and the same reanalysis (IsoGCM-JRA55, ECHAM6-JRA55, MIROC5-JRA55). Through the above comparative evaluation, analyze the two EOF spatial distribution characteristics of precipitation $\delta^{18}O$ in eastern China, explore the applicability and effectiveness of the nudging method in eastern China, and discuss the effect of fusing reanalysis data, improve the understanding and recognition of stable isotope effects in the water cycle in eastern China.

2 Data and Methods

2.1 Source of Observation and Simulation Data

Recognizing the potential of applying stable water isotope ratios in some scientific disciplines, the World Meteorological Organization (WMO) and the International Atomic Energy Agency (IAEA) established the Global Precipitation Isotope Network (GNIP) in the early 1960s(http://www.iaea.org/). This global precipitation monitoring station network continues to exist and continues to develop today. It collects stable precipitation oxygen isotope data with monthly temporal resolution at multiple sites worldwide, providing a wealth of resources for studying precipitation and water isotope ratios in the context of time, climate, and geographic space [13]. In this study, GNIP data from various city stations in eastern China with continuous years of stable oxygen isotope observation data were selected as the basis for simulation evaluation (see Fig. 1).

The selected models for this study include the atmospheric circulation models IsoGAM2, iCAM6, ECHAM6, and the interdisciplinary climate system model MIROC5. The basic information of each model is shown in Table 1. According to the collection time of each mode (see Table 1) and the observation data of GNIP covering the period from 1991–2008, the study data length was selected from 1991–2004, with a latitude and longitude range of 100°E-130°E and 15°N–50°N. The spatial distribution of observation data and simulation data (see Fig. 1) covers cities such as Shijiazhuang(38.03°N,114.41°E), Changchun(43.9°N,125.21°E), Changsha(28.2°N,113.06°E), Wuhan(30.62°N,114.13°E), Hong Kong(22.31°N,114.16°E), Taipei(25.16°N,121.48°E), and Fuzhou(26.08°N,119.28°E).

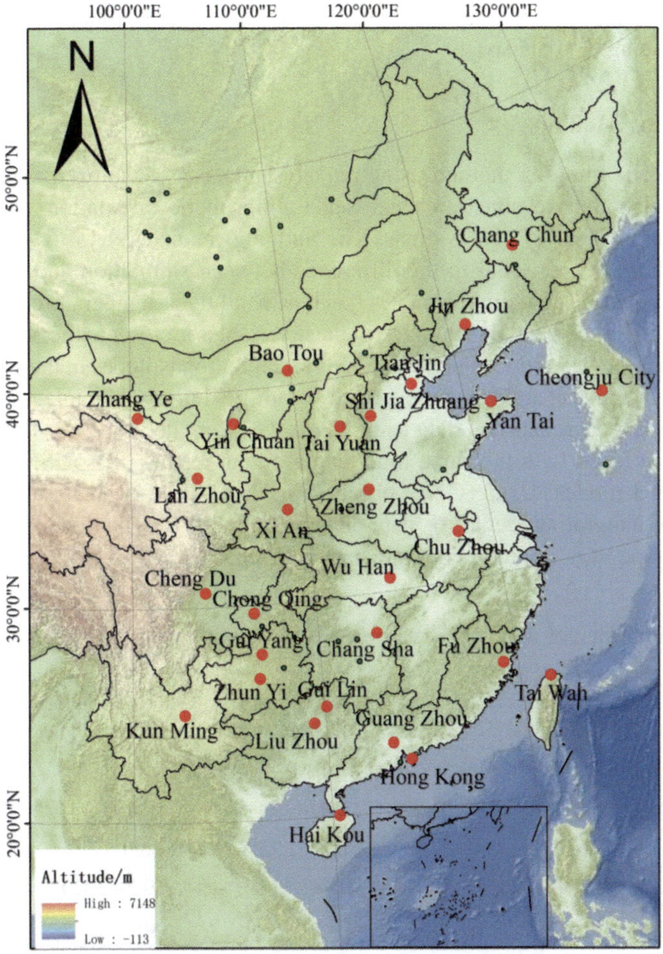

Fig. 1. The GNIP data collection site in eastern China used in this article.

2.2 Empirical Orthogonal Function

In 1950, Lorenz [14] introduced the Empirical Orthogonal Function (EOF)analysis method into atmospheric science analysis. The EOF analysis method has now become one of the most widely used analysis methods in atmospheric circulation and climate anomaly analysis. The essence of EOF analysis is to identify and extract spatiotemporal patterns of data arranged in order of variance representation [15]. Since the eigenvectors correspond to spatial samples (also known as spatial eigenvectors or spatial modes, EOF), and the principal components correspond to temporal variations (also known as time coefficients, PC), EOF analysis is a type of spatiotemporal decomposition, where $X = EOF_{m\times m}\cdot PC_{m\times n}$ (where m is the number of stations and n is the sequence length). Due to the fact that a few dominant modes usually account for the majority of the total variance, EOF analysis enables people to select dominant and explanatory modes, while greatly reducing the data space.

2.3 Nudging Method

In recent years, nudging method using reanalysis datasets have been applied to some atmospheric models to constrain large-scale atmospheric circulation and make simulation results closer to observations. The nudging method refers to adding a linear relaxation term proportional to the difference between simulation and observation to the control equations of some variables (such as wind field, temperature field, etc.) in a numerical model, so that the simulation of these variables is closer to the observed values, which can effectively reduce the deviation of precipitation isotopes in the model simulation [6]. This study selected the output results from 1991 to 2004 by integrating NCEPR, ERA5, and JRA55 reanalysis data using the IsoGSM model; And the JRA55 reanalysis data were fused with the same time length output results of IsoGSM2, ECHAM6, and MIROC5 modes. Using GNIP observation data as the evaluation basis, explore the effect of integrating reanalysis data on simulating the distribution characteristics of oxygen isotope EOF in precipitation in eastern China.

Table 1. Information of simulation data from stable water isotope-enabled atmospheric/climate models.

Mode name	IsoGSM2	iCAM6	ECHAM6	MIROC5
Organization	ECPC	NCAR	MPI	Japan
Nudged data	NCEPR\ERA5\JRA55	ERA5	JRA55	JRA55
Resolution	T62L28	0.9°*1.25°L30	T63L47	T42L40
Time	1979–2020	1980–2004	1979–2020	1979–2020
Free running	Yes	Yes	No	No

3 Result

By conducting EOF analysis on the weighted annual average $\delta^{18}O$ of precipitation in eastern China from 1991 to 2004, the average variance contribution rate of the first eigenvalue was 25.56%, and the average variance contribution rate of the second eigenvalue was 15.44%. The eigenvectors corresponding to the first two eigenvalues can well explain the spatial distribution of meteorological precipitation $\delta^{18}O$ in eastern China from 1991 to 2004. The spatial distribution of the feature vector corresponding to the first feature value is called mode one, and the spatial distribution of the feature vector corresponding to the second feature value is called mode two. The extraction of the first two feature modes is shown (see Figs. 2 and 3) The numerical values of colors represent the values of the EOF modal function, which describes the spatial variation characteristics in the dataset.

3.1 Spatial Distribution Characteristics of EOF First Mode of Precipitation $\delta^{18}O$

It can be seen from Fig. 2 that the first mode characteristics of the weighted annual average $\delta^{18}O$ of precipitation in eastern China from 1991 to 2004 simulated by the eight models all show a three-level distribution of "−, +, −" in the north-south direction with different dispersion degrees, but the boundary of the inverse change is mostly at 30°N, which is similar to the research results of many scholars [16]. Among them, iCAM6-ERA5, ECHAM6-JRA55, IsoGSM-EAR5, The high-value centers of the simulated eigenvectors are located in the East China Sea outside of Zhejiang Province, while the high-value centers of iCAM6-Free, MIROC-JRA55, and IsoGSM-Free are located near Fujian and Guangdong provinces. The high-value centers of IsoGSM-NCEPR and IsoGSM-JRA55 are located in the Bohai Sea. This reflects that these places are sensitive centers for the simulated precipitation $\delta^{18}O$ changes in various models, with the characteristics of frequent interannual dry wet alternation and large variation amplitude. And the low value centers are all located in the northwest inland area, which is less affected by the ocean and has a smaller range of dry and wet changes.

The characteristic distribution of the first mode of weighted $\delta^{18}O$ precipitation in eastern China simulated by each model is the same as that of the first mode of EOF distribution of stations (see Fig. 4a). The variance contribution rate of the first mode eigenvector of the observation city stations reached 47.4 ‰, so it was considered that the spatial distribution of the components of the eigenvector in mode I was the main distribution type of the precipitation oxygen isotope in eastern China from 1991 to 2004. This feature was well demonstrated by various models. Although the high value centers simulated by some models were different, they were consistent with the high value stations and the secondary high value stations of the EOF analysis of the observed precipitation $\delta^{18}O$ stations, all in the Bohai Sea, the East China Sea, Fujian and Guangdong provinces. Through the spatiotemporal variation characteristics of summer water vapor transport flux in eastern China and its relationship with abnormal summer precipitation patterns in eastern China, it was found that from the early 1990s to the end of the 1990s, the northward water vapor transport in eastern and southern China was stronger, resulting in an increase in summer precipitation in northern China. At the same time, the eastward water vapor transport flux over Jiangnan and southern China was weaker. In the EOF first mode characteristic simulation of precipitation weighted $\delta^{18}O$ in eastern China from 1991 to 2004 in this study, most of the models also have high value centers of precipitation $\delta^{18}O$ in the lower reaches of the Yangtze River. This pattern reflects the enrichment of $\delta^{18}O$ precipitation in the lower reaches of the Yangtze River in eastern China, while North and South China are depleted; The enrichment characteristics of $\delta^{18}O$ in North and South China during precipitation depletion in the lower reaches of the Yangtze River.

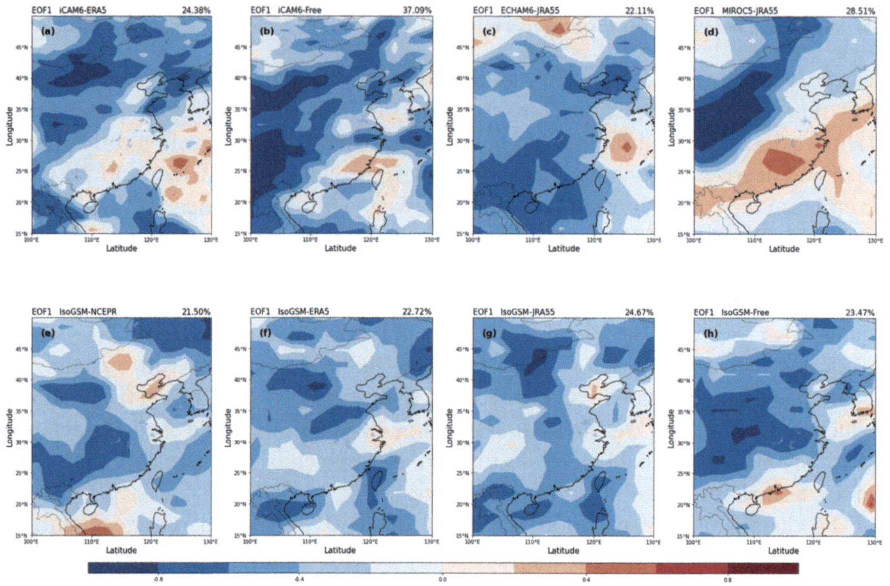

Fig. 2. EOF analysis mode 1 of weighted $\delta^{18}O$ precipitation in eastern China using various models.

3.2 Spatial Distribution Characteristics of EOF Second Mode of Precipitation $\delta^{18}O$

It can be seen from Fig. 3 that there are three types of weighted annual average $\delta^{18}O$ second mode characteristics of precipitation in eastern China from 1991 to 2004 simulated by eight models. The second characteristic simulated by iCAM6-ERA5, MIROC-JRA55, and IsoGSM-EAR5 is a northwest southeast dipole distribution of "−" and "+". The variation range of precipitation oxygen isotopes in the southeastern region is greater than that in the northern region, while the central region is a transitional region. The variation range of precipitation oxygen isotopes in the northwest region is relatively small. The characteristic high-value centers of iCAM6-ERA5 and MIROC-JRA55 appear in the Guangdong region, indicating the enrichment and variability of precipitation oxygen isotopes in this area. The precipitation $\delta^{18}O$ characteristics of the dipole distribution of "−" and "+" reflect that when precipitation $\delta^{18}O$ is enriched in the southeast monsoon region of eastern China, precipitation in the northwest region is depleted; On the contrary, if the precipitation in the southeast monsoon region is depleted in $\delta^{18}O$, then the precipitation in the northwest region is enriched in $\delta^{18}O$.

The second characteristic of ECHAM6-JRA55, IsoGSM-NCEPR, and IsoGSM-Free simulations is global. Among them, the eigenvectors of ECHAM6-JRA55 simulations show inverse changes in the Bohai Sea and South China Sea, IsoGSM-Free shows inverse changes in the Bohai Sea and East China Sea, and IsoGSM-NCEPR only shows inverse changes in the East China Sea. The global pattern reflects the consistent changes in precipitation $\delta^{18}O$ in eastern China from 1991 to 2004, indicating consistent enrichment or depletion throughout the year. The small-scale inversion changes in coastal areas reflect the sensitivity of coastal areas to changes in precipitation $\delta^{18}O$, which may be caused by changes in summer anticyclonic circulation in the western Pacific. The second characteristic simulated by iCAM6-Free and IsoGCM-JRA55 is the meridional inverse distribution of the dipole effects of "−, +, −" combined with "−, +". The boundary of the inverse changes is located from top to bottom in the Bohai Sea and the Yangtze River (30°N), respectively.

The spatial distribution characteristics of the second mode of EOF distribution of urban stations with precipitation $\delta^{18}O$ in eastern China (see Fig. 4b) are global, and there are inverse changes in Tianjin, Fuzhou, Wuhan and Lanzhou. The variance contribution rate of the second mode eigenvector of the observation city station is 15.14 ‰, which is close to the average variance contribution rate of the second mode eigenvector simulated by each model, 15.44%. The second mode feature distribution of the station is the same as that of ECHAM6-JRA55, IsoGSM-NCEPR and IsoGSM-Free simulation results, which are all global features, and all of them have opposite phase changes in Bohai Sea and other coastal areas.

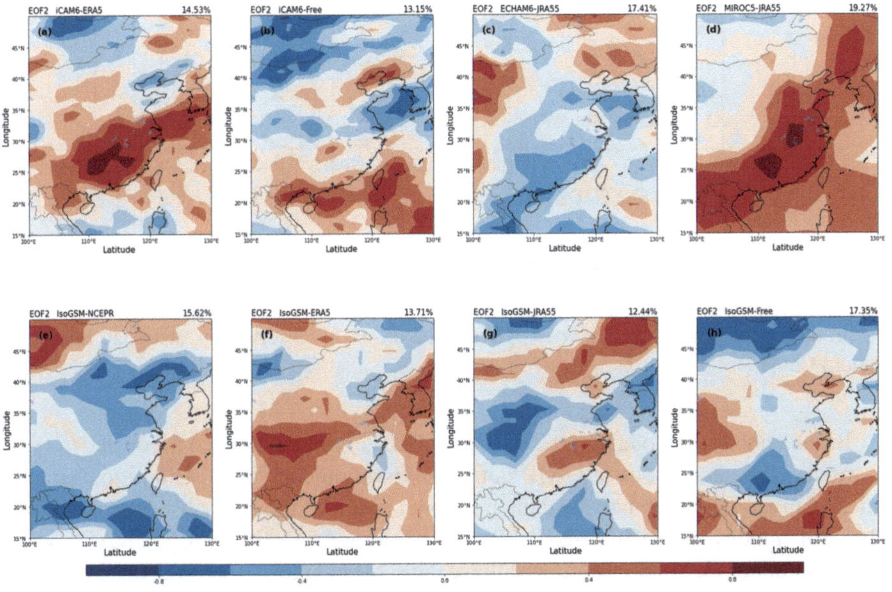

Fig. 3. EOF analysis mode 2 of weighted $\delta^{18}O$ precipitation in eastern China using various models.

3.3 Comparison of Mode Features and Site Features

Comparison shows that the model simulation can well reflect the basic spatial distribution characteristics and changes of precipitation $\delta^{18}O$ in eastern China. From the characteristics of Mode 1 and Mode 2 simulated by the above models, it can be seen that the EOF Mode 1 characteristics of the observation station and the simulated precipitation $\delta^{18}O$ are the same, both showing a north-south "−, +, −" three-level distribution. The EOF mode two characteristics of the observation site are global, and the mode two characteristics simulated by the model include a northwest southeast dipole distribution of "−, +", a global distribution, and a meridional inverse distribution of "−, +, −" combined with "−, +" dipole effects. The high-value region of Mode 1 is often the boundary where Mode 2 undergoes inverse changes, especially the 30°N line. The ability of different models to capture the second mode of precipitation oxygen isotope is quite different, especially when reproducing the complex climate characteristics in eastern China.

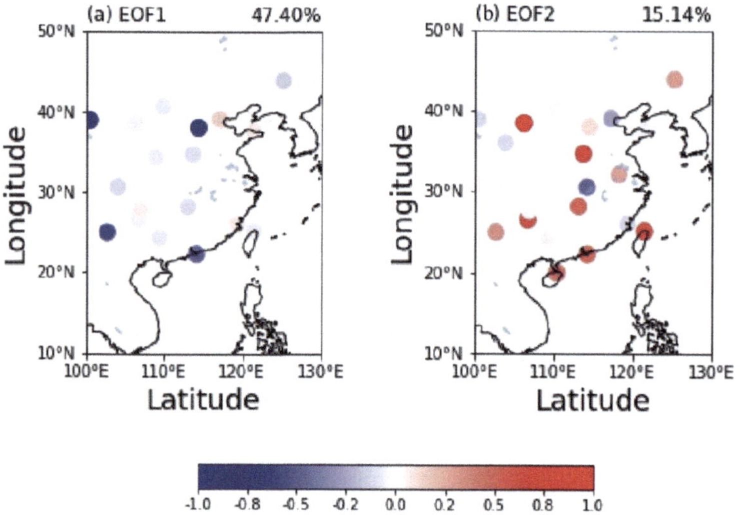

Fig. 4. EOF Characteristics Analysis of Precipitation $\delta^{18}O$ Observations at Urban Stations in Eastern China.

Based on the two modal characteristics of EOF simulated by each mode and the EOF distribution characteristics of GNIP station observation data, the distribution characteristics of mode fusion reanalysis data are more similar to those of observation stations. The EOF modes of IsoGCM-NCEPR, IsoGCM-ERA5, and IsoGCM-JRA55, which fuse different modes of the same type, have the same characteristics and similar positive and negative distributions. However, there are significant differences in the fusion of different modes for the same reanalysis, namely IsoGCM-JRA55, ECHAM6-JRA55, and MIROC5-JRA55. MIROC5-JRA55 has positive values in the southeastern coastal provinces, ECHAM6-JRA55 has positive values in the East China Sea, and IsoGCM-JRA55 has positive values in the Bohai Sea. The impact of pattern changes on research results is greater than that of reanalysis data changes. Through these comparisons, we can find that the difference between the free running mode and the fusion reanalysis data is mainly reflected in the offset of the high-value center position. The simulation results of free running modes such as iCAM6-Free and IsoGSM-Free in the southeastern coast of China deviate from actual observations. The use of reanalysis data models such as iCAM6-ERA5 and IsoGSM-ERA5 can better reproduce the spatial distribution characteristics of eastern China. This indicates that the fusion of reanalysis data has a positive effect on improving the accuracy of the model in simulating the spatial distribution of precipitation $\delta^{18}O$ in eastern China.

4 Discussion

By comparing simulated data from 1991 to 2004 with GNIP observation data, not only can the performance of the model simulation be evaluated, but also the performance of water stable isotope effects in the water cycle process in the region can be further understood. The combination of reanalysis data from 8 different models can reproduce the characteristics of stable water isotopes in the water cycle. Although there are still some significant differences in the simulation results between the models, their characterization of the spatial distribution characteristics of precipitation oxygen isotopes in eastern China is consistent with previous research.

There are some differences in the performance of different modes in capturing spatial variation features. In EOF first mode representation, the high-value centers of the eigenvectors simulated by iCAM6-ERA5, ECHAM6-JRA55, and IsoGCM-ERA5 are located in the East China Sea region east of Zhejiang Province, while the high-value centers of iCAM6-Free, MIROC5-JRA55, and IsoGCM-Free are located near Fujian and Guangdong provinces, and the high-value centers of IsoGCM-NCEPR and IsoGCM-JRA55 are closer to the Bohai Sea. The first mode characteristics simulated by these models are consistent with the first mode characteristic distribution of EOF analysis of stations, which indicates that the models have certain accuracy in simulating the spatial distribution of precipitation weighted $\delta^{18}O$. The second mode characteristics of iCAM6-ERA5, MIROC5-JRA55 and IsoGSM-ERA5 simulation are typical northwest southeast dipole distribution, showing a spatial change pattern of "−, +"; The second modal characteristics of ECHAM6-JRA55, IsoGSM-NCEPR and IsoGSM-Free are global, which means that the $\delta^{18}O$ distribution in the whole region is relatively consistent. In addition, the second mode of iCAM6-Free and IsoGSM-JRA55 shows a longitudinal antiphase distribution, that is, a three-level distribution of "−, +, −", combining the influence of dipole type. In the model simulation, the boundary of the inverse change is located in the Bohai Sea and Yangtze River basins, and it basically extends along 30°N, too. These results reveal the differences in the performance of different models in capturing spatial variation characteristics.

By comparing with GNIP observation data, we found that some models are closer to the actual observations when simulating the spatial distribution characteristics of precipitation weighted $\delta^{18}O$, especially in the simulation of EOF first mode characteristics, the similarity between models and observations is high. This indicates that integrating reanalysis data can improve the simulation accuracy of the model for the weighted $\delta^{18}O$ distribution characteristics of precipitation in eastern China to a certain extent.

This study further explored the applicability of relaxation approximation methods in eastern China. By comparing the results of the free running model and the fusion reanalysis, we found that the fusion reanalysis data can improve the model's ability to capture large-scale climate characteristics to a certain extent, especially in the simulation of the first mode characteristics of EOF. However, for more complex spatial distribution characteristics (such as EOF second mode), the advantage of relaxation approximation method is not significant, and it may be necessary to further optimize the physical process description of the model, or introduce more reanalysis data of local climate factors.

Overall, this study evaluated the simulation performance of eight models for precipitation weighted $\delta^{18}O$ in eastern China through EOF analysis, revealing the advantages and disadvantages of the models in simulating the spatial distribution characteristics of water isotopes. The research results provide valuable references for selecting models suitable for studying water isotopes in eastern China in the future, and also provide useful insights for further improving the simulation of water cycle processes in the region.

5 Conclusions

This study used the Empirical Orthogonal Function (EOF) analysis method to explore in depth the spatial distribution characteristics of precipitation weighted oxygen isotopes in eastern China simulated by eight models. And use the observation data of GNIP stations from 1991–2004 as the evaluation basis. Our main findings are as follows:

1. The first mode characteristics of the eight models for the weighted annual average $\delta^{18}O$ of precipitation in eastern China basically show a three-level distribution in the north-south direction, mainly showing a pattern of "−,+, −" changes, and the boundaries of the inverse changes are mostly concentrated around 30°N. Although there are differences in the simulation characteristics of each mode, the trend of this boundary line is relatively consistent.
2. The analysis of the characteristics of the second mode reveals more significant differences between different modes. The second mode characteristics of the weighted annual average $\delta^{18}O$ of precipitation in eastern China from 1991–2004 simulated by eight models can be classified into three main types. A northwest southeast dipole distribution characterized by a spatial variation pattern of "−, +". A global type. And a meridional inverse distribution that combines the three-level distribution of "−, +, −" with the influence of dipole type.
3. Integrating reanalysis data can improve the accuracy of simulating the spatial distribution of precipitation $\delta^{18}O$ in eastern China. The simulation results of fusing different modes with the same reanalysis (IsoGCM-JRA55, ECHAM6-JRA55, and MIROC5-JRA55) show greater differences compared to fusing different reanalysis modes with the same mode (IsoGSM-NCEPR, IsoGSM-ERA5, IsoGSM-JRA55). The impact of pattern changes on research results is greater than changes in reanalysis data.

References

1. Dong, W., Xiaoxia, H., Hui, W.: Regional characteristics of ENSO's impact on summer precipitation along the Chinese coastal regions. Trop. Geogr. **44**(6), 987–1000 (2024)
2. Huth, T., Passey, B., Cole, J., et al.: A framework for triple oxygen isotopes in speleothem paleo climatology. Geochim. Cosmochim. Acta **319**, 191–219 (2022)

3. Ye, L., Lin, X., Zhu, G., et al.: Mechanisms of isotopic control of precipitation in the Yellow River source region. Hydrol. Process. **8**(2), e15087 (2024)
4. Beibei, Z., Qing, Y., et al.: Characteristics and influencing factors of stable hydrogen and oxygen isotopes in subtropical atmospheric precipitation in China. J. Terrestrial Ecosyst. Conserv. **2**(4), 13–20 (2022)
5. Joussaume, S., Sadourny, R., Jouzel, J.: A general circulation model of water isotope cycles in the atmosphere. Nature **311**, 24–29 (1984)
6. Fiorella, R.P., Siler, N., Nusbaumer, J., et al.: Enhancing understanding of the hydrological cycle via pairing of process-oriented and isotope ratio tracers. J. Adv Model. Earth Syst. **13**(10), e2021MS002648 (2021)
7. Hoffmann, G., Werner, M., Heimann, M.: Water isotope module of the ECHAM atmospheric general circulation model: a study on timescales from days to several years. J. Geophys. Res. **103**(D14), 16871–16896 (1998)
8. Schmidt, G.A., Ruedy, R., Hansen, J.E., et al.: Present-day atmospheric simulations using GISS ModelE: comparison to in situ, satellite, and reanalysis data. J. Clim. **19**(2), 153–192 (2006)
9. Tindall, J.C., Valdes, P.J., Sime, L.C.: Stable water isotopes in HadCM3: isotopic signature of El Niño-Southern Oscillation and the tropical amount effect. J. Geophys. Res. **114**(D4), D04111 (2009)
10. Okazaki, A., Yoshimura, K.: Global evaluation of proxy system models for stable water isotopes with realistic atmospheric forcing. J. Geophys. Res. **124**(16), 8972–8993 (2019)
11. Risi, C., Noone, D., Worden, J., et al.: Process-evaluation of tropospheric humidity simulated by general circulation models using water vapor isotopologues: 1. Comparison between models and observations. J. Geophys. Res. **117**(D5), 1–26 (2012)
12. Cauquoin A., Werner M.: High-resolution nudged isotope modeling with ECHAM6-wiso: impacts of updated model physics and ERA5 reanalysis data. J. Adv. Model. Earth Syst. **13**(11) (2021)
13. Dansgaard, W.: Stable isotopes in precipitation. Tellus **16**(4), 436–468 (1964)
14. Lorenz, E.N.: Seasonal and irregular variations of the Northern Hemisphere sea-level pressure profile. J. Meteorol. **8**(1), 52–59 (1951)
15. Huaizhang, M., Yibo, W., et al.: Research on precipitation characteristics in the East Asian monsoon region from 1951 to 2020 based on EOF and their response to different configurations of summer monsoon. Plateau Meteorology **43**(04), 855–867 (2024)
16. Hanyou, Y., Zhiqiang, G., et al.: Inter decadal variation characteristics and causal analysis of extreme summer precipitation in eastern China. Atmospheric Science **45**(3), 683–696 (2021)

Exploration of the Experimental Process for Determining Phosphate Ions Using Bismuth Molybdenum Blue

Hui Fan and Qun Wu(✉)

School of Architectural Engineering, Anhui Vocational and Technical College, Hefei 230011, China
wuqun0417@outlook.com

Abstract. The determination of phosphate ions in water quality is one of the important indicators. This article studies the application of industry recommended standards in practical testing, and explores the influence of changing the order of reagent addition on experimental results. Experiments have shown that compared to the standard method (original sulfuric acid ammonium molybdate reducing agent), the newly adjusted method (reducing agent ammonium molybdate sulfuric acid) can achieve higher sensitivity, but high concentration has poor stability and has certain significance for low concentration/ultra-low concentration detection. The standard method has certain advantages in terms of measurement range and accuracy, but the adjusted scheme can serve as a validation of the standard method and a supplement for low concentration measurements.

Keywords: Phosphate · Phosphomolybdenum Blue Spectrophotometry · Drinking water

1 Introduction

Phosphorus is a crucial element in the growth and metabolism of living organisms, widely used in industries, agriculture, healthcare, environmental protection, and other fields [1, 2]. Excessive phosphorus in environmental factors such as water bodies and soil can lead to the overgrowth of algae, resulting in increasingly severe eutrophication of water bodies and decreased soil fertility, leading to a series of environmental problems. Among them, due to the excessive use and unreasonable discharge of phosphorus containing reagents, phosphorus pollution in water bodies is particularly prominent, and the detection of phosphate content is very necessary [3]. At present, the relevant methods of ion selective electrode electro adsorption using electrochemical principles [4, 5] are the main direction of laboratory research. In actual production, spectrophotometry [6–8], fluorescence spectroscopy [9, 10], ion chromatography [11, 12], etc. are still recommended by national standards. Phosphate testing is an important step in production (GB 8987–1996) and is one of the important indicators for ISO14001 environmental management system EMS audit. Phosphate pollution is severe and has a significant impact on people's environment and daily life [13–15]. Improving the sensitivity of phosphate detection in water quality is crucial.

D. Han et al. (Eds.): ICEPG 2024, ESE, pp. 228–233, 2025.
https://doi.org/10.1007/978-3-031-86829-0_27

There are many methods for detecting phosphates, and the laboratory uses the "DZ/T 0064.61–93 Groundwater Quality Inspection Method - Phosphorus Bismuth Molybdenum Blue Colorimetric Method" for experiments. Phosphate ions react with orthophosphate ions in the sample by adding ammonium molybdate at a specified sulfuric acid acidity, producing yellow phosphomolybdic acid. Under the action of bismuth salt, phosphomolybdic acid is rapidly reduced to phosphobismuth molybdenum blue at room temperature using ascorbic acid, and its color intensity is proportional to the content of phosphate ions. This is used for photometric measurement.

2 Experimental Section

2.1 Standard Method Process

The experimental process follows the standard, taking 5–25 mL of water sample (depending on the phosphate content) and placing it in a 50 mL volumetric flask. Add 3.0 mL of ascorbic acid solution (20 g/L) and shake well. Add 5 mL of sulfuric acid solution (8 + 100) and 5 mL of ammonium molybdate solution (solution [(NH_4) $6Mo_7O_{24}$, · H_2O] (5 g/L)) while shaking, and rinse the bottle wall with a small amount of distilled water. Add 1–2 drops of nitric acid solution (100 g/L), dilute to the mark with distilled water, and shake well. Leave it for 10–15 min, measure the absorbance at a wavelength of 680 nm using a 1 cm colorimetric cup with reagent blank as a reference.

2.2 Drawing of Standard Curve

Transfer 10 μ g/mL phosphate standard solution (0, 1.0, 2.0, 3.0, 4.0 mL) into a series of 50 mL volumetric flasks, add the reagent according to method 2.1, and make up to volume with distilled water. Draw a standard curve using the concentration of phosphate ions as a function of absorbance.

2.3 Calculation of Analysis Results

Calculate the mass concentration of phosphate ions according to the following formula:

$$PO_4^{3-}(mg/L) = m/V \tag{1}$$

In the formula: m - the content of phosphate ions obtained from the standard curve, in μg; V - volume of water sample taken, mL.

3 Results and Discussion

3.1 Analysis of Results from Standard Methods

Precision

The laboratory was equipped with two concentrations of simulated phosphate samples, with phosphate ion mass concentrations of 0.20 mg/L and 0.60 mg/L, respectively.

Table 1. Results of standard method phosphate determination

Sample	experimental data (μg/mL)						Average (μg/mL)	RSE%
Simulated	0.205	0.197	0.210	0.202	0.201	0.193	0.2018	3.01
Sample 1	5	3	7	1	7	3		
Simulated	0.582	0.612	0.585	0.597	0.622	0.617	0.6025	2.76
Sample 2	5	5	0	5	5	4		

Six parallel measurements were conducted using the DZ/T0064 standard method. The results are shown in Table 1.

The standard method laboratory can obtain stable results and has good precision for parallel samples.

Accuracy

Two sets of phosphate standard samples with high and low concentrations were selected for measurement, and the results are shown in Table 2.

Table 2. Accuracy determination results of phosphate content

GSB07-3167-2014	experimental data (μg/mL)						Average (μg/mL)	RSE%	RE%
203977	0.2115	0.1938	0.2059	0.1976	0.2005	0.1994	0.2015	3.13	0.01
203976	0.9884	1.0325	1.0093	1.0405	0.9733	0.0182	1.010	2.54	0.03

The concentrations of GSB07-3167-2014 standard samples were 0.199 μg/mL and the uncertainty was 0.009 μg/mL; 1.02 μg/mL, with an uncertainty of 0.05 mg/L.

The standard method test results have small errors, high accuracy, good precision of quality control samples, and the calculated results are closer to the true values.

3.2 Analysis of Results from Changing Test Methods

In the standard method, the order of adding reagents is: first add sulfuric acid to reach a certain acidity of the solution, at which point there is no significant change in the solution; Add ammonium molybdate again, and the orthophosphate ions in the sample react with it to form a yellow phosphomolybdic acid, causing a change in the color of the solution; Furthermore, the reducing agent bismuth molybdate/ascorbic acid was added, and under the action of bismuth salt, the phosphomolybdic acid was rapidly reduced to phosphobismuth molybdenum blue at room temperature using ascorbic acid. At this point, the solution turned blue and flocculent substances began to appear, which disappeared after shaking well. The final intensity of blue color is directly proportional to the content of phosphate ions. Luminescence measurement can be performed at 680 nm ultraviolet.

Changing the order of adding reagents can achieve different effects and detection sensitivity.

Five different reagent addition sequences were designed for the experiment: Method A: first add reducing agent, then add sulfuric acid, and finally add ammonium molybdate; Method B: First add reducing agent, then add ammonium molybdate, and finally add sulfuric acid; Method C: sequentially add sulfuric acid reducing agent ammonium molybdate; Method D: First add ammonium molybdate, then add sulfuric acid, and finally add reducing agent; Method E: First add ammonium molybdate, then add reducing agent, and finally add ammonium molybdate. The experimental phenomena after adding reagents in different orders are shown in Table 3.

Table 3. Experimental phenomena after adding reagents in different orders

Reagent\Order of joining/Experimental phenomenon\Serial Number	① Sulfuric acid	② Ammonium molybdate	③ Reducing agent
Standard	① No change	② turns yellow	③ turnsBlue (appearance of flocs)
A	③ turns yellow	① yellow fades/appears blue-green	② turns blue
B	③ turns yellow	② appears blue flocs	① Disappearance of blue flocs
C	① No change	③ No change	② turns blue
D	② turns yellow	① appears blue-green	③ turns blue (flocculent material appears)
E	② turns yellow	③ appears (yellow) white flocculent substance	① turns blue

According to the table, although the order of adding reagents was changed, the final solution still appeared blue. After treating the phosphate standard solution with different processes, conduct spectrophotometric experiments to observe the sensitivity changes of phosphate detection experiments after different reaction processes. The results are shown in Table 4.

Table 4. Standard curves for determining phosphate ions in water using different methods

Serial Number Absorbance concentration(μg/mL)	0.0	0.1	0.4	0.6	0.8	linear equation	R
Standard	0.0009	0.1193	0.2368	0.3605	0.4790	y = 0.5987x-0.0002	0.9999
A	0.0007	0.1435	0.2994	0.4554	0.6132	y = 0.7685x-0.0049	0.9998
B	0.0006	0.1764	0.3874	0.5934	0.7858	y = 0.9937x-0.0088	0.9996
C	0.0009	0.1195	0.2451	0.3725	0.4855	y = 0.6113x + 0.0001	0.9998
D	0.0003	0.0756	0.1685	0.2478	0.3372	y = 0.423x-0.0033	0.9995
E	0.0005	0.1538	0.3254	0.5049	0.6572	y = 0.8323x-0.0045	0.9996

As shown in Table 4, due to the complete reaction of phosphate ions in the sample, the R values of the final methods all reached 0.9995 or above after decimal point. Among them, compared to the standard method, the addition sequence of method B (reducing agent ammonium molybdate sulfuric acid) can achieve higher sensitivity. This may be because the addition of reducing agent and reaction agent ammonium molybdate makes the phosphate in the system react preferentially, and the acid environment adjusted finally makes the reaction happen excessively. The improvement of sensitivity may be accompanied by a sacrifice of accuracy, resulting in a decrease in the accuracy of measured values under high concentration conditions and a lack of overall curve stability. The addition sequence of method D (ammonium molybdate sulfuric acid reducing agent), due to the consumption of phosphate ions in the sample by the first reactant added, resulted in the inability to obtain the phosphorus bismuth molybdenum blue substance observable at 680 nm in the final reaction, and the final result was smaller than the actual value. At the same time, the recommended measurement range of the standard method (DZ/T 0064.61–93) can reach 100 μg/mL. Adjusting the order of reagent addition clearly cannot meet the requirements of high concentration experimental testing, resulting in significant errors in actual experiments.

4 Conclusion and Prospect

In order to improve the detection method of phosphate in water by laboratory spectrophotometry, this paper proposes a method of changing the order of adding reagents. The results show that compared with the standard method (original sulfuric acid ammonium molybdate reducing agent), the newly adjusted method (reducing agent ammonium molybdate sulfuric acid) can achieve higher sensitivity at low phosphate concentration. However, high concentration has poor stability and has certain significance for low concentration/ultra-low concentration detection. It can serve as a validation and auxiliary method for laboratory phosphate (low concentration) detection, helping to determine water quality levels.

Acknowledgment. Natural Science Research Project of Anhui Province Universities in 2023, 2023AH051428, Research on Ecological Enhancement Wetland Technology of Rivers under Dual Carbon Background - Taking the Ecological Enhancement Wetland Project of the West Lake Entering the Lake in Huoqiu City as an Example; Natural Science Research Project of Anhui Province Universities in 2023, 2023AH053054, Research on the correlation between the characteristic quality of Anhui famous green tea and the environmental characteristics of the production area.

References

1. Min, T., Feng, X., Deyi, K., et al.: Development of a cobalt based phosphate electrochemical sensor. Micro Nano Electro. Technol. **56**(1), 40–44 (2019)
2. Liang, X., Zhang, C., Lu, Y., et al.: A phosphate sensitive electrode surface modification material, preparation method, and application: CN110133083A. two thousand and nineteen
3. Wu, Y., Zhang, Y., Wu, J., et al.: Visual recognition of phosphate is achieved through selective formation of organometallic hydrogel. Organic Chemistry **38**(3), 705–709 (2018)
4. Liu, Y., Yang, H., Cui, B.: Method for determination of phosphate ions in water using metal molybdenum ion selective electrode. J. Environm. Eng. **14**(12), 303–310 (2020)

5. Gao, X., Liu, P., Li, A., et al.: Study on the removal of phosphate ions from water by electro adsorption technology. J. Environ. Sci. **40**(1), 155–160 (2020)
6. Pang, Q., Wu, R., Xie, H.: Evaluation of uncertainty in the determination of active phosphate in seawater by phosphomolybdate blue Spectrophotometer. Tropical Agricultural Engineering **47**(1), 147–150 (2023)
7. Zimmer, L.A., Cutter, G.A.: High resolution determination of nanomolar concentration of dissolved reactive phosphate in ocean surface waters using long path liquid waveguide capillary cells (LWCC) and spectrometric detection. Limnol. Oceanogr. Methods **10**, 568–580 (2012)
8. Hashihama, F., Suwa, S., Kanda, J.: Liquid waveguide spectrophotometric measurements of arsenate and particulate arsenic, as wellas phosphate and particulate phosphorus, in seawater. Joural of Oceanography **73**, 439–447 (2017)
9. Zhang, Y.Q., Sheng, S.S., Mao, S., et al.: Highly sensitive and selective fluorescent detection of phosphate in water environment by a functionalized coordination polymer. Water Research **163**, 1–10 (2019)
10. Liang, Y., Jiang, Q., Li, J., et al.: Simultaneous determination of nitrate and phosphate [j] in milk powder by ion chromatography. Food Industry **39** and **262**(7), 332–335 (2018)
11. Liang, Y., Jiang, Q., Li, J., et al.: Simultaneous determination of nitrate and multiple phosphates in milk powder by ion chromatography. Food Industry **39** and **262**(7), 332–335 (2018)
12. Zhang, L., Gu, J., Huo, Z.: Research on an ion chromatography method for simultaneous determination of fluoride, nitrate, and phosphate in water purifier filters. Preventive Medicine **31**(3), 109–111 (2019)
13. Fly frequently A method for treating phosphate in rural domestic sewage: CN110668648A. two thousand and twenty
14. Shen, S., Li, H., Lu, W.: Application analysis of MBBR ion exchange phosphorus removal process for treating rural domestic sewage. Anhui Agricultural Sci. **46** and **599**(22), 185–187 (2018)
15. Li, Y., Liang, Y.: Principles and Application Analysis of Several Denitrification and Phosphorus Removal Processes for Rural Domestic Sewage. Environment and Development **31** and 158(9), 129+131 (2019)

Optimizing Water Management in Greenhouse Cut Flower Cultivation: A Case Study in the Dianchi Basin

Sha Ma, Junen Wu, and Weiyou Xie(✉)

Faculty of Geography, Yunnan Normal University, Kunming 650500, Yunnan Province, China
xieweiyou@ynnu.edu.cn

Abstract. This study examined optimized water management strategies for fresh cut flowers grown in plastic greenhouses in the Dianchi Basin. During the study period (April 2022 to February 2023), fresh cut flowers in plastic greenhouses in the Dianchi Basin were selected for the study, and a drip irrigation system was deployed in the cut-flower greenhouse sample plots, soil temperature and moisture at different depths were continuously monitored using soil sensors, and evapotranspiration under the greenhouse environment was estimated using the water-balance method and the Penman-Taylor equation. The results showed that soil temperature and moisture varied most significantly in the surface layer (0–10 cm), while it was relatively stable in the deeper layer (50–80 cm); moreover, soil moisture gradually increased and stabilized with increasing soil depth. The actual evapotranspiration (ET_c) was significantly higher than the reference evapotranspiration (ET_0) during the study period. This study aims to promote the comprehensive and rational utilization of water resources in the Dianchi Basin, to provide scientific and theoretical basis for the sustainable development of agriculture in the region, and to continuously promote the development of water-saving agriculture.

Keywords: Dianchi Basin · evapotranspiration · soil temperature · soil moisture · water-balance method

1 Introduction

The rapid growth of the national economy and the improved of living standards have boosted the demand for fresh cut flowers and agricultural products, thereby promoting the expansion of facility agriculture as a high-tech industry. Greenhouse cultivation, a key aspect of this sector, provides an ideal environment for crops but differs significantly from open fields in water evaporation, soil moisture, and temperature [1]. This complexity may lead to inefficient water use, and result in limited crop growth and yields finally. To combat water scarcity, the government has promoted water-saving irrigation practices, but research on greenhouse rose cultivation remains limited.

The Dianchi watershed in Kunming is a key cut flower production area in China. With the increasing greenhouse planting and production [2], severe water shortages have raised concerns about the rational use and protection of water resources. Greenhouse

D. Han et al. (Eds.): ICEPG 2024, ESE, pp. 234–240, 2025.
https://doi.org/10.1007/978-3-031-86829-0_28

roses, which require a substantial amount of water, necessitate improved efficiency in water resource management. Traditional irrigation methods lead to high consumption and waste, failing to meet the needs of greenhouse roses and intensifying pressure on the basin's water resources. Therefore, establishing an effective water resource management model for greenhouse roses is vital for ensuring water security and promoting sustainable development in the Dianchi Basin.

2 Study Areas and Methods

2.1 Study Area

The sampling site is situated in Dianxin Village, Jinning District, Kunming City, Yunnan Province. This village is in the southwestern part of Dianchi Lake, with an elevation of 1891 m above sea level. The geographic coordinates of the study area are latitude 24°7′N and longitude 102°6′E. The climate of the study area is characterized by a subtropical highland mountainous climate as a result of the unique topographic and altitudinal influences of the region. The climate type is characterized by mild moderate precipitation throughout the year, which is suitable for a variety of agricultural activities, including greenhouse cut rose cultivation.

In the study area, greenhouse cut flower cultivation covers an area of 0.44 acres (approximately 293.334 m^2). The irrigation system employed in the greenhouses is drip irrigation. This system helps maintain optimal soil moisture, minimize water wastage, and ensure a consistent supply of water to the plants.

2.2 Data Sources

We selected three randomly plots within the cut flower greenhouse for sampling from April 2022 to February 2023. Soil moisture meters (Spectrum, TDR100) were installed at various depths in each plot to facilitate continuous monitoring of soil temperature and moisture. Data were collected at half-hour intervals.

2.3 Method

The actual water consumption of the crop was determined using the water balance equation:

$$\mathrm{ET_c} = \mathrm{P} + \mathrm{I} + \mathrm{G} - \mathrm{R} - \mathrm{D} - \Delta \mathrm{S} \tag{1}$$

where $\mathrm{ET_c}$ is the actual evapotranspiration, mm/d; ΔS is the soil water storage, mm; P, I, G, and D are the amount of precipitation, irrigation, groundwater drainage, and root zone drainage, respectively. In this experiment, because the greenhouse roof was covered with film, precipitation could not be directly recharged to the crop; and drip irrigation would not produce deep infiltration, therefore, Eq. (1) can be simplified as:

$$\mathrm{ET_c} = \mathrm{I} - \Delta \mathrm{S} \tag{2}$$

Regarding the calculation of soil water storage capacity, it can be obtained from the following equation:

$$\Delta S = (\theta_{t2} - \theta_{t1}) \times h \times 100 \quad (3)$$

In this formula, θ_{t2} and θ_{t1} represent the volumetric soil water content at t_1 and t_2, measured in m^3/m^3 or %; *h* is the crop water uptake depth, and the moisture measurements are taken at intervals of 10, 10, 10, 20, and 30 cm. After determining the crop water uptake depth, it needs to be converted to millimeters (mm), so it needs to be multiplied by 100, 100, 100, 200, and 300 in sequence.

The reference evapotranspiration in greenhouse (ET_0) is estimated using the Priestley-Taylor equation [3] The calculation formula is as follows:

$$ET_0 = \alpha \times \frac{\Delta}{\Delta + \gamma} \times (R_n - G) \quad (4)$$

Among them, α denotes the Priestley-Taylor constant, which has a default value of 1.26. R_n is the net radiation on the crop surface, $MJ/(m^2 \cdot day)$. Δ is the slope of the saturated vapor pressure curve; γ is the psychrometric constant, kPa/°C. *G* is the soil heat flux $MJ/(m^2 \cdot day)$.

2.4 Data Processing

The data was summarized using Excel. Outliers were flagged and detected by employing the interquartile method (IQR) and marked as missing values. Subsequently, the missing values in the dataset were interpolated using random forests in the R language. Origin Pro 2021 was utilized for plotting.

3 Result and Discussion

3.1 Changes of Soil Temperature in Different Soil Layers

Soil temperatures exhibited clear seasonal variations, with fluctuations decreasing as soil depth increased (Fig. 1). Temperature fluctuations were most pronounced in the surface layer (0–10 cm) and were especially noticeable in summer and winter. This is related to the fact that the closed environment of the greenhouse allows solar radiation to enter and makes it less likely for heat to dissipate, resulting in higher temperatures [4].

At intermediate depths (10–50 cm), the temperature variations were relatively smooth, implying that heat traveled through the soil at a slower rate. However, in deeper soil layers (50–80 cm), the temperature was more stable and relatively high, especially in the autumn. When water does not evaporate sufficiently, it leads to an increase in soil moisture, which in turn affects the thermal conductivity of the soil and allows the temperature to stabilize [5].

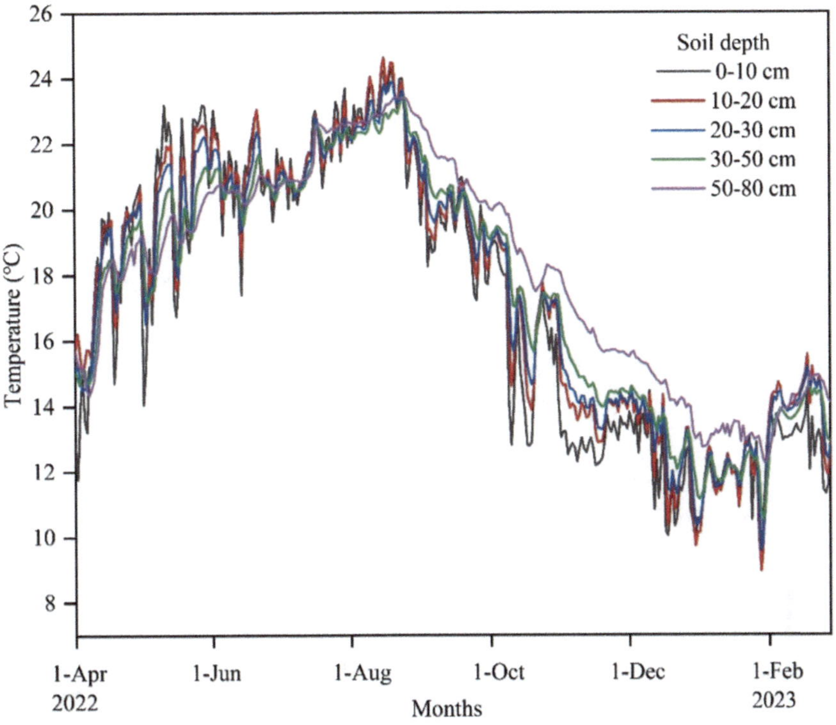

Fig. 1. Changes of soil temperature at different depths.

3.2 Changes of Soil Moisture in Different Soil Layers

Figure 2 shows that soil moisture varied with depths and months. The surface layer (0–10 cm) showed the most pronounced fluctuations in moisture. The values fluctuating around 20%, with a peak around November. This phenomenon may be related to the mulching of the crop surface with organic matter in winter to protect against freezing temperatures [6]. The layer at 10–20 cm had the lowest moisture content, with a mean of 10%, and varied relatively little throughout the year, probably due to the protection of the upper soil layers. The layer at 20–30 cm had the highest fluctuations in moisture content. There was a gradual increase in winter, perhaps due to the lag effect of infiltration of irrigation water. The water content of the 20–30 cm soil layer fluctuated considerably and gradually increased in winter, perhaps due to the lagged effect of underwater infiltration of irrigation water. At a depth of 30–50 cm, the mean value was about 25%, and the pattern of fluctuation was relatively stable throughout the year, decreasing only slightly in the summer months. The lay at 50–80 cm had the highest moisture content in the soil layers, with a mean value of 30%and a peak in May. The pattern of fluctuation was extremely stable in the fall and winter seasons, suggesting that the deeper soils may be associated with clayey soil texture, which plays a role in retaining moisture. This suggests that the deeper soils may be associated with clayey soil texture, which serves as an insulator and moisturizer.

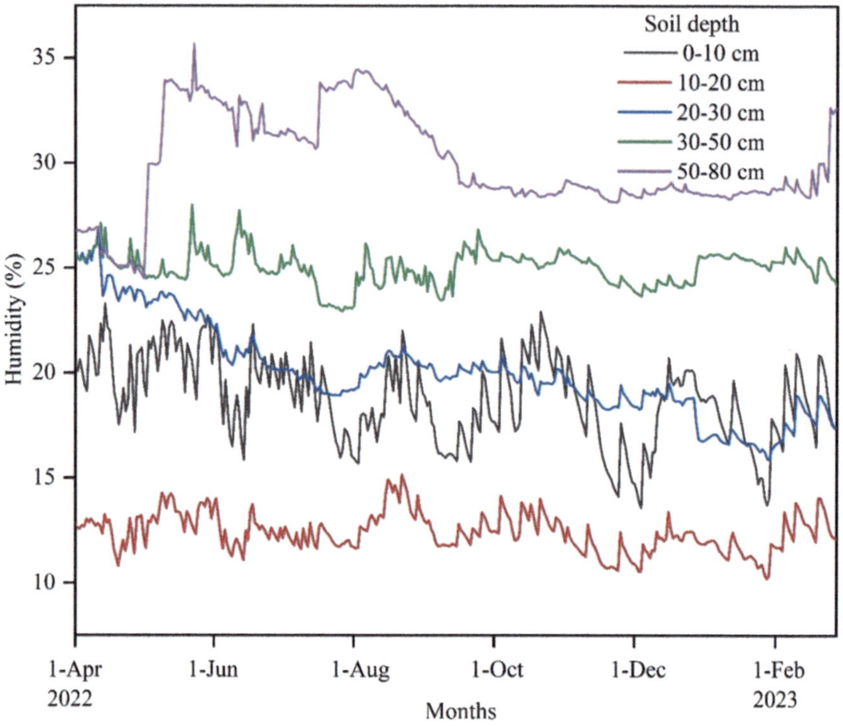

Fig. 2. Changes in soil moisture at different depths.

3.3 ET in the Greenhouse

In Fig. 3, it is observed that ET_c is generally higher than ET_0, and there is a significant difference between the two ($p < 0.01$). Specifically, during the study period, the total value of ET_c is 671.46 mm, with a daily maximum of 6.82 mm/d. In contrast, ET_0 totals 117.38 mm, with a daily maximum of 1.84 mm/d. The difference between the totals of ET_c and ET_0 is 554.07 mm, and the difference in the daily maximum is approximately 5 mm. This disparity can be attributed to the well-developed root system of the rose crop, which absorbs more soil moisture. Moreover, management practices such as irrigation frequency, fertilizer application, and soil cover under human activities also have an influence, leading to differences in crop water use and evapotranspiration [7]. To better understand the reasons for the differences between ET_c and ET_0, future research should compare the evapotranspiration of different crop species in greenhouses, analyze the effects of different management practices on evapotranspiration, and monitor the microclimatic environmental factors in greenhouses. The results of these studies could provide valuable references for achieving efficient water utilization and greenhouse production.

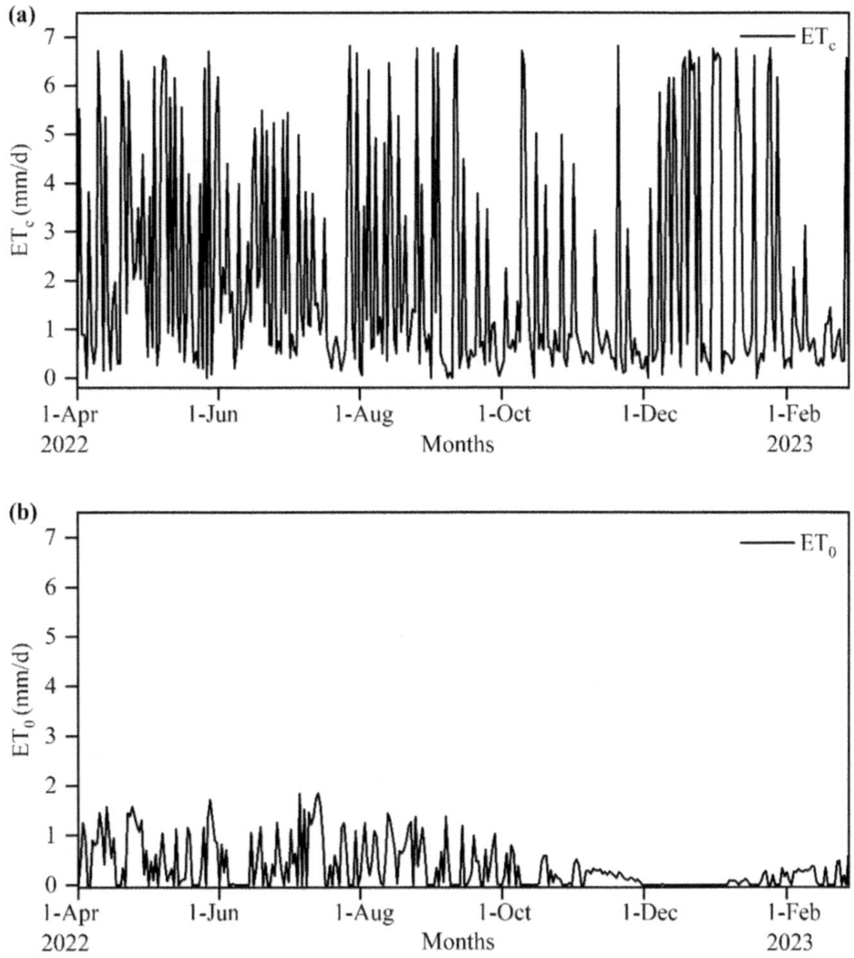

Fig. 3. Seasonal Variation of (a) ET_c and (b) ET_0 in the studied greenhouse environment.

4 Conclusions

In cut flower greenhouse cultivation, the temperature and moisture of the soil layer are continuously monitored to estimate greenhouse ET_c using the water balance method. ET_0 is calculated by the P-T equation using meteorological data. The results show that soil temperature decreases with depth, while moisture increases with soil depth. Consistently, the most significant changes in temperature and moisture are observed in the top layer (0–10 cm). Both ET_0 and ET_c in the greenhouse exhibit significant variability, with total ET_c significantly higher than ET_0. By monitoring the evapotranspiration of water resources in greenhouses, timely monitoring of crop growth water demand can help provide guidance for greenhouse cut flower irrigation and promote water-saving agriculture.

References

1. Butler, D.M., Bates, G.E., Eichler Inwood, S.E.: Tillage system and cover crop management impacts on soil quality and vegetable crop performance in organically managed production in Tennessee. HortScience **51**(5), 1038–1044 (2016)
2. Crago, R., Szilagyi, J., Qualls, R.: What is the priestley-taylor wet-surface evaporation parameter? Testing Four Hypotheses. EGUsphere **2022**(1), 1–26 (2022)
3. Dai, Z., Wang, Q., Jiang, J., et al.: Influence of university agricultural technology extension on efficient and sustainable agriculture. Sci. Rep. **14**, 4874 (2024)
4. Hassani, A.R., Domenighini, P., Belloni, E., et al.: Evaluation of the solar heat gain coefficient of innovative aerogel glazing systems: Experimental campaigns and numerical results. Journal of Building Engineering **62**, 105354 (2022)
5. Qin, K., Chen, F.Y., Ma, L.: Cutting down the travel distance of put systems at Kunming International Flower Auction Market. Int. J. Prod. Res. **53**(3), 3573–3585 (2015)
6. Sun, S., Wang, X., Xu, J., et al.: Study of tomato growth weight-distribution model based on real-time plant weight in a solar greenhouse. Journal of Taibah University for Science **15**, 1027–1037 (2021)
7. Wang, J., Zheng, J., Deng, J., et al.: Effect of moisture content, temperature, and microstructure on the thermal conductivity of loess and its prediction model. Bull. Eng. Geol. Env. **83**, 206 (2024)

Research Progress on the Impact of Reclaimed Water Replenishment on River Environment and Ecology: A Review

Yazhou Dai, Yi Pan(✉), Xiaoqi Yuan, and Jiayi Wei

Xiamen University of Technology, Xiamen 361024, China
2022000116@xmut.edu.cn

Abstract. Reclaimed water from sewage treatment plant is an important water resource. Being discharged into the river channel, it can not only alleviate the water shortage, but also maintain the water cycle of the river ecosystem. Although reclaimed water usually meets water quality standards, its nutrients, trace pollutants and organic compounds may still pose potential risks to the channel environment and ecosystems. This paper summarizes the effects of reclaimed water replenishment on river water quality and ecosystem: reclaimed water replenishment helps to improve the river flow, improving river water quality and diluting pollutant concentrations; nutrients in reclaimed water can promote the growth of aquatic plants and algae, improve the self-purification capacity of the river channel, but it may also lead to eutrophication problems; higher temperatures in reclaimed water may negatively impact aquatic species' survival and reproduction, while excess nutrients can cause overgrowth in riparian vegetation, reducing biodiversity and ecosystem balance; reclaimed water provides essential nutrients for riparian vegetation and wetlands, promoting plant growth and ecosystem stability. Although reclaimed water has broad application prospects, in order to ensure ecological security, it is necessary to strengthen water quality monitoring and risk assessment before discharge to achieve sustainable utilization.

Keywords: Reclaimed water · Ecological water replenishment · Sewage disposal

1 Introduction

Water resources is a vital strategic resource in global sustainable development [1]. Under the background of rapid urbanization, industrialization and global climate change, the shortage of water resources and water environment pollution are increasingly serious, and the river ecosystem is under increasing pressure [2]. As an important unconventional water resource, the reclaimed water in sewage treatment plant can not only effectively alleviate the problem of water shortage by supplementing the river flow, but also promote the restoration of the river ecosystem and the improvement of water quality. The annual reclaimed water usage and sewage treatment volume from 2011 to 2021, demonstrating a steady increase in both metrics over the years as efforts to enhance water reuse have

D. Han et al. (Eds.): ICEPG 2024, ESE, pp. 241–250, 2025.
https://doi.org/10.1007/978-3-031-86829-0_29

expanded. However, although reclaimed water usually meets water quality standards, it still contains certain amounts of nutrients, organic matter, heavy metals, drug residues and trace pollutants, which may pose potential risks to river water quality, ecosystem health and biodiversity [3].

In recent years, with the increasing awareness of river ecosystem protection [4], more and more studies pay attention to the comprehensive impact of reclaimed water discharge from sewage treatment plants on the river environment and ecology. By supplementing the river flow, the reclaimed water can improve the self-purification capacity of the river channel, improve the water quality [5], maintain the water ecological balance, and promote the growth of aquatic organisms to a certain extent. Excessive nutrients and potential pollutants in reclaimed water may also lead to eutrophication, water body biotic imbalance, and structural changes in benthic communities that confer ecological risks [6]. The accumulation of trace pollutants in long-term discharged reclaimed water may adversely affect aquatic organisms and their food chains, and indirectly threaten the stability of the entire ecosystem.

2 Application Status of Reclaimed Water at Home and Abroad

Since the 1980s, both developed countries and some developing countries worldwide have gradually carried out research on reclaimed water discharge and ecological water replenishment [7]. In the field of reclaimed water treatment and ecological water replenishment, the United States, Japan, Israel and other countries are the most typical countries. These countries have established a relatively complete reclaimed water utilization system, especially in arid and semi-arid areas, the technology of reclaimed water for river recharge and agricultural irrigation is relatively mature [8]. By replenishing the river flow, the reclaimed water improves the self-purification capacity of the water bodies and alleviates the problem of water shortage [9].

As a pioneer in the global utilization of reclaimed water, the United States has established a relatively complete regulations, technical standards and application system of reclaimed water. Reclaimed water is not only used for agricultural irrigation, but also widely used in ecological water replenishment and landscape construction of urban river courses [10]. According to statistics, more than 70% of the reclaimed water in the United States is used for environmental restoration and ecological replenishment [5], and a large number of them are used in water restoration and river water replenishment projects. In Japan, despite the abundant rainfall, the water shortage is still severe due to the uneven distribution of resources and the rapid economic development [11]. Since the 1980s, Japan began to promote the reuse of reclaimed water, especially in river water replenishment and landscape water in urban areas. Reclaimed water has become an important supply source for river maintenance. Due to its unique geographical location and extremely lack of water resources, Israel has established the world's most advanced reclaimed water utilization system. In Israel, 72% of urban sewage has been recycled, of which 60% of the reclaimed water is used for ecological water replenishment and groundwater recharge, effectively alleviating the pressure of water resources and realizing the restoration of river and ecosystem.

In China, with the increasingly serious problem of water shortage, the application of reclaimed water is gradually promoted, especially in the river ecological supply and agricultural irrigation has made some progress [7]. In 2021, the Ministry of Ecology and Environment, together with a number of ministries and commissions, issued the Implementation Plan for the Pilot Project of Regional Regenerated Water Recycling, selecting the Beijing-Tianjin-Hebei region, the Yellow River Basin and other areas to carry out the pilot work of reclaimed water recycling, focusing on exploring the application of reclaimed water in the field of river ecological supply. Most of the reclaimed water in China is used for agricultural irrigation and river ecological restoration, especially in the "four rivers" basins (Yellow River, Huaihe River, Haihe River and Liaohe River) with serious water shortage in northern China.

As one of the cities with the most shortage of water resources in China, Beijing has used reclaimed water for river replenishment [12]. In 2021, the reuse rate of reclaimed water in Beijing reached 65%, among which the proportion of reclaimed water replenishing the river has increased year by year, effectively improving the water quality of urban river and improving the health level of water ecosystem. Other cities, such as Shandong and Suzhou, are also vigorously promoting the use of reclaimed water for river ecological water replenishment. In 2021, the amount of reclaimed water in Shandong province reached 1.120 billion m^3, mainly for urban greening, landscape water and river ecological water replenishment. The application of reclaimed water not only alleviates the problem of water shortage to a certain extent, but also effectively maintains the healthy development of river ecological environment.

In general, the research and practice of reclaimed water for river replenishment in China have been deepened, and the technology of reclaimed water has gradually matured and been widely used [13]. However, despite the significant ecological benefits of reclaimed water, its potential environmental risks still exist, and the monitoring and evaluation of the long-term changes of the river ecosystem after the replenishment of reclaimed water need to be strengthened to ensure its safe and sustainable utilization.

3 The Influence of Reclaimed Water Replenishment on the River Channel Environment

Under the background of the increasing shortage of water resources, the reclaimed water of sewage treatment plant, as the alternative water source of river ecological water replenishment [14], shows a broad application prospect. Reclaimed water can not only alleviate the problem of insufficient river flow, but also play a significant role in improving river water quality. There are many successful examples of reclaimed water at home and abroad, and these applications show the positive effect of reclaimed water in improving river water quality and self-purification capacity. However,although reclaimed water has been treated to meet discharge standards, it may still contain residual levels of organic matter, nitrogen, phosphorus, and other trace pollutants, with concentrations often exceeding those found in natural river water [15], which may cause a series of ecological and environmental problems. For example, if the nitrogen and phosphorus and other nutrients rich in the reclaimed water are not effectively controlled, it may lead to the proliferation of algae, thus aggravating the eutrophication of the river water, and then causing ecological problems such as the outbreak of algae bloom. Therefore, when using reclaimed water to supplement the river course, it is necessary to consider the balance between water quality improvement and ecological risk comprehensively, including the water quality characteristics of reclaimed water [16], the long-term impact of reclaimed water on the water body, and the response of the river ecosystem to the replenishment of reclaimed water. Only by reasonably controlling the concentration of pollutants in reclaimed water and combining with the needs of the ecological environment, can the water quality improvement effect of reclaimed water be effectively played and the potential ecological risks can be avoided [16]. The Table 1 below provides an overview of the key impacts of reclaimed water on river environment, along with potential mitigation strategies to address associated risks.

Table 1. Analysis of the impact of reclaimed water on river environment

Impact of Reclaimed Water	Positive Effects	Negative Effects	Solutions
Water quality improvement	Alleviates water shortage, reduces COD and TP	Residual nutrients may cause eutrophication	Regular monitoring and sediment cleaning
Temperature changes	Maintains river flow	Increases river temperature affecting species	Optimize treatment processes, monitor temperatures

3.1 Impact on Potential Water Quality Problems in River Courses

The reclaimed water of sewage treatment plant can not only alleviate the shortage of water resources in the river, but also effectively improve the river water quality and ecological environment. However, long-term reclaimed water discharge may also bring some potential water quality problems, especially due to the impact of residual nutrients and pollutants in the reclaimed water on the river water body [15]. Studies have shown that nitrogen, phosphorus and other nutrients in reclaimed water may cause the risk of water eutrophication and ecological imbalance under certain conditions.

For example, An et al. [9] studied the dilution effect of reclaimed water on the water quality of polluted rivers, and the results showed that after the replenishment of reclaimed water, the concentration of Chemical Oxygen Demand (COD) and Total Phosphorus (TP) in the river significantly decreased. In polluted urban rivers, the COD and TP concentrations were 55.25 mg/L and 0.87 mg/L, respectively. However, after the introduction of reclaimed water into the river, the concentrations of COD and TP decreased to 31.16 mg/L and 0.18 mg/L, respectively. While the replenishment of reclaimed water can improve water quality in the short term by diluting the concentration of pollutants, it still contains certain levels of nitrogen, phosphorus, and other nutrients. These nutrients can easily accumulate in the river sediments and, under certain conditions, may be re-released. Studies indicate that the long-term use of reclaimed water for river replenishment may lead to the release of accumulated nitrogen and phosphorus in the sediments, further exacerbating the problem of water eutrophication [17].

Although the reclaimed water supply can improve the water quality of the river in the short term, if the sediment is not cleaned regularly, more serious eutrophication problems may be caused in the long term [18]. Regular monitoring and sediment cleaning measures should be taken to supplement the river supplemented with reclaimed water, so as to reduce the negative impact of the accumulation of nutrients in reclaimed water on water quality and ensure the ecological health of the river.

3.2 Effect on River Channel Temperature in River Courses

The discharge of reclaimed water not only changes the amount of water in the river but also significantly affects the river's temperature and chemical composition. Studies show that reclaimed water from sewage treatment plants often has a higher temperature due to its source and treatment process. Long-term discharge of reclaimed water into the river may cause the water temperature to rise, especially in the summer or during water shortages. For example, some studies have found that the temperature of reclaimed water is typically 5–8 °C higher than that of natural river water. Under continuous discharge, the river water temperature may increase from its original 18 °C to 25 °C, which can significantly impact the survival, reproduction, and metabolism of species in the river ecosystem [19]. Changes in water temperature can alter the metabolic rate and behavior of fish, invertebrates, and microorganisms in the river, thereby affecting the balance of the entire aquatic ecosystem [20].

Some scholars propose that, due to the difference between reclaimed water temperature and the river environment, the discharge of reclaimed water may not significantly affect the overall river temperature, as the water volume and flow of the river can buffer temperature changes to some extent. For example, in rivers with a higher flow rate, the temperature change is only about 2–3 °C, resulting in a smaller impact. However, in rivers with lower flow rates, the temperature increase can exceed 5 °C. Vliet et al. [21] found that when river discharge decreased by 20% and 40%, river water temperatures increased by an average of 0.3 °C and 0.8 °C, respectively, indicating that reclaimed water discharge or changes in flow can significantly affect river temperatures, especially during dry and warm periods. On the other hand, Woltemade and Hawkins [22] demonstrated through modeling results that higher discharge volumes help mitigate the impacts of increased air temperatures on river temperature, especially under low-flow conditions typical of summer droughts. For example, maintaining stream discharge volume can effectively reduce temperature increases. Additionally, rising temperatures may enhance the biological activity of certain species, increasing their demand for oxygen and nutrients in the water, thereby intensifying the ecological pressure on the river channel [23]. Additionally, the study by Zwolsman and Bokhoven [24] found that the rise in water temperature is more pronounced in areas with lower river flow, especially in regions with concentrated discharge. The increase in temperature, which can exceed 8 °C, limits the dilution of chemical substances and puts significant pressure on the local ecological environment. Therefore, in the process of discharging reclaimed water, changes in water temperature need to be fully considered to avoid adverse impacts on the river ecosystem.

4 Impact of Reclaimed Water Discharge on the River Channel Ecosystem

Using reclaimed water discharge for river recharge can not only effectively improve the water shortage problem [7], but also help to improve the health status of the river ecosystem. The channel supply of reclaimed water has been used in many countries and regions. However, the impact of reclaimed water on the river ecosystem also depends on the treated water quality and the specific environmental conditions of the river course. Reclaimed water replenishment may change the water chemical characteristics of the channel and affect the survival and reproduction of aquatic organisms, especially in the case of long-term replenishment. Exploring the potential impact of reclaimed water discharge on the river ecosystem is crucial to optimize the utilization of reclaimed water and reduce the ecological risk. The following Table 2 outlines key factors to consider in balancing the benefits and risks associated with reclaimed water use in river ecosystems.

Table 2. Impact of reclaimed water on river ecosystems

Impact of Reclaimed Water	Positive Effects	Negative Effects	Solutions
Effects on Aquatic Life	Increases nutrient content, promotes fish growth and biodiversity	Mass algae growth, eutrophication, decline of benthic populations	Mass algae growth, eutrophication, decline of benthic populations
Effects on Riparian Vegetation and Wetlands	Enhances biomass and stability of wetland plants	Overgrowth, increased competition, reduced plant diversity	Plan water replenishment based on local ecological conditions
Chemical and Salinity Changes	Provides essential nutrients for plant growth	Salinization of soil, inhibited growth of salt-sensitive plants	Regularly assess salinity and nutrient concentrations

4.1 Effects on Aquatic Life

Numerous studies, both domestically and internationally, have been conducted on the impact of reclaimed water discharge on aquatic organisms in rivers. The nitrogen, phosphorus, heavy metals, and organic compounds contained in reclaimed water have varying degrees of influence on plankton, benthic organisms, and fish in rivers [25]. For example, some studies have pointed out that high concentrations of nutrients, particularly nitrogen and phosphorus, can lead to massive growth of blue-green algae and green algae, triggering algal blooms. This excessive algal growth not only reduces water transparency but also competes with other phytoplankton, altering the composition of phytoplankton communities. For instance, Mmasseret et al. [26] found that long-term discharge from a waste stabilization lagoon significantly increased the biomass of phytoplankton, such as cyanobacteria, diatoms, and green algae, and altered the phytoplankton community structure by promoting the growth of mixotrophic microorganisms while inhibiting the growth of photosynthetic ones. This change resulted in a decrease in water transparency by about 30%, adversely affecting the living conditions of other aquatic organisms, particularly light-sensitive aquatic plants like submerged macrophytes.

Furthermore, the discharge of reclaimed water also affects the population structure and diversity of zooplankton. In nutrient-rich water bodies, the numbers of zooplankton, such as rotifers and cladocerans, may increase significantly. However, this rapid proliferation of zooplankton is often accompanied by the decline of other species, resulting in shifts in the zooplankton community structure. Nevertheless, some scholars argue that the proper regulation of reclaimed water discharge concentrations can promote the restoration of aquatic biodiversity to a certain extent. For example, Moniz's [27] research indicates that the addition of reclaimed water during dry seasons improves river habitats and water quality by increasing flow velocity and dissolved oxygen. These improvements enhance the nutrient content in the river, providing more resources for benthic organisms such as snails and aquatic insects, thereby boosting the productivity of the river ecosystem. Additionally, many scholars have noted that the impact of reclaimed water on aquatic organisms also depends on the self-purification capacity of the water body and the treatment level of reclaimed water. For instance, when reclaimed water is treated at a higher level, the organic matter and microbial content in the water can promote the recovery of biodiversity in aquatic plants and microorganisms, increasing the overall biodiversity in the river. However, in cases of long-term discharge where the self-purification capacity of the water body is insufficient, the accumulation of harmful substances in the water may lead to a decline in benthic organism populations.

The organic matter and microbes in reclaimed water may also affect the biological community structure of the river channel. Studies have shown that the discharge of reclaimed water increases the content of organic matter in the river, leading to the growth of aerobic microbial and bacterial communities, which, in turn, affects the material cycle and energy flow in aquatic ecosystems [28]. However, the response of aquatic organisms to reclaimed water varies significantly by region, depending on the hydrological characteristics of the river, ecosystem structure, and the amount and frequency of reclaimed water discharge [16]. Therefore, further research on the long-term effects of reclaimed water discharge on riverine aquatic organisms and its ecological impacts is necessary in the future.

4.2 Effects on Riparian Vegetation and Wetlands

The impact of reclaimed water discharge on riparian vegetation and wetlands has also received much attention. The reclaimed water is rich in nitrogen and phosphorus, which provide nutrients for riparian vegetation and wetland plants, but may also negatively affect the ecosystem [29]. Some scholars believe that the use of reclaimed water can promote the growth of wetlands and riparian vegetation and enhance the stability of the ecosystem. For example, the study by Reaves and Croteau-Hartman [30] showed that the long-term supply of reclaimed water can significantly increase the productivity of wetland plants due to the rich supply of nutrients in the reclaimed water. The research found that the biomass of wetland plants increased by approximately 15% to 30%, and vegetation coverage increased by about 20% under the influence of reclaimed water. However, the study also noted that the plant community may shift towards pollution-tolerant species due to the quality of the reclaimed water, potentially leading to a decline in some native species. Nevertheless, these changes still significantly improved the hydrological conditions of the wetlands and enhanced their ecological service functions.

However, it has also been found that higher N & P content in reclaimed water may cause overgrowth of riparian vegetation, increase plant competition, alter plant community structure, and may even trigger a decline in plant diversity. The study by M. van Oorschot [31] et al. showed that the long-term discharge of reclaimed water may accelerate the expansion of nutrient-tolerant plants, thereby suppressing the growth of native vegetation and affecting the balance of plant communities. In addition, the excess nutrients may also trigger eutrophication in wetland waters, leading to abnormal growth of algae and other aquatic plants, further altering the function and stability of the wetland ecosystem [32].

Therefore, although the supply of reclaimed water can provide necessary water and nutrients for riparian vegetation and wetlands, its potential negative effects, such as the accumulation of excess nutrients and increased salinity, may have adverse effects on vegetation growth and wetland ecosystems. Therefore, when using reclaimed water to replenish the channel, the water replenishment and water replenishment frequency should be reasonably planned according to the local ecological environment conditions to minimize the negative impact on riparian vegetation and wetlands [33].

5 Conclusion and Outlook

5.1 Conclusion

The influence of reclaimed water discharged from sewage treatment plant on the river environment and river ecosystem is a multi-dimensional problem.The conclusion is as follows: 1) Reclaimed water replenishment can significantly enhance the river's water quality by reducing the concentration of pollutants such as chemical oxygen demand (COD) and total phosphorus (TP), representing reductions of approximately 43.6% and 79.3%. It can also improve the river's self-purification capacity, contributing to the restoration of the river ecosystem and promoting the growth of aquatic organisms. 2) Despite its positive effects, reclaimed water may contain residual nutrients such as nitrogen and phosphorus, along with other trace pollutants, which can lead to eutrophication. These nutrients may accumulate in the riverbed sediments, potentially re-releasing under certain conditions, exacerbating water quality issues over the long term. 3) The temperature of reclaimed water is typically 5–8 °C higher than that of natural river water. Long-term discharge can lead to an increase in river water temperature, especially in summer or during water shortages, affecting the survival and reproduction of aquatic species. Temperature regulation strategies should be considered to minimize ecological impact. 4) Reclaimed water provides necessary nutrients for riparian vegetation and wetlands, promoting plant growth and ecosystem stability. However, high nutrient content may lead to overgrowth, increased plant competition, and reduced biodiversity. Careful planning of water replenishment is essential to avoid long-term negative effects on vegetation and wetland ecosystems.

5.2 Outlook

The impact of reclaimed water discharge on river ecosystems is influenced by water quality, carrying capacity, discharge frequency, topography, and climate conditions. While reclaimed water offers significant potential for ecological restoration and resource utilization, its environmental risks must be carefully evaluated. Future research should focus on: 1) Assessing the long-term effects of reclaimed water on various river ecosystems by analyzing factors like carrying capacity and climate, and identifying key influences on water quality and ecological balance. 2) Developing advanced sewage treatment technologies, such as bioremediation and membrane filtration, to address emerging pollutants like microplastics, antibiotics, and heavy metals. 3) Conducting long-term ecological monitoring of water quality, biodiversity, and benthic organisms to better understand the risks of reclaimed water discharge. 4) Establishing a robust environmental assessment and monitoring system, with strict discharge standards and policies, to ensure the sustainable use of reclaimed water in river ecological restoration.

References

1. Sophocleous, M.: From safe yield to sustainable development of water resources—the Kansas experience. J. Hydrol. **235**(1–2), 27–43 (2000)
2. Pop, E., Varshney, V., Roy, A.K.: Thermal properties of graphene: fundamentals and applications. MRS Bull. **12**, 1273–1281 (2012)
3. Zhang, H., Cui, B., Liu, C.: Research progress on biological pollutants in agricultural irrigation using reclaimed water. Journal of Henan Agricultural Sciences **53**(3) (2024)
4. Lyu, S., et al.: Wastewater reclamation and reuse in China: Opportunities and challenges. J. Environ. Sci. **39**, 86–96 (2016)
5. National Research Council, Division on Earth, Life Studies, Water Science, Technology Board, & Committee on the Assessment of Water Reuse as an Approach to Meeting Future Water Supply Needs (2012)
6. Ansari, A.A., Gill, S.S., Khan, F.A.: Eutrophication: threat to aquatic ecosystems. Eutrophication: Causes, Consequences and Control, 143–170 (2011)
7. Yi, L., et al.: An overview of reclaimed water reuse in China. J. Environ. Sci. **23**(10), 1585–1593 (2011)
8. Wu, X., et al.: Evaluation of irrigation water use efficiency using remote sensing in the middle reach of the Heihe river, in the semi-arid Northwestern China. Hydrological Processes **29**(9), 2243–2257 (2015)
9. An, S., et al.: Reclaimed water use improved polluted water's self-purification capacity--Evidenced by water quality factors and bacterial community structure. J. Clean. Product. **386**, 135736 (2023)
10. Li, Q., et al.: Optimizing the reuse of reclaimed water in arid urban regions: a case study in Urumqi, Northwest China. Sustainable Cities and Society **51**, 101702 (2019)
11. Oki, T., Kanae, S.: Global hydrological cycles and world water resources. Science **313**(5790), 1068–1072 (2006)

12. Shi, X., et al.: Enhancing water quality and ecosystems of reclaimed water-replenished river: A case study of Dongsha River, Beijing, China. Science of The Total Environment **926**, 172024 (2024)
13. Wang, Y., Zheng, C., Ma, R.: Safe and sustainable groundwater supply in China. Hydrogeol. J. **5**, 1301–1324 (2018)
14. Fito, J., Van Hulle, S.W.H.: Wastewater reclamation and reuse potentials in agriculture: towards environmental sustainability. Environment, Development and Sustainability **23**(3), 2949–2972 (2021)
15. Deng, S., et al.: The utilization of reclaimed water: Possible risks arising from waterborne contaminants. Environmental Pollution **254**, 113020 (2019)
16. Bischel, H.N., et al.: Renewing urban streams with recycled water for streamflow augmentation: hydrologic, water quality, and ecosystem services management. Environmental Engineering Science **30**(8), 455–479 (2013)
17. Wilson, N.J.: Flows of water and nutrients to Lake Tarawera and connected lakes. Diss. The University of Waikato (2022)
18. Kapsalis, V.C., Kalavrouziotis, I.K.: Eutrophication—A worldwide water quality issue. Chemical Lake Restoration: Technologies, Innovations and Economic Perspectives, 1–21 (2021)
19. Sabater, S., Tockner, K.: Effects of hydrologic alterations on the ecological quality of river ecosystems. Water scarcity in the Mediterranean: Perspectives Under Global Change, 15–39 (2010)
20. Siddha, S., Sahu, P.: Impact of climate change on the river ecosystem. Ecological Significance of River Ecosystems, pp. 79–104. Elsevier (2022)
21. Van Vliet, M.T.H., et al.: Global river temperatures and sensitivity to atmospheric warming and changes in river flow. Water Resources Research **47**(2) (2011)
22. Woltemade, C.J., Hawkins, T.W.: Stream temperature impacts because of changes in air temperature, land cover and stream discharge: Navarro River watershed, California, USA. River Res. Appl. **32**(10), 2020–2031 (2016)
23. Ficke, A.D., Myrick, C.A., Hansen, L.J.: Potential impacts of global climate change on freshwater fisheries. Reviews in fish biology and fisheries **17** (2007)
24. Zwolsman, J.J.G., Van Bokhoven, A.J.: Impact of summer droughts on water quality of the Rhine River-a preview of climate change? Water Sci. Technol. **56**(4), 45–55 (2007)
25. Fu, J., et al.: Heavy metals in surface sediments of the Jialu River, China: their relations to environmental factors. Journal of Hazardous Materials **270**, 102–109 (2014)
26. Masseret, E., et al.: Effects of a waste stabilization lagoon discharge on bacterial and phytoplanktonic communities of a stream. Water Environment Research **72**(3) (2000)
27. Moniz, P.: The ecological benefits of using recycled water for streamflow augmentation during the dry season in Marsh Creek, Brentwood, California. Spring, 1–25 (2013)
28. Jiao, G., et al.: Unveiling the hidden impact: How human disturbances threaten aquatic microorganisms in cities. Science of The Total Environment **950**, 175305 (2024)
29. Cai, Y., Feng, M., Zhang, T.: Review of distribution of nitrogen and phosphorus in riparian zones of Chinese inland water bodies (2022)
30. Reaves, R.P., Croteau-Hartman, M.R.: Biological aspects of restored and created wetlands. Proceedings of the Indiana Academy of Science **103**(3–4), 179–194 (1994)
31. Van Oorschot, M., et al.: Modeling invasive alien plant species in river systems: Interaction with native ecosystem engineers and effects on hydro-morphodynamic processes. Water Resour. Res. **53**(8), 6945–6969 (2017)
32. Bornette, G., Puijalon, S.: Response of aquatic plants to abiotic factors: a review. Aquat. Sci. **73**, 1–14 (2011)
33. Patten, D.T.: Riparian ecosytems of semi-arid North America: Diversity and human impacts. Wetlands **18**, 498–512 (1998)

The Fate of Antibiotics and Resistance Genes in Advanced Wastewater Treatment

Zhiguo Li[1,2], Jialu Tang[1,2], Xueting Wang[1,2], Heng Yuan[1,2], Xiaoling Ma[1,2], and Junfeng Wan[1,2](✉)

[1] School of Ecology and Environment, Zhengzhou University, Zhengzhou 450001, People's Republic of China
wanjunfeng@zzu.edu.cn

[2] Henan International Joint Laboratory of Environment and Resources, Zhengzhou University, Zhengzhou 450001, People's Republic of China

Abstract. In recent years, the risk of antibiotics in the environment has increased due to the inefficient treatment of antibiotics in conventional wastewater treatment plants (WWTPs). This has resulted in the emergence of antibiotic resistance (AMR), which is becoming a significant public health concern in modern society. Conversely, wastewater treatment plants (WWTPs) have emerged as a pivotal site for addressing the issue, given their role as reservoirs of antibiotic-resistant bacteria (ARB) and antibiotic-resistant genes (ARG). It has been reported that conventional WWTPs employ specialised disinfection processes to disinfect the target water body, yet the efficacy of removing ARBs and ARGs has been found to be inadequate. This paper presents a review of the efficiency of several advanced oxidation processes (AOPs) for the removal of ARBs and ARGs from wastewater. The review focuses on the combined ultraviolet (UV) and H2O2/chlorine/peroxymonosulfate (PMS)/ozone processes, Fenton/Fenton-like/photo-Fenton, ultrasonic, and multiphase photocatalysis. The objective is to provide an analysis of their bacterial inactivation mechanisms and removal effects. It must be acknowledged that the treatment of actual wastewater still presents certain limitations, and that further comprehensive studies are required to ascertain the full potential of AOPs for the removal of contaminants.

Keywords: Antibiotics · resistance genes · antibiotic-resistant bacteria · advanced oxidation

1 Introduction

Nowadays, with the mass production, overuse, misuse and residual of antibiotics, the antibiotic pressure that develops leads to an increasing number of ARBs/ARGs. Disinfection processes are widely used to remove pathogenic microorganisms from the effluent water, mainly by altering the permeability of biological cell membranes and disrupting the microbial protein structure in order to inactivate the cells. For example, chlorine disinfection is the most common disinfection process in wastewater treatment plants, and then chlorine disinfection also leads to an increase in the abundance of ARGs

D. Han et al. (Eds.): ICEPG 2024, ESE, pp. 251–257, 2025.
https://doi.org/10.1007/978-3-031-86829-0_30

in the effluent water, which may be due to the low concentration of chlorine increases the selective pressure on the bacteria, which alters the structure of the bacterial community and leads to the enrichment of ARGs. Even if ARBs can be inactivated, intracellular and extracellular ARGs may still survive and spread between bacteria via horizontal gene transfer (HGT), thus conferring similar bacterial resistance.

The advanced oxidation process is considered to be a viable technology for AGRs/ARBs removal due to its high efficiency, convenience, wide applicability, and lack of disinfection by-products.AOPs destroy microbial cell membranes, enzymes, and intracellular substances by generating strong oxidising free radicals, such as hydroxyl radical (OH), sulphuric acid radical (SO4), and peroxyl radical (O2), leading to the inactivation of microorganisms. Different AOPs technologies have unique advantages and disadvantages (as listed in Table 1) [1].

Table 1. Advantages and disadvantages of different AOPs processes for treating ARGs.

process	Advantages	disadvantages
Fenton	1. Low cost.2. Simple operation3. Rapid response.4. Wide range of applications	1. Acidic conditions with a pH of 2–5 are required.2.A large amount of sludge will be generated increasing the cost of disposal.3.Contaminants cannot be fully mineralised and converted into certain intermediates
UV/H_2O_2	1. High oxidizing capacity.2. Clean and environmentally friendly	1. The widespread presence of inorganic anions in natural waters has a significant impact on the reaction process.2.Limited by the light transmittance of wastewater
Ultrasound	1. No reagents added.2. Easy to handle.3. Differential or selective degradation	1. a process of high energy consumption
Ozonation	1. No secondary pollution formation	1.Ozone generators are fragile and costly to treat.2. Toxic by-products may be produced during ozone oxidation
Multiphase photocatalysis	1. Thoroughly oxidises and decomposes pollutants2. The photocatalytic process produces no secondary pollution.3. Multiphase photocatalysts are easy to separate and reuse	1.High complexation rates of photogenerated electrons and holes in photocatalysts can limit their efficiency.2. Some catalysts have a narrow range of application to light.3. The problem of light transmission needs to be solved

2 UV Combined with AOP

In order to effectively remove ARGs in the process of HGTs using UV technology combined with different advanced oxidation processes thus forming, oxidation capacity stronger photochemical oxidation process. Common ones are UV/chlorine/H2O2, UV/O3, UV/peroxymonosulfate (PMS) peroxydisulfate (PDS).

Advanced oxidation processes that incorporate UV light similarly utilise in situ generated reactive species (e.g. reactive hydroxyl radicals (OH), Cl-, Cl2 and SO4) to react with target contaminants [2]. Compared to single UV light, more highly reactive radicals can be generated in concert with oxidants, which are able to mineralise or remove ARGs from water more efficiently and also reduce the generation of disinfection by-products (DBPs). These AOPs are widely used to treat hard-to-degrade pollutants in water with better economic viability and disinfection potential. A brief description of these AOPs is provided below.

2.1 UV/H_2O_2

UV/H2O2 is one of the most widely studied UV/AOPs. By irradiating H2O2 in solution, UV light generates hydroxyl radicals through photolysis, which are capable of non-selectively degrading virtually all electron-rich organic pollutants and causing their eventual conversion to water and CO2.

The researchers found that high doses of H2O2 produced hydroxyl radicals that could cause permanent oxidative damage to ARBs. However, higher doses of UV light were needed to completely remove ARGs [3], which showed their persistence even after complete host inactivation. At the same time the competition between H2O2 and ARGs for UV photon adsorption resulted in reduced photolysis of ARGs, causing a limitation in their removal performance.

2.2 UV/Chlorination

In aqueous solution, sodium hypochlorite (NaClO) exists as hypochlorous acid (HClO) and hypochlorite ion (ClO-). Under UV irradiation, HClO and ClO- undergo a photolytic reaction that produces hydroxyl radicals (OH) and chlorine radicals (Cl), and the combination of UV and chlorine, both of which are potent oxidising agents, minimises the inherent drawbacks of the individual processes.

However, different ARBs showed different sensitivities to UV/chlorine. The combination could effectively inhibit the HGT process of ARGs, but it could only damage ARBs to a certain extent, and a short treatment would instead enrich ARGs. Table 2 shows the inactivation effects of UV/chlorine on ARBs and ARGs.

2.3 UV/Ozonation

Considering that ozone has a higher absorption coefficient, the combination of ozone with UV may be stronger than the combination of UV with hydrogen peroxide and hydrogen peroxide alone, the photochemistry of UV improves the oxidation of organics by ozone or by hydroxyl radicals generated by it, and the combination of the two exhibits stronger synergistic effects. The combination with UV significantly increased the removal of ARGs/ARBs compared to ozone alone. However, there is not always a significant difference and some reports have reported that no significant effect was found, which may be due to mass transfer limitations caused by the low solubility of ozone in water and its low photolysis efficiency, which still needs to be studied in depth [6].

Table 2. Inactivation effect of UV/chlorine on ARB and ARG.

process	condition	ARB/ARGs	removal efficiency	References
Secondary wastewater treatment	40 mJ/cm^2 Cl_2:2 mg/L t = 10 min	aacC2, sulII, strB, tetA and tetB	0.41–0.94 log	[4]
	32 mJ/cm^2 Cl_2: 1–2 mg/L t = 20–80 s	Drug-resistant Morgellons and Enterococcus faecalis	1.4 log	
Secondary wastewater treatment	124.8 mJ/cm^2 Chlorine: 30 mg/L t = 30 min	sulI, tetX, tetG	1–1.4 log	[5]
	249.5 mJ/cm^2 Chlorine: 30 mg/L t = 30 min		1.7–2.1 log	

2.4 UV/PMS

The binding of UV light to peroxomonosulfate has also received significant attention. UV light activates the persulfate anion (S2O82-) and produces sulfate (SO42-), an electrophilic free radical with a longer half-life and stronger redox potential than the hydroxyl radical. SO42- adversely affects membrane permeability, intracellular components of bacteria, and cell reproduction. Whereas UV light damages DNA and leads to bacterial inactivation. UV/PMS has been reported to be comprehensively applied for the removal of ARGs, and the results showed superior results than UV/H2O2. However, it is more sensitive to changes in pollutant composition and requires more stringent process parameters than UV/H2O2, and more research is needed to fill the knowledge gap about the effectiveness of UV/PMS in inactivating ARB and reducing the amount of ARGs [7].

3 Fenton and Fenton-Like Methods

In AOPs, Fenton oxidation has been widely used in laboratory and pilot scale to remove ARGs/ARBs and has high removal rate for antibiotics such as 100 per cent reported for ofloxacin. In the removal of ARGs/ARBs, reactive oxygen species are generated by passing hydrogen peroxide and Fe2 + as a catalyst under acidic medium, chain reaction occurs in the presence of Fe2 +, H2O2 is decomposed and Fe2 + reacts with H2O2 to form Fe3 + and -OH with intermediates. Such as superoxide radicals (O2) and monoclinic oxygen (1O2) to oxidatively degrade ARBs/ARGs.ARBs are inactivated during treatment by a decrease in the permeability of the cell membrane followed by swelling and rupture, and eventual loss of cellular constituents, while effluxed ARGs are inactivated by direct oxidation with OH [8].

3.1 Fenton-Like Method

Due to the traditional Fenton process will produce a large amount of iron-containing sludge, difficult to deal with, increase operating costs will also cause secondary pollution, for this reason developed a homogeneous type of Fenton process, the use of other transition metals or metal oxides as a catalyst, to enhance its oxidation capacity, reduce the production of iron-containing sludge, and has a wider PH working range, to overcome some of the shortcomings of the Fenton oxidation..

The oxidative degradation of ARGs is achieved by activation of hydrogen peroxide or persulfate by transition metals or metal oxides to generate strongly oxidising radicals or non-radicals such as monoclinic oxygen.Fenton-like processes have been shown to be effective in reducing the abundance of ARGs, e.g., in the Fe(III):EDDS:H2O2 catalyst system, which can reduce ARBs/ARGs to below the limit of detection in a relatively short time. The detection of ARBs/ARGs in a short time can be reduced below the detection limit. However, limited by the organic composition of the water, the removal performance of the target pollutants may be reduced when other mixtures are included, while the stabilisation and regeneration of the catalysts for the Fenton-like process remains a challenge [9].

3.2 Photo-Fenton Method

Photo-Fenton is the use of a light source to enhance the formation of reactive oxidising substances under UV or sunlight radiation by combining H2O2 with Fe2 + or Fe3 + to produce additional OH [10]. Unlike the Fenton process, in which the iron salts act as photocatalysts and the H2O2 acts as oxidising agent, the iron is able to recycle to react with the H2O2 at a much higher rate, accelerating the generation of hydroxyl radicals and enhancing the disinfection potential. The use of solar radiation in this process also reduces energy requirements. Many drugs, including penicillin, have shown higher depuration efficiencies by the photofenton process. However, due to the influence of pH and light transmission, despite the excellent removal results achieved in the laboratory, the removal rate in real wastewater treatment is still unsatisfactory and further studies are still needed.

4 Ultrasound

The ultrasonic oxidation process is mainly based on the cavitation effect generated by ultrasound, which produces hydroxyl radicals during the treatment process by generating high driving force and high heat to decompose ARBs/ARGs.With the advantages of no additional reagents, easy handling and high differential/selective degradation, it can remove some pollutants that are difficult to remove by conventional disinfection processes, and it has demonstrated a some effectiveness and was found to be an effective pretreatment method for the treatment of ARGs [11]. However, limited by the ultrasound frequency, low frequency and low energy were found to be ineffective in inactivating some microorganisms, and ultrasound is also a high-energy process with a relatively low efficiency of activated oxygen generation, which also restricts its application potential.

5 Heterogeneous Photocatalysis

Multiphase photocatalysis is mainly achieved by electron jump from valence band to conduction band of semiconductor materials such as TiO2 under light conditions to form electron-hole pairs, which are used to generate hydroxyl radicals and other reactive oxygen species (ROS), these ROS are strongly oxidising and capable of mineralising and removing pollutants from water. In addition adsorbed ARBs/ARGs can also be oxidised as indicated by the photocatalyst.

The dosage and nature of the catalyst are important influences on multiphase photocatalysis. Excessive amount of catalyst increases the turbidity of the wastewater, which reduces the light transmittance negatively affecting the subsequent ROS generation. Also catalyst selection and surface area are critical factors in the removal of ARB [12]. The regeneration of ARBs after inactivation and insufficient mineralisation of ARGs is also often faced when used as a disinfection treatment. In addition, multiphase photocatalysis has not been commercialised due to the techno-economic costs.

6 Conclusion

Overall, in this review, we evaluated the removal mechanisms, removal efficacy, and shortcomings of several advanced oxidation processes for the removal of ARBs and ARGs. Conventional wastewater treatment plants for ARBs and ARGs cannot be effectively addressed. Therefore, advanced oxidation processes and their coupling with conventional disinfection are becoming an important area of research in the removal of ARBs/ARGs, and there is still a need for more comprehensive and in-depth studies on the universal removal of ARBs/ARGs and their fate in complex fractions of wastewater/actual wastewater.

References

1. Li, S., Zhang, C., Li F., et al.: Technologies towards antibiotic resistance genes (ARGs) removal from aquatic environment: A critical review. Journal of Hazardous Materials, 411 (2021)
2. Miklos, D.B., Remy, C., Jekel, M., et al.: Evaluation of advanced oxidation processes for water and wastewater treatment – A critical review. Water Res. **139**, 118–31 (2018)
3. Yoon, Y., Chung, H.J., Wen, Di, D.Y., et al.: Inactivation efficiency of plasmid-encoded antibiotic resistance genes during water treatment with chlorine, UV, and UV/H2O2. Water Research **123**, 783–93 (2017)
4. Wang, H., Wang, J., Li, S., et al.: Synergistic effect of UV/chlorine in bacterial inactivation, resistance gene removal, and gene conjugative transfer blocking. Water Research, 185 (2020)
5. Zhang, Y., Zhuang, Y., Geng, J., et al.: Inactivation of antibiotic resistance genes in municipal wastewater effluent by chlorination and sequential UV/chlorination disinfection. Sci. Total Environ. **512–513**, 125–32 (2015)
6. Giannakis, S., Ruales-Lonfat, C., Rtimi, S., et al.: Castles fall from inside: Evidence for dominant internal photo-catalytic mechanisms during treatment of Saccharomyces cerevisiae by photo-Fenton at near-neutral pH. Appl. Catal. B **185**, 150–62 (2016)

7. Rodríguez-Chueca, J., Varella Della Giustina, S., Rocha, J., et al.: Assessment of full-scale tertiary wastewater treatment by UV-C based-AOPs: Removal or persistence of antibiotics and antibiotic resistance genes?. Science of the Total Environment **652**, 1051–61 (2019)
8. Chen, Y.-D., Duan, X., Zhou, X., et al.: Advanced oxidation processes for water disinfection: Features, mechanisms and prospects. Chemical Engineering Journal, 409 (2021)
9. Akbari, M.Z., Xu, Y., Lu, Z., et al.: Review of antibiotics treatment by advance oxidation processes. Environmental Advances, 5 (2021)
10. Michael, S.G., Michael-Kordatou, I., Beretsou, V.G., et al.: Solar photo-Fenton oxidation followed by adsorption on activated carbon for the minimisation of antibiotic resistance determinants and toxicity present in urban wastewater. Appl. Catal. B **244**, 871–80 (2019)
11. Wang, M., Xu, J., Chai, Y., et al.: Differentiation of environmental conditions promotes variation of two quercus wutaishanica community assembly patterns. Forests **11**(1) (2019)
12. Ugwuja, C.G., Adelowo, O.O., Ogunlaja, A., et al.: Visible-light-mediated photodynamic water disinfection @ bimetallic-doped hybrid clay nanocomposites. ACS Appl. Mater. Interfaces. **11**(28), 25483–94 (2019)

Investigating Sediment Composition Distribution in a Reservoir and Assessing the Impact of Static Release on Water Quality

Lu Zhang[1], Yao Du[2], Hailong Li[3](✉), Hanze Li[1], Linlin Hu[1], and Yazhou Fan[1]

[1] Zhejiang Institute of Hydraulics and Estuary (Zhejiang Institute of Marine Planning and Design), Hangzhou 310002, Zhejiang, China
[2] Zhejiang Institute of Communications Co., Ltd., (Zhejiang Institute of Transportation Planning and Design), Hangzhou 310020, Zhejiang, China
[3] Zhejiang Environment Technology Co., Ltd., Hangzhou 310000, Zhejiang, China
569107388@qq.com

Abstract. The reservoir is a vital water source for agricultural and grain production. However, with increasing human activity, it has been continuously receiving substantial amounts of inorganic and organic pollutants from the surrounding watershed. These pollutants accumulate in the bottom sediment of the reservoir through various physical, chemical, and biological processes, resulting in a decline in water quality. In the reservoir's ecosystem, the sediment serves as an exchange medium for elements such as nitrogen, phosphorus, and heavy metals, functioning as either a "source" or a "sink" for these nutrients [1–3]. By sampling and analyzing the concentrations and distributions of carbon, nitrogen, phosphorus, heavy metals, and organic pollutants within the sediment, this study examines the static release dynamics of these key pollutants. The findings reveal that carbon, nitrogen, phosphorus, and heavy metal contamination in the sediment is significant, with external sources identified as the primary contributors. Therefore, implementing pollution control strategies that focus on source management is recommended to protect the aquatic ecosystem and ensure human health.

Keywords: Sediment pollutants · Heavy metals · Carbon · nitrogen and phosphorus · Static release

1 Introduction

The reservoir constitutes a large-scale hydraulic infrastructure project primarily intended for irrigation purposes, while additionally facilitating hydroelectric power generation, flood management, and aquaculture. It encompasses a catchment area of 180 km^2, with a total storage capacity of 170 million m^3, and yields an average annual electricity production of 11 million kWh. The associated irrigation network is among the largest gravity-fed systems in Zhejiang Province, providing water to approximately 300,000 mu (20,000 ha) of cropland primarily used for grain cultivation, as well as 250,000 mu (16,667 ha) of economically valuable forested areas. This reservoir serves as a critical

D. Han et al. (Eds.): ICEPG 2024, ESE, pp. 258–268, 2025.
https://doi.org/10.1007/978-3-031-86829-0_31

water resource for the regions of Qujiang District, Longyou County, Kecheng District, and Lanxi City, thereby playing a pivotal role in supporting agricultural productivity and food security.

As human activities intensify, point source pollution from domestic activities and small-scale industries, along with non-point source pollution from agricultural practices, soil erosion, and housing construction by villagers near the reservoir—facilitated by improved transportation infrastructure—have significantly impacted sediment accumulation, pollution load, and water quality in the reservoir. To address the diverse pollution challenges faced by the reservoir, it is essential to determine and analyze the spatial distribution characteristics of the pollutants. Such analysis provides critical insights for developing effective strategies for controlling water quality pollution in the reservoir, which is of vital importance for safeguarding the health of the population residing within the reservoir's surrounding area.

2 Sample Collection

In accordance with the reservoir's topography, geomorphology, and hydrodynamic characteristics, and adhering to the principle of representativeness [4], seven sediment sampling points were strategically established (see Fig. 1). Undisturbed core sediment samples were collected using an Austrian Uwitec sediment sampler. The contents and spatial distribution of carbon, nitrogen, phosphorus, heavy metals, and organic pollutants in the sediment were assessed using various analytical methods. Furthermore, static release experiments were conducted to examine the kinetics of the primary pollutants' static release within the sediment.

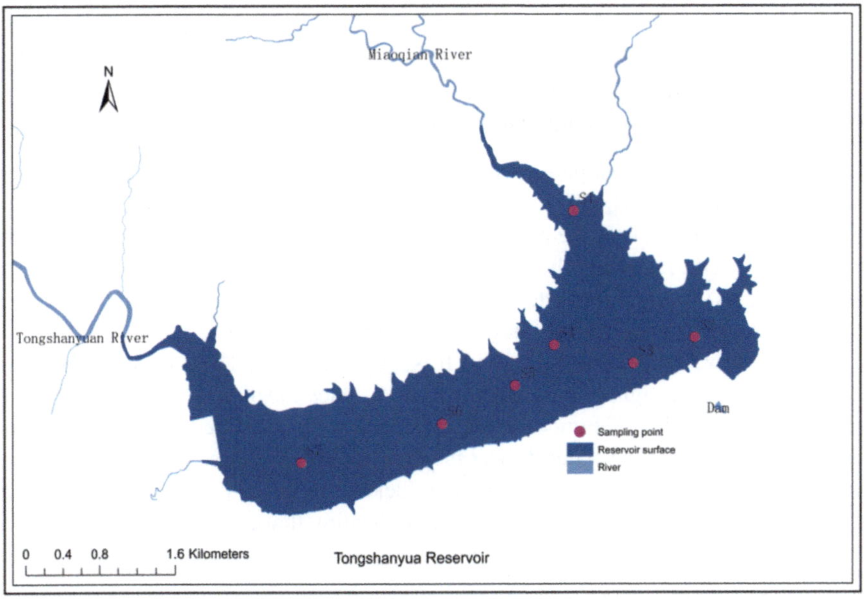

Fig. 1. Reservoir sampling point distribution.

3 Spatial Distribution Patterns of Sediment Pollutants

3.1 CNP Content and Spatial Distribution

(1) Phosphorus Content and Spatial Distribution in Sediments.

Using the SMT method for phosphorus fractionation, 5 cm sediment cores were selected as representative samples. The concentrations of total phosphorus (TP), organic phosphorus (OP), and inorganic phosphorus (IP) were measured at different sediment depths. The IP was further subdivided into iron/aluminum-bound phosphorus (NaOH-P) and calcium-bound phosphorus (HCl-P). The composition and spatial distribution patterns of these phosphorus forms are illustrated in Fig. 2. The results of the analysis are as follows:

1) The vertical distribution of TP across the seven sampling points exhibited a decreasing trend from the surface layer to deeper layers. Phosphorus was predominantly concentrated in the surface sediments, suggesting that its accumulation is strongly linked to human activities.
2) In the surface sediments of the reservoir, the average proportions of OP, iron/aluminum-bound phosphorus, and calcium-bound phosphorus relative to TP showed an increasing trend. This pattern indicates that calcium-bound phosphorus is the dominant phosphorus component in the reservoir sediments. Phosphatase enzymes secreted by microorganisms are critical mediators in the mineralization and degradation pathways of OP. Consequently, microorganisms play a significant role in the mobilization and transformation of phosphorus within the reservoir sediments.
3) The OP content across the seven sampling points ranged from 0.471 mg/g to 0.628 mg/g, while the calcium-bound phosphorus content varied between 0.416 mg/g and 0.554 mg/g, with an average value of 0.467 mg/g. Among the various phosphorus fractions in the sediments, calcium-bound phosphorus is relatively inert, making it less prone to release and less bioavailable.

(2) Nitrogen Content and Spatial Distribution in Sediments.

Using the method developed by Norman et al. to remove nitrate nitrogen from potassium chloride extracts and applying the UV spectrophotometric correction factor method [5], the concentration of nitrate nitrogen in the extract was calculated [6]. An elemental analyzer was used to measure the total nitrogen (TN) content in the sediments. Nitrogen in the sediments was classified into organic nitrogen and inorganic nitrogen, with organic nitrogen accounting for a slightly larger proportion (see Fig. 3).

At the reservoir's central point, which is also the outlet, TN and organic nitrogen concentrations were significantly higher than at other locations. Sampling points S2 and S3 also had higher TN and organic nitrogen levels compared to the other points, possibly due to the higher water flow at the outlet in recent years. These two points are close to the dam, where significant sediment deposition occurs. The TN content in the sediment showed a clear surface enrichment, indicating that external pollution has been the primary contributor to nitrogen accumulation in the water body in recent years.

(3) Total Organic Carbon Content and Spatial Distribution in Sediments.

The spatial distribution of total organic carbon (TOC) across the seven sampling points is shown in Fig. 4. The TOC content in the surface sediments at points S4 to S7 was noticeably lower than at other points. The TOC content was higher in the surface layer and decreased with increasing depth, indicating an increasing trend of organic pollution in the reservoir in recent years. The sediment's water content and TOC levels suggest that the surface sediments have higher organic matter content and are loose, making them more prone to disturbance and thus likely to impact the water quality of the reservoir.

Overall, nitrogen and phosphorus in the reservoir sediments are primarily concentrated in the surface layer. The surface TN and TP load range from 1.40 g/kg to 3.92 g/kg and 1.08 g/kg to 1.47 g/kg, respectively, and gradually increase along the direction of water flow.

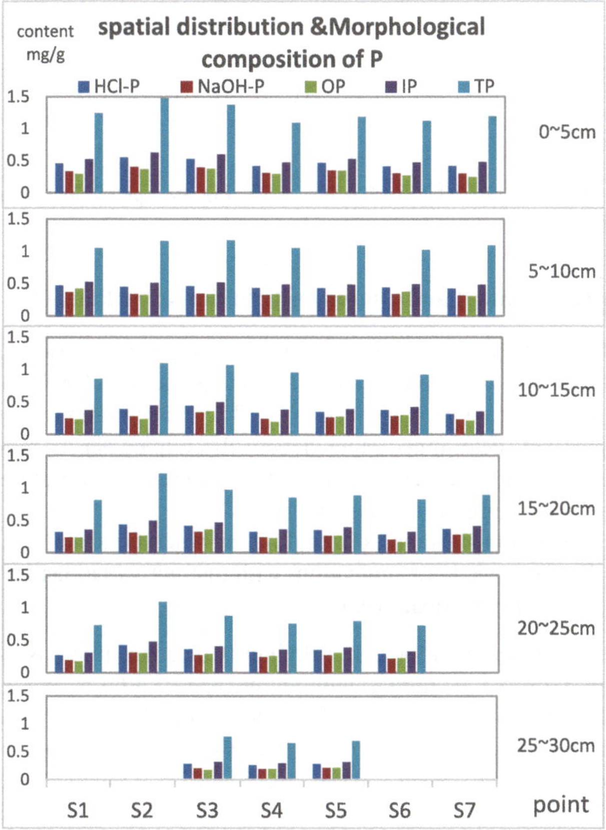

Fig. 2. Morphological composition and spatial distribution of phosphorus content.

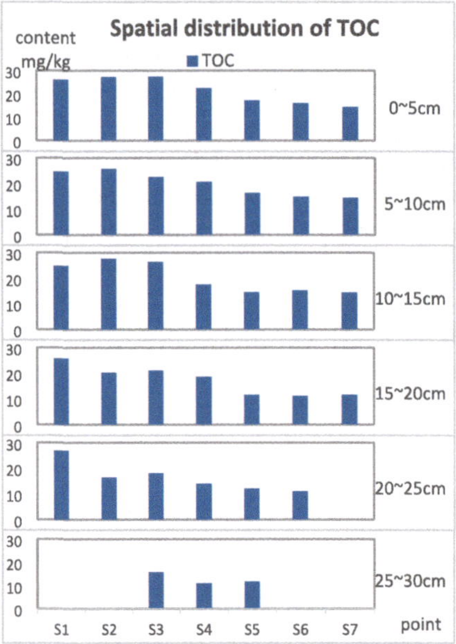

Fig. 3. Spatial distribution of TN & organic N.

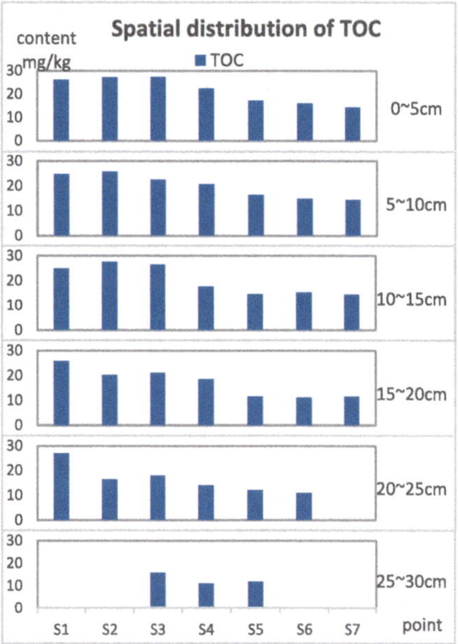

Fig. 4. Spatial distribution of TOC.

3.2 Heavy Metal Content and Spatial Distribution in Sediments

The biotoxicity and bioaccumulation potential of heavy metals pose long-lasting environmental hazards, allowing these elements to accumulate in soil and aquatic sediments, thereby creating significant ecological risks [7]. The concentrations of heavy metals in sediment samples were determined using inductively coupled plasma mass spectrometry (ICP-MS) [8]. The average concentrations of heavy metals at various depths across the sampling points are summarized in Fig. 5.

Analysis of the sediment samples from different reservoir sampling points confirms the presence of heavy metal contamination, with distinct variations in the distribution and pollution levels of individual metals.

1) Among the seven sampling points, cadmium (Cd) concentrations in the sediments ranged from 3.25 to 8.40 mg/kg, with an average of 4.69 mg/kg. Chromium (Cr) levels varied between 51.40 and 178.24 mg/kg, averaging 87.76 mg/kg. Nickel (Ni) concentrations ranged from 39.86 to 80.21 mg/kg, with an average of 56.21 mg/kg. Lead (Pb) concentrations were between 56.16 and 277.32 mg/kg, with an average of 170.90 mg/kg. Mercury (Hg) concentrations varied from 2.51 to 22.68 mg/kg, averaging 7.92 mg/kg, while uranium (U) concentrations ranged from 1.44 to 8.14 mg/kg, with an average of 3.17 mg/kg.
2) Regarding the vertical distribution of heavy metals in the sediments, most metals showed minimal variation between the surface and deeper layers, indicating that heavy metal pollution remains consistently severe in recent years. However, mercury (Hg) and uranium (U) exhibited an increasing trend from the surface to deeper layers,

suggesting that the levels of Hg and U pollution in the reservoir have been effectively controlled in recent years.

3) At sampling point 7, all heavy metals, except for Hg, displayed higher concentrations in the surface layer compared to the deeper layers, indicating a recent intensification of heavy metal pollution at this location.

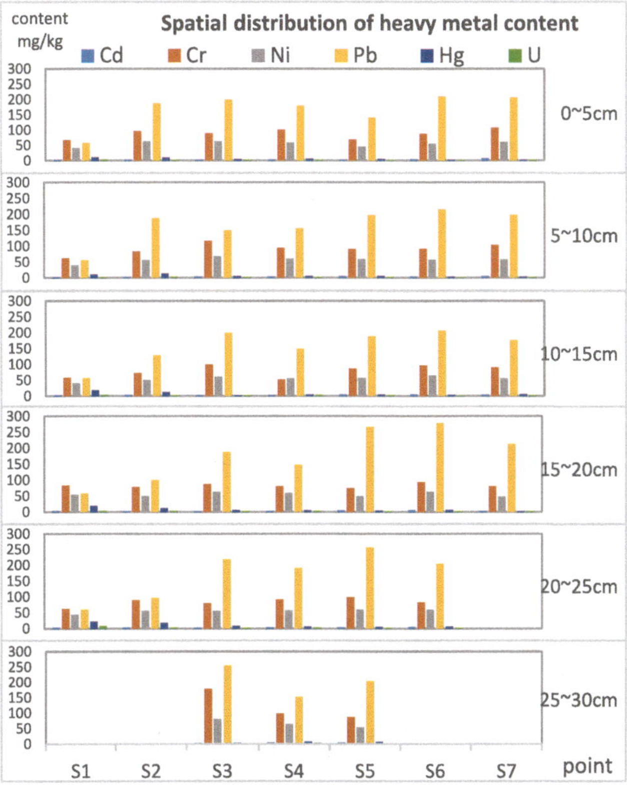

Fig. 5. Heavy metal content at different depths in sediment.

3.3 Persistent Organic Pollutant Content and Distribution in Sediments

The concentrations of DDTs and HCHs in the reservoir sediments were determined using the 3550 C (ultrasonic extraction) and 3620 C (Florisil cleanup column) methods. The spatial distribution of these compounds is illustrated in Fig. 6.

1) DDT was detected at all sampling points, while DDE and DDD were not present. The concentrations of DDT and HCH at sampling points S1 and S2 were significantly higher than those observed at points S3-S7. The highest concentration of DDT was found in the 10–15 cm soil layer at S2, while the highest concentration of HCH was

recorded in the 25–30 cm soil layer at S4. These findings suggest the presence of an active pollution source near S1 and S2, with DDT contamination being particularly severe.

2) The concentration ranges of DDTs and HCHs across the sampling points were 0.7–47.1 μg/kg and 0–25.9 μg/kg, respectively. This indicates that while residues of DDTs and HCHs are present in the sediments, their concentrations remain below the national soil environmental quality level one standard of 50 μg/kg, classifying these sediments as unpolluted and relatively clean [9]. Among the organochlorine pesticides detected, DDT concentrations were significantly higher than HCH concentrations, suggesting recent inputs of DDTs into the reservoir, particularly from upstream pollution sources near S1 and S7.

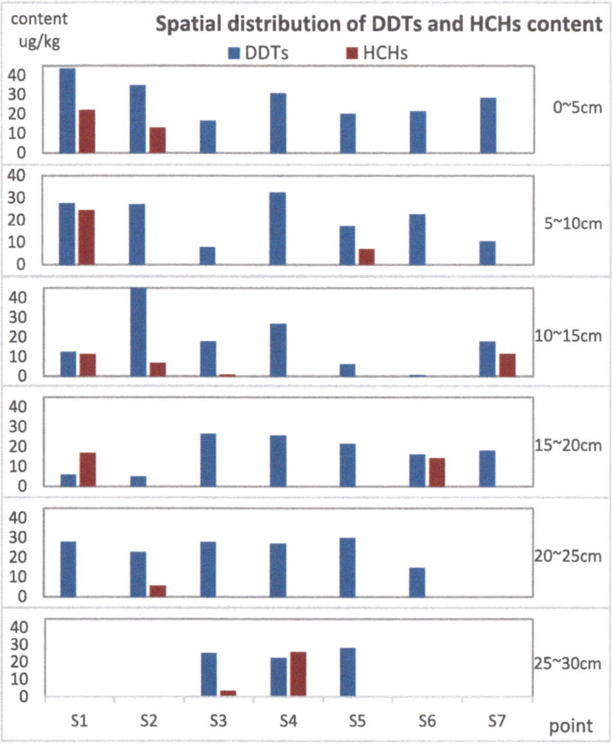

Fig. 6. Spatial distribution of total DDTs and HCHs in sediment.

4 Study of the Static Release Kinetics of Major Pollutants in Sediments

4.1 Nitrogen and Phosphorus Release Kinetics Study

In recent years, the internal nutrient loading of lakes and reservoirs has garnered significant attention, as these bodies are critical sites for the accumulation and regeneration of nitrogen and phosphorus. During events such as wind-induced turbulence and reservoir discharge, sediments can become resuspended, leading to the release of these nutrients. Nitrogen and phosphorus from the sediments enter the water column through molecular diffusion and concentration gradient diffusion [10]. In static release experiments, sedimentary nitrogen consistently showed a release effect on the overlying water. Water samples of 100 ml were collected at intervals of 0, 1, 4, 8, 12, 24, 48, and 96 days to measure total nitrogen (TN) and total phosphorus (TP) concentrations. The result is shown in Fig. 7.

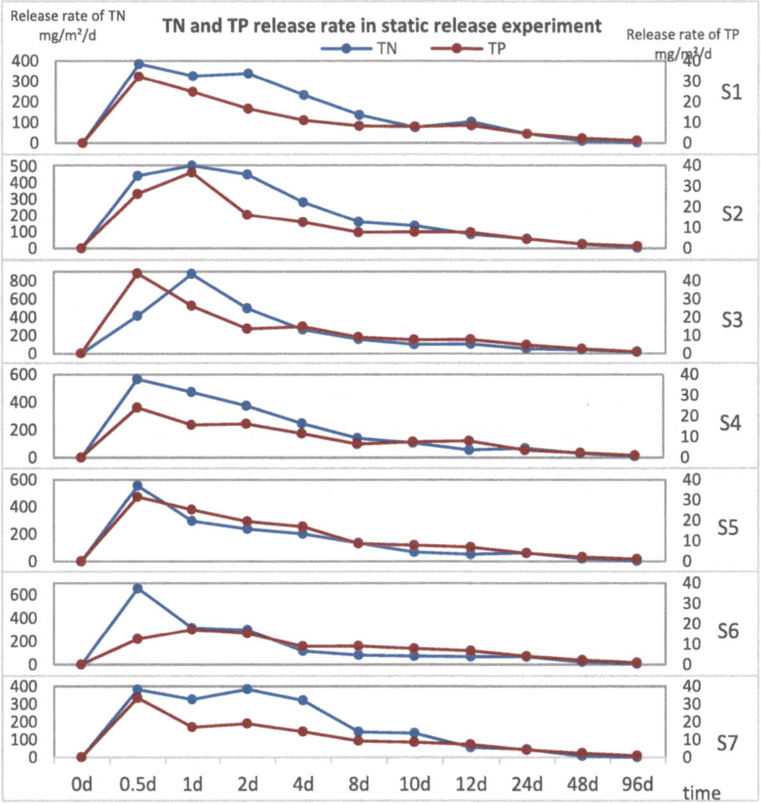

Fig. 7. Change of nitrogen and phosphorus concentration in static release experiment.

The findings from the study of sediment nitrogen and phosphorus release kinetics indicate that sediments serve as a substantial source of nitrogen and phosphorus release into the water. Removing the upper sediment layer, which contains high levels of these nutrients, is a critical strategy for mitigating nitrogen and phosphorus eutrophication in the reservoir.

1) The release rates of TN from the surface sediments at the seven sampling points followed a consistent pattern: a high release rate was observed at the onset of the experiment, which gradually declined over time and eventually stabilized.
2) The release rate of phosphorus in the surface sediments increased rapidly, peaking within the first 12 days, before gradually declining. Across the 96-day period, the average release rate of TP at all seven sampling points remained positive.
3) The release rates of TP and TN from the middle layer sediments mirrored those of the surface sediments. In the early stages, the nitrogen release rate from the middle layer exceeded that of the surface layer; however, the gap between the two gradually narrowed and stabilized over time. The release rates and concentration changes of both TP and TN displayed similar trends, with a gradual decline in release rate during the latter stages of the experiment. Despite this, the concentrations of TP and TN in the overlying water continued to increase.

4.2 Heavy Metal Release Kinetics Study in Sediments

Based on the 30-day simulated release experiment, water samples were extracted from the sediment release devices at 0.5, 1, 5, 10, 20, and 30 days. The concentrations of heavy metals in the water bodies were determined via an inductively coupled plasma mass spectrometer (ICP-MS). The concentrations of heavy metals at different depths are depicted in Fig. 8.

1) During the static release experiment, given the absence of water body disturbance, the water environment remained relatively stable. The majority of the heavy metals in the water bodies could be promptly adsorbed by suspended particles and bottom sediments, accumulated, and settled, with no discernible release process. This indicates that the transfer of heavy metals from the water to the sediment was the predominant process in the experiment. The sediment functioned as the "sink" of heavy metals rather than the "source". Nevertheless, if there were significant disturbances in the water body, it might trigger the release of heavy metals from the sediments into the water, leading to secondary pollution.
2) The concentrations of cadmium (Cd), chromium (Cr), and lead (Pb) dropped below the detection limit by the end of the experiment. The concentrations of heavy metals in the water samples decreased to stable levels, with the exception of uranium (U). The concentration of uranium remained within the range of 0.8–1.0 μg/L throughout the 30-day period, maintaining relative stability.
3) In accordance with the determination results and the national soil quality standard (GB15618–1995), the contents of mercury (Hg) and cadmium (Cd) in the sediment surpassed the third-class standard, belonging to severe pollution and causing relatively severe harm. The contents of other heavy metals were lower than the third-class standard, falling within the realm of relatively clean and light pollution.
4) Concerning the vertical distribution of the sediment, most heavy metals exhibited no significant alteration from the surface to the bottom layer. However, mercury (Hg)

and uranium (U) demonstrated an increasing tendency from the surface to the bottom layer, suggesting that the pollution of mercury (Hg) and uranium (U) in the reservoir has been controlled in recent years.

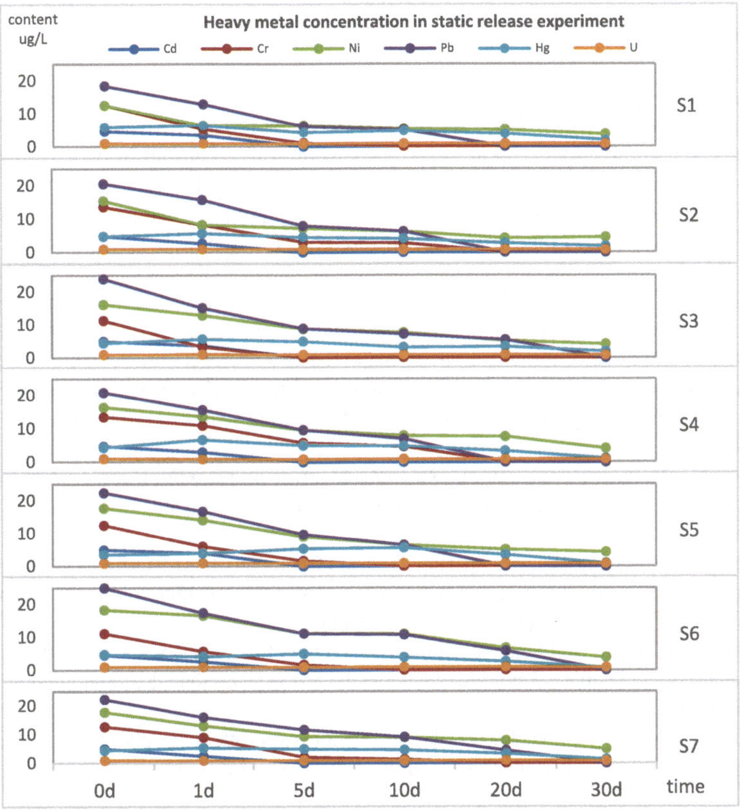

Fig. 8. Change of heavy metal concentration in static release experiment.

5 Conclusion

Analysis of pollutant characteristics and static release experiments reveals that carbon, nitrogen, phosphorus, and heavy metal contamination are significant issues in the reservoir sediments, primarily driven by external pollution sources. Pollutants enter the reservoir through river inflows and subsequently disperse along the flow of water. Therefore, effective water quality management should prioritize source control, adopting a comprehensive approach that addresses point sources, non-point sources, and internal pollution simultaneously. Developing integrated ecological strategies for pollution prevention and disaster mitigation is essential, utilizing both ecological principles and engineering solutions to effectively manage and prevent reservoir pollution.

After achieving effective control of external pollution, dredging measures can be undertaken to remove contaminated sediments. By employing environmentally safe disposal methods, pollutants can be rapidly eliminated from the reservoir, thus protecting the aquatic ecosystem and safeguarding human health.

References

1. Pei, L., Hong, Z., Chao, W., et al.: Research progress of pollutant migration and diffusion in sediment-water system. J. Lake Sci. **30**(6), 1489–1508 (2018). (in Chinese)
2. Shang, L., Sun, R., Jiyuhe, et al.: Risk assessment of heavy metals in river sediments from miyun reservoir. Environ. Sci. Technol. **34**(S2), 344–348 (2011). (in Chinese)
3. Feng, L., Song-Lin, H., Xing, W., et al.: Heavy metal pollution and potential ecological risk assessment of bottom mud in dongguan reservoir. Environ. Sci. Technol. **41**(2), 183–188 (2018)
4. Huifang, C., Shuyun, C., Chunhui, Y., et al.: Internal sediment pollution and release characteristics of hengshan reservoir in Yixing city. Environ. Sci. **41**(12), 5400–5409 (2019)
5. Research group on eutrophication of major lakes and reservoirs in China. Criteria for Investigation of Eutrophication in Lakes, Beijing, China. Environmental Science Press (1987)
6. Rydin, E.: Potentially mobile phosphorus in lake Erken sediment. Water Res. **34**(7), 2037–2042 (2000)
7. Zeng, X., Wang, X., Tu, X., et al.: Research progress on prevention and control technology of heavy metal pollution in farmland soil. J. Ecol. Environ. **28**(9), 1900–1906 (2019). (in Chinese)
8. Lian-Ying, Y., Cheng-Wu, F., Gui-Hua, L., et al.: Characteristics of nutrients and heavy metals in sediment of reservoirs in karst mountainous areas: a case study of Aha reservoir. J. Saf. Environ. Saf. Environ. **17**(5), 2031–5409 (2017)
9. Nanjing Institute of Soil Science, Chinese Academy of Sciences. Soil Environmental Quality Standard of the People's Republic of China GB15618–1995 Shanghai: Shanghai Science and Technology Press (1978)
10. Zhu, G., Chen, Y., Zhou, G., et al.: Simulation experiment on phosphorus release from sediments in the canal (Hangzhou section). J. Lake Sci. **14**(4), 343–349 (2002)

Simulation of the River Water Temperature by Using One-Dimensional Combined Model of Hydrodynamic and Advection-Dispersion

Rongxiang Hu(✉)

Zhejiang Institute of Hydraulics and Estuary (Zhejiang Institute of Marine Planning and Design), Hangzhou 310020, Zhejiang, China
364601233@qq.com

Abstract. The upstream Dam has affected the physical, biological and cultural resources of the river downstream. One of the impacts to downstream physical environment for the aquatic ecosystem is the transformation of water temperature, owing to releases from the upstream reservoir. Temperature is one of the most influential environmental factors affecting irrigation, the growth of fishes and other Species. Adaptive Management Program has considered modifications to release flow and release temperature to increase downstream temperatures. Here, considered the aim river, Zaoshi River, is a wide-Shallow one, we applied one-dimensional combined model of Hydrodynamic (HD) and Advection-Dispersion (AD) to simulate monthly average temperature from the Zaoshi Dam to 17 km downstream. The HD model is based on one-dimensional equation of Saint-venant, the AD model on heat-exchange equation, and the parameters in the combined model need to be determined. In this research, firstly studied many papers, we obtain the main ones from many factors, which is the release volumes and temperature, air temperature, light duration, and relative humidity. Secondly, input the data of the above five items and measured temperature from April 2008 to March 2009 to calibrate the model parameters, and then used the data from April 2009 to March 2010 to check the model built. Compared to measured temperature value, the average relative errors is less than 10%, so the combined model could simulate the water temperature of Zaoshi River very well. In the last, we used the combined model to simulate the water temperature from Apr 2010 to Mar 2011. For Zaoshi the surface water level was affected by the upstream releases and downstream reservoir water level boundary, the temperature of the river was affected by the release mostly. The cause of this result may be the small capacity of the stream for heat storage. So we could change the management of Zaoshi Station to let the temperature satisfy all need of kinds. The modeling approach used here may also prove useful for other systems, particularly the wide-Shallow river below large dams where the capacity of the downstream is not large.

Keywords: water temperature · Hydrodynamic model · Advection-Dispersion model · one-dimensional · Zaoshi River

D. Han et al. (Eds.): ICEPG 2024, ESE, pp. 269–280, 2025.
https://doi.org/10.1007/978-3-031-86829-0_32

1 Introduction

The temperature of river water is an important environmental parameter and has both an ecological and economic significance. On the one hand, water temperature and other water quality factors are closely linked, such as biochemical oxygen demand and re-oxygenation processes related to water temperature; on the other hand, the relationship between water temperature and crop planting and aquatic ecology is closer, such as rice and other plants, water fish and other aquatic organisms are very sensitive to temperature. Recent years, there have been building high dams in domestic. As the building of dams, the flow velocity became slower, the period of water updating longer, and affected by solar radiation and convection and heat transfer, the temperature structure of water body changed. Meanwhile, the water situation of downstream would greatly change. Therefore, changes of water temperature in reservoir and downstream will bring a serious influence on water quality, aquatic growth and so on [1]. Accurate modeling and prediction of water temperature changes of cold water discharge, and effective control, to improve the environment is of great significance to industrial and agricultural production. Therefore, more and more scholars now study this area. They almost firstly studied the water temperature of reservoir, and then simulated the downstream. River water temperatures can be regarded as a function of various hydro-meteorological factors. Primarily, the heat budget of river will be determined by the heat balance processes of radiation, sensible heat flux and evaporation which affect any water surface but, in addition, the changing nature of stream flow will modify the resulting temperatures by purely hydrological factors. In view of the limited availability of historical meteorological data necessary for the application of the model, such as solar radiation, air temperature, discharge and so on, much recent work on river temperature prediction is aimed in the future.

This manuscript is organized as follows. In Sect. 2, we describe the background and the method we used with an emphasis on design of measure water temperature. The combined model is described in Sect. 3. In Sect. 4, the model calibration, validation, and applications are presented. In the last part of this paper, Sect. 5, some conclusion and the future work would be given.

2 Materials and Methods

Research process on river water temperature in China could be divided by three stages:

The first stage is to collect data of water temperature related. In this phase, measure and collect the data of river downstream water temperature per and post of the construction of reservoir, and then used this data to determine and analyze the water temperature influence after the construction.

The second stage is to use empirical formula to simulate it. The use of the previous stage of the water temperature data collected, and established one-dimensional empirical formula to simulate the construction of water temperature downstream after reservoir construction.

The third stage is to use the more accurate method: mathematical model.

With research on the water temperature and technology development, to established one-dimensional, two dimensional, and even three-dimensional mathematical model to simulate the distribution of the river water temperature.

Currently mainly use simulation model and empirical formulas to calculate and predict river water temperature. The former method is to analyze the main factors which affect water temperature, according to the heat, mass balance principle to build the mathematical model which is high precision, suitable for general situation, but it needs more complicated calculation and more data. The latter one is kind of statistical method, analogy and graphical modeling, but it only applies to certain special cases, and its precision is also poor [2].

Empirical formula recommended by regulation for hydrologic computation of water resources and hydropower projects [3] is most commonly used. Much research about it [4, 5] used the former formula to forecast the distribution of downstream water temperature of Reservoir, the impact of its length, etc. LI Ke-feng [6] thought statistical analysis of a lot of meteorological data and water temperature which are the observation data, and established the function to express the relationship between the natural water temperature and air temperature, relative humidity and wind speed. Then the formula was verified by using measured data, finally applied it to river water temperature which was lack of temperature monitoring data.

As the empirical method partly considering water temperature and other elements linked, but mainly get statistics from the measured data of comprehensive, reflecting the statistical rule, but it couldn't consider the effect process to discuss the inherent laws of water temperature, what is more, its calculation process didn't take into account the characteristics of rivers, reservoir operation, etc., its applications have some limitations.

When mentioned modeling, currently the model for one-dimensional vertical have much applications, most of which are vertical temperature model along the river. In practice there are also two-dimensional model to simulate it, but using three-dimensional model is less.

With continuous improvement of the numerical method and computer application technology, vertical one-dimensional mathematical model of water temperature are more applications. In practice, DAI Qun-ying [7] used coupled model of the vertical one-dimensional water model and reservoir vertical water to predict downstream along the Hexi River distribution. Base on the steady flow, Wang Ying [8] simplified the above model to simulated downstream water temperature variation of Nuozhaidu Reservoir and influence on water environment.

The aim river, Zaoshi River, located in east longitude 110°29'~111°23', latitude 29°16'~30°08', in west-north of Hunan Province in china. The whole length of the river is about 17 km. It belongs to the inland mountain area. Its upper boundary is Zaoshi Reservoir, downstream is the Sanjiangkou Reservoir. Controlled basin area of Zaoshi Reservoir is about 3 000 km^2, it occupies 93.7% of Dieshui River. The main task of project development is flood control taking into account the power generation, irrigation, navigation and other utilization requirements. Figure 1 shows the general map of Zaoshi River. Figure 2 shows the level of each section along the downstream.

Obviously Zaoshi Dam has affected the physical, biological and cultural resource of the river downstream. This paper wants to simulate the temperature variation and give some suggestion to prevent from the bad influence.

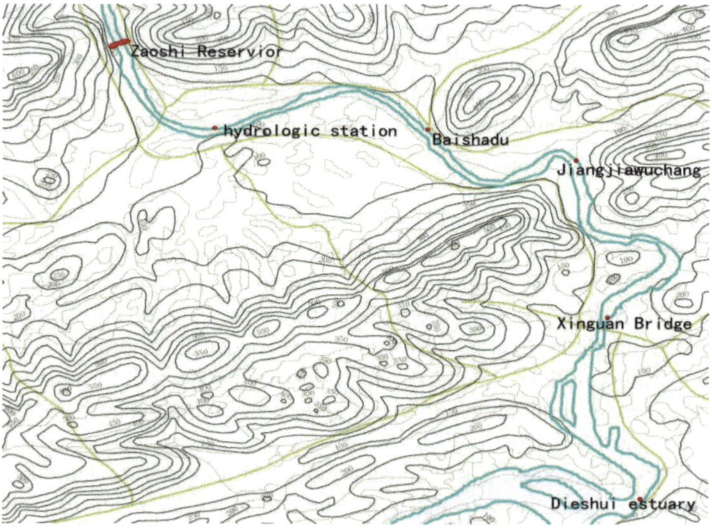

Fig. 1. The general map of Zaoshi River

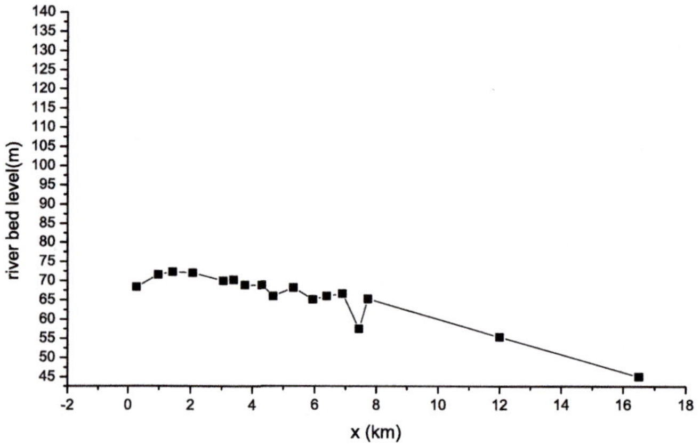

Fig. 2. The level of each section along the downstream

The one-dimensional unsteady flow temperature combined model was chosen for the objective river. It was composed by the one-dimensional Saint-Venant equations and the temperature Advection-Dispersion equation. As the model is the mechanism of formation in the water temperature based,when apply it, need more field data in order to achieve more accuracy than the empirical method. Therefore, data collection before becomes exceptionally important. Generally, categories of information required as follows:

1) Terrain information. Pretreatment of river terrain is an important part of Dynamics simulation [9];
2) Meteorological data. Synthesis of existing research results, obtained the major influencing factors of river downstream, such as weather and water temperature are two factors in the upstream inflow. Meteorological factors include temperature, solar radiation, cloud cover, wind speed, relative humidity;
3) Boundary conditions. For Upstream boundary conditions usually is flow and temperature, it is to say the discharge value of exports from power stations and water temperature value. For downstream boundary conditions is water level, and often use the value of the end section;
4) Initial conditions. Time-period initial discharge and temperature values of discharge, river water level and the initial temperature of the river;
5) Validation data. To validate the results of simulation need flow and temperature measured. So how to measure the temperature is important for validation process. In this study, Fig. 3 shows the kind of thermometer to measure of water temperature. Its measuring accuracy is 0.2°C, measuring range –7–40 °C.So the river was divided by 4 controlled sections, at 10:00 and 15:00 on every 8, 18, and 28 of each month.

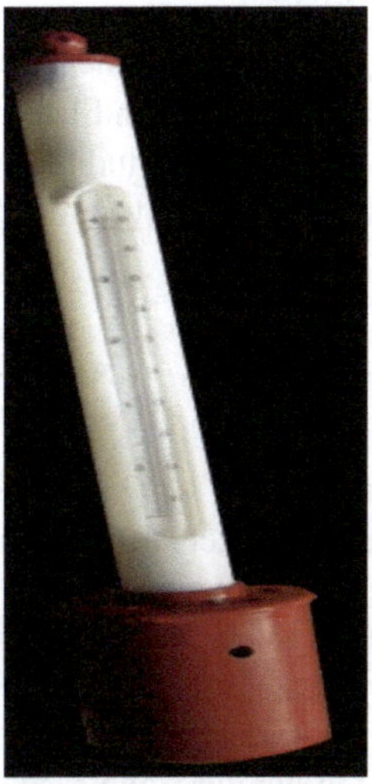

Fig. 3. The measurement of water temperature

3 Combined Model

In this paper, we used two basic models to combine and calculate the water flow volume and water temperature.

(1) Hydrodynamic model (HD).

In the form of z, Q variable, without considering tributaries, the flow equation is de Saint-Venant system of equations, as follows:

$$B\frac{\partial z}{\partial t} + \frac{\partial Q}{\partial x} = 0 \tag{1}$$

$$\frac{\partial Q}{\partial t} + \frac{2Q}{A} \times \frac{\partial Q}{\partial x} + (gA - \frac{BQ^2}{A^2})\frac{\partial z}{\partial x}\bigg|_z + gA\frac{Q^2}{K^2} = 0 \tag{2}$$

where z is the section water level, m; Q is the section flows, m3/s; A is the area of flow section, m2; B is the width of the river, m; n is bed roughness, according to experiences; K is discharge modulus, m3/(s•m2); t is the time step, s; x is the length step; g is the acceleration of gravity, m/s2. For HD model, first apply Difference Method to discrete above equation, and use pursuit method to solve the discrete equation, you can get more detail in master paper [6].

(2) Advection-Dispersion model (AD).

Neglected the heat exchange between water and river bottom, on assumption of unsteady state conditions, complete vertical and lateral mixing in river cross-sections, advection dominating over dispersion (longitudinal) and a simplified expression for surface heat exchange, the model of the advection-dispersion derives from the heat balance equation as follows:

$$\frac{\partial(AT)}{\partial t} + \frac{\partial(QT)}{\partial x} = \frac{\partial}{\partial x}(AE_L\frac{\partial T}{\partial x}) + \frac{B_S S}{\rho C_P} \tag{3}$$

where T is the section water temperature,°C; Q is the section flow, m^3/s; A is the area of flow section, m^3/s; B_s is the width of water surface of the section, m; E_L is longitudinal dispersion coefficient; ρ is the water density, kg/m^3; C_P is the specific heat of water, 4184 J/(kg°C) = 4184 W/(s kg°C); t is the time step, s; x is the length step; S is the net heat exchange flux unit according the water surface, it included the heat exchange between the external (sun and air) and the water, W/m^2, it is calculated by a set of empirical equations that account for air temperature, wind speed, solar radiation, atmospheric long wave radiation, long wave back radiation, evaporation/condensation, conduction, solar reflection and atmospheric reflection. You can read this paper [10] to know more details.

In general, heat in the river affected by the vertical diffusion role is far less than the plug-flow dispersion role. Heat exchange unit included the water surface, the riverbed, the internal heat exchange, etc. However, the river bed and internal heat exchange is also very small in the calculation of heat exchange, so we usually only consider the heat surface exchange. We could also get discrete equation using Difference Method. So the left of the equation is

$$\frac{\partial (QT)}{\partial x} = \frac{Q_{i+1}T_{i+1} - Q_{i-1}T_{i-1}}{2\Delta x} \tag{4}$$

The right of the equation is

$$\begin{aligned} \frac{\partial}{\partial x}\left(AD_L\frac{\partial T}{\partial x}\right) &= \frac{\partial}{\partial x}\left(AD_L\frac{T_i - T_{i-1}}{\Delta x}\right) \\ &= \frac{\left(AD_L\frac{T_i - T_{i-1}}{\Delta x}\right)_{i+1} - \left(AD_L\frac{T_i - T_{i-1}}{\Delta x}\right)_i}{\Delta x} \\ &= \frac{A_{i+1}D_{L\,i+1}(T_{i+1} - T_i) - A_iD_{L\,i}(T_i - T_{i-1})}{\Delta x^2} \end{aligned} \tag{5}$$

And

$$\frac{B\varphi}{\rho C_p} = \frac{B_i\varphi_i}{\rho C_p} \tag{6}$$

At last, the left equals to the right, we can get as follows:

$$\begin{aligned} \frac{Q_{i+1}T_{i+1} - Q_{i-1}T_{i-1}}{2\Delta x} &= \frac{A_{i+1}D_{L\,i+1}(T_{i+1} - T_i) - A_iD_{L\,i}(T_i - T_{i-1})}{\Delta x^2} \\ &\quad + \frac{B_i\varphi_i}{\rho C_p} \\ \Delta x(Q_{i+1}T_{i+1} - Q_{i-1}T_{i-1}) &= 2A_{i+1}D_{L\,i+1}(T_{i+1} - T_i) \\ &\quad - 2A_iD_{L\,i}(T_i - T_{i-1}) + 2\Delta x^2\frac{B_i\varphi_i}{\rho C_p} \\ 2A_{i+1}D_{L\,i+1}T_i + 2A_iD_{L\,i}T_i &= -\Delta x(Q_{i+1}T_{i+1} - Q_{i-1}T_{i-1}) \\ &\quad + 2A_{i+1}D_{L\,i+1}T_{i+1} + 2A_iD_{L\,i}T_{i-1} + 2\Delta x^2\frac{B_i\varphi_i}{\rho C_p} \end{aligned} \tag{7}$$

The solution process of combined model of one-dimensional river water temperature is as follows: First, base on the upstream and downstream boundaries of the hydrodynamic model calculate a time step, then substituted those into the thermal diffusion equations, calculated the same time period water temperature, and then calculated the next step evolution of flow and water temperature conditions, until the end of computation time.

4 Results and Discussion

The modeled flow and temperatures were compared with the observed record in order to fit the combined model from Apr 2008 to March 2009, using solar radiation (Ra, W/m^2), air temperature (Ta, °C), cloud cover (C, %), wind speed (U, m/s), relative humidity (RH, %), and to verify the combined by using the data from April 2009 to March 2010, at last apply the model to predict the temperature of the next year and After analysis from April 2008 to March 2010 the minimum discharge is 19 m^3/s, and the maximum one is 215 m^3/s. The below Fig. 4 shows the process of different water surface flow (the end section was effected by Sanjiangkou Reservoir, Normal water level H = 65 m) via different release volume. It shows when the water level constant, surface water level of river did not vary with the flow significantly, especially the distance above 12 km.

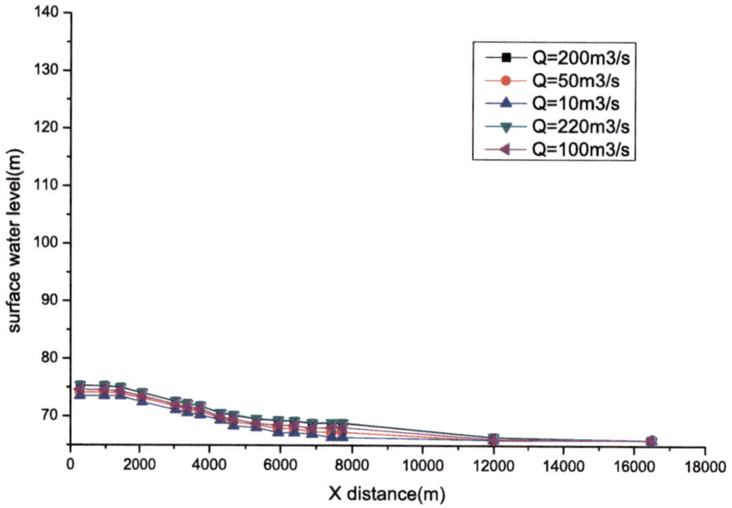

Fig. 4. Surface water level of river when the different discharges happened

From the figure above, the water lever maximum difference between the biggest discharge and the smallest one is about 1.5 m. The Zaoshi Dam could adjust the flow very well comparatively the nature river. It has less bad influence to the downstream river.

Solutions of hydrodynamic model included the area of each section (A, m^2), width of water surface (B, m) and so on. Through substitution the above result to water temperature, modified the difference between the observed data and simulate result from Apr 2008 to Mar 2009, we get some important parameters: main river bed roughness n = 0.3–0.4, beach n = 0.6–0.8. The more the discharge, the larger value we use. The longitudinal dispersion coefficient can range from 30–50.

Figure 5 shows the numerical is very near to the measured value. Also, in Table 1 the difference between numerical and measured is manly low to 0.2°C, just one value to 0.5°C, and the relative error is manly less 2% from Table 2. So we thought the model built can simulate this river very well. We can find that the river water temperature did not change a lot along the river. In spring and autumn, the different value is less than other two seasons. What is most important is that the river temperature is equal to the release. So we can conclude the release from the dam is the most main factor.

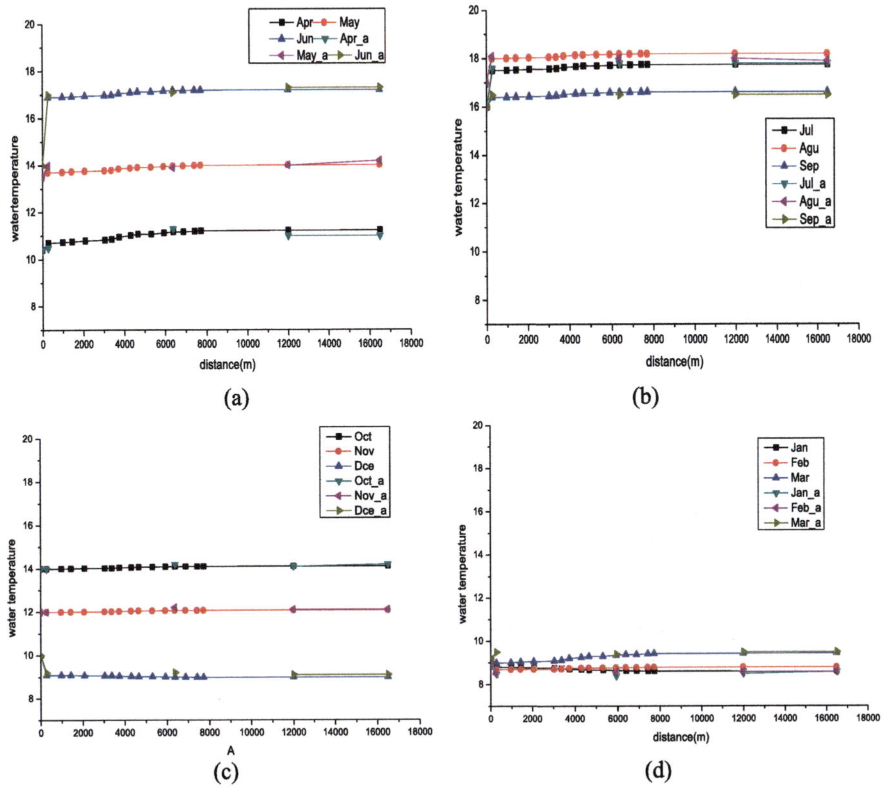

Fig. 5. Simulate results for the application of the combined model from Apr 2009 to Mar 2010(E.g. Apr means simulate value, Apr_a means actual value)

Table 1. Difference of the numerical divided by the measured value at 4 measured sections °C

Distance	Month											
	Jan	Feb	Mar	Apr	May	Jun	Jul	Agu	Sep	Oct	Nov	Dce
280	–0.3	–0.2	0.5	–0.2	0.3	0.1	0.1	0.1	0.1	–0	–0	0.1
5940	–0.2	–0.2	0.0	0.1	–0.1	–0.1	0.08	–0.2	–0.1	0.11	0.14	0.2
12000	–0.1	–0.2	0.06	–0.2	–0	0.09	0.06	–0.2	–0.1	–0	0.03	0.12
16500	0	–0.2	0.05	–0.2	0.19	0.09	0.06	–0.3	–0.1	0.09	0.03	0.12

Table 2. Relative error between the numerical and measured value at 4 measured sections

Distance	Month											
	Jan	Feb	Mar	Apr	May	Jun	Jul	Agu	Sep	Oct	Nov	Dce
280	–4%	–2%	5%	–2%	2%	1%	1%	1%	1%	0%	0%	1%
5940	–3%	–2%	0%	1%	–1%	0%	0%	–1%	–1%	1%	1%	2%
12000	–1%	–2%	1%	–2%	0%	1%	0%	–1%	–1%	0%	0%	1%
16500	0%	–2%	1%	–2%	1%	1%	0%	–2%	–1%	1%	0%	1%

We use the reservoir water temperature model [11] to get the vertical temperature before the dam from Apr 2010 to Mar 2011, as the upper boundary of AD. After we solved combined model of HD and AD using the recently decade average flow and meteorological data, the simulate result is the Fig. 6. The initial condition is that the whole river water temperature is the same, Given 9.6 °C. Table 3, Table 4, Table 5 gave the other calculate boundary.

Table 3. The discharge and water temperature of Zaoshi reservoir for each month

month	1	2	3	4	5	6	7	8	9	10	11	12
Q	68.3	88	94.2	109.2	158	206.8	207.6	58.1	55.1	58.5	50.3	41.6
T	9.6	8.8	10.7	11.9	15.0	15.0	19.0	19.6	15.5	16.3	15.5	12.3

Table 4. The decade average air temperature and relative humidity of Zaoshi

month	1	2	3	4	5	6	7	8	9	10	11	12
Ta,°C	4.8	6.3	10.5	16.4	21.3	25	28.2	27.2	23.1	17.8	12.2	7
RH,%	49	54	58	62	66	74	76	75	76	66	55	49

Table 5. Water level of Sanjiangkou Reservoir in each month for the down boundary

month	1	2	3	4	5	6	7	8	9	10	11	12
H,m	68	68	68	68	68	68	68	68	68	68	68	68

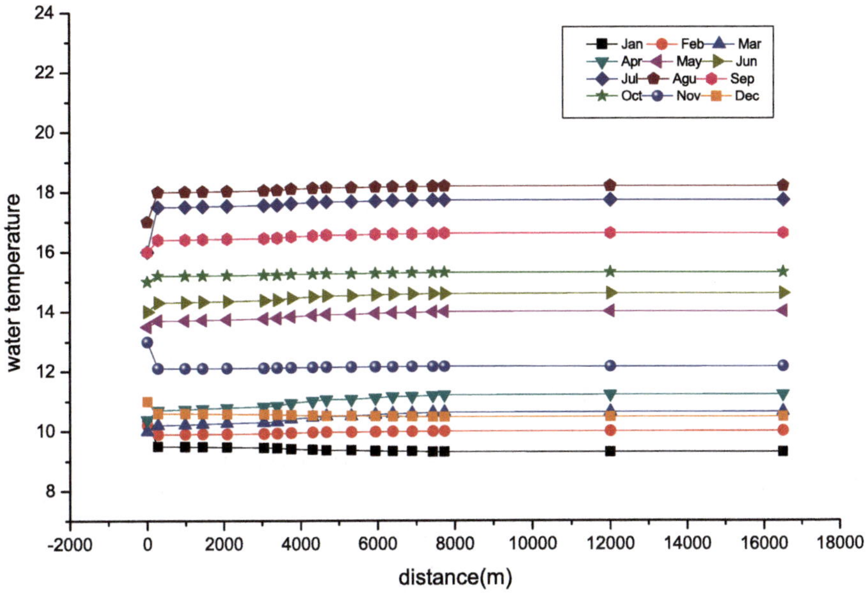

Fig. 6. Simulate result of the combined model after validation form Apr 2010 to Mar 2011

From the above figure, we can know the water temperature of downstream is near o the release, and the value is no significant as the distance. The difference between "0" and "280" is largest than other section. The water temperature value is above 15°Cfrom Jul to Sep, it has a little bad influence to irrigation. So it doesn't need any measurement to adjust the influence now. It also shows that water temperature value increases in the start river section from May to Oct, while in other months the value decreases and then increases via the distance. On the basis of analysis on the average air temperature and water temperature, we initially deduced it may be caused by the Meteorological factors.

5 Conclusions

The result of the model provides an ideal tool for testing and optimizing man-made interventions and management solutions to alleviate existing problems in reservoirs. Similar it is the perfect tool during the design phase or operating period for carrying out Environmental impact Assessment of reservoirs as well as the impact upstream and downstream.

In the beginning of this research, firstly studied many papers, we obtain the three stages of the water temperature simulation and the main factors, which is the release volumes and temperature, air temperature, light duration, and relative humidity. Secondly, input the data of the above five items and measured temperature from April 2008 to March 2009 to calibrate the model parameters, and then used the data from April 2009 to March 2010 to check the model built. We used the combined model composition of Hydrodynamic model and Advection-Dispersion model to predict the Zaoshi very well. In the last we used the model to simulate the water temperature from Apr 2010 to

Mar 2011.In the modeling process, Meteorological conditions, including air temperature, solar radiation, wind, and humidity,Hydraulic parameters, including in stream flow rate, inflow temperature, river width and length, bottom slope and bed roughness are needed. For Zaoshi the water surface level of the river was affected by the up discharge boundary and down Sanjiangkou reservoir water level boundary, the temperature of the river was affected by the release water temperature mostly. The cause of this result may be the small capacity of the stream for heat storage. We should collect these factors very accurately before the simulation. The simulate result figure shows water temperature value increases in the start river section from May to Oct, while in other month the value decreases and then increases. After Analysis we initially deduced it may be caused by the Meteorological factors. Form Apr 2010 to Mar 2011 the water temperature value is above 15°C from Jul to Sep, it has a little bad influence to irrigation. It doesn't need any measurement to adjust the influence now.

Next status we could change the management of Zaoshi Station to let the temperature satisfy all need of kinds. It is to say is to do more research on how to simplify the model for Zaoshi environment, and simulate more accurate. We also try to calculate the maximum and minimum temperature of each month by obtaining more information.

Acknowledgements. Foundation Item: Zhejiang Province Basic Public Welfare Research Program (LG22E090001); Water Conservancy Science and Technology Program (RC2119); Zhejiang Institute of Marine Planning and Design (ZIHE23Y007).

References

1. CAO, Y.Z.: Research on the water temperature simulation for Longyangxia reservoir and Influence on temperature of river downstream. Xi'an University of Technology, 3 (2008)
2. Wang, T.A.N.G., Li, W.A.N.G., Quan, Q.U.A.N.: Water temperature forecast for the section downstream of heishanxia reservoir on the yellow river. Power Syst. Clean Energy **24**(4), 71–75 (2008)
3. Regulation for hydrology computation of water resource and hydropower projects. SL278–2002
4. Xiao-qun, Z.H.A.N., Jian, C.H.E.N., Jian-jun, H.U.: Water temperature calculation for shankouyan reservoir and influence on temperature of river downstream. Water Res. Prot. **21**(1), 30–35 (2005)
5. La-hua, J.I.N., Feng-jun, X.U.: Water Environment Simulation and Visualization Technology. Chemical Industry Press, Beijing (2004)
6. LI, K.F., HAO, H.S., ZHUANG, C.Y.: A new method for predicting water temperature of river by using meteorological factors. J. Sichuan Univ. (Eng. Sci. Ed.) **38**(1), 1–4 (2006)
7. Qun-ying, D.A.I.: Study on water temperature forecasting of the reservoir and its Lower rivers, p. 5. HoHai University, Nanjing (2006)
8. Wang, Y., Zang, L., Zhang, X.: Water temperature of river model and its forecast for downstream flow of nuozhadu reservoir. J. Xian Univ. Technol. **19**(3), 235–239 (2003)
9. GU, L.Z., GUAN, K., ZHENG, G.D.: Discussion about terrain pre-treatment techniques in MIKE21. Guangdong Water Res. Hydropower 8 (2008)
10. Roy, R., Gu, Li, Y.T.: River temperature sensitivity to hydraulic and meteorological parameters. J. Environ. Manage. **66**, 43–56 (2002)
11. ZHANG, H.l., ZHOU, X.D., CHEN, P.C.: Research on precipitation methods in reservoirs temperature. Water Conservancy Sci. Technol. Econ. (2008)

Analysis of Treatment Methods for Substation Effluent

Lijun Fu(✉), Yanling Guan, Liang Zhang, Bing Wei, Guoxin Li, Han Wang, and Jingyuan Chi

Power Science Research Institute of Heilongjiang Electric Power Co., Ltd,, Harbin, Heilongjiang, China
584085737@qq.com

Abstract. It is the social responsibility of power grid enterprises to actively implement the national environmental protection policy. The classification of substation effluent, the present situation of oil-water treatment in substation, the environmental risk of substation effluent, the treatment methods of production effluent and domestic effluent are described in detail. Biological contact oxidation method is usually used to treat domestic sewage of substation, and the treatment measure of production wastewater is "High-efficiency adsorption + electric oxidation" integrated accident oil pool, and the oil-water separation device is installed inside the accident oil pool, when the oily wastewater flows into the accident oil pit, the oil-water separation is carried out by the oil-water separation facilities, and the waste water is discharged after being treated by the oil-water separation device.

Keywords: Substation · Effluent · Environmental protection · Treatment

1 Introduction

With the rapid development of our country's economy, the power consumption load is increasing day by day, and environmental problems are constantly arising with the operation high-voltage substations. During the normal operation of the substation, on-duty personnel will generate a certain amount of domestic sewage; the main transformer within the high-voltagestation, when converting electrical energy into magnetic energy, the iron core will heat up under various effects, and a large amount of transformer oil is injected into the shell to ensure the safe and stable operation of the transformer and avoid accidents.The maintenance personnel will produce waste oil, wastewater or a mixture of both during the repair or maintenance process, which will affect the water environment around thestation [1, 2]. Therefore, the wastewater discharged from the substation needs to be treated to meet the clean emission standards before being discharged.

D. Han et al. (Eds.): ICEPG 2024, ESE, pp. 281–288, 2025.
https://doi.org/10.1007/978-3-031-86829-0_33

2 Classification and Current Situation of Wastewater Discharged from Substations

2.1 Classification of Wastewater Discharged from Substations

The wastewater generated by the substation during operation is divided into domestic sewage and industrial wastewater according to its source.

The wastewater produced during the production process of the substation is industrial wastewater, for example: the wastewater generated when staff rinse insulating porcelain bottles radiators (including the main transformer cooling fans) during equipment maintenance, the small amount of oily wastewater produced when oil-containing equipment releases oil, and the oil leakage drainage caused by sudden equipment failures, all of which will be discharged into the accident oil pool, with the main pollutants being oil and SS.

Domestic sewage refers to the sewage and domestic washing water produced by the staff of the substation during the operation period, with pollution factors such BOD5, COD, SS, ammonia nitrogen, and E. coli. As the scale of power enterprises gradually expands and the number of personnel increases, the amount domestic sewage produced is also increasing. If it is not treated and directly discharged, it will cause certain environmental pollution [2–4].

2.2 Current Status of Oil and Water Treatment in Substations

In daily operation, the accident oil pit of the substation collects not only the oil leaking from the equipment but also some of the rainwater produced the substation. In daily operation, the accident oil pit of the substation collects not only the oil leaking from the equipment but also some of the rainwater produced within the substation. Once there is heavy rain, these untreated or incompletely treated oil-water mixtures will be discharged from the station with the rainwater, causing pollution water bodies and soil environment [5, 6].At present, the oily wastewater generated by substations is mainly collected, settled, and separated from oil and water by the accident oil pool, and naturally overflowed according to the water level in the pool to form direct discharge. The accident oil pool is a traditional concrete structure, which has the following problems in operation and treatment effect:

(1) The accident oil pool only has the function of natural physical stratification of oil and water, with poor treatment effect, and the effluent cannot meet standards.
(2) The accident oil pool lacks water quality monitoring instruments, making it impossible to effectively verify the water quality during use.
(3) The substation duty personnel lack professional knowledge in wastewater treatment.

3 Environmental Risk of Wastewater Discharge from Substations

At present, the accident oil pool in substation design has a single function, which is only for wastewater storage and simple sedimentation. The oily is naturally discharged directly according to the water collection height in the pool without any treatment. The components of oily wastewater discharged directly are complex and contain various toxic organic compounds, with high COD values, posing a serious risk of ecological and environmental. The harm of oily wastewater mainly lies in its pollution of rivers, lakes, and seas as well as the land. Scientific research shows that if oily wastewater is discharged into rivers and lakes, it will cover the water surface, thus isolating the exchange between gases the water and the atmosphere, leading to a sharp decrease in the oxygen content of the water. The decrease in oxygen content in water bodies can directly affect the growth of aquatic organisms, leading to the death of aquatic plants and animals, causing a in water quality, and directly impacting the utilization of water resources [7]. More seriously, if oily wastewater directly contaminates drinking water sources, it can lead to large-scale human illnesses and even directly cause group food poisoning, significant harm.

In addition, when oily wastewater is dumped on the ground without treatment, it will also pollute the soil. The oil will adhere to the leaves plants, blocking the normal photosynthesis of plants; the sediment of oily wastewater will affect the normal growth of plant roots, leading to large-scale death of plants.

Some pollutants in wastewater are not toxic themselves, but they are harmful to water bodies due to their large quantity or high concentration. For example, when the amount of organic matter discharged into the water body exceeds the allowable amount, the water body will experience anaerobic corruption. When a large amount of inorganic matter flows in, it will increase the salt concentration in the water, causing a change in osmotic pressure which has an adverse effect on living things (plants and animals and microorganisms).

4 Wastewater Treatment Methods and Control Measures for Substations

4.1 Methods for Treating Production Wastewater

In the technical field of oily wastewater treatment, there are mainly oil separation, chemical demulsification, coagulation and flotation, biological degradation andorption, oxidation and other processes. Among them, the oil separation process is only suitable for preliminary pretreatment of oily wastewater and cannot meet the requirements of deep treatment. The techniques such as demulsification, coagulation and flotation, and biological degradation are not suitable for the operation conditions of substations due to complex operation, difficult maintenance, need for chemical reagents, and large equipment size, and cannot be used as the treatment technology for oily wastewater from substation accidents. Only adsorption, oxidation and other process technologies are suitable for advanced treatment, simple operation and maintenance, and unattended operation, and are suitable as selected technology for the new accident oil pit of the substation.

The process flow of the new integrated emergency oil pit technology is that the oil-water mixture from the main transformer pit naturally flows into the isolation chamber the integrated equipment, where it undergoes preliminary purification through collection and static stratification. Then, the wastewater below the oil layer is pumped into the equipment chamber the inlet pump. The equipment chamber consists of adsorption and electro-oxidation units. The wastewater first undergoes oil removal treatment in the adsorption unit, naturally overflows into the electro-oxidation unit for advanced purification treatment. After being tested by the detection equipment to meet the standards, it is discharged outward by outlet pump. The entire process is automatically controlled and intelligently monitored [7, 8]. The operation process is shown in Fig. 1.

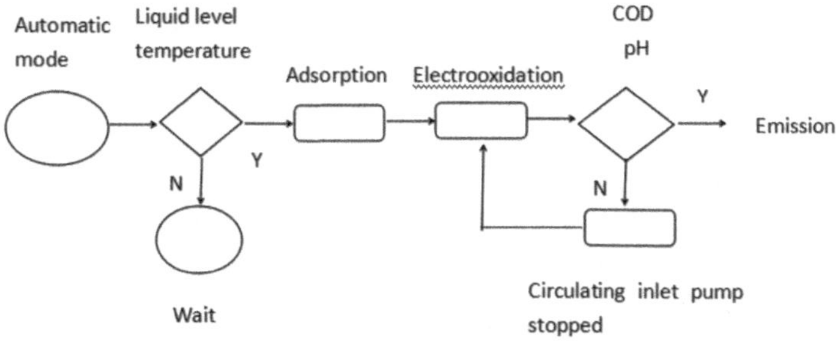

Fig. 1. Operation process

The integrated sewage treatment equipment is adopted, and the automatic control system is configured to merge and treat the construction wastewater, domestic wastewater, and rain generated by the project. The sewage treatment efficiency is high, and the treated water quality meets the construction requirements and is used on-site, effectively solving the problem of and sewage recycling at the construction site, saving resources, reducing construction costs, and continuously improving the construction of 3 C green power grid.

The special construction site slurry treatment and debugging agent is used. This agent is divided into A and B agents. It automatically mixes two liquids in proportion and adds them to the equipment for use. They are stored separately to ensure that their effectiveness will not decrease before use. Compared to wastewater treatment that use a variety of chemical reagents to treat wastewater, the process is complex and time-consuming, making this technology more suitable for wastewater treatment in engineering projects.

In the early stage, BIM technology was used to plan and set up water storage tanks and reasonably arrange pipelines. The reclaimed water pipe and municipal were set in parallel, allowing users to freely choose between reclaimed water and municipal water from the same outlet, making water use more convenient and flexible. A check valve installed on the municipal pipeline to prevent reclaimed water from flowing back into the municipal pipeline and contaminating the municipal water.

The centrifugal pump and the self-priming pump are used in combination. The centrifugal pump is used as a booster pump to supply water the toilet and the building floors, ensuring the water pressure meets the daily flushing requirements. The self-priming pump is used for the water storage tank at the site entrance for washing cars and water guns, ensuring the water pressure meets the minimum water pressure required for green construction vehicle washing. The technical principles are shown in Fig. 2.

Fig. 2. Technical principles

4.2 Treatment of Domestic Wastewater

Biological contact oxidation is usually used to treat domestic sewage in substation. The process flow is shown in Fig. 3. This is because in many sewage treatment processes, biological contact oxidation process because of high removal rate, less sludge, strong biological activity, low engineering costs, low operating costs. No need to add nutrient salt, stable effluent quality and convenient operation and management. It has been widely used in various wastewater treatment units.

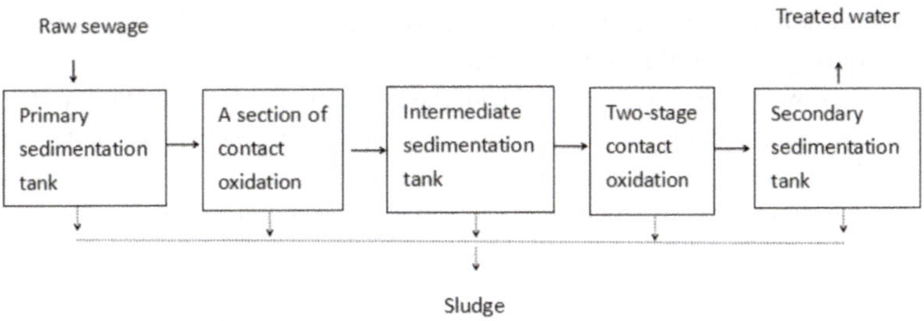

Fig. 3. Process flow of biological contact oxidation treatment technology

The 330 kV substation is divided into a production area and an office area. The office area has a green space of 3,000 square meters, with 6 staff members during the day and 3 at night, as well as 5 security guards and cen staff. The station is equipped with 5 toilets, 1 electric water heater, 1 fully automatic washing machine, and 1 canteen. The wastewater discharged is domestic sewage. The domestic sewage in the station is collected through pipelines into the septic tank, then flows into the regulating pool, and is lifted 2 sewage pumps to the station's infiltration well for discharge. The water quality is shown in Table 1.

Table 1. Wastewater quality discharged from a 330 kV substation

Project	Adjust the water quality of the regulating pool	Comprehensive emission level one	Urban water quality for greening
pH	6.0–9.0	6.0–9.0	6.0–9.0
COD_{Cr}/(mg/L)	350	100	–
Suspended solids/(mg/L)	44.8–155.4	70	10(NTU)
BOD_5/(mg/L)	28.0–119.5	20	20
Oil content/(mg/L)	20	10	–
Ammonia nitrogen/(mg/L)	5.6–26.3	15	20
Total phosphorus/(mg/L)	0.8–3.3	0.5	–

Therefore, for substations that require domestic sewage treatment, the following points must be considered: 1) Rainwater and sewage pipelines should be; 2) Add oil separators for oily catering wastewater; 3) Fully consider the capacity of sewage treatment equipment, and design the integrated A/O process treatment equipment with a sewage treatment capacity of 0.5m3/h, with a retention time of 26.5h (greater than the ~ 16h specified in the outdoor drainage design code); 4) Add a sewage return pump from the aerobic tank to the anoxic tank to enhance denification effects and prevent the death of biofilm during sewage flow interruptions.

Since the flow of domestic sewage from the substation is small and the water quality is simple, and the green area of the office is large enough digest the recycled and purified sewage, it is decided to use the existing sewage treatment facilities as the basis, adopt the underground integrated A/O process sewage treatment and the station area greening sprinkler irrigation pipeline network to use the purified water for office area greening sprinkler irrigation.

The existing regulating tank can be used. The existing lift pump has a processing flow of 10.0 m/d and cannot be utilized so two additional lift pumps with a flow rate of 0.5 ~ 2.5 m/d are added, along with an oil separator for the catering wastewater outlet an improved integrated A/O process sewage treatment equipment, and a landscaping irrigation network system. The regulating tank has a volume of 60 m and a retention of 300 ~ 150 h. The data after A/O process treatment is shown in Table 2.

The sewage treatment facilities are mainly made up of grille well, regulating tank, sedimentation tank, contact oxidation tank, outlet tank, sludge tank, rainwater well and fan. Two-stage contact oxidation process is adopted for domestic sewage treatment. A stereoscopic composite packing is arranged in the oxidation tank, which can attach and

Table 2. Effluent water quality of the clear water tank

Project	Adjust the water quality of the regulating pool	Clear water pond water quality	Comprehensive emission level one	Urban water quality for greening
pH	6.0–9.0	6.0–9.0	6.0–9.0	6.0–9.0
COD_{Cr}/(mg/L)	350	48	100	–
Suspended solids/(mg/L)	44.8–155.4	27	70	10(NTU)
BOD_5/(mg/L)	28.0–119.5	8	20	20
Oil content/(mg/L)	20	2	10	–
Ammonia nitrogen/(mg/L)	5.6–26.3	2	15	20
Total phosphorus/(mg/L)	0.8–3.3	0.5	0.5	–

grow a large number of active biofilms. The microorganism carries on the degradation to the organic matter in the water unceasingly, the microorganism unceasingly metabolism. A high quality variable hole aeration system is installed at the bottom of the tank, and the aeration is carried out by a fan to ensure the oxygen requirement for the normal growth of aerobic microorganisms in the biofilm. With the continuous degradation of organic matter in wastewater by microorganisms, the biofilm becomes thicker and thicker. As the inner layer of the membrane thickens and can not get oxygen, it loses its viscosity and falls off, flowing with water into the secondary sedimentation tank. The new biofilm is formed on the packing again, so that the organic matter in the wastewater is degraded continuously. Contact oxidation process can not only remove organic pollutants in wastewater, but also has the function of nitrification and denitrification [8, 9].

5 Conclusion

The wastewater from substations, especially those far from urban areas, meeting the standards for discharge is a new environmental protection requirement under the current new situation. Each high-voltage substation should be equipped with a closed small sewage treatment system, including primary clarifier, secondary clarifier, sludge tank and disinfection tank, etc. For the discharge of waste oil, the accident oil pool should be built in the high-voltage substation. When the transformer oil leaks or is replaced, it should be collected completely before being imported into the accident oil pool to avoid the pollution of water quality and environment. Substation managers should regularly arrange personnel to check the use of the accident oil pool, whether there is leakage, and water and oil separation process, recycling. Other wastes, such as oil residue, oily wastewater, containing dangerous substances, should be recycled by a specialized agency, substation staff shall not be their own treatment, to prevent accidents.

Acknowledgements. This paper studies is funded by the science and technology project "Research and application of substation wastewater treatment technology" (Project number: 52243724000W) of State Grid Heilongjiang Electric Power Company Limited.

References

1. Yu, F., Wu, Z., Tu, F.: Research on clean emission design technology of substation. Clean Up World **37**(4), 100–102 (2021)
2. Huang, L.: Construction technology of integrated rain and sewage treatment for green power grid project. Civ. Archit. Guangdong **30**(4), 94–95,98 (2023)
3. Xu, Y., Yang, D., Wang, F.: Application of electroadsorption technology in wastewater treatment of electric power industry. Clean Coal Technol. **37**(3), 138–146 (2021)
4. Qiu, B., Li, G., Zhang, H.: Study on application of hydraulic cavitation technology for oil-bearing wastewater from oil pool. Hunan Electr. Power **42**(3), 106–109 (2022)
5. Liu, X., Li, H., Guo, Z.: Study on mobile integrated oil-bearing wastewater treatment device in substation. Green Technol. **116**(2), 109–110 (2020)
6. Xu, K.: Research status and prospect of oily wastewater treatment. Chem. Ind. Manage. **24**, 21–22 (2021)
7. Pan, Z., Huang, B., Liang, Y.: Technical reform and practice of anti-maloperation for oil discharge and nitrogen injection fire extinguishing device. Hunan Electr. Power **41**(1), 44–47 (2021)
8. Suss, M.E., Porada, S., Sun, X., et al.: Water desalination via capacitive deionization. Energy Environ. Sci. **8**(8), 2296–2319 (2015)
9. Porada, S., Schipper, F., Aslan, M., et al.: Capacitive deionization using biomass-based microporous salt-templated heteroatom-doped carbons. Chemsuschem **8**(11), 1867–1874 (2015)

Evaluation of Treatment Processes for Removal of Microplastics in Wastewater

Yanjing Zhu[1], Sijia Yang[1], Kai Yang[1], and Meizi Zhao[2](✉)

[1] IVL Swedish Environmental Research Institute Beijing Representative Office, Beijing 100005, China
[2] TianJin Eco-Environmental Monitoring Center, Tianjin 300191, China
zhaomz_2004@126.com

Abstract. Microplastic in waste water and performance of WWTP (Wastewater treatment plant) in China is presented and compared with other studies mainly from Sweden. The commonly used treatment processes in municipal wastewater treatment plants are introduced and compared regarding to microplastic removal. Literature review has been carried out mainly from Chinese and Swedish study and different microplastic removal processes and techniques are discussed to prevent and reduce emissions of microplastics. Results show that treatment combinations of screen-grit chamber+coagulan+ pre-sedimentation, MBR, combinations of coagulation and sand filtration gave the best microplastic removal in primary, secondary and tertiary process. In addition, it is also found that coagulation can effectively remove microplastics or strengthen the removal of microplastics, which could apply in the primary, secondary and tertiary treatment processes of WWTP in China. This study can further provide technical and data support for the removal of microplastics from actual wastewater.

Keywords: microplastic pollution · wastewater treatment plant · microplastic removal

1 Introduction

Microplastics are usually defined as plastics and rubber particles with a diameter between 1 μm and 5 mm [1]. Nanoplastic is the name for plastic/rubber particles that are even smaller. Microplastics have been found in various regions, environmental media (air, water and soil) and biological tissues (marine life, human being etc.) [2, 3]. Due to its non-biodegradable, long persistent properties and joint action with other chemical substances, microplastics have potential toxic effect [4]. Therefore, it attracts lots of attention to do research on toxicity of microplastics, microplastic pollution distribution and removal technologies.

Wastewater is one of the main carriers of microplastics migration and transformation. Microplastics in municipal wastewater are mainly from personal care products and laundry but with combined sewer systems also road wear particles and plastic litter particles are frequently found in wastewater [5, 6].Traditional wastewater treatment plant

D. Han et al. (Eds.): ICEPG 2024, ESE, pp. 289–305, 2025.
https://doi.org/10.1007/978-3-031-86829-0_34

(WWTP) can remove part of the existing microplastics through filtration, settling, flotation, adsorption etc. Previous studies show that microplastic removal efficiency varied from 50 -100% in municipal wastewater treatment plants [7]. However, due to the different characteristics of microplastics in wastewater and different treatment processes adopted in WWTPs, the removal of microplastics in WWTP are significantly different. As a result, the effluent from WWTP is still an important source of microplastics to water environment.

2 The Status of Current Microplastic Pollution in China, Swedish and Finland

2.1 Microplastic Pollution in China

China is one of the biggest countries for plastic production and consumption. Microplastic pollution is therefore serious in Chinese water environments. The average concentration of microplastics in Tai Lake is 3.4–25.8 particle/liter [8]; water channels connecting to Yangtze river contains microplastics 1.8–2.4 particle/liter and water channels in the textile industrial zone has 2.1–71 particle/liter [9]. Microplastic concentration is 0.0003–0.9 particle/liter in the Chinese sea surface water [10], which is an average pollution level compared to other countries.The distribution of the size of WWTPs (designed capacity) were large (2000–5000 m^3)–34.1%, medium (1000–2000 m^3)–20.8% and small (1000 m^3)–22.9% (Fig. 1).

The most common used wastewater treatment processes in Chinese WWTPs are divided into primary treatment, secondary treatment and tertiary treatment. Primary treatment mainly includes screen, grit chamber and primary sedimentation. Organic matter and nitrogen biological removal is in the secondary treatment, e. g. activated sludge (CAS, AAO, AO, SBR, Oxidation ditch, etc.), biofilm and MBR. The distribution of different treatment process in secondary treatment process is activated sludge (83.76%), biofilm (5.46%) and MBR (0.59%) (data based on the number of installations in the WWTPs). Coagulant sedimentation, air flotation, adsorption, biological filter and membrane filtration are widely applied in tertiary treatment.

Several studies have been carried out to investigate the microplastic properties in Chinese WWTPs. The results show that the composition, size and color of microplastics in municipal wastewater vary with different regions in China. Microplastic properties in influent from 25 WWTPs located in 8 cities (Shanghai, Beijing, Wuhan, Wuxi, Xiamen, Guilin, Changzhou, Nanjing) have been studied [9, 11, 16]. Studies show that fiber was the majority of the microplastics in the wastewater in Shanghai, Beijing and Changzhou, while in Wuxi and Guilin, most of the microplastics were in fragment form. The composition of the microplastics also varied between the cities, in Wuhan, 54.8% of the microplastics were polyamide (PA = nylon) [9] while in Beijing 42.3% of the microplastics were Polyethylene terephthalate (PET) [15]. A high percentage of the microplastics had light color, transparent or white in most of the cities, but in Guilin, 70% of the microplastics had red and green color. A comparison has been made to show the different properties of microplastics in different WWTPs in China (Fig. 2.A-C).

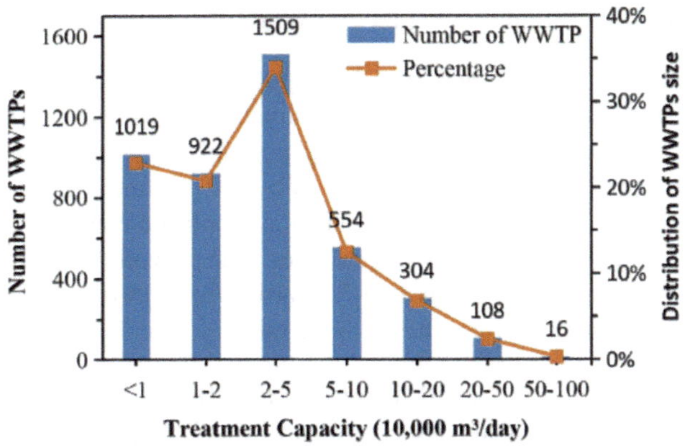

Fig. 1. Treatment capacity of municipal WWTPs in China

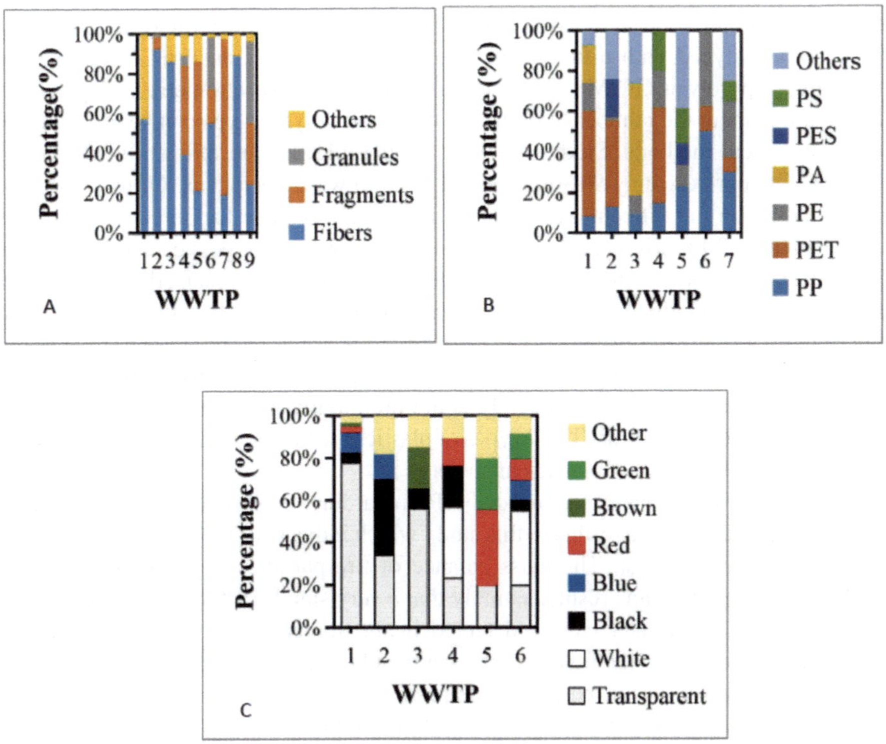

Fig. 2. Microplastic characteristics in WWTP of China. (A) Shapes; (B) Type of polymer; (C) Color [9–18]. (PP: Polypropylene, PET: Polyethylene terephthalate, PE: Polyethylene, PA: Nylon, PES: Polyester, PS: Polystyrene)

2.2 Microplastic Pollution in Sweden and Finland

Sweden is one of the first countries started monitoring microplastics. Gewert shows that the level of microplastics in the Stockholm archipelago, which is part of the Baltic Sea, was 0.0014 particle/liter [19]. Concentrations of microplastics in the four largest Swedish lakes have also been investigated. In the size of > 300 μm fraction, the variation in the levels of microplastics was relatively low. The microplastic concentration is less than 0.6 particle/m3 among 80% of the water samples. The exceptions apply to the samples taken in close connection to cities where 7.2 particle/m^3 was found. This is consistent with the fact that larger cities are often singled out as a significant source of release of microplastics.

Two of the largest WWTPs in Sweden (Käppala in Stockholm and Gryaab in Gothenburg) together with Aalborg University (Denmark) analyzed microplastics including tire particles in the influent, effluent and points among the different treatment processes in WWTPs [21]. Comparisons with previous studies at Sjölunda WWTP in Malmö, Sweden and a study of 10 WWTPs in Denmark have made it possible to draw some general conclusions about microplastic pathways in wastewater. All plastic particles in this project have been analyzed with different types of FTIR (Fourier Transform Infrared Spectroscopy) or pyrolysis GC-MS (gas chromatography-mass spectrometry). Very detailed information about the number of particles, mass and polymer type can be attained with these analytical methods. Studies show that microplastics are efficiently removed at Gryaab, by approximately 99% of microplastics to water recipients. Of the amount of microplastics in the influent wastewater (after 20 mm coarse bar screens) to Gryaab, approximately 30% originates from domestic wastewater (2 g/pe, year) and 70% originates from stormwater, infiltration and other sources (5 g/pe, year). About 30% of the sewer system in Gothenburg is combined which is the reason for substantial parts of microplastics from stormwater.

Domestic wastewater in Stockholm and in Gothenburg, contains 3–10 g microplastics/person/year and contains relatively common polymer types: mainly polyethylene, polyester and polypropylene. These types of plastics are the most abundant in all samples and reflects well the use of plastics in society.

Previous assumptions by the Swedish EPA regarding tire particles are approximately 100 times higher than the results in this study which indicates a load to WWTPs of 1.2 g tire particles/person, year. The mass balance of tire particles at Gryab and Sjölunda WWTPs showed the same results, namely that a very small fraction of the tire particles in the influent wastewater can be detected in the sludge or the effluent wastewater. However, the analytical methods are uncertain and under development.

In Finland a study showed that wastewater influent contained 180 textile fiber and 430 synthetic particles/liter. The effluent contained 4.9(±1.4) textile fiber and 8.6(±2.5) synthetic particles/liter [22]. Another study investigated 4 WWTP in Finland and it showed that fiber was the major MP in wastewater, 39–81% and 29–100% in the influent and effluent [23]. The wastewater contained polyethersulfone (PES) (60%), polyeten (PE) (14%), polyacrylate (7%), polyVinyl Chloride (PVC) (5%), polystyrene (PS) (4%) and polypropen (PP) (3%) [24].

In another comparative study of the removal of microplastics by conventional activated sludge method and MBR process it was found that granular microplastics and fiber microplastics accounted for 17.8% and 82.2% of the total microplastics, respectively. The polymer types of microplastics were mostly polyester (PES) fiber and polyethylene [25].

2.3 Microplastic Removal in Wastewater Treatment Plant

Microplastic concentration in the wastewater was much higher than the values in the river, lakes and sea. Treatment technology can remove microplastics from wastewater and reduce the input to water environment. In recent years, investigations have been made to remove microplastics in the wastewater. Most of microplastics in the wastewater had a size range of 0.05–0.1mm [26]. The size and color distribution of the microplastics were related to the people living style. Study from three wastewater treatment plants in Shanghai showed that the microplastic concentration were 117–226.37 particle/liter and 52–83.16 particle/liter in inflow and outflow respectively, removal rate was 55.6–63.25% [11, 12]. Seven wastewater treatment plant in Xiamen obtained microplastics 1.57–13.69 particle/liter in inflow and 0.2–1.73 particle/liter in effluent, which gave a removal efficiency of 79.3–97.8% [14]. Baresel investigated wastewater treatment plants in Sweden and results showed that influent contained 13–65 particle/liter and removal efficiency was 96% [27]. Effluent from Ryaverket treatment plant only contained microplastics 0.008 particle/liter in the influent and treatment process gave 99.89% removal efficiency. The great removal difference is due to the variation in the treatment process, wastewater quality, the other solid particle size and property of microplastics in the wastewater. Moreover, the different methods of analysis would give uncertainty on the results also (Table 1).

Table 1. Removal efficiency of MP in WWTPs in China and Sweden.

Location	Scale (10,000m^3/d or PE)		WWTP number	Finest mesh (μm)	MP concentration (n/L)		Removal (%)	Ref
					Influent	Effluent		
China		280	1	75	117	52	55.6	[12]
		> 100	2	80	171.89–226.27	69.03–83.16	59.8–63.3	[11]
		100	1	50	12.03	0.59	95.1	[15]
		2	1	20	79.9	28.4	64.4	[9]
		50	1	10	4.72	0.33	93	[13]
		12	2	25	0.28	0.05–0.13	53.6–82.1	[20]
		30	1	43	1.70	0.32	80.9	[17]
		30	1	25	0.7	0.07	90.0	[29]
		0.1–30	11	25	79.33–342.67	3.63–10.59	89.2–97.2	[18]

(continued)

Table 1. (*continued*)

Location	Scale (10,000m^3/d or PE)		WWTP number	Finest mesh (μm)	MP concentration (n/L)		Removal (%)	Ref
					Influent	Effluent		
		--	7	43	1.57–22.91	0.2–1.73	79.3–97.8	[14]
Sweden		12000	1	300	15.1	0.008	99.9	[27]
		740000	1	300	7.3	0.008	99.9	[42]
		14000	1	300	12.1	0.023	99.7	[42]
		Pilot	1	20	12.03	0.59	100	[27]
		--	436	20	13–65	--	> 96	[28]

3 Microplastic Removal in Different Treatment Processes in WWTP

3.1 Primary Treatment

The most used technologies in the primary treatment include screening, grit chamber, flotation,pre-sedimentation to remove big particles, sand and oil. Apart from the commonly used the treatment process mentioned above, primary filtration can effectively remove microplastics, such as drum filter system. Filtration is a process the remove particles in a certain range of size from water. It gives higher removal efficiency as it provides better control mechanism and requires much less space. Depending on the mesh size, MP can be removed. In this process, MP-density is not a driving factor as in the other primary treatment process.

Most of the studies did microplastic removal evaluation, however, not for each treatment process, but for the whole primary treatment. Table 2 shows the removal effect of microplastics by the primary treatment process of sewage treatment plants in China, Sweden and other countries. The removal rate of microplastics in the primary treatment process is 16–98.4%, and the average removal rate is 58.6%.

In Chinese sewage treatment plants, the removal efficiencies of microplastics by the primary treatment process were between 24.3–58.8%, and the average removal rate was 43.1% [9, 11, 15, 30]. It shows that the removal efficiency of a single aerated grit chamber and pre-sedimentation on microplastics was low, only 24.25% and 35.29%, while the removal rate of microplastics by combined aerated grit chamber and pre-sedimentation process increased to 46.5% [17]. Screen - aerated grit chamber - pre-sedimentation and screen-rotational flow grit chamber– pre-sedimentation was commonly used in primary treatment processes in WWTP in China. Both treatment line gave a high removal efficiency for microplastics, 58.8% and 49.7% respectively [11, 15].

Baresel and Olshammar in Sweden studied the removal of microplastics (20 μm -5 mm) in primary treatment by screen- pre-sedimentation/coagulants - drum filter. The whole removal treatment process reached 88.2% for microplastic removal. The addition of coagulant helped the sedimentation and removal of microplastics. The drum filter with a pore size of 1 mm and 2 mm had a removal rate of 89% and 55% for microplastics

Table 2. Removal effect of microplastics by the primary treatment process

Location	Process	Finest mesh (μm)	MP concentration* (n/L)		Removal (%)	Ref
			Influent	Effluent		
China	Aerated GC	43	2.17	1.64	24.25	[17]
	Screen-Aerated GC	80	171.89	86.84	46.5	[11]
	High efficient PS	43	1.598	1.034	35.29	[30]
	Aerated GC + PS	43	1.598	0.854	46.56	[30]
	Screen + Aerated GC + PS	20	79.9	47.4	40.7	[9]
	Screen + Aerated GC + PS	50	12.03	4.96	58.8	[15]
	Screen-rotational flow GC + PS	80	226.27	113.77	49.7	[11]
Sweden	Screen(6mm)	20	126	106	16%	[34]
	Screen + PS&Coagulants				88.2	
	Drum filter(1mm)	20	370	41	89%	
	Drum filter(2mm)	20	370	166.5	55%	
USA	Screen + GC	20	85.6	55.6	35.1	[31]
Australia	Screen + GC	1.5	11.8	5.23	55.7	[32]
Spain	PS	25	451	171	62.1	[35]
Australia	Screen + PS	25	12	2.22	83.2	[25]
	Screen + GC + PS	25	12	1.54	87.5	
	Screen + GC + PS	25	12	1.44	88	
Korea	Screen + GC + PS	1.2	4200	1568	62.7	[34]
	Screen + GC + PS	1.2	31400	12580	56.8	
	Screen + GC + PS	1.2	5840	2080	64.4	
Finland	Screen + Aerated GC + PS	20	430	290.7	32.4	[22]
France	Screen + GC&Oil removal + PS	100	293	90	69.3	[43]
Scotland	Screen + GC&Oil removal + PS	11	15.7	3.4	78.3	[36]
USA	Screen + GC + PS&Oil removal	20	133	21.1	84.1	[31]
	Screen + GC + PS&Oil removal	20	92.9	10.8	88.4	[31]
Finland	Screen + GC&$FeSO_4$ + PS	20	380–686.7	9.9–14.2	97.4–98.4	[24]

*MP measurement was carried out based on the mesh size on the left column. GC: Grit chamber; PS: Primary sedimentation

in sewage respectively. The smaller the pore size of the filter, the higher the removal efficiency of microplastics reached.

The removal rates of microplastics by the primary treatment process in the United States, Australia, South Korea and Finland were between 32.4–98.4%, and the average removal rate was 69.7%, which is much higher than the primary treatment of the Chinese sewage treatment plants. This may be due to the quality of the influent water and the operating conditions of the process. The average value of influent microplastics in the seven sewage treatment plants in China was 70.8 per liter, while the average value of influent microplastics in sewage treatment plants in Europe and the United States (excluding Sweden and South Korea) was 160.7 per liter [22, 23, 31, 32, 43]. In addition, precipitation chemical (for example, hepta) is often dosed before the pre-sedimentation step in WWTPs of Europe and America, which helps to remove microplastics. In China, precipitation chemical is usually added in the tertiary treatment rather than primary treatment.

In the primary treatment process, screen-grit chamber-pre-sedimentation and screen-pre-sedimentation had a removal rate of 35.1–83.2% for microplastics. To remove microplastics, pre-sedimentation was better than grit chamber, but both single treatment process were not as good as the combined processes [25, 31, 32, 35]. Ziajahromi investigated three combined treatment processes: screen-pre-sedimentation, screen-grit chamber-pre-sedimentation and screen-grit chamber-pre-sedimentation to remove microplastics, and the removal rates of microplastics were between 83.2–88% [25]. When the concentration of microplastics in the influent was as high as 4 200–31 400 particle/L, screen-grit chamber-pre-sedimentation combined process can still maintain the microplastic removal rate in the range of 56.8–64.4% [34]. These show that the screen-grit chamber-pre-sedimentation can effectively remove microplastics in sewage.

In addition, because some microplastics have lower density than water and have high affinity with grease and coagulants, the addition of degreasing regent or coagulants can enhance the removal of particles together with microplastics in the primary treatment process. By adding oil removal skimming in grit chamber, screen-grit chamber (with oil removal skimming) – pre-sedimentation could remove 69.3–78.3% of microplastics [36, 43]. Moreover, another study with screen-grit chamber-pre-sedimentation (with oil removal skimming) gave 84.1–88.4% removal for microplastics [31]. Talvitie showed that screen-grit chamber + FeSO4 + pre-sedimentation could remove microplastics as high as 97.4–98.4%, which indicates that adding coagulant could effectively strengthen removal of microplastics [23].

To summarize, the single primary treatment process had a low removal rate of microplastics, among them, pre-sedimentation gave better removal than screen and grit chamber. The treatment line: screen-grit chamber-pre-sedimentation and screen-grit chamber-pre-sedimentation (with oil removal skimming) can effectively remove microplastics from wastewater. Moreover, the addition of coagulate could significantly improve the microplastics removal.

3.2 Secondary Treatment

The secondary treatment refers to the biological treatment, which focuses on removing of organic matter and nitrogen and sometimes phosphorus. The biological treatment can be achieved by activated sludge (AAO, AO, SBR, CAS, modified AS, Oxidation dich, etc.), biofilm sludge or MBR system. In the biological treatment step, sludge can adsorb or behavior as a net to capture the microplastics from the water phase.

Activated Sludge Related Process

Activated sludge (AS) has a certain adsorption and aggregation effect on microplastics. At the same time, the microorganisms in the sludge can also take in part of the microplastics, and then through sedimentation or filtration, the microplastics in the sewage can be separated from the water phase. Many studies showed the microplastics could be removed by activated sludge system (3.6%-42.9%) [37, 38]. Film shape with similar size as suspended solids of wastewater (less than 20um) and fiber shape microplastics were easily to be adsorbed by extracellular polymeric substance (EPS) and settled along with sludge in secondary sedimentation [9, 18, 39]. In fact, any shapes of microplastic could also be wrapped by sludge flocs and then separated from the water phase during settling [39]. 17% microbeads and 13% foam were captured in activated sludge [9].

Biofilm Related Process

Biofilm technology is a biological wastewater treatment process in which solid media are added to suspended growth reactors to offer attachment surfaces for biofilms, in order to increase the concentration of the micro-organism concentration and retention time. The study suggested that adsorption and fixation were the major reason of microplastic removal. Hidayaturrahman and Lee observed that the microplastic removal efficiency in biofilm based process was less than that in activated sludge based process.

MBR Process

Due to the high effluent quality, MBR process is becoming more and more popular for the future treatment plant [40]. The membrane used in MBR is normally microfiltration and ultrafiltration [39] with pore size is 0.1–50 um and 0.001–0.1 um, respectively. Due to the small pore size, the most of the microplastics would be captured by membrane. Baresel et al. showed that MBR with UF (nominal pore size 0.04 μm) could remove 100% of microplastics with the membrane pore size of 0.2 um. Other studies also confirmed that MBR gave a high microplastic removal [23, 31, 38, 41].

General Treatment Efficiency by Secondary Treatment Process

Table 3 shows the removal effect of the secondary treatment process of some municipal sewage treatment plants on microplastics. The removal rate of microplastics in the secondary process was 25.1–99.9%, with an average value of 66.6%.

The secondary treatment processes in Chinese sewage treatment plants are mainly activated sludge (conventional and modified AS, AO, AAO) and biofilm (biological aerated filter). The removal rates of microplastics were 19.9–91.4%, and the average removal rate was 46.4%. The removal rates of microplastics by conventional activated sludge (CAS), AO and AAO were low (19.9–54.5%), and the removal rate of microplastics by biological aerated filter was 76.7%. It is showing that the removal rate of microplastics by the modified activated sludge (AS) process could reach 91.4%, indicating that changing the operating conditions of the secondary treatment process could effectively improve the removal of microplastics [17].

The secondary treatment processes of wastewater treatment plants in Europe, America and South Korea mainly include activated sludge (CAS, AAO), biofilm (biological trickling filter, activated biological filter) and MBR. The removal rates of microplastics were 42.1–99.9%, and the average removal rate was 76%. The membrane module in the MBR process can effectively retain microplastics, and the removal rate of microplastics was > 99%, which is much higher than other secondary sewage treatment processes. The removal rate of microplastics by a single CAS process was maintained between 42.1–87.6%, and the average removal rate was 66.5%. By adding coagulant, the removal rate of microplastics by CAS process could reach 92.6%, while the removal rate of microplastics by AAO + SS was 93.7%, which is much higher than the CAS and AAO process of China's sewage treatment plant (19.9–54.5%). This means that the CAS and

AAO process can be optimized to achieve high-efficiency removal of microplastics. The removal rates of microplastics by aerated biofilm filter + CAS + SS and Biofilter were 48.4% and 61.6%, respectively, which are not significantly different from CAS process.

In the secondary treatment process of the sewage treatment plant, the MBR process can have a removal rate of more than 99% for microplastics. The modified AS and AAO processes, such as adding coagulants, extending hydraulic retention time and settling time, etc., can also effectively separate microplastics from sewage water.

Table 3. Removal efficiency of MP by secondary treatment process in WWTPs

Location	Process	Finest mesh (μm)	MP concentration*(n/L)		Removal (%)	Ref
			Influent	Effluent		
China	CAS-SS	10	1.75	1.24	29.1	[13]
	Modified AS + SS	43	1.64	0.14	91.4	[17]
	AO + SS	80	86.84	69.6	19.9	[11]
	AAO + SS	20	47.4	34.1	28.1	[9]
	AAO + SS	50	4.96	2.26	54.5	[15]
	AAO + SS	80	113.77	85.17	25.1	[11]
	BAF	43	0.944	0.22	76.7	[17]
Finland	CAS + SS	20	290.7	68.6	76.4	[22]
Finland	CAS + SS	20	12.1	1.5	87.6	[24]
USA	CAS + SS	20	21.1	5.9	60.9	[31]
Korea	CAS + SS	1.2	1568	710	54.7	[34]
Korea	CAS + SS	1.2	12580	7863	42.1	[34]
Korea	CAS + SS	1.2	2080	433	77.3	[34]
Australia	CAS + SS	25	1.44	0.48	66.7	[25]
Scotland	CAS + SS(Add polymer)	11	3.4	0.25	92.6	[36]
Australia	ABF + AS + SS	1.5	5.23	2.7	48.4	[32]
Spain	AAO + SS	25	171	10.7	93.7	[35]
France	Biofilter	100	90	35	61.1	[43]
Canada	Trickling filters + Solid contact tank + SS	1–65	2.6	0.5	80.8	[44]
Finland	MBR	250	5	0.4	99.2	[38]
Finland	MBR	20	6.90	0.005	99.9	[23]
USA	MBR	20	55.6	0.5	99.1	[31]

*MP measurement was carried out based on the mesh size on the left column. AS (AAO, AO, CAS, SBR, modified AS, Oxidation ditch, etc.): Activated sludge; CAS: Conventional activated sludge; SS: Secondary sedimentation; BAF: Biological aerated filter; ABF: Activated bio-filter.

3.3 Tertiary Treatment

Tertiary treatment, which is a polishing step after the biological treatment, normally includes filtration and disinfection, for instance coagulation, activated carbon, advanced oxidation (ozone, chlorine and ultraviolet disinfection), membrane technologies, to remove the emerging contaminates.

Table 4 shows the removal effect of microplastics by a tertiary treatment process with two or more treatment units. The removal rates of microplastics in the tertiary treatment process of sewage treatment plants in China, Sweden, Australia and South Korea were between 66.7–96.2%, and the average removal rate was 82.9%. In addition, in Swedish sewage treatment plants, activated sludge is combined with different tertiary treatment units, and the removal rate of microplastics was all above 98%.

Table 4. Removal of MP by combined tertiary treatment processes in WWTPs

Location	Process	Finest mesh (μm)	MP concentration* (n/L)		Removal (%)	Ref
			Influent	Effluent		
China	Denitrification + UF + O_3 + UV	50	2.26	0.59	71.7	[15]
	Air flotation + UF + RO + Chlorination	0.45	368	17	95.4	[33]
	Coagulation + MF + RO + O_3	0.45	291	25	91.4	[33]
	Coagulation + V-filter filtration + Chlorination	0.45	494	125	74.7	[33]
	High efficiency clarification + Sand-filter + Chlorination	0.45	1030	312	69.7	[33]
Sweden	AS + Sand filter	20			98.4	[34]
	AS + UF	20			99.4	[34]
	AS + pond	20			98.7	[34]
	AS high HRT&MF	20			99.8	[34]
	MF-O3-GAC	20	6000	1340	77.8	[34]
Australia	Flocculation + Chlorination + UF + RO	25	0.48	0.21	66.7	[25]
Korea	Coagulation + O_3	1.2	710	33	95.4	[34]
Korea	Coagulation + Membrane disc-filter	1.2	7863	297	96.2	[34]
Korea	Coagulation + Rapid sand filter	1.2	433	66	84.8	[34]

*MP measurement was carried out based on the mesh size on the left column. AS: Activated sludge; UF: Ultrafiltration; MF: Microfiltration; UV: Ultraviolet; RO: Reverse osmosis; GAC: granular activated carbon filter.

Moreover, some sewage treatment plants use a single treatment unit in tertiary treatment. In order to better understand the effect of a single treatment unit on the removal of microplastics, various treatment technologies have also been investigated and analyzed. The results are shown in Table 5 below.

Among the tertiary treatment processes applied in Chinese sewage treatment plants, air flotation had the highest removal rate of microplastics, reaching 47.3–73.4%. The removal rates of microplastics by coagulation were between 19–48.3%, and the average removal rate was 32.5%. V-filter filtration and Sand-filter also had removal rates of

Table 5. Removal of MP by single tertiary treatment process in WWTPs

Location	Process	Finest mesh (μm)	MP concentration* (n/L)		Removal (%)	Ref
			Influent	Effluent		
China	Coagulation	0.45	494	400	19.0	[33]
	Coagulation	0.45	291	203	30.2	[33]
	Coagulation	0.45	1030	533	48.3	[33]
	Air flotation	10	1.24	0.33	73.4	[13]
	Air flotation	0.45	368	194	47.3	[33]
	V-filter filtration	0.45	400	181	54.8	[33]
	Sand-filter	0.45	533	351	34.1	[33]
	UF	0.45	194	178	8.2	[33]
	MF	0.45	203	130	36.0	[33]
Korea	Coagulation	1.2	710	328	53.8	[34]
Korea	Coagulation	1.2	7863	1444	81.6	[34]
Korea	Coagulation	1.2	433	215	47.1	[34]
Finland	Air Flotation(Add PAX)	20	2	0.1	95	[23]
Finland	Membrane disc-filter (20μm)	20	2	0.03	98.5	[23]
Korea	Membrane disc-filter(10μm)	1.2	1444	297	79.4	[34]
Korea	Rapid sand filter	1.2	215	66	73.8	
Italy	Sand filter	10	0.9	0.4	55.6	[45]
Finland	Rapid sand filters	20	0.7	0.02	97.1	[34]
USA	Granular sand filter	20	9.5	2.6	72.7	[31]
USA	Sand filter	125			15	[46]
Finland	Biological filter	20	68.6	8.6	87.5	[22]

*MP measurement was carried out based on the mesh size on the left column. PAX: Polyaluminium chloride; UF: Ultrafiltration; MF: Microfiltration; UV: Ultraviolet; RO: Reverse osmosis.

more than 30% for microplastics. Generally, the pore size of UF and MF membrane is generally less than 1μm, and almost all microplastics should be removed after UF or MF process. However, UF and MF can only remove 8.2% and 36% of microplastics, respectively. The low removal of microplastics by UF and MF may be attributed to the following reasons: (1) The effluent after membrane filtration was contaminated; (2) The detection method was not accurate enough; (3) Membrane was broken.

In the sewage treatment plants in Finland and South Korea, the removal rates of membrane disc-filter (20 μm and 10 μm) for microplastics were 98.5 and 79.4%, respectively, and the removal rate of drum filter(1 mm) for microplastics in sewage was 89% [22, 27, 34], which shows that membrane with disc-filter and drum filter had a good removal effects on microplastics, and the pore size of the filter had a decisive effect on the removal of microplastics [24, 34]. Coagulation, air flotation and sand filter are commonly used in tertiary treatment. In Korean sewage treatment plants, the removal rates of microplastics by coagulation were 47.1–81.6%, with an average value of 60.8%. Studies have proved that aluminum salt, iron salt and PAM coagulant could effectively remove microplastics in sewage [39]. By adding polyaluminum chloride (PAX), the removal rate of microplastics by the air flotation unit was as high as 95%, which is much higher than the range of 47.3–73.4% in Chinese sewage treatment plants [23]. The removal rates of sand filter for microplastics were between 15–97.1% [33, 34, 45, 46]. Only the sand filter treatment unit of a water treatment plant in New York had a poor removal effect on microplastics (15%) [46]· The general removal rates of microplastics from sand filter were 55.6–97.1%, with an average value of 74.8%, which shows that sand filtration had a good removal effect on small fraction microplastics from the effluent of secondary treatment process [33, 34, 45]. In addition, the removal rate of microplastics by biological filter was 87.6%, indicating biological filter as one of tertiary treatment processes may has good effect and potential in microplastics removal from wastewater.

To summarize, membrane disc-filter had a high removal rate of microplastics due to the fact that the pore size of filter membrane was smaller than the size of MPs. Coagulation, air flotation, sand filter and their combination process not only had high removal rate of microplastics, but also were widely applied in tertiary treatment processes.

3.4 Summarization

(1) In primary treatment, combined technologies could obtain a good microplastics removal efficiency, for instance screen-grit chamber-pre-sedimentation process gave removal range between 32.4–88% in China, Australia, South Korea and Finland, and the average removal rate was 62.4%. After adding coagulant before pre-sedimentation, screen-grit chamber + FeSO4 + pre-sedimentation could remove microplastics as high as 97.4–98.4%. Therefore, screen-grit chamber +coagulant+ pre-sedimentation process is the preferred primary process for removing microplastics. Both of screen-grit chamber-pre-sedimentation and coagulant-sedimentation processes have been widely applied for suspended solids removal in WWTPs in the world. Coagulant-sedimentation as an effectively wastewater treatment technology, has the advantages of simple installment, low investment and operational cost. Adding chemical coagulant before pre-sedimentation is commonly used in Swedish WWTPs This is an important reason why the removal rate of microplastics in Swedish WWTPs is higher than that in China. Therefore, adding coagulant in screen-grit chamber + pre-sedimentation process would increase the microplastic removal rate effectively.

(2) In the secondary treatment, the MBR process gave the best removal efficiency which was higher than 99% of removal based on the current studies. Apart from MBR process, 25.1–87.6% of microplastics can be removed by activated sludge (CAS and AAO), and the average removal rate was 54.8%. The modified AS and AAO process can remove 91.4% and 93.7% of microplastics. The removal efficiency of microplastics by AS, biofilm and MBR mainly depends on the sludge separation equipment. The removal efficiency of MBR for microplastics is high due to its small pore size filter membrane, which can intercept sludge and microplastics simultaneously. Some methods may help to improve the removal of microplastics by AS (CAS, AAO,AO, SBR, Oxidation ditch, etc.), such as adding flocculants to enhance the settling performance of sludge and microplastics, extending the residence time of suspended solids in sedimentation tank, and using membrane filtration with smaller pore size.

(3) In the tertiary treatment, membrane filtration system has a high removal rate of microplastics due to the fact of the filter pore size is less than the size of microplastic. However, due to the high investment and operational cost, membrane is not widely used. Coagulation and sand filter are commonly used in tertiary treatment, which also have high removal rates of microplastics. The removal rate of microplastics by coagulation and sand filter was 46.36% and 58.05%.

4 Conclusion and Recommendations

4.1 Conclusion

This paper investigated different municipal wastewater treatment technologies and microplastic removal efficiency from treatment process was compared. Results showed that in primary treatment, screen-coagulant- pre-sedimentation gave a high removal efficiency. In the secondary treatment, MBR achieved the highest removal. The AS (improved activated sludges and AAO) processes can also effectively remove microplastics. The removal efficiency of microplastics by secondary treatment process (AS, biofilm and MBR) depends on the sludge separation equipment. In tertiary process, coagulation and sand filter and their combination processes have high removal rates of microplastics.

4.2 Recommendations

(1) The study is based on the municipal wastewater treatment, the microplastic properties in municipal wastewater is different compared the ones in the industry wastewater. Thus, the treatment technology might be different in the industrial WWTPs.
(2) It is one of the main problems to study microplastics because of the disunity of MPs monitoring technology. Therefore, it is important to modify and standardize the MPs monitoring (sampling and analysis) technologies.
(3) This paper focuses on the microplastic separation from wastewater. However, the separated microplastic will end up in the sludge and future needs better microplastic removal in the sludge management.

Acknowledgements. This sresearch is a deliverable in the project "Identify, prevent and reduce microplastics load to the Yellow Sea and Bohai Sea– A Source-to-Sea approach (IPR-MP)" that has been financed by the Swedish Environmental Protection Agency (grant number NV-00907–20).

References

1. Wang, Y.Q., Wang, H.D., Yan, Z.K., et al.: Research progress on pollution status and removal technology of microplastics in freshwater. Mod. Chem. Ind. **43**, 1–10 (2023)
2. Zhou, Y.W.J., Zou, M., Jia, Z., et al.: Microplastics in soils: a review of methods, occurrence, fate, transport, ecological and environmental risks (review). Sci. Total. Environ. **748**, 141368 (2020)
3. Han, W., Yang, Q., Liu, X., et al.: Research progress on the sources of microplastics in WWTPs. Water Purif. Technol. **42**(4), 23-30,150 (2023)

4. Kögel, T.B.Ö., Toto, B., Bienfait, A.M., Sanden, M.: Micro- and nanoplastic toxicity on aquatic life: determining factors. Sci. Total. Environ. **709**, 136050 (2020)
5. Acharya, S.,Rumi, S.S.,Hu, Y.,et al.: Microfibers from synthetic textiles as a major source of microplastics in the environment: a review.Text. Res. J. **91**(17/18), 2136-2156 (2021). 10,1177/0040517521991244
6. Zhang, X.F., Huang, Y.Q., Li, Q., et al.: Research progress on emerging microplastics pollutant in municipal sewage treatment plants. Mod. Chem. Ind. **43**(2), 17–21 (2023)
7. Han, S., Fu, Y., Zhang, Y., et al.: Characteristics, detection and removal of microplastics in urban sewage. Ind. Water Wastewater **52**(5), 1–5 (2021)
8. Huang, J.: Occurrences, fluxes of microplastics and the carrier function for PPCPs in typical waters of Taihu Lake Basin. Dissertation for the Master Degree, Jiangnan University, China (2024)
9. Liu, X., Yuan, W., Di, M., Li, Z., Wang, J.: Transfer and fate of microplastics during the conventional activated sludge process in one wastewater treatment plant of China. Chem. Eng. J. **362**, 176–182 (2019)
10. Anuj, S., Pritam, P., Pandit, R., Chopade, L.: Eradication of microplastics in wastewater treatment: overview. B. Res. Appl. Chem. **233** (2023). https://doi.org/10.33263/BRIAC133.223
11. Jia, Q.L., Chen, H., Zhao, X., Li, L., Nie, Y.H., Ye, J.F.: Removal of microplastics by different treatment processes in shanghai large municipal wastewater treatment plants. Huan Jing Ke Xue **40**, 4105–4112 (2019)
12. Bai, M.: Occurrence characteristics of microplastics in yangtze river estuary and plastics flux into the sea, China. Dissertation for Doctoral Degree, East China Normal University, China (2022)
13. Hao, X., Zhai, X., Wu, Y., Liu, R.: Evolution and fate of microplastics in sewage treatment. China Water Wastewater **35**, 20–26 (2019)
14. Long, Z., et al.: Microplastic abundance, characteristics, and removal in wastewater treatment plants in a coastal city of China. Water Res. **155**, 255–265 (2019)
15. Yang, L., Li, K., Cui, S., Kang, Y., An, L., Lei, K.: Removal of microplastics in municipal sewage from China's largest water reclamation plant. Water Res. **155**, 175–181 (2019)
16. Huan, J., Tuo, J., Wang, L., Liu, J.: Abundance, characteristics and seasonal variation of microplastics in a domestic sewage treatment plant in Nanjing, China .J. Water Process Eng. **55** (2023). https://doi.org/10.1016/j.jwpe.2023.104200
17. Wang, W.: Study on characteristics distribution of microplastics in Xiamen urban wastewater and its control measures. Master Ministry Nat. Res. (2019)
18. Xu, X., Jian, Y., Xue, Y., Hou, Q., Wang, L.: Microplastics in the wastewater treatment plants (WWTPs): occurrence and removal. Chemosphere **235**, 1089–1096 (2019)
19. Gewert, B., Ogonowski, M., Barth, A., Macleod, M.: Abundance and composition of near surface microplastics and plastic debris in the stockholm archipelago, baltic sea. Mar. Pollut. Bull. **120**, 292–302 (2017)
20. Lv, X., Dong, Q., Zuo, Z., Liu, Y., Huang, X., Wu, W.-M.: Microplastics in a municipal wastewater treatment plant: fate, dynamic distribution, removal efficiencies, and control strategies. J. Clean. Prod. **225**, 579–586 (2019)
21. Tumlin, S.B.C.: Kartläggning av mikroplaster – till, inom och från avloppsreningsverk. Svenskt Vatten utveckling. Rapport Nr 2020–8 (2020)
22. Talvitie, J., Heinonen, M., Pääkkönen, J.P., et al.: Do wastewater treatment plants act as a potential point source of microplastics? preliminary study in the coastal gulf of finland, baltic sea. Water Sci. Technol. **72**, 1495–1504 (2015)
23. Talvitie, J., Mikola, A., Koistinen, A., Setala, O.: Solutions to microplastic pollution - removal of microplastics from wastewater effluent with advanced wastewater treatment technologies. Water Res. **123**, 401–407 (2017)

24. Talvitie, J., Mikola, A., Setälä, O., Heinonen, M., Koistinen, A.: How well is microlitter purified from wastewater? - a detailed study on the stepwise removal of microlitter in a tertiary level wastewater treatment plant. Water Res. **109**, 164–172 (2017)
25. Ziajahromi, S., Neale, P.A., Rintoul, L., Leusch, F.D.L.: Wastewater treatment plants as a pathway for microplastics: development of a new approach to sample wastewater-based microplastics. Water Res. **112**, 93–99 (2017)
26. Jiang, L., Chen, M., Ying, H. et al.: Effects of different treatment processes in four municipal wastewater treatment plants on the transport and fate of microplastics. Sci. Total Environ. **831**, 1–12 (2022)
27. Magnusson, K., Noren, F.: Screening of microplastic particles in and down-stream a wastewater treatment plant. IVL Swed. Environ. Res. Inst. (2014)
28. Baresel, C., Olshammar, M.: Quantification of microplastic discharge from wastewater treatment plants, including sanitary sewer overflow events, to the baltic sea. IVL Swed. Environ. Res. Inst. (2019)
29. Xie, Y., et al.: Impact of sewage treatment plant on the accumulation of microplastics in freshwater organisms in the guilin urban section of the lijiang river. Environ. Sci. 1–13 (2020)
30. Wang, W., Long, Z., Yu, X., Zhang, J., Lin, H.: Microplastics characteristic in yundang wastewater treatment plant of xiamen. Mar. Environ. Sci. **38**, 205–210 (2019)
31. Michielssen, E.A.: Fate of microplastics and other small anthropogenic litter (SAL) in wastewater treatment plants depends on unit processes employed. Environ. Sci. Water Res. Technol. **2**, 1064–1073 (2016)
32. Raju, S., et al.: Improved methodology to determine the fate and transport of microplastics in a secondary wastewater treatment plant. Water Res. **173**, 115549 (2020)
33. Wang, R., Ji, M., Zhai, H., Liu, Y.: Occurrence of phthalate esters and microplastics in urban secondary effluents, receiving water bodies and reclaimed water treatment processes. Sci. Total. Environ. **737**, 140219 (2020)
34. Hidayaturrahman, H., Lee, T.-G.: A study on characteristics of microplastic in wastewater of South Korea: Identification, quantification, and fate of microplastics during treatment process. Mar. Pollut. Bull. **146**, 696–702 (2019)
35. Edo, C., González-Pleiter, M., Leganés, F., Fernández-Piñas, F., Rosal, R.: Fate of microplastics in wastewater treatment plants and their environmental dispersion with effluent and sludge. Environ. Pollut. **259**, 113837 (2020)
36. Murphy, F., Ewins, C., Carbonnier, F., Quinn, B.: Wastewater treatment works (WwTW) as a source of microplastics in the aquatic environment. Environ. Sci. Technol. **50**, 5800–5808 (2016)
37. Chen, S., Yang, L., Zhonghong, L.: Research advances in microplastics nanoplastics/removal technology in municipal wastewater treatment systems China. Environ. Eng. **41**(S2), 38–48 (2023)
38. Lares, M., Ncibi, M.C., Sillanpää, M., Sillanpää, M.: Occurrence, identification and removal of microplastic particles and fibers in conventional activated sludge process and advanced MBR technology. Water Res. **133**, 236–246 (2018)
39. Zhang, X., Chen, J., Li, J.: The removal of microplastics in the wastewater treatment process and their potential impact on anaerobic digestion due to pollutants association. Chemosphere **251**, 126360 (2020)
40. Andersson, S.L., Westling, K., Andersson, S., Lindblom, E.: Long term trials with membrane bioreactor for enhanced wastewater treatment. IVL (2019)
41. Ogonowski, M., Gerdes, Z., Gorokhova, E.: What we know and what we think we know about microplastic effects – a critical perspective. Curr. Opin. Environ. Sci. Health **1**, 41–46 (2018)
42. Magnusson, K., Jorundsdottir, H., Noren, F., Lloyd, H., Talvitie, J., Setala, O.: A Nordic perspective on waste water treatment plants as pathways for microscopic anthropogenic particles to marine systems. IVL Swed. Environ. Res. Inst. (2016)

43. Dris, R., Gasperi, J., Rocher, V., Saad, M., Renault, N., Tassin, B.: Microplastic contamination in an urban area: a case study in Greater Paris. Environ. Chem. **12**, 592–599 (2015)
44. Gies, E.A., et al.: Retention of microplastics in a major secondary wastewater treatment plant in Vancouver, Canada. Mar. Pollut. Bull. **133**, 553–561 (2018)
45. Magni, S., et al.: The fate of microplastics in an Italian wastewater treatment plant. Sci. Total Environ. **652,** 602–610 (2019)
46. Mason, S.A., et al.: Microplastic pollution is widely detected in US municipal wastewater treatment plant effluent. Environ. Pollut. **218**, 1045–1054 (2016)

Study on the Optimal Dosage of Sodium Percarbonate for Treating Severe Cyanobacterial Blooms

Zhichong Zhang[1,2], Jiangchuan Tu[1,2], Mulan Zhu[1,2](✉), and Ping Chen[1,2]

[1] School of Environmental Science and Engineering, Xiamen University of Technology, Xiamen 361024, China
292890310@qq.com

[2] The Key Laboratory of Water Resources Utilization and Protection of Xiamen, Xiamen 361024, China

Abstract. In recent years, with the increase in agricultural activities, point-source and non-point-source pollution during the process have led to excessive levels of nutrients such as nitrogen and phosphorus in small rural water bodies, causing frequent cyanobacterial blooms. Sodium percarbonate, as a new environmentally friendly oxidant, has gained increasing attention for its application in controlling cyanobacterial blooms. However, existing studies lack research on optimizing its dosage from the perspective of cost-effectiveness. To better guide practical applications, this study simulated severe cyanobacterial bloom conditions and conducted comparative experiments on the effectiveness of sodium percarbonate in treating cyanobacterial blooms at eight different dosages: 0, 2, 3, 4, 5, 6, 7, and 8 mg/L. The results showed that, for severe cyanobacterial blooms, considering treatment efficacy, cost, and ecosystem safety, 6 mg/L was the optimal choice. At this dosage, after 48 h of treatment, the algal cell density decreased from an initial 60 cells/L to 1.5 cells/L, achieving a removal rate of 97.5%. The bloom severity was reduced from severe to no visible bloom. These findings provide a reference for determining the optimal dosage of sodium percarbonate for practical applications in treating severe cyanobacterial blooms.

Keywords: Cyanobacterial blooms · Sodium percarbonate · Optimal dosage

1 Introduction

With the increase in agricultural activities, point-source and non-point-source pollution has exacerbated eutrophication problems in small rural water bodies [1, 2]. Eutrophication often leads to cyanobacterial blooms, which not only cause mass fish deaths but also produce harmful substances such as microcystins (MCs), odorous compounds, and other algal organic matter (AOM), posing a serious threat to aquatic environments [3, 4]. According to the "Technical Specifications for Remote Sensing and Ground Detection

Z. Zhang and J. Tu—Co-first author

D. Han et al. (Eds.): ICEPG 2024, ESE, pp. 306–311, 2025.
https://doi.org/10.1007/978-3-031-86829-0_35

of Algal Blooms" (HJ 1098–2020) issued by China's Ministry of Ecology and Environment, algal blooms are categorized into five levels: no significant bloom, mild bloom, light bloom, moderate bloom, and severe bloom. These levels are defined based on algal cell density. The boundary for moderate and severe blooms is defined by an algal cell density (D, cells/L) of $5.0x10^7 < D < 1.0 \times 10^8$ and $D > 1.0 \times 10^8$, respectively. Generally, blooms at or below the light level are considered to be in a healthy state and do not require special treatment.

Currently, there are three main approaches to controlling cyanobacterial blooms: physical, biological, and chemical methods. Physical methods, such as harvesting, may have immediate effects but are expensive and prone to recurrence, limiting their practicality in large-scale applications. Biological methods, such as using aquatic organisms to suppress algal growth, offer ecological benefits but are slow and may introduce invasive species. Chemical methods, including copper sulfate, permanganates, ozone, and polyaluminum chloride (PAC), may cause algal cells to release toxins upon death, leading to secondary pollution, and some chemicals can accumulate in organisms, adversely affecting ecosystems [5].

Sodium percarbonate ($Na_2CO_3 \cdot H_2O_2$), also known as solid hydrogen peroxide, is an eco-friendly oxidant. Compared to liquid hydrogen peroxide, it is cost-effective, easy to store and transport, and its non-toxic, odorless, and non-polluting properties have gained attention in recent years. When added to water, it reacts rapidly to form sodium carbonate and hydrogen peroxide without generating additional pollutants. In aqueous solution, hydrogen peroxide produces oxidative radicals such as hydroxyl radicals ($\cdot OH^-$) and superoxide radicals ($\cdot O_2{}^-$), which have strong oxidative effects on algae [6], disrupting photosynthesis and protein synthesis, ultimately killing algal cells.

Sinha et al. [7] demonstrated that adding 4.0 mg/L of sodium percarbonate to pond water with an algal cell density of 1.1×10^9 cells/L (severe bloom) effectively removed algae, achieving a removal efficiency of 81.8%, and degrading algal toxins. Chen Chao et al.[8] found that in cyanobacterial waters with a total Chl-α concentration of 200 μg/L (severe bloom), adding 5 mg/L of sodium percarbonate resulted in a total Chl-α removal rate of approximately 81.9%. Greenfield et al.[9] discovered that in artificially cultivated water containing 4 × 105 cells/L of Anabaena (mild bloom), the addition of 3.64 mg/L of sodium percarbonate achieved a removal rate of 55%. Previous studies by our research group have shown that treatment with 4.5 mg/L of sodium percarbonate for 123 h effectively managed a severe cyanobacterial bloom with a Chl-α concentration of 352.33 μg/L, achieving a removal rate of 86.22%, without causing lethal effects on fish in the water [10]. However, these studies primarily focus on the effect of a specific dosage of sodium percarbonate under particular algal cell concentrations, and there is limited research comparing the algal removal effects of different dosages of sodium percarbonate in severe cyanobacterial blooms.

To address this gap and improve the application of sodium percarbonate in treating cyanobacterial blooms, this study aims to conduct comparative experiments under eight different dosages of sodium percarbonate (0, 2, 3, 4, 5, 6, 7, and 8 mg/L) in severe cyanobacterial bloom conditions. The results will provide a reference for selecting the optimal dosage for practical applications in treating severe cyanobacterial blooms.

2 Experimental Materials and Methods

The algae species used in this experiment is Microcystis aeruginosa, a cyanobacterium commonly found in algal blooms. The algal strain (FACHB-905) was purchased from the Wuhan Institute of Hydrobiology, Chinese Academy of Sciences, and cultured indoors using BG-11 medium under controlled lighting conditions.

A microcosm simulating severe cyanobacterial blooms was prepared by adding an appropriate amount of distilled water to the concentrated Microcystis aeruginosa culture obtained from the expansion process. To investigate the optimal dosage of sodium percarbonate for treating severe cyanobacterial blooms, eight microcosms with a volume of 2 L each were set up. The initial total Chl-α concentration in all microcosms was the same, at 227.9 μg/L (a severe bloom level). Sodium percarbonate was added to the eight experimental water bodies at dosages of 0 (blank control), 2, 3, 4, 5, 6, 7, and 8 mg per liter of water, respectively. The sodium percarbonate used in this experiment was provided by Wuhan Lullaby Pharmaceutical Chemical Co., Ltd.

The experiment lasted for 48 h, and during the experiment, four water quality parameters—algal cell density, Chl-α concentration, dissolved oxygen (DO), and pH—were monitored every 8 h.

3 Results and Discussion

3.1 Algal Cell Density and Total Chl-*α* Removal Efficiency

Under the conditions of applying sodium percarbonate at dosages of 0, 2, 3, 4, 5, 6, 7, and 8 mg/L per liter of experimental water, the changes in algal cell density and total Chl-α content over 48 h of treatment for severe cyanobacterial blooms are shown in Fig. 1 and Fig. 2.

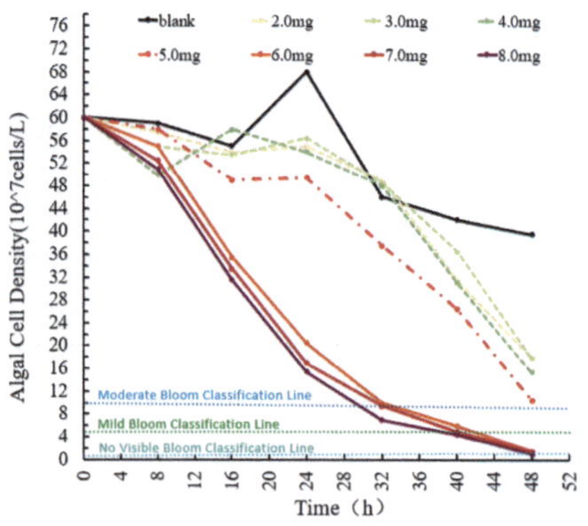

Fig. 1. Variation of Algal Cell Density Over Time at Different Dosages

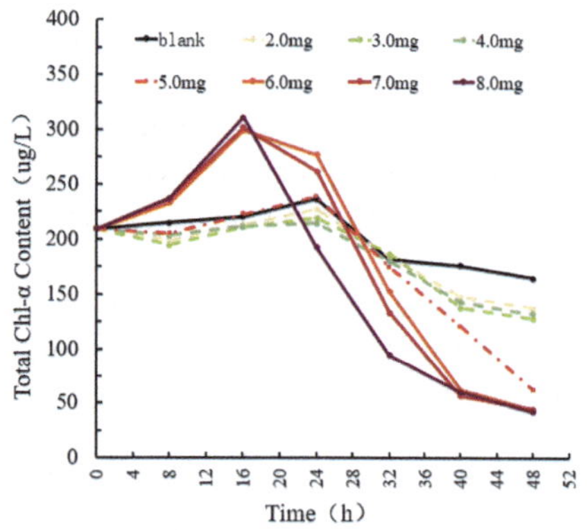

Fig. 2. Variation of Algal Cell Density Over Time at Different Dosages

From Fig. 1, it can be observed that the removal efficiency of algal cell density by sodium percarbonate is distinctly divided at a dosage of 5 mg/L (indicated by the red dashed line in the figure). When the dosage is ≤ 5 mg/L, the decline in algal cell density is slow within the first 24 h. After 48 h of treatment, while the bloom severity is noticeably reduced, it still remains in the severe bloom category. However, when the dosage is > 5 mg/L, the algal cell density begins to decrease rapidly. After 48 h of treatment, the severe bloom can be effectively reduced to a state of no visible bloom. Additionally, with 5 mg/L as the dividing point, when the dosage is > 5 mg/L (i.e., 6, 7, and 8 mg/L), the removal rates of algal cells are 97.5%, 97.5%, and 98.3%, respectively, which are significantly higher than the removal rates of 28.3%, 70%, 70%, 74.2%, and 82.5% for dosages ≤ 5 mg/L (i.e., 0, 2, 3, 4, and 5 mg/L). Furthermore, from Fig. 1, it is evident that the removal efficiency of algal cells at dosages of 6, 7, and 8 mg/L does not show a significant difference. Therefore, considering the removal efficiency, cost, and minimizing ecological impact, the optimal dosage is 6 mg/L.

From Fig. 2, it is evident that the total Chl-α content in the water does not decrease in sync with the reduction in algal cell density but instead shows an initial increase followed by a decrease. This may be due to the large amount of Chl-α released by the cyanobacteria upon their death, which was not immediately oxidized and removed, indicating a certain delay in removal [11].

Figure 2 also shows that the removal efficiency of total Chl-α content is similarly divided by the sodium percarbonate dosage of 5 mg/L (indicated by the red dashed line). At dosages ≤ 5 mg/L, the total Chl-α content increases slightly in the initial phase and then decreases slowly after 24 h. When the dosage is > 5 mg/L, the total Chl-α content rises sharply in the initial stage but decreases rapidly after 24 h. The experimental results further reveal that at dosages > 5 mg/L (i.e., 6, 7, and 8 mg/L), the removal rates of total Chl-α after 48 h are 77.6%, 76.2%, and 72.7%, respectively, which are significantly better than the removal rates of 30.1%, 37.1%, 41.5%, 37.9%, and 70.2% at dosages of 0, 2, 3, 4, and 5 mg/L. Moreover, similar to the removal of algal cell density, there is no significant difference in the removal efficiency of total Chl-α content among the 6, 7, and 8 mg/L dosages.

Therefore, considering the removal rates of algal cell density and total Chl-α content, along with the cost and ecological safety, the optimal dosage of sodium percarbonate is determined to be 6 mg/L.

3.2 The Variation of pH and Dissolved Oxygen (DO) Levels

During the 48-h experiment, the variations in pH and dissolved oxygen (DO) levels in the heavily cyanobacterial bloom test water at sodium percarbonate dosages of 0, 2, 3, 4, 5, 6, 7, and 8 mg/L are shown in Fig. 3 and Fig. 4, respectively.

From Fig. 3 and Fig. 4, it is evident that the threshold for effective impact on algal cell activity is at a dosage of 5 mg/L of sodium percarbonate. When the dosage is ≤ 5 mg/L, the effect on algal cell viability is not significant, and the DO and pH in the algal water exhibit the typical oscillation associated with normal algal cell activity, showing fluctuations in a 24-h cycle. For instance, during the 16th to 40th hours of the experiment, which corresponds to 7 AM to 7 AM the following day, a complete 24-h oscillation cycle is observed.

In contrast, when the dosage exceeds 5 mg/L, sodium percarbonate effectively and rapidly kills algal cells, resulting in a significant and sustained decline in both DO and pH levels without fluctuations.

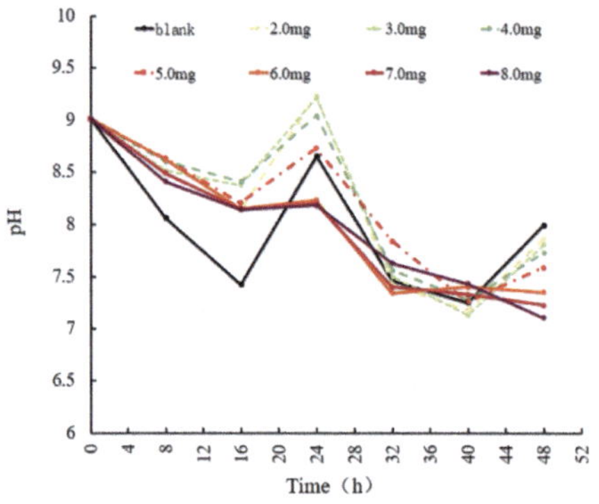

Fig. 3. Variation of Algal Cell Density Over Time at Different Dosages

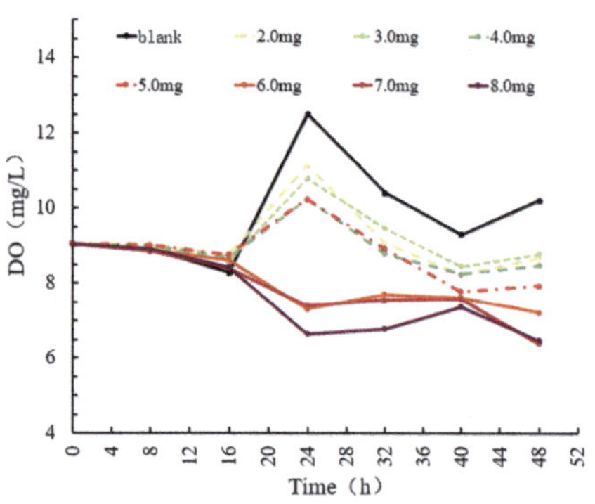

Fig. 4. Variation of Algal Cell Density Over Time at Different Dosages

It is important to note that the 24-h cycle observed in normal bloom conditions is due to the photosynthetic activity of algal cells during daylight hours, where CO2 is converted into O2, leading to an increase in dissolved oxygen and pH. Conversely, at night, the algal cells undergo respiration, converting O2 back into CO2, which results in decreased dissolved oxygen and a corresponding drop in pH [12].

4 Conclusion

This study conducted a comparative experiment on the effects of sodium percarbonate on heavily cyanobacterial blooms at eight different dosages: 0, 2, 3, 4, 5, 6, 7, and 8 mg/L. Throughout the experiment, algal cell density, total Chl-α content, dissolved oxygen (DO), and pH levels were monitored every 8 h. After 48 h of experimentation, the following main conclusions were drawn:

(1) **Threshold Effect of Sodium Percarbonate**: For heavily cyanobacterial blooms, the effectiveness of sodium percarbonate treatment is delineated at a dosage of 5 mg/L. When the dosage exceeds 5 mg/L (i.e., 6, 7, or 8 mg/L), the removal rates for algal cell density reached 97.5%, 97.5%, and 98.3%, respectively, while the removal rates for total Chl-α were 77.6%, 76.2%, and 72.7%. Under these three dosage conditions, the heavily polluted water could be reduced to a state with no obvious blooms. In contrast, when the dosage is ≤ 5 mg/L (i.e., 0, 2, 3, 4, or 5 mg/L), the removal rates for algal cell density were only 28.3%, 70%, 70%, 74.2%, and 82.5%, while the removal rates for total Chl-α were 30.1%, 37.1%, 41.5%, 37.9%, and 70.2%, respectively. Despite significant improvement in bloom levels under these five dosages after 48 h, the water still remained within the heavily polluted category.

(2) **Lack of Difference Among Higher Dosages**: When the dosage is > 5 mg/L (i.e., 6, 7, or 8 mg/L), there is no significant difference in the removal rates of algal cells or total Chl-α among the three dosages. Therefore, considering treatment efficacy, cost, and ecosystem safety, the optimal dosage of sodium percarbonate for managing heavily cyanobacterial blooms is 6 mg/L.

(3) **Trends in pH and DO**: The trends in pH and DO levels in the water also exhibit a threshold effect at 5 mg/L. When the dosage exceeds 5 mg/L (i.e., 6, 7, or 8 mg/L), algal cells are severely harmed, leading to a stable decline in both pH and DO levels as the cells lose their ability to photosynthesize. Conversely, when the dosage is $\leq$ 5 mg/L, algal cells maintain a certain level of activity, resulting in pH and DO values that still resemble those of a normal bloom, exhibiting the typical 24-h oscillation pattern.

Acknowledgements. This work was financially supported by "Fujian Provincial Science and Technology Project (2023I0042)".

References

1. Freedma, N.B.: Environmental ecology. Academic Press, San Diego (2002)
2. Qin, B.Q., Gao, G., Zhu, G.W.: Lake eutrophication and its ecosystem response. Chin. Sci. Bull. **58**(10), 855–864 (2013)
3. Li, L., Gao, N., Deng, Y., et al.: Characterization of intracellular & extracellular algae organic matters (AOM) of microcystic aeruginosa and formation of AOM-associated disinfection byproducts and odor & taste compounds. Water Res. **46**(4), 1233–1240 (2012)
4. Li, L., Gao, N.Y., Yin, D.D., et al.: Treatment processes in waterworks for removal of algae and algae toxins from raw Water. China Water Wastewater **24**(6), 20–24 (2008)
5. Fang, Y.B., Wang, C.Y., Tang, W.W., et al.: Comparison of advantages and disadvantages of algae removal technology, application status and new technology progress. Ind. Water Treat **40**(9), 1–6 (2020)
6. Ren, J.: The Inhibition mechanism and effect of UV/H2O2 process on microcystis aeruginosa and its degradation mechanism to microcystin-LR..Fudan. Univ. (2011)
7. Sinha, A.K., Green, W.R.: Mitigating cyanobacterial blooms and cyanotoxins in hypereutrophic ponds following the application of a granular hydrogen peroxide-based algaecide. Arkansas Bull. Water Res. 28–38 (2018)
8. Chen, C., Fan, F., Shi, X.L., et al.: Research on the control of chemical oxidants over water-blooming cyanobacteria. China Environ. Sci. **38**(11), 4307–4313 (2018). https://doi.org/10.19674/j.cnki.issn1000-6923.2018.0481
9. Greenfield, D.I., Duquette, A., Goodson, A., et al.: The effects of three chemical algaecides on cell numbers and toxin content of the cyanobacteria Microcystis aeruginosa and Anabaenopsis sp. Environ. Manage. **54**, 1110–1120 (2014)
10. Tu, J.C., Zhu, M.L., Lin, Z. et al.: Comparative study on the removal effect of cyanobacteria by two algae removal schemes using new environmental protection oxidant. ICRER 2022. Environmental Science and Engineering. Springer, Cham (2023). https://doi.org/10.1007/978-3-031-31808-5_3
11. Gao, Z.Q., Meng, C.X.: Progress in cyanobacteria water bloom. J. Anhui Agric. Sci. **37**(16), 7597–7598 (2009). https://doi.org/10.13989/j.cnki.0517-6611.2009.16.019
12. Lin, Z., Zhu, M.L., Tu, J.C., Jiang, H., Han, Y.L.: Study on the correlation between the variation of DO and pH value and chlorophyll-a concentration in water body. ICRER 2022. Environmental Science and Engineering. Springer, Cham (2023). https://doi.org/10.1007/978-3-031-31808-5_1

Three-Dimensional Numerical Analysis of Bionic Structure of Integrated Wastewater Treatment Equipment Optimisation

Feiyu Lin, Dongfeng Li(✉), Shengnan Wu, Chengzhuo Wang, Haonan Li, Xiahui Chen, Zimiao Chen, Luobin Wu, Zihao Fang, and Fuqing Bai

Zhejiang University of Water Resources and Hydropower, Xuelin Street 508, Hangzhou, China
lidf@zjweu.edu.cn

Abstract. The current design of integrated wastewater treatment equipment features a smooth deflector plate that, while capable of guiding water flow to some degree, suffers from significant flow disturbances at its junctions. This leads to localized low flow rates, the development of circulating vortices, and an uneven flow distribution, which can result in sludge reflux and deposition—conditions detrimental to the efficient circulation of mixed liquids. To mitigate these issues, we propose a biomimetic approach inspired by the convex knot structure found at the leading edge of the humpback whale's pectoral fin. By constructing a three-dimensional computational fluid dynamics (CFD) model incorporating this biomimetic structure, we conducted comparative analyses that revealed a more stable internal flow field and enhanced uniformity in liquid mixing. The biomimetic model demonstrated superior vortex suppression capabilities, effectively minimizing sludge reflux and deposition. These findings underscore the potential for practical application in optimizing the operational efficacy of integrated wastewater treatment systems through advanced numerical simulations.

Keywords: Fluent · Wastewater treatment · Bionic · Optimisation analysis

1 Introduction

The integrated sewage treatment equipment is characterized by its optimal layout of reaction units, resulting in a compact design that fosters close interconnections between pools. This configuration significantly reduces the lag time typically associated with traditional sewage treatment processes. Current research into various integrated treatment systems reveals a reliance on conventional methodologies, often modifying pool structures to enhance reaction conditions and fluid dynamics. This integration of multiple reaction units within a single reactor not only streamlines the treatment process but also contributes to substantial savings in both initial investments and long-term operational costs. Consequently, integrated equipment emerges as the most economically efficient solution across a range of wastewater treatment applications.

Currently, both domestic and international scholars primarily focus on enhancing process design and optimizing equipment structure in integrated systems [1–4]. Research

D. Han et al. (Eds.): ICEPG 2024, ESE, pp. 312–319, 2025.
https://doi.org/10.1007/978-3-031-86829-0_36

indicates that employing diverse process combinations can significantly reduce energy consumption and costs, while simultaneously improving treatment efficiency by lowering pollutant levels. However, our investigation reveals that the majority of studies prioritize process optimization, with limited attention given to structural enhancements of the equipment itself. This paper aims to address these research gaps in the optimization of integrated wastewater treatment equipment structures. We employ a realizable k-ε multiphase flow model to simulate the internal flow field and analyze the impact of baffles, both with and without bionic structures, on flow dynamics.

2 Simulation Methodology

2.1 Modelling

The integrated equipment studied in this paper primarily consists of an anaerobic tank, an anoxic tank, an aerobic tank, and additional components. By appropriately simplifying the corresponding geometric model with dimensions of 4 m in length, 2.5 m in width, and 2.7 m in height, the model features an orifice spacing of 0.15 m. Revit drawing software was then used to create a complete three-dimensional geometric model of the integrated equipment, as shown in Fig. 1.

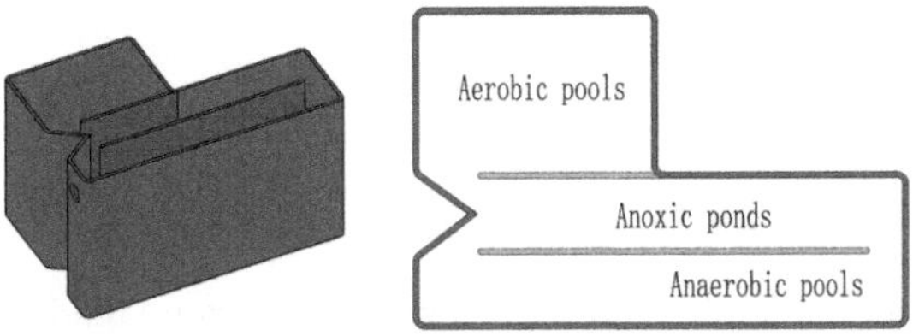

Fig. 1. Three-dimensional geometric model of integrated equipment

A bionic convex structure, inspired by the leading edge of the humpback whale's pectoral fin, has been incorporated into the internal baffle of the equipment. The design of this biomimetic structure was rendered using Revit drawing software, as illustrated in Fig. 2.

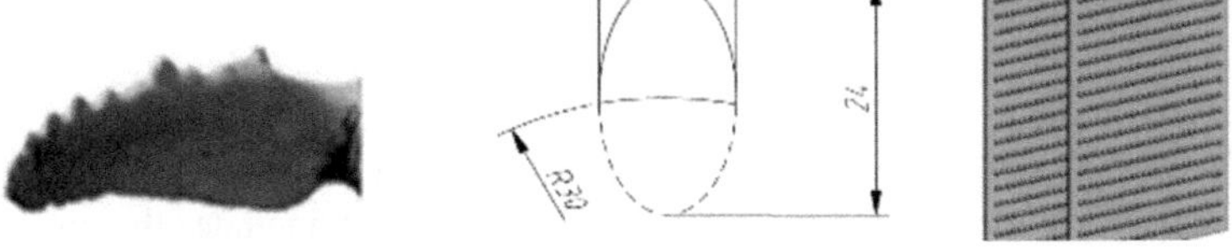

Fig. 2. Three dimensional geometric model of the convex structure and baffle of the leading edge of the pectoral fin of the bionic humpback whale

2.2 Grid Division

Utilizing the meshing module of ANSYS Fluent, we selected appropriate distance and curvature parameters and activated the advanced sizing function for mesh refinement. The computational domain of the integrated equipment model comprises approximately 687,000 mesh elements, with a maximum mesh size of 50 mm and a minimum size of 2 mm. The structural mesh configuration is illustrated in Fig. 3, demonstrating the precision and complexity of the modeling approach employed in this study.

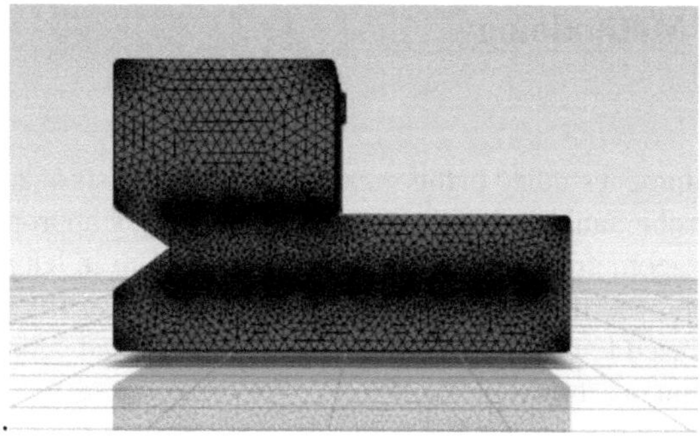

Fig. 3. Structural meshing diagram

2.3 Parameter Settings for Model Solving

1) In this paper, the simulation object is designed as a general fluid, so the velocity inlet and free outflow are used, and the conditions are set as shown in Table 1. The outlet flow rate is an unknown value before the flow problem is solved.

Table 1. Boundary condition settings for model solution

Boundary	Boundary type	application condition	numerical value	direction
Inlet	velocity - inlet	non-pressure flow	0.6–0.7	+ X
Liquid phase	velocity outlet	outflow Unpressurized flow	Unknown value	+ X
Gas-phase	Pressure outlet	Compressible flow	Unknown value	+ Y

2) Solving with the FLUENT solver [5].

Check the mesh and define the length unit; check the mesh information attributes as well as the model setup parameter information, which includes the mesh volume, size range, number of nodes, and so on. Verify whether they are correct, and if there are any errors, modify them immediately.

Creation of computational model: Consulting the data and analysing the comparisons, it can be determined that the Realizable k-ε model can achieve the goal of convergence in this numerical simulation, and all other settings are adopted as default values. In the multiphase flow model, the Mixture model is used for two-phase flow calculations because the model has strong coupling.

Set the fluid material properties and boundary conditions: this paper focuses only on the state without considering aeration and the distribution of the water flow field in the pool, so the fluid material is set to water, with the density, temperature, and other physical properties taken as default values. The inlet boundary condition is a velocity inlet, the outlet is free outflow, and the other boundaries are all wall surfaces, with the default setting being a no-slip condition.

3) Post-processing of calculation results:

Create equivalent surfaces and display the calculation results on these surfaces. Boundary surfaces are automatically defined as faces in FLUENT, such as the inlet and outlet boundary surfaces. Plot the flow field cloud map and the trace map, among others.

3 Numerical Solution and Result Analysis

3.1 Computational Convergence Analysis

Selected the simulation process in Fig. 4 for calculating the iterative residual curve graph. The curve shows a downward trend, the data region is stable, and it essentially meets the convergence conditions. Throughout the calculation process, there are no significant changes, indicating good convergence of the model's calculation results.

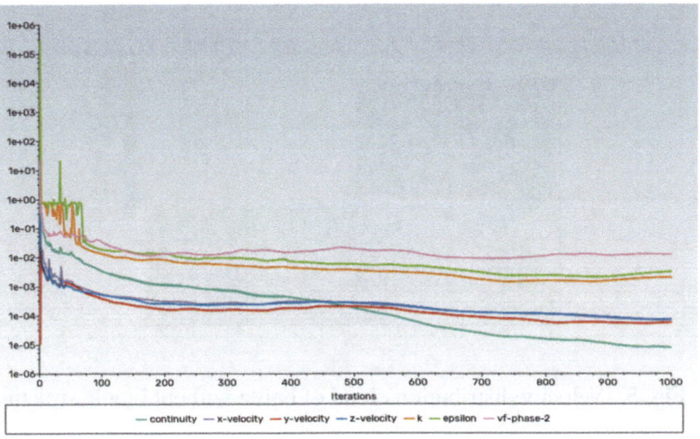

Fig. 4. Iterative residual plot

3.2 Flow Field Analysis of Integrated Equipment Under Different Working Conditions

To investigate the optimal structure of the integrated equipment, a detailed study and analysis of its internal structure was conducted. The effective depth of the equipment is 2 m. A cross-section is taken at 1 m from the ground, representing the midpoint of the effective depth, and the flow field inside the equipment is analyzed under various internal structures.

Flow Field Analysis of Baffle Without Bionic Structure

It is generally believed that to achieve effective circulation, it is necessary to reduce the high friction resistance of the baffle surface on water flow rates. Through simulation calculations, we obtained the velocity distribution cloud diagram of the flow field in the integrated equipment. As shown in Fig. 5. When the baffle surface lacks a bionic structure, the internal structure of the integrated equipment allows for relatively uniform mixing and flow in the anaerobic and anoxic pools. However, in the aerobic tank, the flow rate of the mixed liquor is low, which can lead to sludge deposition, adversely affecting the stable operation of the biochemical system. Figure 6 illustrates the flow rates at each stage, indicating that only 19% of the overall calculation domain has mixed liquid flow rates greater than 1 m/s, while 50% of the area has flow rates less than 0.7 m/s. Figure 7 shows that the high resistance of the baffle surface decreases the overall flow rate, leading to issues such as sludge reflux and sedimentation. This hampers the circulation of the mixed liquor within the equipment. As a result, the mixed liquor flow rate along the direction of water movement gradually levels off, with an average flow rate of less than 0.7 m/s in the aerobic pool, which seriously impacts the treatment effectiveness in this zone. The velocity vector diagram in Fig. 7 and the velocity streamline diagram in Fig. 8 reveals that dead zones and vortices are likely to occur at the corners of the anaerobic and anoxic zones, further affecting the overall mixed liquid flow rate and resulting in a significant low-velocity area at the back of the aerobic zone.

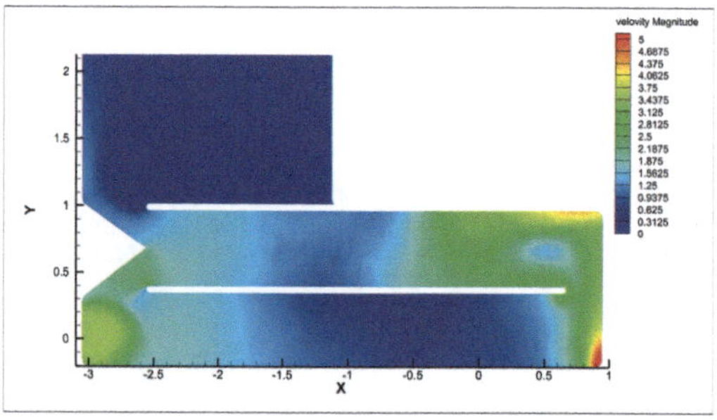

Fig. 5. Velocity distribution cloud of baffle without bionic structure

Flow Field Analysis of Baffle Surface Humpback Whale Bionic Structure

This study examines the impact of an integrated wastewater treatment equipment's internal structure on its flow field, specifically by incorporating a humpback whale-inspired bionic structure into the baffle. CFD simulations resulted in the data presented in Figs. 9 and 10. Analysis of Fig. 10 reveals that the proportion of the mixture with flow rates below 0.7 m/s decreases to approximately 35%, a reduction of 15% compared to the original design. Conversely, the proportion exceeding 1 m/s increases to about 45%,

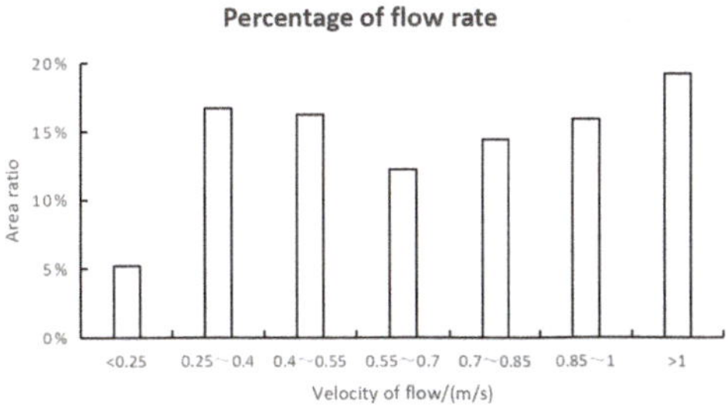

Fig. 6. Baffle without bionic structure mixed liquid flow rate percentage

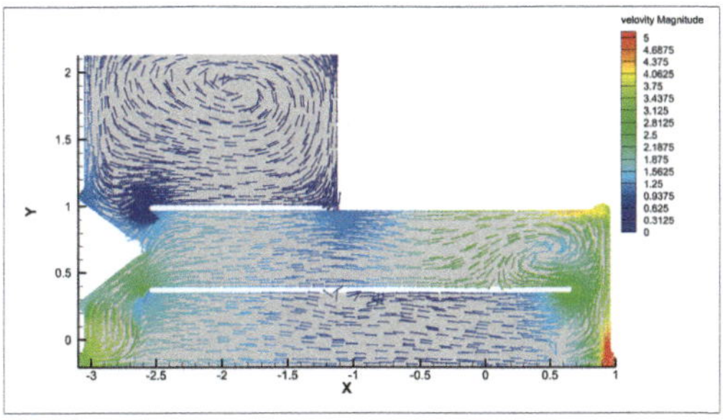

Fig. 7. Velocity vector diagram of baffle without bionic structure

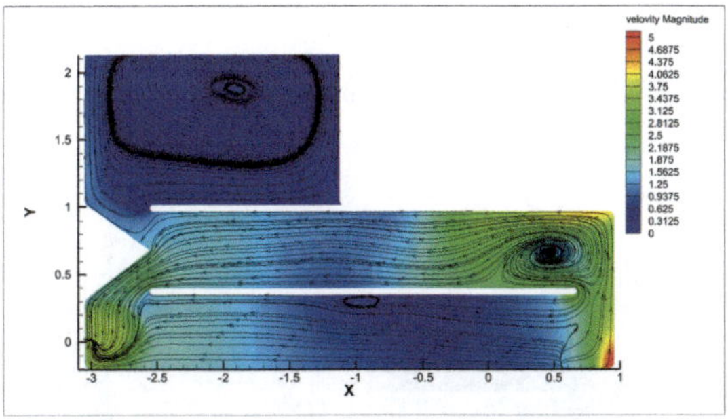

Fig. 8. Streamline diagram of Baffle without Bionic Structure

reflecting a 26% improvement. The aerobic zone exhibits lower flow rates with even mixing, while the corners of the anoxic and anaerobic pools show no significant low-velocity zones, facilitating better mixture promotion and reducing sludge deposition. Overall, the flow rate stabilizes, resulting in a more uniform flow field.

Additionally, the efficiency of the integrated equipment is influenced by internal vortices. Due to the limitations in visualizing vortex distribution through velocity profiles, this study employs velocity vector and fluid trajectory diagrams for analysis. Figures 11 and 12 demonstrate that the bionic structure reduces vortex areas compared to the smooth baffle, resulting in smoother circulation of mixed liquids in the anoxic and anaerobic tanks, with no prominent vortices at the corners. Overall, the humpback whale-inspired design enhances fluid stability and ensures thorough mixing across all regions.

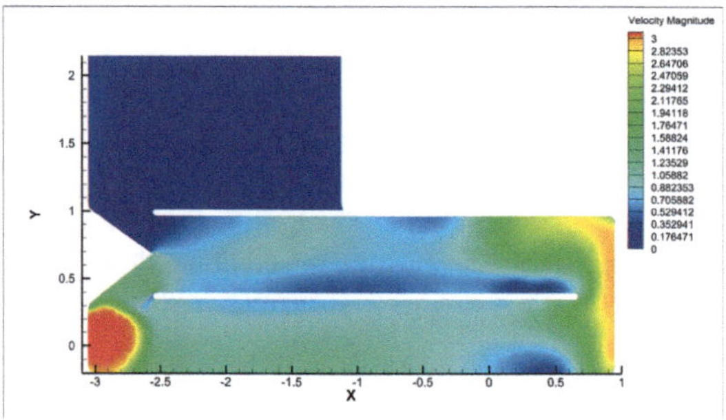

Fig. 9. Velocity distribution cloud of humpback whale-like bionic structures

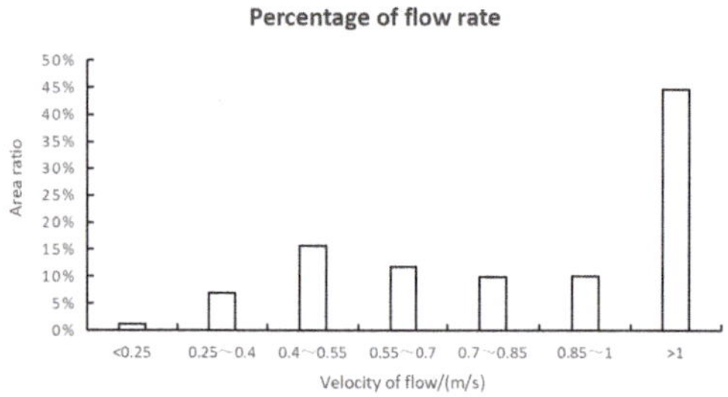

Fig. 10. Humpback whale-like bionic mixture structure flow rate percentages

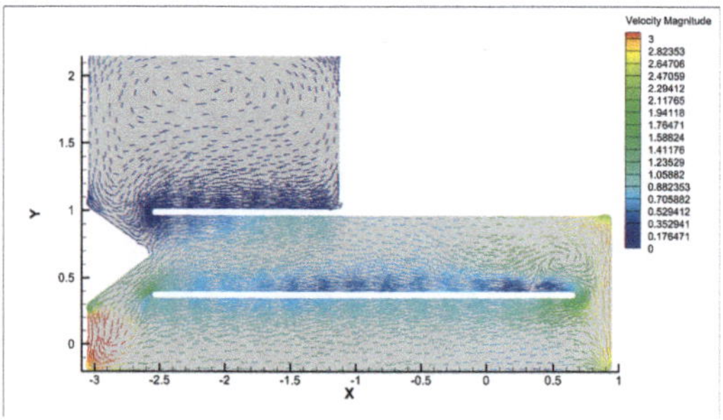

Fig. 11. Velocity vectors of humpback whale-like bionic structures

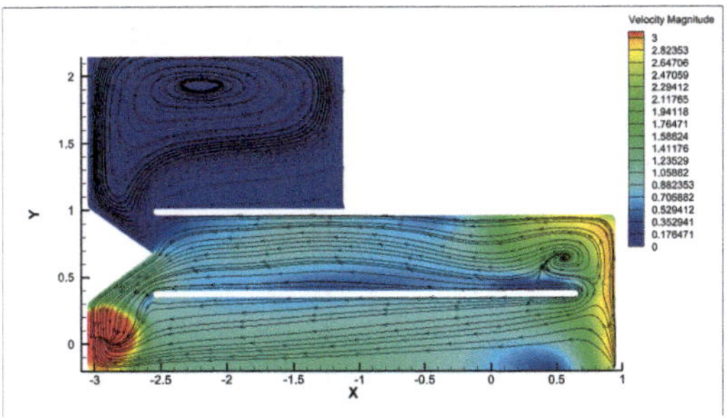

Fig. 12. Streamline diagram of humpback whale-like bionic structures

4 Conclusion

This study employs computational fluid dynamics (CFD) with the Realizable k-ε multiphase flow model to analyze and compare the internal flow field of integrated wastewater treatment devices. The goal is to optimize internal circulation and enhance pollutant treatment efficacy. Key findings include:

1) The baffle surface structure significantly influences sewage treatment efficiency. An optimized baffle can enhance the average flow rate of mixed liquids, promote uniform internal processes, reduce eddy currents, and improve treatment efficiency. Transitioning from a smooth baffle to a humpback whale-inspired bionic structure leads to a marked increase in the average flow rate, with the proportion of flow exceeding 1 m/s rising from 19% to 45%. Concurrently, the flow rate below 0.7 m/s decreases from 35% to 15%, resulting in reduced disparity between overall and local flow rates, thus enhancing circulation and pollutant treatment.
2) At a given flow rate, the sewage treatment rate correlates directly with the average flow rate of the mixture. Faster flow rates facilitate the transport of greater volumes of sewage per unit time, mitigating issues such as sludge deposition and achieving higher treatment efficiencies.

References

1. Wei, B.: Comparison of common integrated domestic wastewater treatment facility processes. Low Carbon World **14**(09), 31–33 (2024). https://doi.org/10.16844/j.cnki.cn101007/tk.2024.09.061
2. Ahmed, T., Bram, K.: Treatment of an aerobically pre-treated domestic sewage by a rotating biological contactor. Water Res. **36**(1), 147–155 (2002)
3. Zhang, R., Ziang, Y.: Design and application of integrated wastewater treatment system for water conservation and emission reduction. Equip. Manag. Maint. Repair (15), 181–184 (2024). https://doi.org/10.16621/j.cnki.issn1001-0599.2024.08.54
4. Xia, S.B., Liu, J.X.: An innovative integrated oxidation ditch with vertical circle (IODVC) for wastewater treatment. J. Environ. Sci. (China) **16**(3), 367–370 (2004)
5. Fan, D.Q.: Research on integrated treatment equipment for high concentration aquaculture wastewater based on CFD simulation. Guilin University of Electronic Science and Technology (2022)

Identification of Critical Source Areas and Preferred Best Management Practices (BMPs) for Total Phosphorus in the Fuzhou River Watershed

Huazhi Zhang[1,2], Yuannan Wang[1,2](✉), and Xiaojing Li[1,2]

[1] Changjiang Schinta Software Technology Co., Ltd, Wuhan 430010, China
913864366@qq.com

[2] Internet+ Intelligent Water Conservancy Key Laboratory of Changjiang Water Resources Commission, Wuhan 430010, China

Abstract. To better manage total phosphorus (TP) pollution from nonpoint sources in the Fuzhou River Basin, this study constructed a Soil and Water Assessment Tool (SWAT) model for the basin. Based on the calibrated SWAT model, we identified the critical source areas (CSAs) of TP pollution and evaluated the reduction effects of individual Best Management Practices (BMPs) and their combinations. Additionally, we optimized the selection of BMPs by balancing their environmental cost-effectiveness. The findings reveal that the SWAT model exhibits robust applicability in the Fuzhou River Basin, with the CSAs accounting for 37.61% of the total basin area and contributing 60.65% of the total TP load. These CSAs are primarily concentrated in agricultural and orchard zones, where high rates of nutrient loss per unit area are prevalent. Combined BMPs outperform individual BMPs in mitigating nonpoint source pollution loads, and there is a positive correlation between the number of combined BMPs and the average reduction rate. However, given a fixed implementation area within the basin, an optimal threshold for the reduction efficiency exists beyond which incremental BMP combinations yield diminishing returns. Among the individual BMPs, FS (filter strips) stands out as the most cost-effective measure for reducing TP loads in the Fuzhou River Basin. In contrast, the combination of FS+GW+FR10 (filter strips, grassed waterways, and 10% phosphate fertilizer reduction) proves to be the most cost-effective among the evaluated BMP combinations.

Keywords: Fuzhou River · Total Phosphorus · SWAT Model · Best Management Practices (BMPs) · Cost-Benefit Analysis

1 Introduction

Rivers, serving as essential freshwater resources and key components of the Earth's hydrological cycle, play a crucial role in societal development and the integrity of ecosystems. In recent years, however, the quality of river water has significantly deteriorated due to anthropogenic activities such as urbanization and rapid industrial and agricultural

D. Han et al. (Eds.): ICEPG 2024, ESE, pp. 320–334, 2025.
https://doi.org/10.1007/978-3-031-86829-0_37

expansion within watersheds [1–3]. River water pollution is generally categorized into two types: point source pollution and non-point source pollution (NPS) [4]. While point source pollution is relatively straightforward to identify and manage, and has received considerable attention with effective controls implemented across many regions in China [5], NPS pollution presents more substantial challenges due to its inherent randomness, complex mechanisms, ambiguity, widespread sources, extensive impact ranges, and the difficulty in pinpointing critical source areas [6, 7], influenced by factors such as hydrological and meteorological conditions and land use patterns. Currently, NPS pollution stands out as the primary contributor to surface water quality degradation globally [8]. According to the Second National Census of Pollution Sources, agricultural NPS emissions account for 67.22% of total phosphorus (TP) emissions in China.

Best Management Practices (BMPs) constitute a comprehensive prevention and control system aimed at protecting water resources from pollution [9]. These practices include both engineering and non-engineering measures, which have been shown to effectively control non-point source (NPS) pollution and the transport of sediments into receiving water bodies [10–12]. Model simulations serve as a potent tool for the quantitative study of NPS pollution, facilitating its control and reduction [13]. The Soil and Water Assessment Tool (SWAT) model, developed by Dr. Jeff Arnold at the Agricultural Research Service (ARS) of the United States Department of Agriculture (USDA), is a physically-based, continuous-time, watershed-scale model capable of predicting the long-term impacts of diverse land management practices on hydrology and nutrient dynamics within watersheds [14]. This model is extensively utilized by scholars worldwide as a hydrological tool for evaluating BMPs [15, 16].

This study focuses on the Fuzhou River Basin and constructs a SWAT hydrological and water quality model to identify critical source areas (CSAs) of total phosphorus (TP) pollution. It evaluates the reduction efficiency of different best management practices (BMPs) and their combinations under various scenarios, aiming to achieve an optimal balance between reduction efficiency and costs. The findings aim to provide scientific guidance for the prevention and control of non-point source pollution in the basin.

2 Study Area and Methodology

2.1 Overview of the Study Area

The Fuzhou River is the second largest river in Dalian City, spanning a total length of 129.4 km with a drainage area of 1,648 square kilometers (Fig. 1). It originates from the southern slopes of Laomao Mountain in Tongyi Township, Pulandian City, and flows through multiple townships in Pulandian and Wafangdian Cities before emptying into the Bohai Sea at Laoyangtou, Xilanqi Village, Santaizi Township, Wafangdian City. This study focuses on the region downstream of the Songshu Reservoir and the estuary, located at latitudes 39°30′ to 39°59′ N and longitudes 121°34′ to 122°21′ E. The region is characterized by a warm-temperate sub-humid climate, with uneven rainfall distribution, where 60.2% of the precipitation occurs during the flood season (July to September). The annual average precipitation is 654 mm, and the average annual runoff volume is 345.31 million m^3. The annual mean temperature ranges from 8.6 °C to 10.5 °C. Land use within the basin primarily consists of farmland (45%) followed by forestland (28%).

The study area is located within the administrative region of Wafangdian City, where the midstream and downstream areas are predominantly agricultural, with corn being the main crop. Livestock production, primarily involving chickens, pigs, and cattle, is concentrated below the Dongfeng Reservoir. Non-point source pollution mainly originates from agriculture, livestock breeding, and rural domestic sewage discharge.

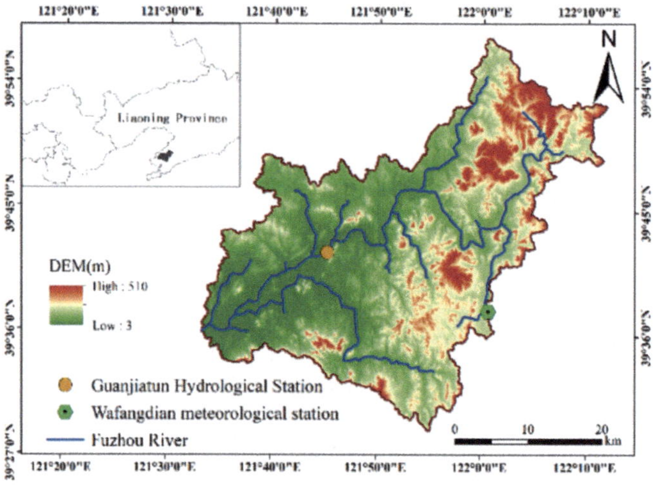

Fig. 1. Overview map of the Fuzhou River Basin

2.2 SWAT Model Input Data

The data required for model construction include digital elevation data, meteorological data, hydrological data, land use data, and soil spatial data. The descriptions and sources of these data are listed in Table 1.

Table 1. Basic data and sources

Data Name	Data Description	Data Sources
DEM	GDEMV3 digital elevation data with 30-m resolution	Geospatial Data Cloud (https://www.gscloud.cn/)
Land Use	Remote sensing data at 30-m resolution for monitoring current land use in China	Center for Resource and Environmental Sciences and Data, Chinese Academy of Sciences (https://www.resdc.cn/)

(*continued*)

Table 1. (*continued*)

Data Name	Data Description	Data Sources
Soil	World Soil Database (HWSD) China Soil Dataset: 1 km by 1 km	National Cryosphere Desert Data Center (http://www.ncdc.ac.cn/)
Meteorological	Daily-scale meteorological data from Wafangdian Meteorological Station, 2008–2020	National Meteorological Information Center (https://m.data.cma.cn/)
Hydrological	Monthly runoff data from Guanjiatun Hydrological Station, 2008–2020	Wafangdian Ecological Environment Branch
Water Quality	TP monitoring data from the Guanjiatun section, 2018–2020	China Environmental Monitoring Center (CEMC) (https://www.cnemc.cn/)
Agricultural Management	Data related to tillage, fertilization, etc.	Second Pollution Census

2.3 Model Calibration and Validation

Model calibration involves estimating model parameters by comparing model outputs with observed data under a set of assumed parameters. Validation, on the other hand, ensures that the calibrated model can accurately simulate conditions outside the calibration period. In this study, a warm-up period and a simulation period were defined within the SWAT model [17]. Twenty-three runoff parameters and 14 water quality parameters were selected, and the SUFI-2 algorithm in SWAT-CUP was applied for calibration and validation. Model performance was evaluated using the coefficient of determination (R^2) and the Nash-Sutcliffe Efficiency (NSE). Typically, NSE values above 0.75 indicate excellent model performance, while values between 0.50 and 0.75 are considered satisfactory [18]. Similarly, R^2 values above 0.75 indicate excellent simulation results, whereas values between 0.36 and 0.75 are deemed satisfactory [19].

The coefficient of determination (R^2) is calculated as:

$$R^2 = \left\{ \frac{\sum_{i=1}^{n} (S_i - S_{ave}) \times (Q_i - Q_{ave})}{\left[\sum_{i=1}^{n} (S_i - S_{ave})^2\right]^{0.5} \times \left[\sum_{i=1}^{n} (Q_i - Q_{ave})^2\right]^{0.5}} \right\}^2 \tag{1}$$

The Nash-Sutcliffe efficiency coefficient is calculated as:

$$NSE = 1 - \frac{\sum_{i=1}^{n} (Q_i - S_i)^2}{\sum_{i=1}^{n} (Q_i - Q_{ave})^2} \tag{2}$$

where Q_i is the measured data, m^3/s; Q_{ave} is the average value of the measured data, m^3/s; S_i is the simulated data, m^3/s; S_{ave} is the average value of the simulated data, m^3/s.

2.4 Identification of Critical Source Areas for Total Phosphorus Pollution Loads

Critical source areas (CSAs) refer to regions within a watershed that generate significant pollutant outputs despite their relatively smaller size. Identifying CSAs provides a targeted basis for the deployment of Best Management Practices (BMPs). Based on the annual average outputs of TP from the SWAT model for the period 2018 to 2020, the TP loading intensity per unit area was categorized into five levels: light, moderate, high, severe, and extremely severe, using the natural breaks classification method [20]. Sub-watersheds with severe and extremely severe TP loading intensities were designated as CSAs. The calculation formula is as follows:

$$P_i = \frac{PT_i}{A_i} \tag{3}$$

P_i is the load intensity of the sub-basin during time period i, kg/ha; PT_i is the total load generated by the sub-basin during time period i, kg; A_i is the sub-basin area.

2.5 BMPs Setup Method

In response to the non-point source pollution scenario in the Fuzhou River Basin [21, 22], appropriate BMPs were selected for scenario simulations. Five non-structural measures were devised, including contour tillage (CT), strip cropping (SC), and phosphorus fertilizer reduction measures at 10% (FR10), 30% (FR30), and 50% (FR50) reduction rates. Additionally, three structural measures were considered: terracing (TR), grassed waterways (GW), and filter strips (FS). The specific parameter settings for these BMPs were derived from a thorough review of the SWAT model input-output manual [23] and relevant literature on non-point source pollution mitigation using the SWAT model [24–28]. These detailed measures and their parameters are summarized in Table 2.

Table 2. Measure Parameter Setting Table.

BMPs	Abridge	parameterization
Phosphorus Fertilizer Reduction 10%	FR10	Adjustment of the fertilizer.dat module parameters: phosphorus content is 90% of the original value
Phosphorus Fertilizer Reduction 20%	FR20	Adjustment of the fertilizer.dat module parameters: phosphorus content is 80% of the original value

(continued)

Table 2. (*continued*)

BMPs	Abridge	parameterization
Phosphorus Fertilizer Reduction 30%	FR30	Adjustment of the fertilizer.dat module parameters: phosphorus content is 70% of the original value
Grassed Waterways	GW	GWATL defaults to the length of the HRU side of the square, and GWATW is set to 300 m
Vegetated Filter Strips	FS	FILTER_RATIO is 5, FILTER_CON is 0, FILTER_CH is 0, and the filter w parameter is set to default values
Terracing	TR	TERR_CN is the original value of CN2 minus 5; TERR_SL parameter values are shown in Table 3, TERR_P parameter values are shown in Table 4
Contour Tillage	CT	Adjust CONT_CN to the original value of CN2 minus 2; CONT_P parameter values are shown in Table 4
Strip Cropping	SC	STRIP_N is 0.19, STRIP_CN is the original value of CN2 minus 3, STRIP_C is 0.2, STRIP_P values are shown in Table 4

Table 3. Design Values for Terrace Widths at Different Slopes.

Slope (°)	Field width (m)	Width of the field bank (m)
0–5	10–15	0.3
5–15	2–10	0.3
15–25	3–6	0.3
>25	1.5–3	0.3

Table 4. Values of USLE_P Parameter Settings for the Simulation of Contour Tillage and Terracing Scenarios.

Slope (°)	USLE_P	
	CT	TR
1–2	0.6	0.12

(*continued*)

Table 4. (*continued*)

Slope (°)	USLE_P	
	CT	TR
3–8	0.5	0.1
9–12	0.6	0.12
13–16	0.7	0.14
17–20	0.8	0.16
21–25	0.9	0.18

2.6 Comprehensive Benefits Assessment of BMPs

To conduct an exhaustive evaluation of the environmental benefits and comprehensive cost-effectiveness of reducing total phosphorus (TP) loads in the Fuzhou River Basin, eight individual best management practices (BMPs) and their combinations were employed. Since the three phosphorus fertilizer reduction measures cannot be implemented concurrently, a total of 101 scenarios were simulated.

The formula for determining the environmental benefit reduction efficiency is as follows:

$$R(\%) = \frac{p_{BAS} - p_{BMP}}{p_{BMP}} \times 100\%, \tag{4}$$

where: R is the pollution load reduction efficiency, %; p_{BAS} is the pollution load under the initial setting, kg; p_{BMP} is the pollution load output from the model under the management measures, kg.

By considering relevant literature and standards, it was assumed that the duration for implementing the measures would be 10 years. The annual costs for each BMP were then determined, with specific values presented in Table 5. These costs were obtained through extensive literature reviews [20, 24, 25, 28–30].

Table 5. Costs of measures in the Fuzhou River Basin [¥/(ha-a)].

FR10	FR25	FR50	FS	GW	TR	CT
204.1	510.2	1020.2	1150	2060	760	750

A comprehensive benefit analysis was performed using a cost-effectiveness metric, calculated as follows:

$$CE = \frac{p_{BAS} - p_{BMP}}{\cos t}, \tag{5}$$

Where CE is the pollution load reduction per ten thousand yuan, i.e., the cost-effectiveness value, kg/ten thousand yuan; $\cos t$ is the cost required for the arrangement of management measures, ten thousand yuan; p_{BAS} is the pollution load under the initial setup, kg; p_{BMP} is the pollution load under the management measures, kg.

3 Results and Analysis

3.1 Model Calibration and Validation Results

The calibration and validation results of the SWAT model for the Fuzhou River Basin are depicted in Fig. 2. For streamflow, the R^2 and NSE values during the calibration period were 0.95 and 0.84, respectively, while during the validation period, they were 0.90 and 0.80. For total phosphorus (TP), the corresponding R^2 and NSE values during the calibration period were 0.81 and 0.76, and 0.80 and 0.52 during the validation period. Both the R^2 and NSE values for streamflow and TP during the simulation and validation periods exceeded 0.5, indicating the excellent applicability of the model in the Fuzhou River Basin.

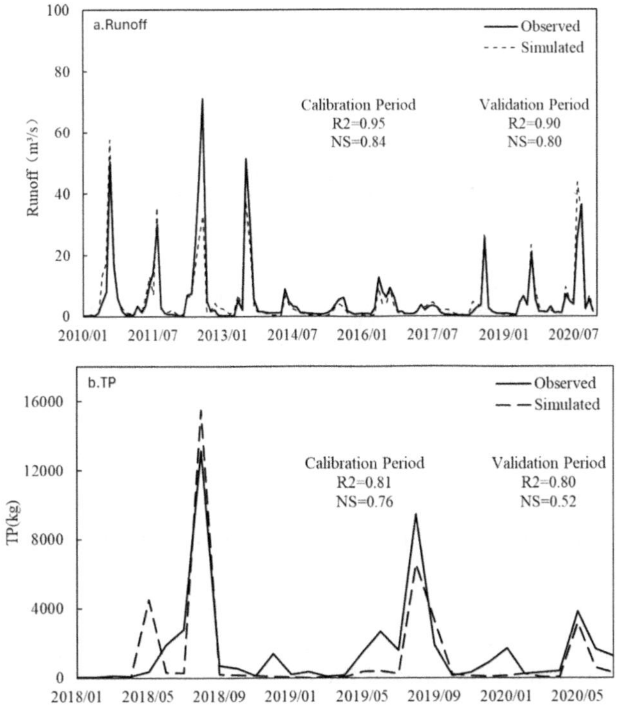

Fig. 2. Calibration and validation plots of the SWAT model for monthly runoff (a) and total phosphorus (b) load.

3.2 Critical Source Areas for Total Phosphorus Pollution Loads

Figure 3 illustrates the annual average unit area load of total phosphorus (TP) in the Fuzhou River Basin from 2018 to 2020. The results reveal significant spatial variations in TP load intensity, with higher loads observed in the midstream and downstream regions, as well as in Sub-basin 6 located upstream. This is attributed to the predominance of cultivated land in these areas, particularly in Sub-basin 6, where land use is dominated by orchards. Intense agricultural activities lead to the accumulation of non-point source pollution within the soil. In contrast, the upstream regions, primarily covered by forests, generate less pollution. The annual average TP load across all sub-basins ranged from 0.07 to 0.60 kg/ha, with a mean value of 0.36 kg/ha.

Based on the model classification, Sub-basins 8, 17, 6, 15, 4, 16, 9, 10 (extremely heavy pollution load) and 14, 21, 24, 26, 29 (heavy pollution load) were identified as critical source areas. These areas, covering 491.21 hectares and accounting for 37.61% of the total basin area, contribute 60.65% of the TP load. Therefore, the implementation of best management practices (BMPs) should be prioritized in agricultural lands within these critical source areas.

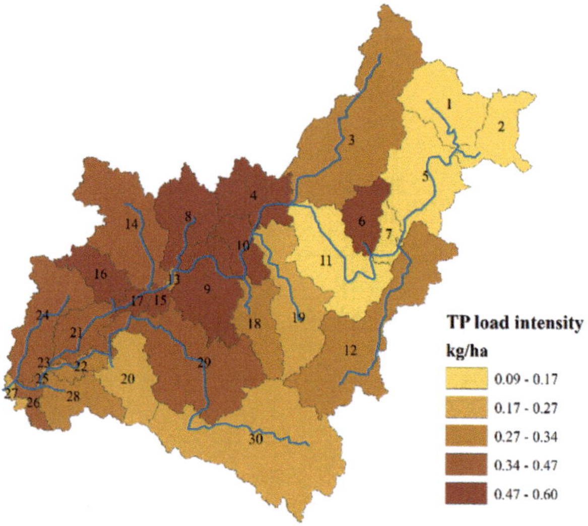

Fig. 3. Spatial distribution of total phosphorus loading intensity in the Fuzhou River Basin

3.3 Evaluation of the Efficiency of BMPs

The reduction efficiencies of individual Best Management Practices (BMPs) on total phosphorus (TP) load are presented in Fig. 4 and Table 6. The reduction rates ranged from 0.53% to 38.97%, demonstrating that all measures effectively mitigated TP load. Specifically, the order of effectiveness was as follows: FS (38.97%) > GW (26.40%) > TR (20.80%) > FR50 (17.08%) > SC (14.67%) > CT (8.72%) > FR30 (8.34%) > FR10 (0.53%). Research indicates that runoff-induced erosion and soil transport are primary pathways for phosphorus entry into water bodies, with water volume, sediment load, and total phosphorus flux exhibiting a positive and synchronous relationship [31, 32]. The notable variations in TP load reduction among different BMPs stem from their distinct mechanisms.

FS and GW, both classified as end-of-pipe treatment measures, exhibit significant reduction effects by directly intercepting sediment and other pollutants through reduced flow velocities [33–35]. TR, an intermediate interception measure, functions as a dam in the field to retain runoff and prevent erosion [36]. CT and SC, representing contour farming practices, reduce the transport of sediment with runoff by increasing surface roughness and field water storage capacity [23, 37, 38], also demonstrating considerable reduction rates. FR10, FR30, and FR50, belonging to source control measures, effectively manage TP load by curbing phosphorus inputs at the source through reduced fertilizer application [39].

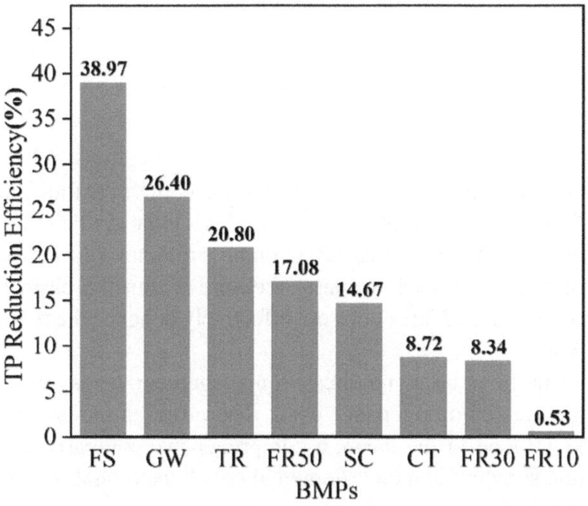

Fig. 4. Efficiency of single measures in reducing TP pollution in the watershed.

The reduction efficiency of total phosphorus (TP) load in the Fuzhou River Basin by all Best Management Practices (BMPs) adopted in this study is shown in Fig. 5. The results indicate that all BMPs have a certain effect on reducing the TP load. The reduction efficiency ranges from 8.92% to 41.46% for the combination of two measures (average: 26.45%), from 14.38% to 45.25% for the combination of three measures (average: 34.05%), from 16.84% to 46.42% for the combination of four measures (average: 38.99%), from 30.39% to 46.18% for the combination of five measures (average: 42.77%), and from 44.59% to 46.18% for the combination of six measures (average: 45.49%). The BMPs with the highest reduction efficiency among the combinations of one to six measures are FS+FR50, FS+TR+FR50, FS+TR+FR50+GW, FS+TR+FR50+GW+SC, and FS+TR+FR50+GW+SC+CT, respectively.

Compared to single BMPs, combined BMPs exhibit higher efficiency in reducing the TP load in the basin, and the average reduction efficiency gradually increases with the increasing number of measures combined. However, there are significant differences in the reduction efficiency among different scenario combinations. For example, some combinations of two measures show lower reduction efficiency than a single measure,

Table 6. Reduction rates of TP loads by different BMPs.

Type of measure	Scenarios	Total phosphorus load (kg)	Reduction rate (%)
Baseline scenario	BAS	1523.46	0.00
Non-engineering measures	CT	1523.43	8.72
	SC	1407.33	14.67
	FR10	1337.55	0.53
	FR25	1448.225	8.34
	FR50	1341.38	17.08
Engineering measures	TR	1433.63	20.80
	GW	1287.00	26.40
	FS	1355.825	38.97

while the maximum reduction efficiency of scenarios with three, four, and five measures is close to that of the combination of six measures. This indicates that the reduction effect of combined scenarios on TP load is not a simple summation of individual measures. When the allocated area for each scenario measure within the basin is fixed, simply increasing the number of measures does not effectively enhance the reduction efficiency of TP pollution load.

The deployment of a large number of management measures correspondingly increases construction and maintenance costs. Therefore, in the process of controlling and reducing the TP load in the basin, a comprehensive comparison should be made based on the actual situation and environmental cost-benefit analysis of the basin.

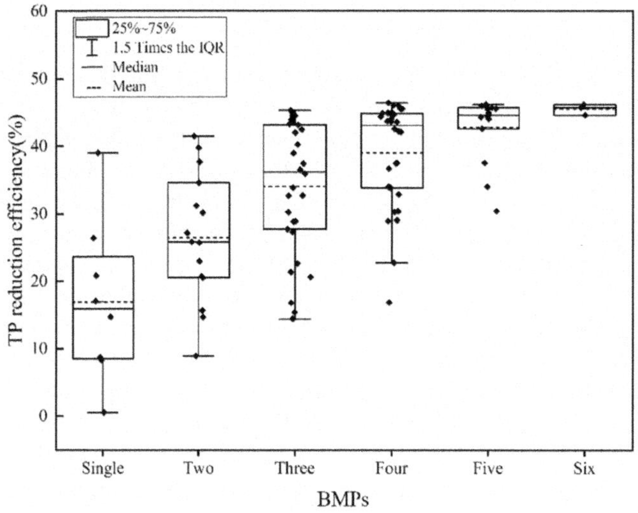

Fig. 5. Reduction efficiency of TP pollution in the watershed by combinations of measures.

3.4 Environmental Cost-Effectiveness Assessment of BMPs

The cost-effectiveness (CE) of individual BMPs, as depicted in Table 7, reveals that GW attains the highest CE value among the standalone BMPs, standing at 241.25 kg/10,000 yuan, followed closely by FS with a CE value of 34.54 kg/10,000 yuan. The deployment length of GW is equivalent to the side length of the square HRU where the measure is placed, while FS covers 10% of the field area. The significantly smaller deployment areas for these two measures result in lower overall investments, hence leading to substantially higher CE values compared to other interventions. In summary, the average CE value for non-structural measures stands at 1.96 kg/10,000 yuan, whereas the average for structural measures reaches 92.86 kg/10,000 yuan, indicating that structural measures exhibit superior environmental cost-effectiveness in the Fuzhou River Basin.

Table 7. Reduction rates of total phosphorus loads by different BMPs

Type of measure	Scenarios	Total phosphorus load (kg)	Cost-effectiveness value (kg/million dollars)
Baseline scenario	BAS	1523.46	/
Non-engineering measures	CT	1523.43	1.18
	SC	1407.33	3.99
	FR10	1337.55	0.30
	FR25	1448.225	2.37
	FR50	1341.38	1.95
Engineering measures	TR	1433.63	2.79
	GW	1287.00	241.25
	FS	1355.83	34.54

Table 8 highlights the top ten combinations of BMPs based on their comprehensive environmental cost-effectiveness. Notably, the CE values of these multi-scenario measures exceed those of all single-scenario measures except GW and FS. As evident from the BMPs efficiency assessment in Sect. 2.3, combined scenarios exhibit greater efficiency in reducing TP loads in the basin compared to standalone measures. Consequently, in watersheds with specific non-point source pollution reduction targets, multi-scenario approaches are more suitable. Among the top ten combinations in terms of comprehensive environmental cost-effectiveness, FS+GW+FR10, FS+GW+FR30, FS+SC+GW+FR10, and FS+SC+FR10 stand out as particularly effective strategies, highlighting the potential of strategic combinations tailored to the unique characteristics of the Fuzhou River Basin.

Table 8. Top 10 BMPs in terms of combined environmental cost-effectiveness

Serial Number	BMPs	Total phosphorus load (kg)	Cost-effectiveness value (kg/million dollars)
1	FS+GW+FR10	19786.67	14.18
2	GW+FR10	25500.00	13.85
3	FS+FR10	21433.33	13.06
4	FS+GW+FR30	19283.33	9.25
5	FS+FR30	20716.67	8.56
6	GW+FR30	24013.33	8.33
7	FS+SC+GW+FR10	19136.67	6.64
8	FS+SC+FR10	19963.33	6.39
9	GW+SC+FR10	23980.00	5.46
10	FS+SC+GW+FR75	18743.33	5.40

4 Conclusion

(1) The Soil and Water Assessment Tool (SWAT) model has demonstrated robust applicability within the Fuzhou River Basin, achieving satisfactory simulation results for both runoff and total phosphorus (TP). Notably, the identified critical source areas, encompassing 37.61% of the total basin area, contribute a substantial 60.65% of the TP load. These areas are primarily concentrated in agricultural and orchard zones, where high rates of nutrient loss per unit area are prevalent.

(2) Within the Fuzhou River Basin, individual Best Management Practices (BMPs) exhibited a range of effectiveness in mitigating TP loads, spanning from 0.53% to 38.97%. Notably, FS (filter strips) and GW (grassed waterways) emerged as highly potent strategies, significantly reducing TP loads. Furthermore, the integration of multiple BMPs surpassed single measures in terms of efficiency, revealing a positive correlation between the number of combined BMPs and the average reduction rate. However, given a fixed implementation area within the basin, an optimal threshold for the reduction efficiency exists beyond which incremental BMP combinations yield diminishing returns.

(3) In terms of cost-effectiveness analysis, FS stands out as the most economically viable single BMP for mitigating TP loads in the Fuzhou River Basin. Additionally, the combined approach of FS+GW+FR10 offers the optimal balance between cost and effectiveness among the various BMP combinations.

References

1. Xia, S., Lu, W., Yang, J., et al.: Characteristics of spatial and temporal distribution of water quality in Suzhou rivers and analysis of influencing factors. Environ. Sci. 1–14 (2024)

2. Hua, Z., Li, R., Yang, Z., et al.: Spatial and temporal trends of nutrient status and its causes in the Erhai Sea from 2017 to 2022. Lake Sci. 1–13 (2024)
3. Smith, V.H., Tilman, G.D., Nekola, J.C.: Eutrophication: impacts of excess nutrient inputs on freshwater, marine, and terrestrial ecosystems. Environ. Pollut. **100**(1), 179–196 (1999)
4. Zhou, H.: A review of research on the impacts of surface source pollution and land use in river basins. Mod. Agric. Sci. Technol. (18), 101–105+109 (2024)
5. Zhang, D., Zhang, X., Zhang, J.: Integrated research and evaluation on nonpoint source pollution in shanghai suburbs. Acta Agricul Turae Shanghai **50**(5), 976–985 (1997)
6. Liu, R., Xu, F., Zhang, P., et al.: Identifying nonpoint source critical source areas based on multi-factors at a basin scale with SWAT. J. Hydrol. **533**, 379–388 (2016)
7. Keliang, C., Xiaodong, Z., Xianghua, W., et al.: Study on management and control of nonpoint source pollution from urban surface runoff. Chin. J. Popul. Resour. Environ. **5**(1), 39–44 (2007)
8. Zheng, Y., Han, F.: Markov Chain Monte Carlo (MCMC) uncertainty analysis for watershed water quality modeling and management. Stoch. Environ. Res. Risk Assess. **30**(1), 293–308 (2016)
9. Sun, Q.-Q., Zhang, C.-P., Yu, X.-X., et al.: Progress of best management practices for agricultural surface pollution in China. J. Ecol. **32**(3), 772–778 (2013)
10. Wang, H., Zou, M., Fang, W., et al.: Definition of critical source areas and control strategies for nonpoint source pollution in Wuqiangxi watershed based on SWAT model. J. Agric. Eng. 1–11 (2024)
11. Liu, W., Chen, L., Li, J., et al.: Watershed-scale pollution traceability simulation-optimization of prevention and control methods: a case study of Shun'an River Basin in Tongling City. Environ. Sci. **44**(7), 3835-3845 (2023)
12. Panagopoulos, Y., Makropoulos, C., Mimikou, M.: Reducing surface water pollution through the assessment of the cost-effectiveness of BMPs at different spatial scales. J. Environ. Manag. **92**(10), 2823–2835 (2011)
13. Xu, Z.-X., Zheng, H., Ma, J.-F.: A review on the application of hydrological modeling in the assessment of ecosystem hydrological services. J. Water Ecol. 1–18 (2023)
14. Arnold, J.G., Srinivasan, R., Muttiah, R.S., et al.: Large area hydrologic modeling and assessment part I: model development 1. JAWRA J. Am. Water Resour. Assoc. **34**(1), 73–89 (1998)
15. Wang, X., Qiao, F., Lei, K., et al.: Research and application progress of SWAT model in China. China Rural Water Conserv. Hydropower (5), 109–113 (2015)
16. Du, B., Wu, L.: Analysis of the current status of SWAT model research based on bibliometrics. People's Pearl River **44**(12), 44–53 (2023)
17. Refsgaard, J.C.: Parameterization, calibration and validation of distributed hydrological models. J. Hydrol. **198**(1–4), 69–97 (1997)
18. Moriasi, D.N., Arnold, J.G., Van Liew, M.W., et al.: Model evaluation guidelines for systematic quantification of accuracy in watershed simulations. Trans. ASABE **50**(3), 885–900 (2007)
19. Motovilov, Y.G., Gottschalk, L., Engeland, K., et al.: Validation of a distributed hydrological model against spatial observations. Agric. For. Meteorol. **98–99**, 257–277 (1999)
20. Huang, W.: Research on non-point source pollution characterization, prevention and control in Zhupi River Basin based on SWAT model. Huazhong University of Science and Technology (2022)
21. Cai, S.: Research on the current situation analysis of water pollution and treatment countermeasures of Fuzhou River. Dalian University of Technology (2020)
22. Dong, L., Jia, A., Chen, X.: Evaluation of water environment in watersheds based on gray water footprint. China Rural Water Conserv. Hydropower (2), 43–50 (2021)

23. Arnold, J.G., Kiniry, J.R., Srinivasan, R., et al.: Soil & water assessment tool (2010)
24. Ruan, B.N., Wu, L.L., Guo, J.W., et al.: Evaluation and screening of best management practices in the Jinghe River Basin taking into account the comprehensive benefits of environmental costs. China Environ. Sci. 1–11 (2023)
25. Wang, X., Hao, F., Zhang, X.: Optimization of best management practices for nonpoint source pollution in Danjiangkou reservoir basin. China Environ. Sci. **33**(07), 1335–1343 (2013)
26. Teshager, A.D., Gassman, P.W., Secchi, S., et al.: Simulation of targeted pollutant-mitigation-strategies to reduce nitrate and sediment hotspots in agricultural watersheds. Sci. Total. Environ. **607–608**, 1188–1200 (2017)
27. Kaini, P., Artita, K., Nicklow, J.W.: Optimizing structural best management practices using SWAT and genetic algorithm to improve water quality goals. Water Resour. Manag. **26**(7), 1827–1845 (2012)
28. Huang, K.: Research on the best management measures of surface source pollution in Danjiang River Basin based on SWAT model. Xi'an University of Technology (2020)
29. Xu, H., Xu, Z., Liu, P.: Estimation of nonpoint source pollution loads and preference of best management measures in Zhangwei South Canal Basin. Environ. Sci. **34**(3), 882–891 (2013)
30. Wang, Y.: Cost analysis of mechanized straw return technology. Jiangsu Agric. Mech. **6**, 49–50 (2010)
31. Lou, B., Qiu, G.: Changes in total phosphorus fluxes and spatial sources of the Jialing River (2002–2019). China Environ. Sci. 1–11 (2023)
32. Liu, J., Chen, Q., Xu, Y., et al.: Prediction of phosphorus input fluxes and their reduction in the Dongting Lake area of the Yangtze River Basin based on a hybrid model. Environ. Sci. 1–19
33. Chen, X., Liu, T., Cheng, J., et al.: Effectiveness and influencing factors of grassed filter strips on runoff, sediment and phosphorus deterrence and control in vegetable fields in South China. Res. Soil Water Conserv. **26**(2), 377–383 (2019)
34. Huang, S., Wu, Y., Mao, A.: Effect of Riparian buffer zone vegetation on phosphorus retention in runoff water. For. Sci. Res. **37**(1), 73–81 (2024)
35. Zhang, P., Li, Q., Kan, H., et al.: Simulation study on the control effect of grass-planted river on nonpoint source pollution based on SWAT model. J. Agric. Environ. Sci. **33**(6), 1204–1209 (2014)
36. Du, X., Li, S., Peng, Y.: Study on the benefits of plant hedges and Shikan terraces in improving sloping arable land. China Soil Water Conserv. (9), 39–41+69 (2010)
37. Li, L.H.: Study on nitrogen and phosphorus pollution in Qingyi River Basin based on SWAT model. Sichuan Agricultural University (2020)
38. Yang, C.-H., Li, Y., Wang, X., et al.: Effects of sugarcane contour planting and new planting allocation ratio on nitrogen and phosphorus nutrient loss due to slope gully erosion. Soil Water Conserv. Bull. **43**(5), 69–78 (2023)
39. Wang, Y., Chu, C., Liu, X., et al.: Analysis of nonpoint source pollution and simulation of fertilization measures in rate water basin. J. Water Resour. Water Eng. **30**(4), 6–13 (2019)

Air and Gas Pollution Control

Numerical Simulation Study on Oxygen-Enriched Combustion Process of Municipal Sludge

Jialin Li, Jinxu Jiang(✉), and Lei Huang

Hefei General Machinery Research Institute Co. Ltd, Hefei, China
jjx13146056060@163.com

Abstract. In this paper, a comprehensive municipal sludge oxy-fuel combustion numerical simulation model taking various processes, including combustion, heat and mass transfer into account was established. The model's precision has been validated through the inversion analysis of experimental data from sludge oxy-fuel combustion tests. A quantitative examination was conducted to assess the impact of oxygen concentration on the thermogravimetric curves of sludge particles and the temperature distribution within the reaction zone. The numerical inversion results closely match the experimental outcomes, with a maximum relative error of 7.9% and an average relative error of 6.07%. During the drying phase of sludge particles, the particle weight exhibited a slight reduction as the reaction time increased. The effect of oxygen concentration on the weight loss curve was found to be negligible. In the volatile ignition and combustion phase, the sludge particles experienced the most rapid weight loss. At this stage, the influence of oxygen concentration on the weight loss curve was particularly pronounced. An increase in oxygen concentration was found to substantially enhance the combustion rate of the sludge. The surface temperature of sludge particles experiences a modest increase in response to higher oxygen concentrations. Meanwhile, the maximum temperature within the gas phase reaction zone of tube furnace exhibits a slight downward trend as oxygen concentration rises. In the carbon combustion stage, the weight loss curve of the sludge declines as the oxygen concentration increases.

Keywords: municipal sludge · oxygen-enriched combustion · numerical simulation · temperature field

1 Introduction

Urbanization and industrialization are advancing relentlessly, resulting in an annual rise in China's sludge production. Achieving sludge reduction, harmless treatment, and resourceful utilization has become a pivotal issue in the environmental protection sector. Current established methodologies for sludge disposal predominantly encompass landfilling, composting, natural desiccation, and thermal incineration. Oxy-combustion technology stands out for its ability to minimize sludge volume, reduce pollutant emissions, and improve resource recycling efficiency.

D. Han et al. (Eds.): ICEPG 2024, ESE, pp. 337–345, 2025.
https://doi.org/10.1007/978-3-031-86829-0_38

A series of studies focusing on the combustion process of sludge. In the field of combustion characteristic research, small-scale laboratory experiments are predominantly conducted. Utilizing experimental methods such as thermogravimetric analysis and tube furnace combustion tests, these studies have unveiled the combustion reaction mechanisms of sludge [1–4]. In the field of numerical simulation of the sludge combustion process, the research primarily extends to the modeling of sludge combustion and the co-combustion modeling of sludge with other solid particulate materials [5–7].

In general, investigations on the deployment of oxy-combustion techniques for sludge incineration remains comparatively limited. Given the distinct composition and combustion properties of sludge compared to conventional solid fuels, there is a pressing need for further comprehensive studies to demystify the oxy-combustion mechanisms specific to sludge and to delineate its combustion behavior within an enriched oxygen atmosphere. A numerical model for the oxy-combustion of sludge was established and verified in this study. The processes of drying, pyrolysis, volatile combustion, and char combustion of sludge within a tube furnace under oxy-fuel conditions were simulated. The temperature distribution within sludge particles and the tube furnace at various oxygen levels was investigated using parameterized simulations.

2 Material

The sludge was collected from the municipal wastewater treatment plant in Zhengzhou New District, Zhengzhou, Henan Province. The proximate analyses and ultimate analyses of the sludge sample are detailed in Table 1.

Table 1. Proximate analyses and ultimate analyses of Sludge

Proximate analyses[a] (wt.%)					ultimate analyses[a](wt.%)			
C	*H*	*O*	*N*	*S*	*V*	*FC*	*A*	*M*
29.85	6.78	12.13	5.86	0.21	49.32	4.32	40.23	3.21

[a]air drying base; V-Volatile; FC-Fixed Carbon; A-Ash; M-Moisture

3 Numerical Simulation Model Development

3.1 Geometric Model

The modeling is based on the oxy-combustion experiment of sludge in a tube furnace. The model's dimensions include the tube furnace's diameter D, height H and the cylindrical sludge's diameter d and height h. As depicted in Fig. 1, the sludge particle's central axis coincides with that of the tube furnace, and the model's geometric dimensions and boundary conditions are symmetrical about the central axis. To reduce numerical simulation costs, a two-dimensional symmetrical model was established.

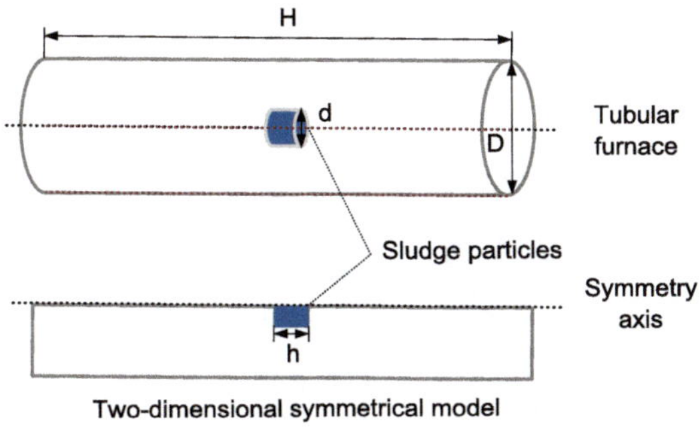

Fig. 1. Schematic diagram of geometric model

3.2 Mesh and Boundary Condition Setting

The left boundary of the model serves as the gas inlet, allowing the establishment of various oxy-fuel conditions within the tube furnace by adjusting the oxygen mass fraction in the mixed gases. The model's upper boundary is an isothermal, set to the tube furnace wall's preset temperature. The volatilization process at the sludge-gas interface (as shown in red in Fig. 2) is modeled by applying a mass source term. Due to the distinct heat and mass transfer at the gas-solid interface, significant changes in temperature and concentration are observed in this region. To accurately capture the temperature and concentration distributions within the tube furnace, localized mesh refinement was applied at the gas-solid interface positions in the model.

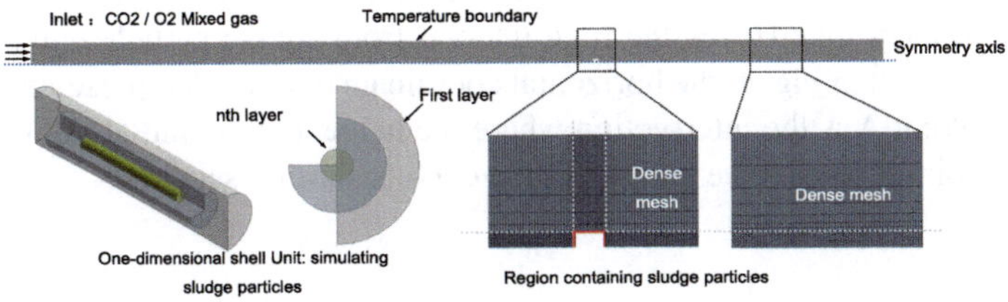

Fig. 2. Model mesh and boundary condition setting

3.3 Simulation of the Sludge Oxy-Fuel Combustion Process

The tube furnace has a diameter of 80 mm and a height of 1500 mm. Its quartz tube's heating zone spans 1100 mm (as shown in red in Fig. 3). The cylindrical sludge particles are 10 mm in both diameter and height. The tube furnace inlet introduces a CO_2 and O_2 mixture with an oxygen mass fraction of 30%, at a flow rate of 10 L/min. The temperature and concentration fields within sludge particles during drying, pyrolysis, and char combustion are calculated using MATLAB programming. The combustion of pyrolysis gases in the fluid domain is simulated numerically via FLUENT. Heat and mass transfer between solid sludge particles and the gas phase is simulated with a custom C-language subroutine.

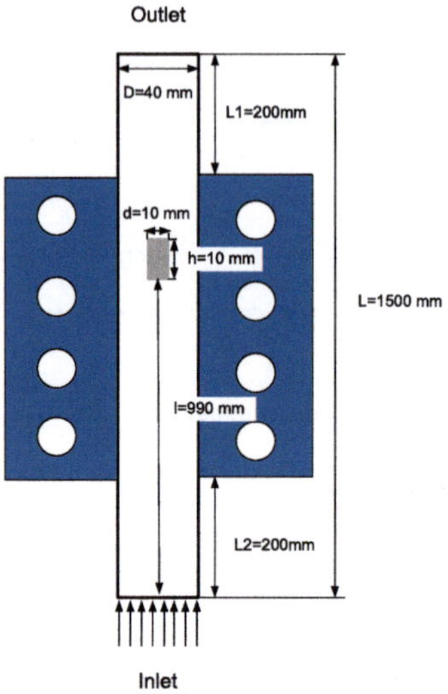

Fig. 3. Geometric dimensions schematic diagram of tube furnace and sludge particles

Prior to the introduction of solid sludge particles, the furnace air is preheated. The heating section's wall is maintained at a constant 1073.15 K, and a 30% O_2/CO_2 mixture is fed at 0.033 m/s. It's worth noting that temperature, oxygen concentration, and inlet gas velocity are all adjustable variables in the numerical model.

The thermal gravimetric curve, illustrating the sludge particles' weight loss over time, helps characterize the various stages of sludge's oxygen-enriched combustion. The ignition delay time (t_d) is the time interval from sludge particle entry to volatile ignition. As shown in Fig. 4, the horizontal coordinate of point A represents the ignition delay time. Point A is the intersection where the tangent at the initial high weight loss rate point (Point B) meets the parallel to the curve's starting segment.

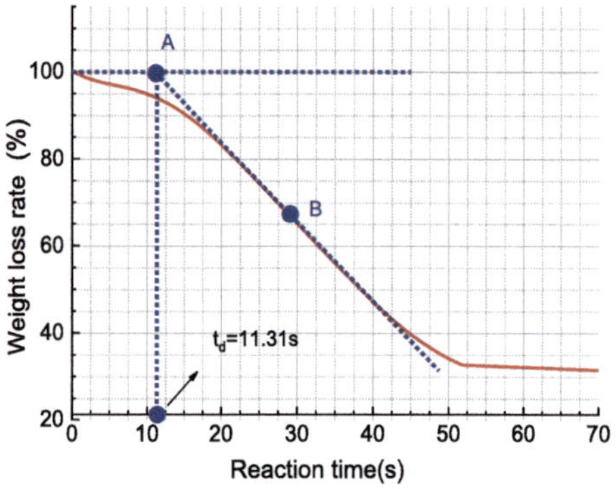

Fig. 4. Sludge particle thermogravimetric curves

Figure 5 displays the temperature field distribution during different reaction stages in the tube furnace's gas phase. Upon initial entry of sludge particles into the tube furnace (at $t = 1$ s), the surface temperature of the particles is markedly lower than that of the furnace interior. Convective heat transfer between the gas phase and solid sludge particle surfaces markedly enhances the evaporation of moisture and the pyrolysis process within the particles. With the progression of pyrolysis and a rising concentration of volatiles, a vigorous chemical reaction with the furnace's oxygen initiates the combustion of the volatiles. During the stable combustion phase of volatiles (at $t = 30$ s), the temperature in the ignition zone significantly rises, while the internal temperature of the sludge particles remains overall lower than that of the gaseous reaction domain within the tube furnace. In the char combustion phase ($t = 60$ s), temperatures around the particles stabilize.

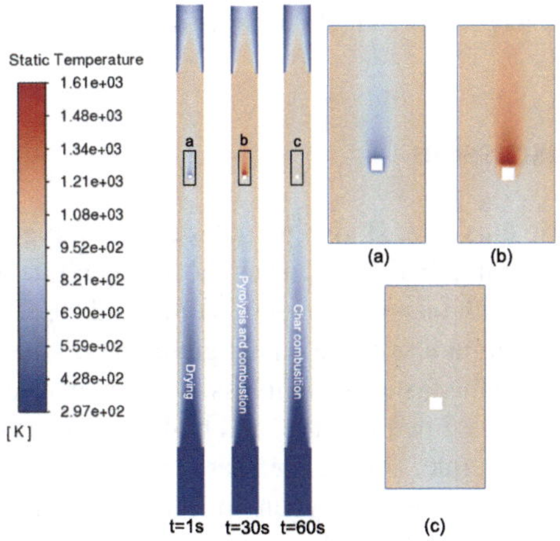

Fig. 5. Temperature distribution map in tube furnace

3.4 Verification of Numerical Model

A parametric analysis compared the experimental and numerical results of sludge particle center temperatures at volatile ignition for oxygen mass fractions of 30%, 40%, 50%, and 80%. As depicted in Fig. 6, the numerical model of sludge oxy-combustion established in this study can accurately calculate and obtain the sludge particle center temperature under different oxygen concentrations. The numerical inversion results for the sludge ignition temperature under various oxygen-enriched atmospheres show a maximum relative error of 7.9% and an average relative error of just 6.07% compared to experimental results, thereby validating the accuracy of the sludge oxy-combustion numerical simulation model.

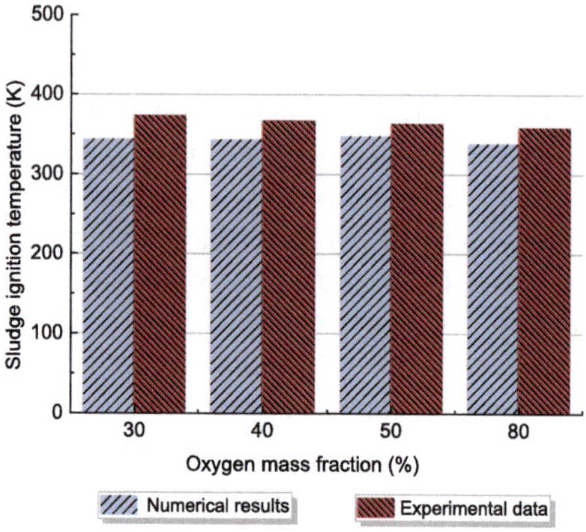

Fig. 6. Experimental and numerical results of sludge particle center temperature at varying oxygen concentrations

4 Results and Discussion

The thermal gravimetric curves of sludge particles at four oxygen concentrations (0.3, 0.4, 0.5, and 0.8) are illustrated in Fig. 7. During the drying phase of sludge particles, the weight decreases slightly with increasing reaction time, and the effect of oxygen concentration variations on the weight loss curve is negligible. In the volatile combustion phase, sludge particles experience the fastest weight loss, with oxygen concentration exerting a pronounced effect on the gravimetric curve, and higher oxygen concentrations notably accelerating the combustion rate. During the char combustion phase, the sludge weight loss curve decreases with increasing oxygen concentration, and the rate of mass loss for sludge particles increases under higher oxygen concentrations. Oxygen concentration enhances the rate of char combustion, with this promotional effect becoming more pronounced as the combustion process advances.

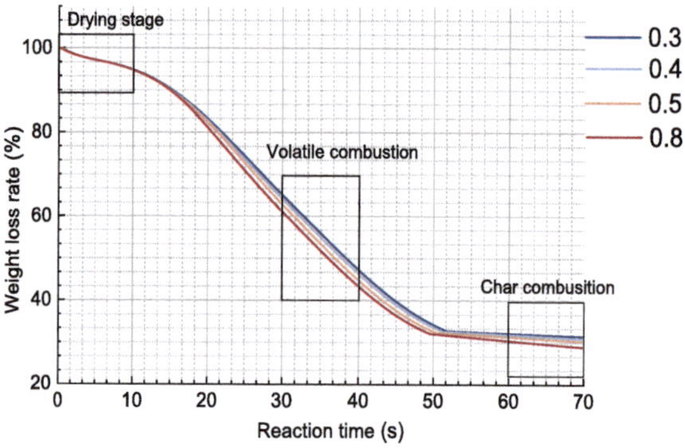

Fig. 7. Thermogravimetric curves of sludge particles under different oxygen concentration ratios

The study of temperature field distribution in sludge oxy-combustion can reflect the thermophysical behavior of sludge at different reaction stages and reveal the heat exchange mechanism between the sludge particles and the gases inside the tube furnace. The surface temperature changes of sludge particles have been studied at three characteristic moments corresponding to the various reaction stages of sludge oxy-combustion: during the drying phase at $t = 4$ s, the stable combustion of volatiles at $t = 35$ s, and the char combustion phase at $t = 65$ s. Under different oxygen concentrations, the surface temperature of sludge particles during the volatile combustion phase is significantly higher compared to the drying phase. As shown in Fig. 8, Sludge particle surface temperature during the volatile combustion phase is significantly higher compared to the drying phase. In the char combustion phase, the rise in particle surface temperatures is modest, mainly attributed to the heat released from the internal combustion of char. Variations in oxygen concentration exert a negligible effect on the surface temperatures of sludge particles across the different stages.

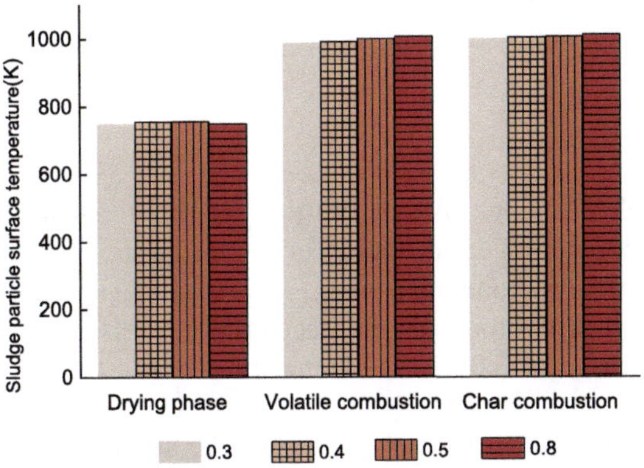

Fig. 8. Sludge particle surface temperature variations vs diverse oxygen levels

In the volatile combustion phase, the maximum temperature in the tube furnace's gas phase tends to decrease with higher oxygen concentrations, yet the change remains relatively minor (as shown in red in Fig. 9). The temperature within the gaseous reaction domain is primarily affected by the heat released from the combustion of volatiles. Under conditions of high oxygen concentration, the ignition delay for volatiles in the tube furnace is shortened. During the stable combustion phase, a decrease in volatile concentration can limit the combustion reaction rate.

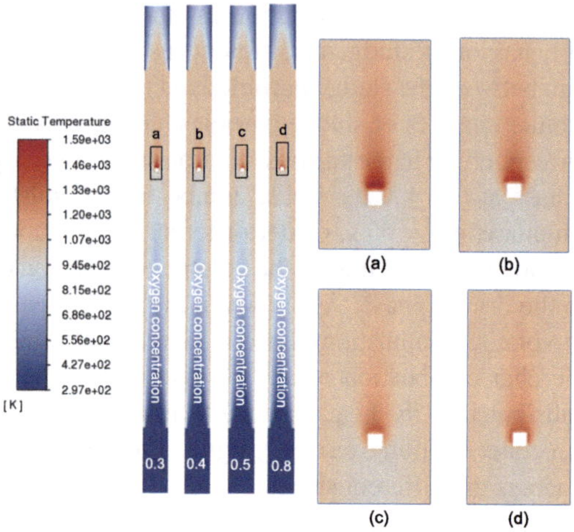

Fig. 9. Temperature distribution cloud of gas phase reaction in tube furnace under varied oxygen levels

5 Conclusions

A numerical model for sludge oxy-combustion has been established, and numerical simulations of sludge particle combustion in a tube furnace under O_2/CO_2 enriched conditions have been carried out in this study. The numerical inversion of sludge particle ignition temperatures, across four oxygen concentration levels, exhibits a maximum error of 7.9% and an average of 6.07%, validating the model's suitability for engineering applications.

The thermal gravimetric analysis of sludge particles at various oxygen concentrations indicates that during the drying phase, the weight of the particles decreases slightly with reaction time, and the impact of oxygen concentration variations on the weight loss curve is negligible. In the volatile combustion phase, sludge particles experience rapid weight loss, with oxygen concentration significantly influencing the gravimetric curve and higher concentrations markedly accelerating the combustion rate. During the char combustion stage, the weight loss curve of sludge decreases with increasing oxygen concentration.

During the volatile combustion phase, the surface temperature of sludge particles significantly increases compared to the drying phase. In contrast, the temperature rise during the char combustion phase is relatively minor. In the volatile combustion phase, the peak temperature in the tube furnace's gas phase tends to decrease with higher oxygen concentrations, yet the overall change is minimal.

Acknowledgment. This research has been co-financed by Anhui Provincial Key Research and Development Project (Grant No. 2022010672); Doctoral Science and Technology Foundation of Hefei General Machinery Research Institute Co., Ltd (Grant No. 2024011188).

References

1. Wang, M.Q., Yu, H.L., Chen, M.J.: Experimental study on the pyrolysis and combustion of municipal sewage sludge. J. Univ. Shanghai Sci. Technol. **36**(02), 185–188+193 (2014)
2. Dai, C.H., Hou, J.W., Fan, W.S.: Study on combustion performance and influencing factors of the pyrolytic carbon derived from low-temperature fast pyrolysis of municipal sewage sludge. Environ. Eng. **41**(S2), 558–561+567 (2023)
3. Wang, C.X.: Investigation on characteristics of sewage sludge combustion and pollutant emission. University of Science and Technology of China, Anhui (2021)
4. Bien, J., Jurand, D.: Conversion of sulfur and nitrogen to gaseous components from sewage sludge combustion under oxy-firing conditions. Desalin. Water Treat.: Sci. Eng. **57**(3), 1223–1229 (2016)
5. Dottei, A., Holtz, D., Müller, K.: Numerical investigation of sewage sludge combustion in a fluidized bed reactor: a comparison of 2D and 3D simulations. Powder Technol. **428**, 118834 (2023)
6. Max,S., Christian, S., Günter, S.: Modelling and simulation of fluidized bed steam-oxygen gasification of sewage sludge using thermochemical equilibrium and experimental data. Fuel **341** (2023)
7. Zhang, Z.L.: Research and prospect of coal-fired and sludge-coupled technologies in CFB boilers. Therm. Power Gener. **49**(5), 7–13 (2020)

Spatiotemporal Characteristics and Genetic Analysis of Cyanobacteria Blooms in Hulun Lake Based on CIE Chromaticity Space

Qingzhu Zhang(✉)

School of Geomatics and Urban Spatial Informatics, Beijing University of Civil Engineering and Architecture, Beijing, China
2108570022110@stu.bucea.edu.cn

Abstract. Cyanobacterial blooms are one of the most significant global challenges affecting aquatic ecosystems. Both their outbreaks and dissipation pose serious threats to the ecological environment of the affected water bodies and the safety of surrounding drinking water sources. As one of China's five largest freshwater lakes, Hulun Lake plays a crucial role in maintaining the stability of the local ecosystem. This study monitors and identifies cyanobacterial blooms in Hulun Lake by converting the surface reflectance of blooms in Landsat imagery into the CIE-xy chromaticity space. A dataset covering cyanobacterial bloom outbreaks in Hulun Lake from June to September for each year from 2003 to 2023 was generated. The study finds that wind speed and temperature are among the key factors driving the cyanobacterial blooms, with their effects becoming apparent 1 to 5 days after these conditions occur.

Keywords: Cyanobacteria blooms · Hulun Lake · Influencing factor

1 Introduction

A "bloom" typically refers to the phenomenon where the biomass of phytoplankton significantly exceeds the average levels in water bodies, aggregating at the surface to form visible algal mats [1]. Cyanobacterial blooms are one specific type of bloom, and their outbreaks cause the water to appear green. The occurrence and dissipation of cyanobacterial blooms are essentially manifestations of the cyanobacteria's vertical movement, either floating and aggregating or sinking and mixing [2]. The outbreak of cyanobacterial blooms is the result of the combined influence of physical, chemical, and biological factors in lakes [10]. Therefore, it is essential to study the driving factors behind cyanobacterial blooms in Hulun Lake, incorporating environmental conditions such as wind speed and precipitation, along with water quality parameters such as nitrogen and phosphorus concentrations and chlorophyll-a levels, to enable early prediction of cyanobacterial bloom events and inform relevant policy measures [5–7].

Once a cyanobacterial bloom occurs, it can severely impact the local ecosystem and compromise the safety of drinking water sources. Moreover, the decay of cyanobacteria

D. Han et al. (Eds.): ICEPG 2024, ESE, pp. 346–353, 2025.
https://doi.org/10.1007/978-3-031-86829-0_39

releases noxious odors and toxic compounds, exacerbating water pollution. With the increasing influence of climate change and human activities, the eutrophication of lakes has intensified, leading to more frequent and severe cyanobacterial blooms and their associated consequences.

The short cycle of cyanobacteria bloom outbreak, the range is uncertain and there is no obvious early warning, even within a few hours will occur large-scale migration movement, which makes the monitoring and prediction of cyanobacteria bloom very difficult. Traditional monitoring methods mainly rely on water samples collected in the field. Although this method has high accuracy and can effectively monitor the changes of various indicators in the water body, it is difficult to carry out monitoring work in a large range and cannot collect data with long time series and high frequency [12]. Remote sensing data has the advantages of wide monitoring range, fast speed and long period, which effectively make up for the shortcomings of traditional monitoring methods, and provide the possibility for the monitoring of cyanobacteria blooms in large areas and long time series.

Currently, there are numerous methods for monitoring cyanobacterial blooms using remote sensing, including vegetation index approaches, hyperspectral remote sensing, and machine learning techniques [13]. The vegetation index method involves establishing vegetation indices or new band combinations based on the spectral reflectance characteristics of cyanobacterial blooms, and then using threshold segmentation to detect and monitor blooms. However, the accuracy of this method is heavily influenced by the thresholding technique used [3]. Hyperspectral remote sensing operates on a similar principle. Nonetheless, spectral curves vary across different water bodies and within different areas of the same water body, and the cost of obtaining hyperspectral data is relatively high. Machine learning, on the other hand, requires large datasets to achieve high accuracy, which is not always feasible, as not all water bodies have access to extensive historical monitoring data.

Hulun Lake is the fifth largest freshwater lake in China and the largest lake in Inner Mongolia [3]. It plays a crucial role in regulating the local climate and supporting the surrounding ecological environment. In recent years, due to the impact of climate change and human activities, the frequency and severity of cyanobacterial blooms in Hulun Lake have been increasing. Investigating the causes and patterns of cyanobacterial bloom outbreaks in Hulun Lake is of great significance for mitigating the associated risks and preventing future occurrences.

This study begins by analyzing the color characteristics of cyanobacterial blooms, converting their reflectance in the red, green, and blue bands into the CIE chromaticity space [8, 9]. The boundary conditions of cyanobacterial blooms in Hulun Lake are then calculated within this chromaticity space, forming the basis for a long-term monitoring and identification model for cyanobacterial blooms. The model achieved an accuracy of 80% when compared to in-situ measurements. Additionally, this study explored the drivers of cyanobacterial bloom outbreaks in Hulun Lake by analyzing daily meteorological data from nearby weather stations. The results indicate that wind speed is the most significant factor influencing changes in cyanobacterial bloom area.

2 Materials and Methods

2.1 Study Area

Hulun Lake is located between the New Barag Right Banner and New Barag Left Banner on the Hulunbuir grassland in Inner Mongolia (Fig. 1).The annual average temperature ranges from −0.6 °C to 1.1 °C, and the average annual precipitation is less than 300 mm [11]. Hulun Lake experiences a semi-arid grassland climate, characterized by cool but short summers and long autumn and winter seasons with lower temperatures.

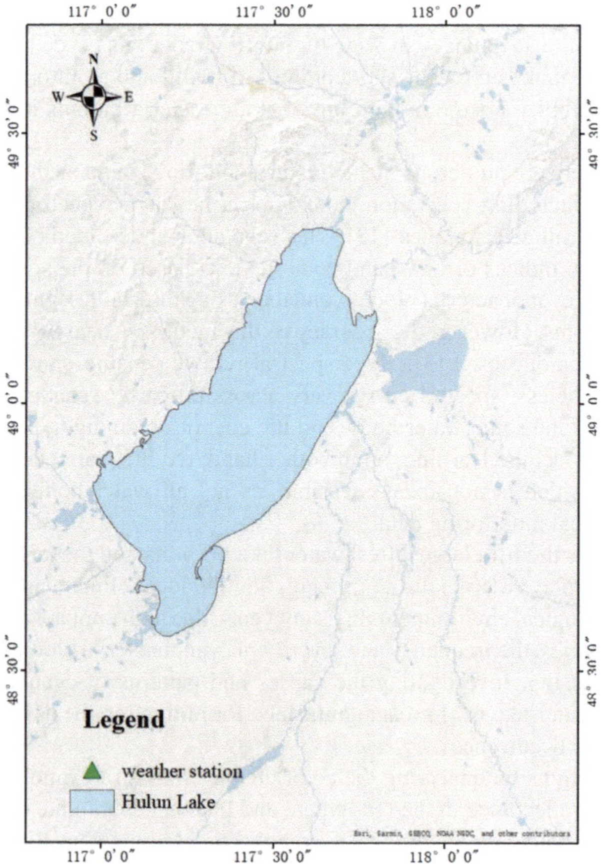

Fig. 1. Hulun Lake

2.2 Sample Collection and Data Source

Satellite Dataset

Landsat 5, 8, and 9 images, taken under clear or mostly clear conditions from June to September each year between 2013 and 2023, were obtained from the UGGS official website. The Normalized Difference Vegetation Index (NDVI) was used to assist in the visual interpretation of cyanobacterial blooms, resulting in a sample set for Hulun Lake. Using all available Landsat imagery from 2013 to 2023, a long-term time series identification of cyanobacterial blooms in Hulun Lake was conducted on the Google Earth Engine (GEE) platform.

Cyanobacteria Bloom Detection Algorithm

This study utilizes Landsat 5, 7, 8, and 9 imagery from 2013 to 2023 to identify pixels containing algal blooms in Hulun Lake during the months of June to September, establishing an algal bloom sample database for the Hulun Lake basin. The dataset was divided into training and validation sets in a 7:3 ratio. To assess the model's accuracy, we compared the number of cyanobacterial bloom pixels identified by the model with those in the training set within the same region, using the ratio between the two as the measure of model accuracy. The surface reflectance values of all pixels in the sample database were converted from the red, green, and blue bands into the CIE-xy color space. The conversion formula is as follows:

$$X = 2.7389R + 1.7517G + 1.1302B \tag{1}$$

$$Y = 1.0000R + 4.5907G + 0.0601B \tag{2}$$

$$Z = 0.0000R + 0.0565G + 5.5943B \tag{3}$$

$$x = \frac{X}{X + Y + Z} \tag{4}$$

$$y = \frac{Y}{X + Y + Z} \tag{5}$$

The x-axis of the CIE-xy color space was divided into intervals with a step size of 0.002, determining the corresponding maximum and minimum values of y. These points were then used to perform the fitting, resulting in the boundary condition formula for cyanobacterial blooms in Hulun Lake.

Environmental Parameters

Due to the limited number of meteorological stations near Hulun Lake and their considerable distance from the lake, this study selected only the closest station's data, including atmospheric pressure, wind speed, temperature, relative humidity, and precipitation, as the meteorological input for Hulun Lake. Using the Granger causality test, the meteorological factors were screened to determine the causal relationships between different meteorological variables and cyanobacterial bloom outbreaks.

3 Results and Conclusion

3.1 CIE-xy Feature Space

The surface reflectance values of the pixels in the sample database were converted into the CIE-xy color space, resulting in the color-based feature space for cyanobacterial blooms in Hulun Lake. The boundaries of this feature space are shown in the figure (Fig. 2). The region above the boundary represents the color characteristics of cyanobacterial blooms in Hulun Lake.

$$y = 158.68723456x^2 - 105.89265448x + 18.01397501 \tag{6}$$

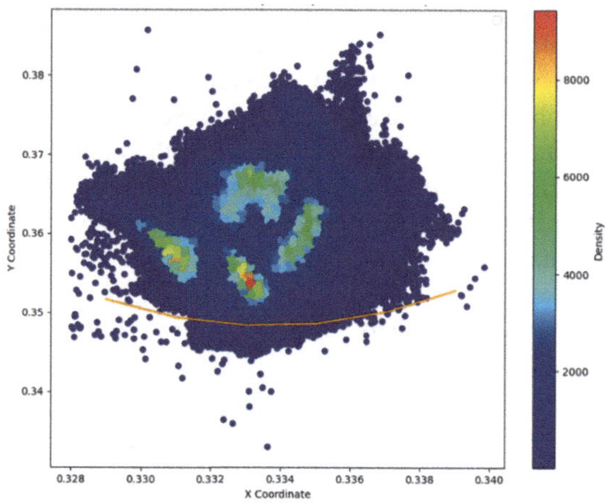

Fig. 2. CIE color space of cyanobacteria blooms in Hulun Lake

3.2 Temporal and Spatial Characteristics of Cyanobacteria Bloom

The time series of cyanobacterial bloom outbreaks in Hulun Lake from June to August for each year between 2003 and 2023 is shown in the figure below (Fig. 3). Compared with the training set, the model achieved an accuracy of 80%. Hulun Lake is most prone to cyanobacterial blooms in July and August each year. Between 2003 and 2023, there were three instances where the bloom area exceeded 80% of the total lake area, occurring in 2007, 2009, and 2010. In particular, the cyanobacterial bloom in 2007 covered 2,010 square kilometers, accounting for 96% of the total lake area.

3.3 Meteorological Factors Affecting the Outbreak Area of Cyanobacteria Blooms

Using meteorological data from the nearest weather station to Hulun Lake for the period 2003 to 2020, a heatmap of the Pearson correlation coefficients between meteorological

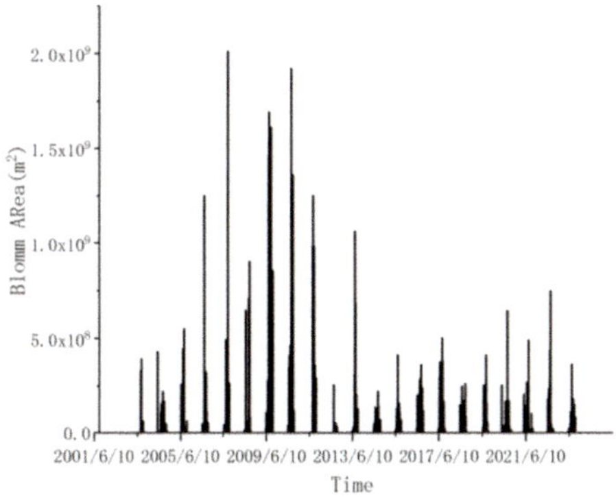

Fig. 3. The outbreak area of cyanobacteria blooms in Hulun Lake from June to September in 2003–2023

factors (Fig. 4). The results indicate that the correlation between cyanobacterial bloom area and changes in the surrounding environment is relatively weak. However, based on the p-values at different lag periods (Table 1), it is evident that there is a certain causal relationship between environmental factors and cyanobacterial bloom outbreaks, suggesting a delayed response of cyanobacterial blooms to climate variability. Specifically, the causal relationship between average wind speed and bloom area is most pronounced with a one-day lag, while for temperature, the strongest causal relationship occurs with a lag period of 3 to 5 days.

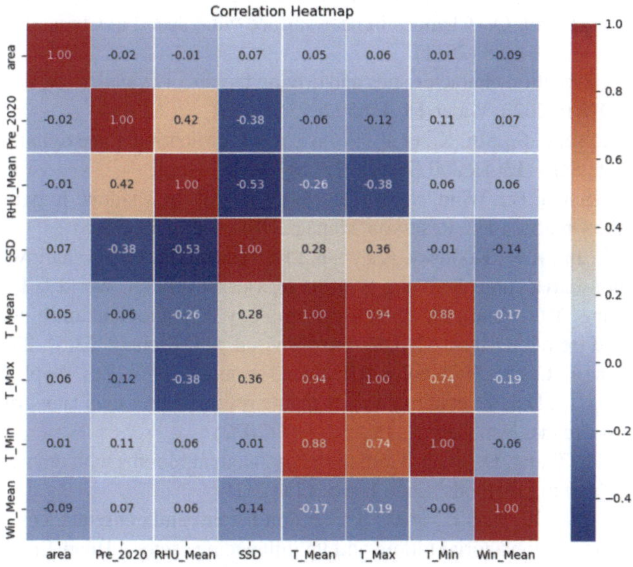

Fig. 4. Pearson correlation coefficient thermal map of cyanobacteria bloom outbreak area and meteorological factors in Hulun Lake

Table 1. Causal relationship between cyanobacteria bloom outbreak area and meteorological factors

	Pre	RHU_Mean	SSD	T_Mean	T_Max	T_Min	Win_Mean
1	0.77	0.24	0.05	0.26	0.56	0.22	0.05
2	0.88	0.11	0.13	0.40	0.62	0.21	0.08
3	0.96	0.21	0.18	0.26	0.65	0.01	0.13
4	0.99	0.31	0.16	0.25	0.14	0.03	0.07
5	0.47	0.11	0.21	0.27	0.12	0.04	0.07

This study focuses on the color characteristics of cyanobacterial blooms during their outbreak, utilizing the CIE-xy chromaticity space to extract the color boundaries of cyanobacterial blooms in Hulun Lake. Using these boundary conditions, cyanobacterial bloom outbreaks over the past 20 years in Hulun Lake were monitored and extracted. The study further demonstrated that there is a lag of 1 to 5 days between meteorological conditions and cyanobacterial bloom outbreaks. This research provides new insights for the monitoring and management of cyanobacterial blooms in Hulun Lake.

References

1. Qin, B.Q., Zhou, J.J., Elser, J.J., et al.: Water depth underpins the relative roles and fates of nitrogen and phosphorus in lakes. Environ. Sci. Technol. **54**(6), 3191–3198 (2020)
2. Guo, H., Xu, M., Hu, Q.: Changes in near-surface wind speed in China: 1969–2005. Int. J. Climatol. **31**(3), 349–358 (2011)
3. Li, X., et al.: Evolution characteristics and driving factors of cyanobacterial blooms in Hulun lake from 2018 to 2022. Water **15**, 3765 (2023)
4. Zhang, Y., Loiselle, S., Shi, K., et al.: Wind effects for floating algae dynamics in eutrophic lakes. Remote Sens. **13**(4), 800 (2021)
5. Kann, J., Welch, E.B.: Wind control on water quality in shallow, hypereutrophic upper Klamath lake Oregon. Lake Reservoir Manag. **21**(2), 149–158 (2005)
6. Havens, K.E., Jin, K.R., Rodusky, A.J., et al.: Hurricane effects on a shallow lake ecosystem and its response to a controlled manipulation of water level. Sci. World J. **1**, 44–70 (2001)
7. Shi, K., Zhang, Y.L., Zhang, Y.B., et al.: Phenology of phytoplankton blooms in a trophic lake observed from long-term MODIS data. Environ. Sci. Technol. **53**(5), 2324–2331 (2019)
8. Te Robertson, A.R.: CIE 1976 color-difference formulae. Color. Res. Appl. **2**, 7–11 (1977)
9. Hou, X., Feng, L., Dai, Y., et al.: Global mapping reveals increase in lacustrine algal blooms over the past decade. Nat. Geosci. **15**, 130–134 (2022)
10. Yu, P., Gao, R., Zhang, D., et al.: Predicting coastal algal blooms with environmental factors by machine learning methods. Ecol. Indic. **123** (2021)
11. Ouyang, T., Zhao, L., Ji, L.L., et al.: Succession pattern and consequences of the dominant species during cyanobacterial bloom and its influencing factors. Huan Jing Ke Xue **43**(10), 4480–4488 (2022). https://doi.org/10.13227/j.hjkx.202112249

12. Wen, C.C., Huang, T.L., Kong, C.H., et al.: Analysis of mechanism and start-up thresholds of seasonal algal blooms in a northern eutrophic stratified reservoir. Huan Jing Ke Xue **44**(3), 1452–1464 (2023). https://doi.org/10.13227/j.hjkx.202205012
13. Ma, J., Jin, S., Li, J., et al.: Spatio-temporal variations and driving forces of harmful algal blooms in Chaohu lake: a multi-source remote sensing approach. Remote Sens. **13**(3) (2021). https://doi.org/10.3390/rs13030427

Spatiotemporal Variations and Influencing Factors of Grey Water Footprint of Planting Industry in Zhejiang Province, China

Yiling Weng, Jiayan Song, and Hua Zhu(✉)

School of Geomatics, Zhejiang University of Water Resources and Electric Power, Hangzhou 310018, China
zhuhua18@mails.ucas.ac.cn

Abstract. Investigating the status of water pollution in the planting industry and its influencing factors can help promote the intensive utilization of agricultural water resources and the control of water pollution. The grey water footprint of the planting industry (PGWF) in Zhejiang and its influencing factors from 2010 to 2021 were analyzed using the GWF model and the Logarithmic Mean Divisia Index (LMDI) model. The results indicate that: (1) The PGWF in Zhejiang exhibited a downward trend from 2010 to 2021, decreasing from 24.50×10^8 m^3 in 2010 to 11.08×10^8 m^3 in 2021. (2) From 2010 to 2021, the spatial pattern of PGWF shifted from being 'higher in the northeastern regions and lower in the southwestern parts' to 'higher in the central area and lower around the periphery'. (3) Economic development and technological progress drive up the PGWF in Zhejiang, whereas adjustments in industrial structure, reduced water intensity, and population act as constraints on its expansion. Notably, water intensity exerts the most significant negative influence on the variation of the PGWF, whereas economic development has the strongest positive effect.

Keywords: Planting Industry · Grey Water Footprint · LMDI · Zhejiang

1 Introduction

Water resources play a vital role in maintaining global ecological balance and ensuring economic well-being. Still, they face severe challenges, especially in regions with rapid economic development [1]. With the accelerating urbanization, the problems of water scarcity and pollution have become increasingly severe, posing a major obstacle to sustainable development [2]. Zhejiang is well-developed in both agriculture and economy. However, irrational irrigation practices and excessive use of fertilizers and pesticides consume a significant amount of water resources and increase pollution discharge. Therefore, conducting an in-depth analysis of the grey water footprint of the planting industry (PGWF) and promoting its development towards a green and sustainable direction are urgent issues that need to be addressed. This significantly influences the development of ecological civilization in Zhejiang, as well as nationwide.

D. Han et al. (Eds.): ICEPG 2024, ESE, pp. 354–361, 2025.
https://doi.org/10.1007/978-3-031-86829-0_40

Initially proposed by Hoekstra et al. in 2008 [3], the grey water footprint (GWF) is defined as the quantity of freshwater necessary to dilute pollutants to safe concentrations, as stipulated by water quality monitoring standards [4]. The GWF model provides a new approach for water pollution assessment, effectively combining water quantity and water quality to more intuitively demonstrate the degree of water pollution [5]. Initially, studies primarily focused on the GWF related to nitrogen and phosphorus elements [6–10], but subsequently broadened their scope to include the GWF of industries, households, and agricultural sectors. For instance, Yang Fan conducted a study on the spatial autocorrelation of GWF intensity across Shandong from 2001 to 2014, measuring the negative environmental impact on water resources amidst economic growth [11]. Meng et al. utilized the Logarithmic Mean Divisia Index (LMDI) method to analyze the factors influencing the changes in agricultural GWF in the Pearl River Basin from 2006 to 2020 [12]. Guo Wei et al. employed the GWF model, the Theil index, and the LMDI to investigate the spatiotemporal evolution of agricultural GWF efficiency and its driving factors in the Yellow River Basin from 2000 to 2021 [13].

Over the past decade, Zhejiang has made significant progress in agricultural water pollution control. However, there has been limited research focusing on the PGWF and its driving factors. Consequently, this paper selects Zhejiang as the study area to investigate the spatiotemporal variations in the PGWF and quantify the contributions of its influencing factors. The objective is to offer scientific insights for agricultural water pollution control in Zhejiang.

2 Materials and Methods

2.1 Study Area

Located on the southeast coast of China and in the southern wing of the Yangtze River Delta, Zhejiang covers a land area of 105,500 km^2. Between 2010 and 2021, the average total water consumption was 19.217 billion m^3, of which agricultural use comprised roughly 40%. In 2021, the agricultural GDP reached CNY 216.9 billion, representing 2.95% of the total GDP. Meanwhile, the agricultural workforce numbered 22.35 million people, accounting for 43.85% of the total population.

2.2 Data Sources

The datasets employed in this study, encompassing annual nitrogen fertilizer application amounts, GDP, GDP of the planting industry, and the population employed in this industry, were compiled from the "Zhejiang Provincial Statistical Yearbook" and municipal statistical yearbooks between 2010 and 2021. The data on water consumption were obtained from the water resources bulletins of various prefecture-level cities, as well as the "Zhejiang Water Resources Bulletin" for the years 2010 to 2021.

2.3 Methods

Estimation of GWF in the Planting Industry

Water pollution of planting industry is primarily caused by nitrogen (N), and therefore, this paper selects nitrogen fertilizer as the indicator of water pollution. The PGWF in Zhejiang is estimated based on the nitrogen content entering water bodies, which is determined by the nitrogen fertilizer leaching rate [13]. The calculation formula is as follows:

$$GWF_{pl} = \frac{\alpha \bullet N}{C_{max} - C_{nat}} \tag{1}$$

where GWF_{pl} represents the PGWF (m^3); α is the nitrogen fertilizer leaching rate (%), with a national average value of 7% taken from reference [14]; N represents the annual nitrogen fertilizer application amount (kg); C_{max} is the standard concentration of pollutants in water quality (kg/m^3), which is set at 0.015kg/m^3, according to the "Environmental Quality Standards for Surface Water (GB3838-2002)"; and C_{nat} is the natural concentration of water (kg/m^3), which is assumed to be 0 kg/m^3 based on reference [4].

LMDI Model

Based on the extended Kaya identity [15], this study utilizes the LMDI multiplicative model [16] to quantify the contributions of driving factors to the PGWF in Zhejiang. The decomposition of the PGWF is as follows:

$$GWF_{pl} = \frac{GWF_{PL}}{WU} \times \frac{WU}{GDP} \times \frac{GDP}{TVA} \times \frac{TVA}{APS} \times PS = TL \times WE \times IS \times EE \times PS \tag{2}$$

where WU stands for total water consumption; GDP is the gross domestic product; TVA denotes the total agricultural GDP; APS represents the agricultural population; The technological level (TL) represents the technical effect, which indicates the GWF per unit of water used in the planting industry and is inversely proportional to technological level; The WE stands for the water intensity effect, representing the efficiency of water resource utilization; The IS is the industrial structure effect, which is the reciprocal of the proportion of the agricultural GDP; The EE signifies the economic effect, representing the per capita agricultural GDP and indicating the agricultural economic development level; and The PS denotes the population effect.

3 Results Analysis

3.1 Temporal Evolution of PGWF

As shown in Fig. 1, the PGWF in Zhejiang exhibited a fluctuating decline between 2010 and 2021, indicating a gradual reduction in water pollution from the planting industry. Except for the years 2019 to 2021, the PGWF in Jiaxing and Shaoxing was notably higher than that in other regions of Zhejiang. Notably, the peak values for them in 2010 amounted to 3.66×10^8 m^3 and 3.36×10^8 m^3, respectively. From 2010 to 2021, both Jiaxing and Shaoxing witnessed a substantial decrease in their PGWF. Notably, Jiaxing experienced the most significant reduction, with its PGWF decreasing to 2.60×10^8 m^3 in 2021. This significant drop underscores the notable achievements in water pollution control within its planting industry. The PGWF in Hangzhou, Ningbo, Jinhua, Wenzhou, Taizhou, Huzhou, Quzhou, and Lishui all exhibited similar downward trends. Zhoushan maintained a relatively low PGWF, ranging from 0.07×10^8 m^3 to 0.14×10^8 m^3, which was the lowest across cities and showed minimal fluctuation.

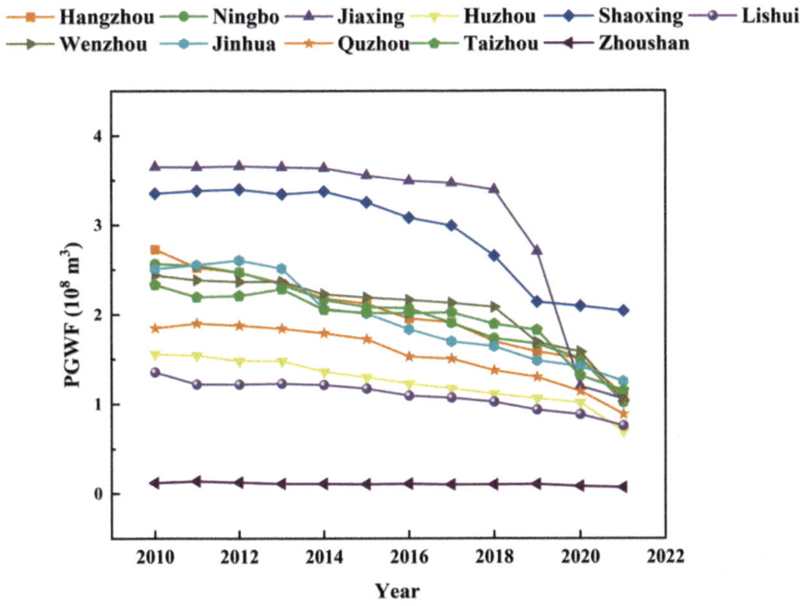

Fig. 1. PGWF in various cities of Zhejiang from 2010 to 2021.

3.2 Spatial Distribution Characteristics of PGWF

Spatial visualization of the PGWF in Zhejiang was conducted for the years 2010 and 2021, as shown in Fig. 2. In 2010, the spatial pattern of the PGWF generally showed a trend of "higher in the northeast and lower in the southwest", except Huzhou. This was due to the low and flat terrain in northern Zhejiang, which is rich in water resources and has a relatively large-scale planting industry with higher nitrogen fertilizer application rates. In contrast, the southern region features complex terrain with more mountainous areas, resulting in a smaller-scale planting industry and lower nitrogen fertilizer application rates. Compared to 2010, the PGWF in all prefecture-level cities in 2021 saw a significant reduction. Moreover, the spatial distribution of the PGWF evolved into a pattern of "higher in the central regions and lower in the surrounding areas", although there was little change along the eastern coast. Shaoxing emerged as the city with the highest PGWF, with a value of 2.04×10^8 m^3. From 2010 to 2021, the distribution range of high PGWF across Zhejiang tended to decrease, while the distribution range of low PGWF gradually expanded. This reflects the good progress made by Zhejiang in water pollution control and water resources management of planting industry.

3.3 Factors Influencing the PGWF

As shown in Fig. 3, in terms of absolute value, the average effect of technological change between 2011 and 2021 was 0.953. The absolute values in 2012, 2014, and 2015 were significantly greater than 1, indicating a positive impact of technological advancements on the PGWF during these years. In contrast, the absolute values for the other years were

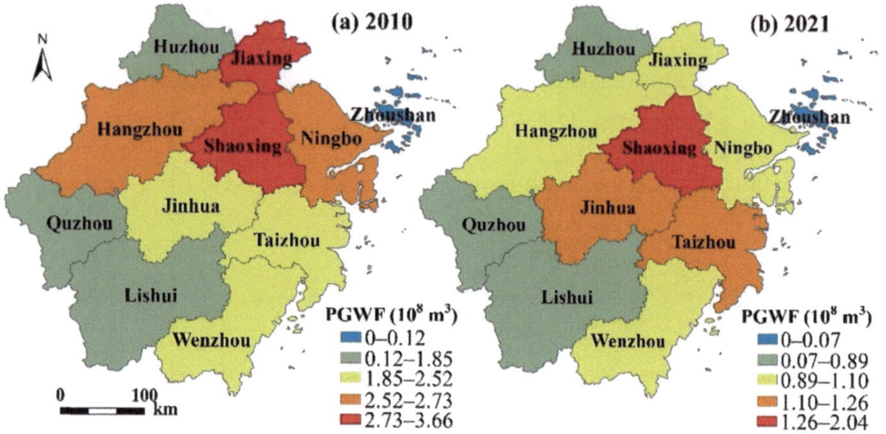

Fig. 2. Distribution of PGWF in various cities of Zhejiang in 2010 and 2021.

significantly less than 1, suggesting a negative influence of technological advancements on the PGWF. As seen in Table 1, in terms of relative quantity, the average annual contribution rate during the study period was −0.525, indicating that the technological level was a major factor contributing to the decrease in the PGWF.

The WE is represented by the volume of water used per unit of agricultural GDP. In terms of absolute value, the average effect of water intensity over the 11 years was 0.897, continuously contributing negatively to the PGWF during the study period. In terms of relative quantity, the water intensity effect was the dominant factor driving the negative growth of the PGWF between 2011 and 2021, with an average annual contribution rate of −3.518.

The IS effect is determined by the reciprocal of the agricultural GDP's proportion within the overall GDP. In terms of absolute value, it showed a positive correlation with the PGWF in all years except 2012 and 2020, with an average value of 1.056. This suggests a trend where, as the proportion of high-pollution industries in agriculture decreased, the PGWF also tended to decrease. In terms of relative influence, the annual average contribution of the IS effect during the research period stood at 1.051. The peak of its contribution occurred in 2013, when it emerged as the foremost factor fueling the growth of the PGWF.

The EE is represented by the per capita agricultural GDP. In terms of absolute value, the EE generally had a positive impact on the PGWF over the 11 years, with a mean annual effect of 1.075. In terms of relative quantity, the mean annual contribution rate of the economic effect between 2011 and 2021 reached 2.885, making it the dominant factor driving the positive growth of the PGWF. Its contribution rate peaked in 2015, reached a trough in 2016, and then rebounded in the following years.

The PS is represented by the population employed in the planting industry. In terms of relative quantity, the average effect during the study period was 0.971. In 2015, the absolute value was significantly less than 1, indicating the largest negative impact on the PGWF. Over the 11 years, the average annual contribution rate was −0.892, making it a major factor contributing to the decrease in the PGWF.

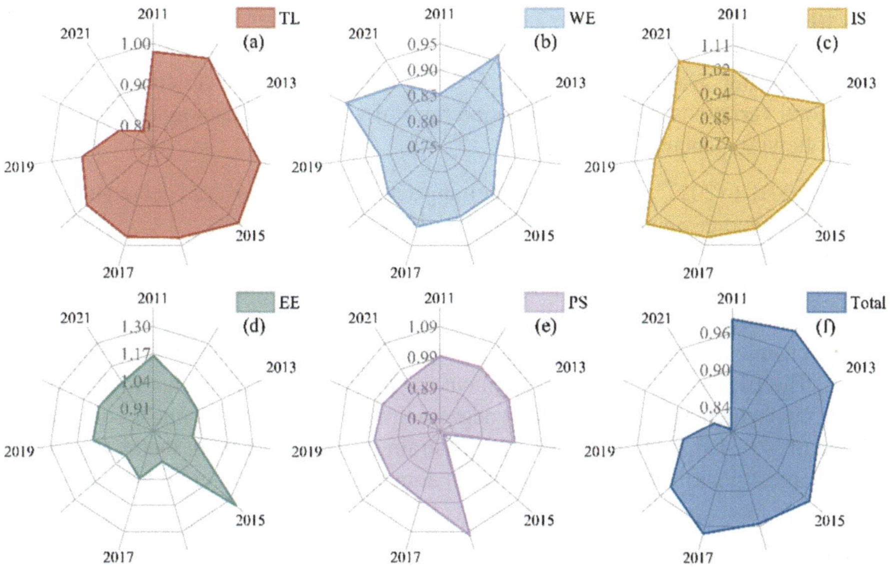

Fig. 3. Absolute contribution rates of factors influencing PGWF from 2010 to 2021.

Overall, the variations in the PGWF in Zhejiang are predominantly shaped by water intensity and economic factors. Economic development and technological advancements contribute to an expansion in the PGWF, whereas industrial structure adjustments, water use efficiency improvements, and population dynamics act as constraints on this growth.

Table 1. Standardized contribution rates of factors influencing PGWF from 2010 to 2021.

Year	TL	WE	IS	EE	PS
2010–2011	−1.148	−8.805	1.051	8.246	−0.344
2011–2012	0.695	−5.932	−5.102	9.702	−0.363
2012–2012	−2.373	−8.019	6.871	2.599	−0.077
2013–2013	0.359	−2.573	1.319	−0.086	−0.020
2014–2013	1.056	−4.005	1.024	10.096	−9.171
2015–2014	−0.451	−2.506	0.963	−1.105	2.099
2016–2014	−0.761	−3.181	2.544	1.137	−0.739
2017–2015	−0.535	−1.819	2.226	−0.375	−0.497
2018–2015	−0.620	−1.102	0.245	0.712	−0.235
2019–2016	−0.943	−0.248	−0.052	0.462	−0.219
2020–2016	−1.057	−0.512	0.471	0.343	−0.245

4 Conclusion

After conducting an assessment of the PGWF in Zhejiang, this paper delves into its spatiotemporal variations and the contribution rates of various influencing factors, arriving at the following conclusions:

(1) From 2010 to 2021, the PGWF in Zhejiang demonstrated an overall decreasing trend, with an average annual value of 20.05×10^8 m^3.
(2) The spatial distribution of the PGWF in Zhejiang transformed from being "higher in the northeast and lower in the southwest" in 2010, to "higher in the center and lower around the edges" in 2021.
(3) The changes in the PGWF in Zhejiang are primarily influenced by the WE and the EE. The EE and the IS tend to augment the PGWF, while the technological effect, the WE (which exhibits the largest negative impact with a mean annual contribution rate of −3.518), and the PS tend to constrain its growth. Among these, the EE has the largest positive impact, with a mean annual contribution rate of 2.885.

Funding. This research was supported by the National College Students' Innovative Entrepreneurial Training Plan Program (grant no. S202311481039) and the Joint Funds of the Zhejiang Provincial Natural Science Foundation of China (grant no. LZJWY23E090004).

References

1. Zhang, Y., et al.: Coupling coordination analysis of China's water resources utilization efficiency and economic development level. J. Clean. Prod. **373**, 133874 (2022)
2. Redhu, S., Jain, P.: Unveiling the nexus between water scarcity and socioeconomic development in the water-scarce countries. Environ. Dev. Sustain. **26**(8), 19557–19577 (2024)
3. Hoekstra, A.Y., Chapagain, A.K.: Globalization of Water: Sharing the Planet's Freshwater Resources. Wiley (2011)
4. Hoekstra, A.Y.: Water footprint assessment in supply chains. In: Bouchery, Y., Corbett, C.J., Fransoo, J.C., Tan, T. (eds.) Sustainable Supply Chains. SSSCM, vol. 23, pp. 77–98, Springer, Cham (2024). https://doi.org/10.1007/978-3-031-45565-0_4
5. Yu, C., et al.: A hybrid water-quality-index and grey water footprint assessment approach for comprehensively evaluating water resources utilization considering multiple pollutants. J. Clean. Prod. **248**, 119225 (2020)
6. Fu, T., et al.: Measurement and driving factors of grey water footprint efficiency in Yangtze River Basin. Sci. Total. Environ. **802**, 149587 (2022)
7. Zhang, L., et al.: China's provincial grey water footprint characteristic and driving forces. Sci. Total. Environ. **677**, 427–435 (2019)
8. Zhang, Y., et al.: Improving potential of nitrogen linked gray water footprint in China's intensive cropping systems. J. Clean. Prod. **269**, 122307 (2020)
9. Wu, M., et al.: Formation mechanism and step effect analysis of the crop gray water footprint in rice production. Sci. Total. Environ. **752**, 141897 (2021)
10. Wu, M., et al.: Blue, green, and grey water footprints assessment for paddy irrigation-drainage system. J. Environ. Manag. **302**, 114116 (2022)
11. Fan, Y.: The measure and spatial pattern analysis of the municipal grey water footprint in Shandong province. Water Saving Irrig. **2**, 69–75 (2017). In Chinese

12. Meng, Y., et al.: Study on agricultural grey water footprint and its influencing factors in the pearl river basin. Ecol. Econ. **39**(12), 126–133 (2023). In Chinese
13. Guo, W., Hu, X.: Spatial-temporal pattern evolution and driving factors of agricultural grey water footprint efficiency in yellow river basin. Bull. Soil Water Conserv. **44**(2), 437–445 (2024). In Chinese
14. Wang, C., et al.: Reducing soil nitrogen losses from fertilizer use in global maize and wheat production. Nat. Geosci. **17**(10), 1008–1015 (2024)
15. Hu, P., et al.: Decomposition analysis of industrial pollutant emissions in cities of Jiangsu based on the LMDI method. Environ. Sci. Pollut. Res. **29**(2), 2555–2565 (2022)
16. Jiang, P., et al.: Research on spatial and temporal differences of carbon emissions and influencing factors in eight economic regions of China based on LMDI model. Sci. Rep. **13**(1), 7965 (2023)

Deep Learning for Interpreting Regional Surface Ozone Spatio-Temporal Prediction

Canju Zheng[1,4], Hengqing Shen[2], Jianan Sun[3], Guangliang Liu[1,4](✉), Haowei Cao[1,4], and Jie Zhang[1,4]

[1] Key Laboratory of Computing Power Network and Information Security, Ministry of Education, Shandong Computer Science Center, Qilu University of Technology (Shandong Academy of Sciences), Jinan, China
guangliangliu@163.com, caohw@sdas.org

[2] Environment Research Institute, Shandong University, Qingdao 266237, Shandong, China
hqshen@sdu.edu.cn

[3] Ningbo Yonghuanyuan Environmental Engineering and Technology Co., LTD, Ningbo 315012, Zhejiang, China

[4] Shandong Provincial Key Laboratory of Computer Networks, Shandong Fundamental Research Center for Computer Science, Jinan, China

Abstract. Against the backdrop of improving urban particulate matter ($PM_{2.5}$) pollution in China, ground-level ozone (O_3) has emerged as the second most critical air pollutant. Given China's vast geographic expanse, O_3 pollution exhibits strong regional characteristics. Exploring the key influencing factors in different regions can aid government agencies in developing targeted O_3 prevention and control strategies. In this study, we develop a regional short-term O_3 spatio-temporal prediction transformer model (RO3former) to accurately and efficiently forecast ground-level O_3 concentrations in the Beijing-Tianjin-Hebei (BTH), Yangtze River Delta (YRD), Pearl River Delta (PRD), Fenwei Plain (FWP), and Chengdu-Chongqing (CC) regions. Additionally, we quantify the contributions of the driving factors through factor elimination experiments in these different regions. The results demonstrate the model's strong performance, with correlation coefficients stabilized between 0.85 and 0.92 and normalized mean bias (NMB) ranging from −5.87% to 2.77% across the regions and seasons. The input factor ablation experiments reveal that the key forcing factors vary by region. Overall, the historical O_3 concentration has the most significant effect on short-term O_3 concentration prediction in China, followed by wind speed (u, v) and temperature. The spatial differences in drivers contribute to the heterogeneity of O_3 concentrations and exposure risks across China. Therefore, future O_3 management strategies should involve region-specific control policies tailored to the identified key influencing factors.

Keywords: O_3 concentration · Spatio-temporal prediction · Different regions · Driving factors

C. Zheng and H. Shen—Have contributed equally to this work.

D. Han et al. (Eds.): ICEPG 2024, ESE, pp. 362–374, 2025.
https://doi.org/10.1007/978-3-031-86829-0_41

1 Introduction

In recent decades, China has experienced rapid urbanization and energy-intensive development, leading to increasingly severe air pollution problems [1]. Especially in urban agglomerations such as Beijing-Tianjin-Hebei (BTH), Yangtze River Delta (YRD) and Pearl River Delta (PRD), Fenwei Plain (FWP), and Chengdu-Chongqing (CC) [2]. As a pollutant with strong oxidizing properties, O_3, when its concentration reaches high levels near the ground surface, can have adverse effects on the human respiratory system, immune system, tissues, and organs such as the skin, thus threatening people's health and even life [3]. In addition, the presence of O_3 has adverse effects on social production and daily life. For example, it affects the normal growth of plants, leading to a decrease in crop yields and losses in agricultural production [4].

Since 2013, a series of stringent clean air measures have been implemented. These clean air policies can be divided into two phases, 2013–2017 and 2018–2020, which are summarized in two major plans: the Action Plan for Prevention and Control of Air Pollution (2013–2017) and the Three-Year Action Plan for Winning the Battle for Blue Sky (2018–2020) [5]. Phase I policies focused on reducing fine particulate matter ($PM_{2.5}$) and achieved significant results. Phase II policies continued the efforts of Phase I and further reduced SO_2, NO_X, and $PM_{2.5}$ emissions. The implementation of the two phases of clean air policies has significantly improved China's air quality. According to data from the Bulletin of China's Ecological and Environmental Conditions, from 2017 to 2022, the national annual average $PM_{2.5}$ concentration has achieved year-on-year reductions of 43 $\mu g/m^3$, 39 $\mu g/m^3$, 36 $\mu g/m^3$, 33 $\mu g/m^3$, 30 $\mu g/m^3$, and 29 $\mu g/m^3$, respectively. At the same time, however, the national annual average ozone concentration remained at a high level, with specific values of 149 $\mu g/m^3$, 151 $\mu g/m^3$, 148 $\mu g/m^3$, 138 $\mu g/m^3$, 137 $\mu g/m^3$, and 145 $\mu g/m^3$. Against the backdrop of the year-on-year improvement in particulate matter pollution in China's cities, O_3 pollution partially offsets the health benefits of the improvement in $PM_{2.5}$ and has become one of the most serious air pollutants in China [6].

O_3 is typically formed from nitrogen oxides ($NO_X = NO + NO_2$) and volatile organic compounds (VOC_S) through a series of complex photochemical reactions [7]. In clean air, there is a photochemical equilibrium between NO_X, NO_2, and O_3. As shown in reactions (1)–(3), reaction (3) is known as the titration reaction, which plays the role of eliminating O_3.

$$NO_2 + hv(< 420nm) \rightarrow NO_2^* \rightarrow NO + O \tag{1}$$

$$O + O_2 + M \rightarrow O_3 + M \tag{2}$$

$$NO + O_3 \rightarrow NO_2 + O_2 \tag{3}$$

Once biological or anthropogenic VOC_S are emitted into the atmosphere, their oxidation products, including HO_2, RO_2, and other reactive intermediates, react with NO:

$$RO_2 + NO \rightarrow NO_2 + RO \tag{4}$$

$$HO_2 + NO \rightarrow NO_2 + OH \tag{5}$$

The above two reactions initiate the conversion of NO to NO_2 and form a competitive reaction with the titration reaction [8]. In polluted environments, due to the presence of excess VOC_S, the NO_2-NO-O_3 system continuously consumes NO and replenishes the loss of NO_2, thereby promoting the production of O_3. This process disrupts the dynamic equilibrium in the absence of VOC_S, leading to a gradual accumulation of O_3 to uncontrollable concentrations. The formation of O_3 in the actual atmospheric environment is much more complex.

China has a land area of more than 9.6 million square kilometers, a latitude of nearly 50° from north to south, complex and variable meteorological and economic conditions, and a topography that is high in the west and low in the east. In addition, O_3 pollution is closely related to human socioeconomics, resulting in complex O_3 sources [9]. Population density, economic development, topography, and energy structure all have important impacts on O_3 [10, 11].

Many previous studies have investigated how meteorological factors and anthropogenic emissions affect surface O_3 concentrations by modulating photochemical reaction rates [12]. For example, Yin et al. found that solar radiation has a significant effect on the long-term changes of O_3 in Guangzhou [13]. Sun et al. applied the GEOS-Chem model and found that meteorological drivers had a greater influence than emissions on summer surface O_3 changes in China from 2003 to 2015 [14]. In addition, Wang et al. concluded that the main meteorological factors for surface O_3 concentrations at the national scale show significant regional differences [11].

However, few studies have investigated the factors leading to O_3 variations in different regions of China. Although some studies have applied air quality modeling and statistical analysis to investigate the effects of meteorology and emissions on O_3 concentrations in China [3, 15], air quality modeling and statistical analysis methods have limitations due to the complex mechanisms of O_3 formation, the incompleteness of the VOC_S monitoring system, and the high computational cost. Deep learning is gradually applied to O_3 concentration prediction due to its advantages, such as lower learning cost for network construction and faster training speed. In this paper, we adopt a deep learning-based spatio-temporal series prediction model, RO3former, i.e., which is designed to utilize multi-routine observation data with a well-developed observation system instead of imperfect O_3 precursor data to efficiently and accurately predict the air quality concentrations in Beijing-Tianjin-Hebei, Yangtze River Delta, Pearl River Delta, Chengdu-Chongqing and Fenwei regions, and to quantify the contributions of driving factors in different regions through the factor ablation experiment, so as to improve the overall prediction level of O_3 concentration and enhance the interpretability of the model.

2 Methodology

2.1 Study Areas and Data

As shown in Fig. 1, this study covers the Beijing-Tianjin-Hebei (BTH), Yangtze River Delta (YRD), Pearl River Delta (PRD), Fenwei Plain (FWP), and Chengdu-Chongqing (CC) regions due to the obvious regional differences in the main meteorological factors affecting the ground-level O_3 concentration. In order to investigate the effects of different factors on O_3 in different regions, we compiled a data set containing 12 different types of data for each region, including $PM_{2.5}$, PM_{10}, SO_2, NO_2, CO, O_3, wind speed (u, v), air pressure, solar ultraviolet radiation (UVB), relative humidity, and temperature. Where O_3, $PM_{2.5}$, PM_{10}, SO_2, NO_2, CO, wind speed (u, v), air pressure, relative humidity, and temperature data were obtained from the China Air Quality Reanalysis Dataset (CAQRA) (https://doi.org/https://doi.org/10.11922/sciencedb.00053, last accessed April 25, 2024). Solar ultraviolet radiation (UVB) data were obtained from the fifth-generation European Reanalysis Dataset (ERA5) of the European Center for Medium-Range Weather Forecasts (ECMWF) (https://cds.climate.copernicus.eu/cdsapp#!/home, last accessed April 25, 2024). The datasets are divided into a training set (2013–2016), a validation set (2017) and a test set (2018).

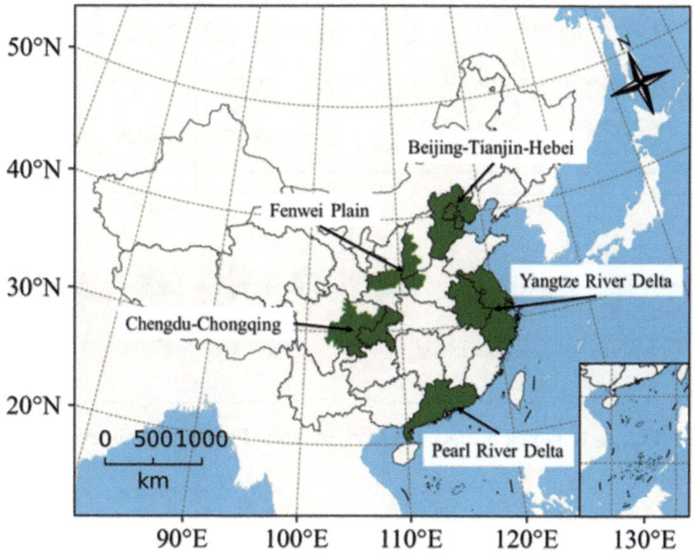

Fig. 1. Study areas.

2.2 RO3former

The RO3former model was developed based on Earthformer, a hierarchical Transformer codec model based on a cuboid Attention block [16]. In the cuboid attention layer, the input spatio-temporal tensor is decomposed into non-overlapping rectangles by a local or extended decomposition strategy. Then, self-attention is applied in each attention layer of the cube to limit the computational complexity to small rectangles, thus reducing the overall computational complexity. Features in the cube can focus on both local and global vectors. In this way, the local cubes capture the dynamics of the entire system and exchange information with each other. The model solves the high complexity problem of the Transformer architecture in obtaining high-dimensional spatio-temporal observational data applications, which is potentially advantageous in the field of O_3 concentration prediction.

The RO3former model is shown in Fig. 2. It requires the input of 12 in drivers within different regions, including $PM_{2.5}$, PM_{10}, SO_2, NO_2, CO, O_3, wind speed (u, v), air pressure (psfc), relative humidity (RH), temperature (temp), and solar ultraviolet radiation (UVB) for the previous 24 h. The model then generates O_3 predictions for the next 24 h within the region. Table 1 describes the multi-channel Earthformer model parameters used for training.

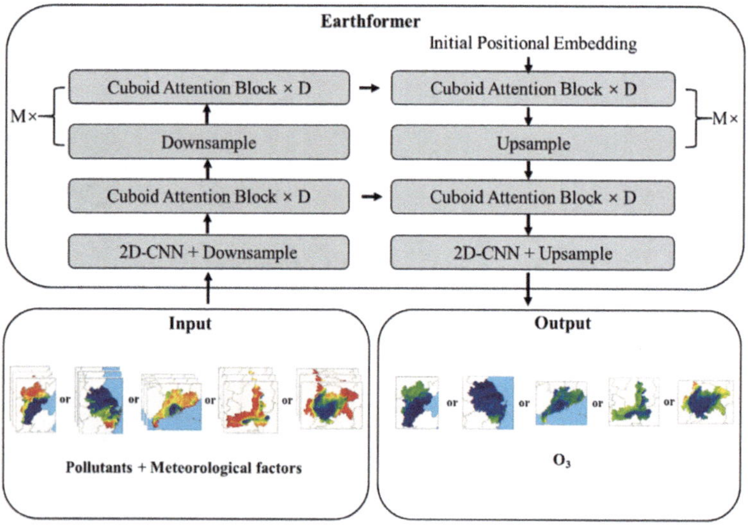

Fig. 2. Regional Surface O_3 Concentration Prediction Framework.

2.3 Evaluation Indicators

The performance of the model was evaluated using statistical metrics, including Root Mean Square Error (RMSE), Pearson's correlation coefficient (Pearson's r), and Standardized Mean Bias (NMB), as described in Eqs. (6)–(8).

$$RMSE = \sqrt{\frac{1}{n}\sum_{i=1}^{n}(\hat{y}_i - y_i)^2} \tag{6}$$

Table 1. .

RO3former	Model Parameters
parameters (M)	2.3
optimizer	Adam
learning rate	0.0005
batch size	1
base_units	64
epochs	30

$$Pearson's\, r = \frac{\sum_{i=1}^{n} \left(\hat{y}_i - \bar{\hat{y}}_i\right)\left(y_i - \bar{y}_i\right)}{\sqrt{\sum_{i=1}^{n} \left(y_i - \bar{\hat{y}}_i\right)^2 \sum_{i=1}^{n} \left(\hat{y}_i - \bar{y}_i\right)^2}} \tag{7}$$

$$NMB = \frac{\sum_{i=1}^{n} (\hat{y}_i - y_i)}{\sum_{i=1}^{n} (y_i)} \times 100 \tag{8}$$

where y_i is the true value of O_3, $\hat{y}_i$ represents the predicted value, and n is the number of samples.

3 Results and Discussion

3.1 Trends in Ozone Concentrations

Figure 3 shows the trends of annual average O_3 concentrations of BTH, YRD, PRD, FWP, and CC from 2013 to 2018. From the figure, it is clearly seen that the O_3 concentrations of BTH, YRD, and FWP show an annual increase during these years. In contrast, the CC region showed a decrease in O_3 concentration until 2015 and then started to show an increasing trend after 2015. PRD, on the other hand, showed a decrease in O_3 concentration until 2016, but it has gradually increased since 2016. The phenomenon of decreasing O_3 concentration in CC and PRD areas is possibly the reason for the Clean Air Action Plan launched in China in 2013. Under the influence of this program, the emission of nitrogen dioxide (NO_2) was effectively controlled and thus decreased rapidly, which indirectly led to a decrease in O_3 concentration [17, 18]. However, the O_3 concentration still shows a gradual increase in various regions, and the problem of O_3 pollution needs to be urgently solved. By analyzing the key influencing factors in different regions, new ideas for regional O_3 control can be provided.

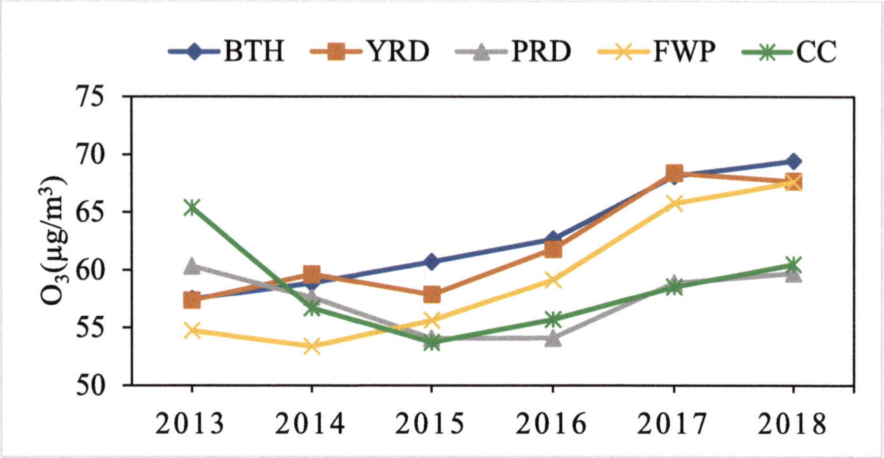

Fig. 3. Annual average concentrations of O_3 in the BTH, YRD, PRD, FWP, and CC regions between 2013 and 2018.

3.2 Evaluations of Model Performance

The model shows reasonable simulation results in all five regions (Fig. 4). The correlation coefficients of each region in different seasons are stable between 0.85 and 0.92, while the NMB varies from −5.87% to 2.77%. The model performs best in winter (RMSE = 10.88 μg/m^3, MAE = 7.74 μg/m^3) and worst in summer (RMSE = 26.23 μg/m^3, MAE = 18.34 μg/m^3). The performance of the RO3former model is comparable to the recent numerical and machine learning models [19–21].

The predictions were generally better in the autumn than in other seasons. This may be related to the relatively high temperatures and favorable meteorological conditions in the autumn [21]. The O_3 concentration is higher in the autumn and winter than in the spring and summer, and these changes are largely influenced by temperature and solar radiation [22]. In spring, the O_3 production is favored by the gradual increase of temperature and solar radiation. In summer, although high temperatures and intense solar radiation provided favorable conditions for photochemical reactions, high humidity also affected O_3 production to some extent. In winter, low temperatures are not conducive to photochemical reactions of O_3, and O_3 concentrations are generally low across the country [20].

The ability of the model to predict O_3 concentrations varies by region and is influenced by seasonal and regional characteristics. In general, the predictive ability of the model is better in the PRD and CC regions than in the BTH, YRD, and FWP regions. The bias is usually large in summer, especially in the BTH region (NMB = −5.87%, RMSE = 26.23 $\mu g/m^3$, MAE = 18.34 $\mu g/m^3$), as reported by [23], except that the error in the PRD region was relatively large in autumn (NMB = 2.77%, RMSE = 14.54 $\mu g/m^3$, MAE = 10.02 $\mu g/m^3$). This phenomenon is caused by the fact that O_3 concentrations peak in June during summer in most regions except the PRD due to high temperatures and intense solar radiation, which accelerate photochemical reactions and provide an ample supply of precursors for O_3 production [24]. The high summer O_3 concentrations in the BTH, YRD, and FWP regions resulted in model underestimation with NMBs of −5.87%, −5.24%, and −4.55%, respectively [25, 26]. In contrast, O_3 concentrations in the PRD region are relatively low in summer and high in autumn, mainly due to the influence of the Asian monsoon [27]. In addition, the FWP also showed an underestimation of O_3 concentration in winter (NMB = −4.17%) and spring (NMB = −3.40%), which may be related to anthropogenic emissions such as residential heating and power plants in northern cities [28]. The CC, on the other hand, also showed a larger NMB in winter and spring, which was mainly influenced by VOC_S [29]. In conclusion, the RO3former model shows good simulation results in different regions, proving its good applicability and wide usefulness.

3.3 Input Factor Ablation Experiment

The formation of O_3 is the result of a complex interaction of several drivers rather than a dominant single key factor. In most cases, the historic O_3 level has the most significant influence on the next day's O_3 prediction, while meteorological factors such as wind speed and temperature also play an important role.

Analyzing from the perspective of seasonal variations, the O_3 concentration at historical moments in spring (Fig. 5a and e) and summer (Fig. 5b and f) contributed the most to the O_3 concentration, followed by wind speed, temperature, solar UV radiation, air pressure, relative humidity, and conventional air pollutants, in that order. In autumn (Fig. 5c and g), temperature decreases and its effect on O_3 decreases, and the effect of solar UV radiation increases [30]. In winter (Fig. 5d and h), temperature is no longer the key factor dominating the O_3 concentration due to the lower temperature. At this time, the contribution of wind speed to O_3 exceeds that of historical O_3 and becomes the most dominant factor influencing O_3 concentration [31].

The BTH region experiences more severe O_3 pollution in summer compared to other regions. The high temperature and abundant sunshine in summer provide favorable conditions for O_3 formation, and the densely populated, industrialized, and anthropogenic emissions in the BTH region [32], coupled with the influence of southerly winds, make the region prone to O_3 accumulation. Regional O_3 pollution in the YRD region occurs mainly in spring, and most of the MDA8 peak ozone in its subregions occurs in April-May [33]. The main factors affecting O_3 in the YRD region are wind speed and temperature. The seasonal trend of O_3 in the PRD region is relatively weak. This is mainly due to the perennial high temperatures in the region, as well as unfavorable factors such as cloudy and rainy weather, oceanic air, and strong convection brought by the Asian summer

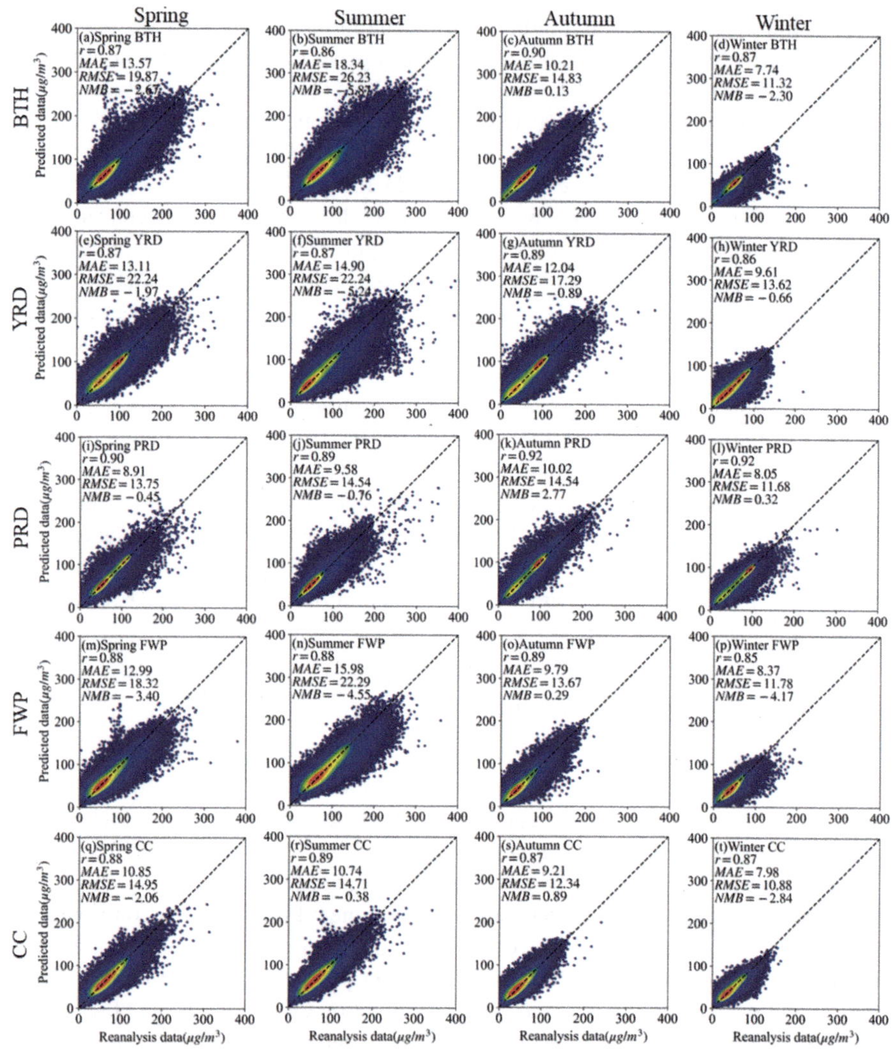

Fig. 4. Density scatterplots of BTH, YRD, PRD, FWP, and CC regions in different seasons.

monsoon, which is not conducive to the formation and accumulation of O_3 [34]. Solar radiation is the most critical factor in the FWP region in spring, while temperature has the greatest effect in the FWP region in summer [35]. In the CC region, the increase in surface O_3 is mainly influenced by changes in anthropogenic emissions, with a weaker effect of meteorological conditions due to high emission levels and basin topography [12].

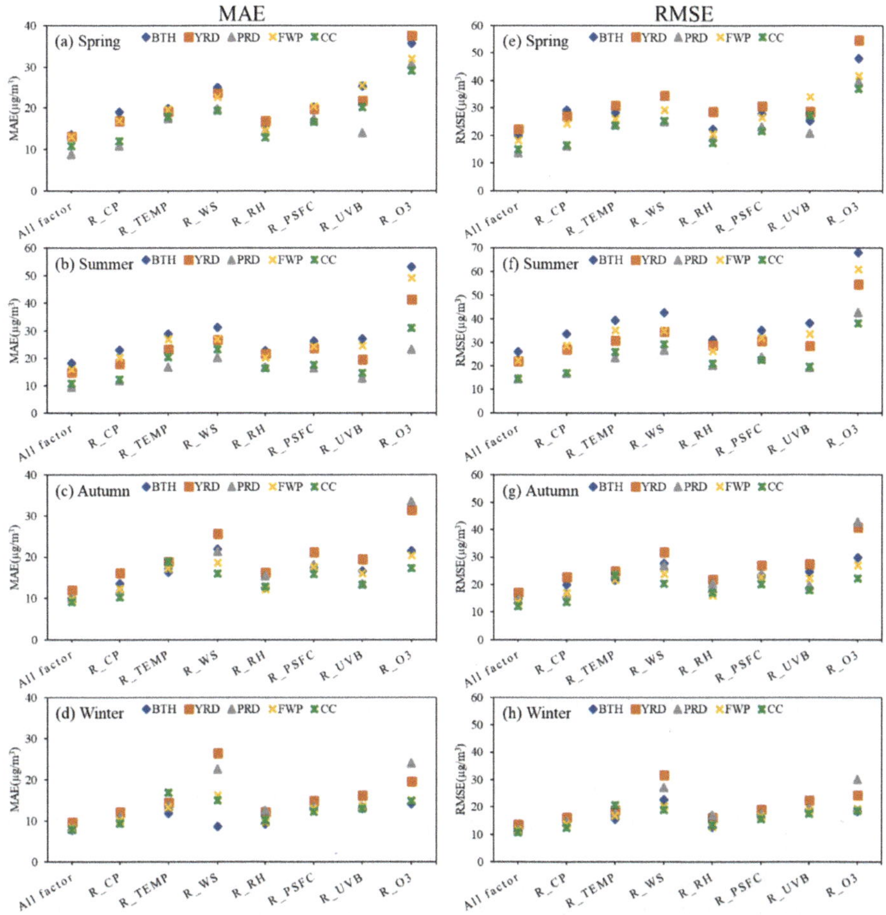

Fig. 5. MAE (a-d) and RMSE (e-h) generated by the ablation factors in different seasons. The removed factors in different scenarios: All factors–nothing; R_CP–conventional pollutants; R_TEMP–temperature; R_WS (u, v)–wind speed (u, v); R_RH–relative humidity; R_PSFC–pressure; R_UVB–downward UV radiation at the surface; R_O3–historical ozone.

4 Conclusion

In this study, we developed the RO3former model for short-term O_3 prediction in the Beijing-Tianjin-Hebei, Yangtze River Delta, Pearl River Delta, Fenwei Plain, and Chengdu-Chongqing regions. The model performs well in different regions, with correlation coefficients stabilized between 0.85 and 0.92, and the normalized mean bias (NMB) remains relatively low between -5.87% and 2.77%.

Through multifactorial ablation experiments, we have thoroughly investigated the key factors influencing O_3 prediction in different seasons in the above five regions. It is found that the formation of O_3 is not dominated by a single key factor but is the result of a complex interaction of multiple drivers. In most cases, the historic O_3 level has the most significant influence on the next day's O_3 prediction, while meteorological factors such as wind speed and temperature also play an important role.

The results of this study fully demonstrate the great potential and value of deep learning techniques for short-term spatial and temporal prediction of surface O_3 in the Beijing-Tianjin-Hebei, Yangtze River Delta, Pearl River Delta, Fenwei Plain, and Chengdu-Chongqing regions, and reveal a strong relationship between regionalized O_3 pollution and seasonal variations in meteorological conditions.

Acknowledgments. This study is financially supported by the Key R&D Program of Shandong Province, China (No. 2022CXGC020106), Jinan Science and Technology Bureau (No. 202228034) and the Major Innovation Special Project for the Integration of Science, Education, Industry of Qilu University of Technology (Shandong Academy of Science) (No. 2023JBZ02 and 2023HYZX01), and the China Postdoctoral Science Foundation (No. 2021M691921). We extend our gratitude to the Institute of Atmospheric Physics, Chinese Academy of Sciences (IAP/CAS) and the European Centre for Medium-Range Weather Forecasts (ECMWF) for providing atmospheric pollutant data and meteorological data for this study. We would like to express our gratitude to the anonymous reviewers for their valuable feedback, which has contributed to the enhancement of this paper.

References

1. Shu, L., Zhu, L., Bak, J., et al.: Improving ozone simulations in Asia via multisource data assimilation: results from an observing system simulation experiment with GEMS geostationary satellite observations. Atmos. Chem. Phys. **23**(6), 3731–3748 (2023)
2. Guan, Y., Xiao, Y., Rong, B., et al.: Long-term health impacts attributable to PM2.5 and ozone pollution in China's most polluted region during 2015–2020. J. Clean. Prod. **321**, 128970 (2021)
3. Yu, R., Lin, Y., Zou, J., et al.: Review on atmospheric ozone pollution in China: formation, spatiotemporal distribution, precursors and affecting factors. Atmosphere **12**(12), 1675 (2021)
4. Feng, Z., Sun, J., Wan, W., et al.: Evidence of widespread ozone-induced visible injury on plants in Beijing, China. Environ. Pollut. **193**, 296–301 (2014)
5. Liu, Y., Geng, G., Cheng, J., et al.: Drivers of increasing ozone during the two phases of clean air actions in China 2013–2020. Environ. Sci. Technol. **57**(24), 8954–8964 (2023)
6. He, C., Wu, Q., Li, B., et al.: Surface ozone pollution in China: Trends, exposure risks, and drivers. Front. Public Health **11**, 1131753 (2023)
7. Liu, C., Shi, K.: A review on methodology in O3-NOx-VOC sensitivity study. Environ. Pollut. **291**, 118249 (2021)
8. Yang, J., Zhao, Y.: Performance and application of air quality models on ozone simulation in China – a review. Atmos. Environ. **293**, 119446 (2023)
9. Yang, G., Liu, Y., Li, W., et al.: Association analysis between socioeconomic factors and urban ozone pollution in China[J]. Environ. Sci. Pollut. Res. **30**(7), 17597–17611 (2023)
10. Pei, Z., Chen, X., Li, X., et al.: Impact of macroeconomic factors on ozone precursor emissions in China. J. Clean. Prod. **344**, 130974 (2022)
11. Wang, T., Xue, L., Brimblecombe, P., et al.: Ozone pollution in China: a review of concentrations, meteorological influences, chemical precursors, and effects. Sci. Total Environ. **575**, 1582–1596 (2017)

12. Ma, D., Wang, T., Wu, H., et al.: The effect of anthropogenic emission, meteorological factors, and carbon dioxide on the surface ozone increase in China from 2008 to 2018 during the East Asia summer monsoon season. Atmos. Chem. Phys. **23**(11), 6525–6544 (2023)
13. Yin, C., Deng, X., Zou, Y., et al.: Trend analysis of surface ozone at suburban Guangzhou, China. Sci. Total Environ. **695**, 133880 (2019)
14. Sun, L., Xue, L., Wang, Y., et al.: Impacts of meteorology and emissions on summertime surface ozone increases over central eastern China between 2003 and 2015. Atmos. Chem. Phys. **19**(3), 1455–1469 (2019)
15. Ge, Q., Zhang, X., Cai, K., et al.: Ozone pollution in Chinese cities: spatiotemporal variations and their relationships with meteorological and other pollution factors (2016–2020). Atmosphere **13**(6), 908 (2022)
16. Gao, Z., Shi, X., Wang, H., et al.: Earthformer: exploring space-time transformers for earth system forecasting. Adv. Neural Inf. Process. Syst. **35**, 25390–25403 (2022)
17. Li, K., Jacob, D.J., Liao, H., et al.: Anthropogenic drivers of 2013–2017 trends in summer surface ozone in Chin. Proc. Natl. Acad. Sci. **116**(2), 422–427 (2019)
18. Lin, C., Lau, A.K.H., Fung, J.C.H., et al.: Removing the effects of meteorological factors on changes in nitrogen dioxide and ozone concentrations in China from 2013 to 2020. Sci. Total Environ. **793**, 148575 (2021)
19. Cheng, M., Fang, F., Navon, I.M., et al.: Spatio-temporal hourly and daily ozone forecasting in china using a hybrid machine learning model: autoencoder and generative adversarial networks. J. Adv. Model. Earth Syst. **14**(3), e2021MS002806 (2022)
20. Xiong, K., Xie, X., Huang, L., et al.: Improved O3 predictions in China by combining chemical transport model and multi-source data with machining learning techniques. Atmos. Environ. **318**, 120269 (2024)
21. Yan, X., Guo, Y., Zhang, Y., et al.: Combining physical mechanisms and deep learning models for hourly surface ozone retrieval in China. J. Environ. Manage. **351**, 119942 (2024)
22. Chen, Z., Li, R., Chen, D., et al.: Understanding the causal influence of major meteorological factors on ground ozone concentrations across China. J. Clean. Prod. **242**, 118498 (2020)
23. Ye, X., Wang, X., Zhang, L.: Diagnosing the model bias in simulating daily surface ozone variability using a machine learning method: the effects of dry deposition and cloud optical depth. Environ. Sci. Technol. **56**(23), 16665–16675 (2022)
24. Lu, X., Zhang, L., Chen, Y., et al.: Exploring 2016–2017 surface ozone pollution over China: source contributions and meteorological influences. Atmos. Chem. Phys. **19**(12), 8339–8361 (2019)
25. Shipeng S, Meng F a. N, Jinhua T a. O, et al. Estimating ground-level ozone concentration in China using ensemble learning methods[J]. National Remote Sensing Bulletin, 2023, 27(8): 1792–1806
26. Cheng, M., Fang, F., Navon, I.M., et al.: Assessing uncertainty and heterogeneity in machine learning-based spatiotemporal ozone prediction in Beijing-Tianjin- Hebei region in China. Sci. Total Environ. **881**, 163146 (2023)
27. Li, S., Wang, T., Huang, X., et al.: Impact of east asian summer monsoon on surface ozone pattern in China. J. Geophys. Res.: Atmos. **123**(2), 1401–1411 (2018)
28. Ren, H.H., Cheng, Y., Wu, F., et al.: Spatiotemporal characteristics of ozone and the formation sensitivity over the Fenwei Plain. Sci. Total Environ. **881**, 163369 (2023)
29. Huang, D., Li, Q., Han, Y., et al.: Biogenic volatile organic compounds dominated the near-surface ozone generation in Sichuan Basin, China, during fall and wintertime. J. Environ. Sci. **141**, 215–224 (2024)
30. Zhou L, Sun L, Luo Y, et al. Air pollutant concentration trends in China: correlations between solar radiation, PM2.5, and O3. Air Q. Atmos. Health **16**(8), 1721–1735 (2023)
31. Gao, M., Gao, J., Zhu, B., et al.: Ozone pollution over China and India: seasonality and sources. Atmos. Chem. Phys. **20**(7), 4399–4414 (2020)

32. Guan, Y., Liu, X., Zheng, Z., et al.: Summer O3 pollution cycle characteristics and VOCs sources in a central city of Beijing-Tianjin-Hebei area, China. Environ. Pollut. **323**, 121293 (2023)
33. Tong, L., Xiao, H., Yi, H., et al.: Spatial regionalization on surface ozone in the Yangtze River delta of China[J]. Asia-Pac. J. Atmos. Sci. **58**(2), 207–218 (2022)
34. Ma, X., Yin, Z.: Dipole pattern of summer ozone pollution in the east of China and its connection with climate variability. Atmos. Chem. Phys. **21**(21), 16349–16361 (2021)
35. Wang, S., Ren, Y., Xia, B.: PM2.5 and O3 concentration estimation based on interpretable machine learning. Atmos. Poll. Res. **14**(9), 101866 (2023)

A Carbon Emission Reduction Approach Based on Improved Life Cycle Assessment Algorithm and Mixed Integer Linear Programming

Dongge Zhu[1(✉)], Jia Liu[1], Zhenhua Yan[1], Hongwei Han[2], Rui Ma[1], Fangzhou Dang[3], and Xiaowen Huang[3]

[1] Ningxia Electric Power Research Institute, Yinchuan 750002, China
zhu1230701@163.com

[2] State Grid Ningxia Electric Power Co., Ltd., Yinchuan 750001, China

[3] State Grid Xintong Yili Technology Co., Ltd., Xiamen 361001, China

Abstract. Due to the complex composition of carbon emission driving factors of products/services, it is difficult to achieve carbon emission reduction effects. Therefore, this paper proposes a study on carbon emission reduction methods based on improved life cycle assessment algorithms and mixed integer linear programming. When using the mixed integer linear programming (MILP) model of linear approximation for specific emission reduction planning, define integer variables to represent whether a certain emission reduction technology is adopted, continuous variables to represent the resource usage and clean energy substitution at each stage, and the objective function of constructing the MILP model is to minimize the total carbon emissions throughout the life cycle. When the linear expression between the carbon emissions corresponding to the product/service and the nonlinear factors reaches convergence in the objective function, a carbon emission reduction plan is output. In this test result, the maximum reduction in carbon emissions during the life cycle reached 16.76%, and the overall carbon emissions dropped by 11.95%.

Keywords: Improved Life Cycle Assessment Algorithm · Mixed Integer Linear Programming · Carbon Emission Reduction · Carbon Emission Drivers · Carbon Emissions · Linear Approximation · Objective Function

1 Introduction

Economic development is the main driver of carbon emission growth. With the advance of industrialization and urbanization, energy consumption has risen sharply, especially the use of fossil fuels. Therefore, in the process of carbon emission reduction, it is necessary to balance the relationship between economic development and carbon emission control to avoid excessive restrictions leading to economic stagnation [1, 2], at the same time, we are actively exploring ways of resource utilization of waste to achieve resource recycling [3].

D. Han et al. (Eds.): ICEPG 2024, ESE, pp. 375–380, 2025.
https://doi.org/10.1007/978-3-031-86829-0_42

In the carbon emission reduction method of fuel cell vehicles based on the life cycle perspective [4], although the carbon emissions of different hydrogen production methods are analyzed in detail through the life cycle assessment (LCA) method, other environmental problems that may be encountered in the use of fuel cell vehicles cannot be fully considered. In the carbon emission reduction method [5] that considers the transformation performance characteristics of resource-based cities, the transformation performance characteristics of resource-based cities and their impact on carbon emissions are explored. However, this study lacked in-depth discussion on the carbon emission reduction paths and strategies of non-resource-based cities. In the enterprise carbon emission reduction method based on peer effect [6], the role of peer effect in enterprise carbon emission reduction is discussed.However, this study failed to fully analyze the specific mechanism and influencing factors of peer effects.

On the basis of the above, this paper proposes the study of carbon emission reduction methods based on the improved life cycle assessment algorithm and mixed integer linear programming, and carries out comparative tests to analyze and verify the performance of the designed methods.

2 Carbon Reduction Methodology Design

2.1 Carbon Emissions Analysis Based on Improved Life Cycle Assessment Algorithms

In this paper, based on the improvement of the whole life cycle assessment methodology, the carbon emissions of products/services are categorized into several major stages.

This paper carries out the analysis of specific carbon emission factors from five stages, namely, material acquisition, manufacturing, transportation and distribution, use and maintenance, and disposal and recycling [7]. Based on the above stage division, combined with the corresponding products/services carbon emission drivers, carbon emissions are analyzed and calculated, which can be expressed as follows

$$Q = \sum q_i w_i + \lambda * e + \sum y_j l_j + \Delta w * T - \gamma * B \tag{1}$$

Among them, Q indicates the total carbon emissions corresponding to the drivers of carbon emissions of products/services over the life cycle. q_i represents the carbon emission factor of energy type i, w_i represents the consumption of energy type i, λ denotes the carbon emission factor of the production process. e indicates the efficiency coefficient of the production process. y_j represents the carbon emission intensity parameter of j transportation mode, l_j indicates the transportation distance corresponding to j transportation mode, Δw represents the carbon emissions corresponding to the energy use efficiency of the product during the use process. T indicates the total length of time the product has been in use, i.e., the effective life of the product. γ indicates that disposal reduces the amount of wasted resources. B indicates the carbon emission factor corresponding to the waste disposal resource.

2.2 Carbon Reduction Based on Mixed Integer Linear Programming

Combined with the effective control of carbon emissions corresponding to products/services in the full life cycle of Part 1.1, this article uses a linear approximation mixed integer linear programming (MILP) model to carry out specific emission reduction planning [8].

First, collect specific data on the driving factors of carbon emissions at each stage based on Sect. 2.1. In the decision variable definition stage, this paper defines an integer variable to represent whether to adopt a certain emission reduction technology (0 means not to use it, 1 means to use it), and a continuous variable to represent the resource usage and clean energy substitution amount at each stage [9]. On this basis, the objective function of the constructed MILP model is to minimize the total carbon emissions throughout the life cycle, which can be expressed specifically as

$$f(x) = Min \sum_{t \in Stages} \sum_{a \in Activitiest} (Q_{t,a} \times E_{t,a} \times (1 - \sum_{r \in \mathrm{Re}ductiont} R_{t,r} \times \delta_{t,r} \times x_{t,r})) \quad (2)$$

Among them, $f(x)$ denotes the objective function, $Q_{t,a}$ represents the quantity of activity a in stage t, $E_{t,a}$ denotes the corresponding carbon emission factor, $R_{t,r}$ represents the emission reduction rate of emission reduction technology r in stage t, $\delta_{t,r}$ represents the effectiveness index of technology r at stage t (1 if applicable, 0 otherwise), $x_{t,r}$ represents a binary decision variable indicating whether to adopt technology r.

Based on this, the way to solve the objective function using the MILP solver can be expressed as

$$cor(\Gamma, \Lambda_x) = Lf(x)_{[n-1]}f(x)_{[n-2]} \cdots f(x)_{[1]}f(x)_{[0]}Q \quad (3)$$

Among them, $cor(\Gamma, \Lambda_x)$ represents the linear expression between the corresponding carbon emissions of the product/service and the nonlinear factor, L denotes the row vector in the dataset corresponding to the product/service corresponding to the driver of carbon emissions. $f(x)_{[n-1]}$ indicates the intermediate value of the jump process Mubai function.

The linear expression is approximated in such a way that

$$cor(\Gamma, \Lambda_x) = cor(\Gamma * (x + Q) \oplus \Lambda_x) \quad (4)$$

When the linear expression between the products/services corresponding to carbon emissions and the nonlinear factors reaches convergence in the objective function under the premise that the linear approximation correlation coefficient is a constant value [10], the corresponding result is taken as the final carbon emission reduction scheme.

Through the above steps, the linear approximation MILP model is used to effectively reduce the carbon emissions of the product/service throughout its life cycle.

3 Application Testing and Analysis

3.1 Test Environment

For the specific test environment, a food waste treatment system is used as the specific test object. The system is applicable to all kinds of catering establishments, canteens and communities, and is capable of treating a variety of food waste, including food waste, food waste, fruit and vegetable waste. The system is divided into four main stages: pre-treatment, biodegradation, post-treatment and resource utilization. The details are shown in Fig. 1.

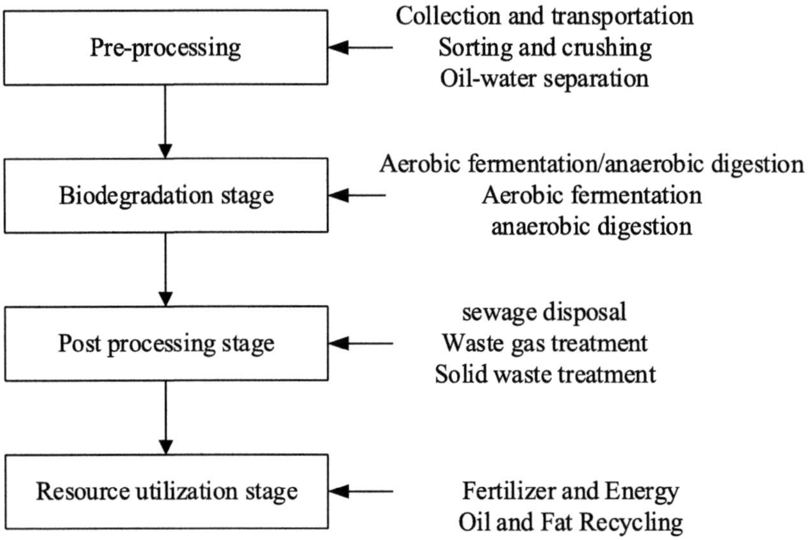

Fig. 1. Flow of food waste treatment system

The current carbon emissions of the kitchen waste treatment system are statistically analyzed, and the corresponding data results are shown in Table 1.

Table 1. Statistical table of carbon emission data from the operation of the food waste treatment system.

Stage	Carbon emissions/kg/t
pre-processing	62.60
Biodegradation stage	108.60
Post processing stage	223.00
Resource utilization stage	85.40
total	479.60

Combining the information shown in Table 1, three methods were used for carbon abatement treatment, and the application effects of different methods were compared.

3.2 Test Results and Analysis

On the basis of the above test environment, the application effects of the three different methods were counted.

The application effects of three different carbon emission reduction methods are analyzed, Among them, under the carbon emission reduction method based on the life cycle perspective, the reduction in carbon emissions in different stages is relatively balanced, and the overall. The carbon emission reduction is 31.02kg/t. Under the carbon emission reduction method that considers performance characteristics, the carbon emission reduction in the post-processing stage is the most obvious. The carbon emission reduction effect in other stages has not been fully reflected, but the overall corresponding Carbon emission reduction has also reached 35.52kg/t. In the test results of the design method in this article, not only the reduction in carbon emissions at different stages reached more than 6.0%, but also the carbon emissions in the post-processing stage dropped by 16.76%, and the overall carbon emissions dropped by 11.95%. Compared with the control group, the corresponding carbon emission reductions increased by 26.29 kg/t and 24.79 kg/t respectively.

Combining the above test results, it can be concluded that the carbon emission reduction method designed in this paper can effectively realize the control of carbon emissions in the overall life cycle and achieve good carbon emission reduction.

4 Conclusion

This article proposes research on carbon emission reduction methods based on improved life cycle assessment algorithms and mixed integer linear programming, comprehensively assessing the resource consumption, environmental impact and carbon emission stages of products/services throughout their life cycles. By improving the LCA algorithm, a large number of practical applications are combined data, through model simulation and data analysis, can more accurately quantify carbon emissions at each stage. With the help of MILP, under the premise of satisfying a series of linear constraints, the optimal solution of the objective function containing integer variables. By quantitatively assessing the emission reduction effects and economic costs of different policies, we can more accurately formulate emission reduction targets and measures. Companies can also use this method to conduct their own carbon footprint assessments and optimize emission reduction strategies. By improving production processes, optimizing energy structure, improving resource utilization efficiency and other measures, we can reduce carbon emissions and improve market competitiveness.

References

1. Yan, G., Song, Q., Ni, Y., et al.: Pricing, carbon emission reduction and recycling decisions in a closed-loop supply chain under uncertain environment. J. Ind. Manage. Optim. **19**(1), 224–245 (2023)
2. Meng, L., Zhou, W., Li, Y., et al.: Building a low-carbon energy community based on energy multi type networks. Comput. Simul. **39**(12), 108–115 (2022)
3. Wu, H., Wang, S.: Design and optimization of intelligent orchard frost prevention machine under low-carbon emission reduction. J. Clean. Prod. **433**(Dec.25), 139808.1–139808.16 (2023)
4. Teng, Z., Tan, C., Liu, P., et al.: Analysis on carbon emission reduction intensity of fuel cell vehicles from a life-cycle perspective[J]. Energy Frontiers **18**(1), 16–27 (2024)

5. Xu, W., Zheng, J., Zhou, J., et al.: Transformation performance characteristics of resource-based cities and their carbon emission reduction effects. J. Nat. Res. **38**(1), 39–57 (2023)
6. Guo, J., Chang, S., Guo, M.: The impact of peer effect of industrial robot application on enterprise carbon emission reduction. Sci. Rep. **14**(1), s4159 (2024)
7. Yan, J., Jing, J., Li, Y.: Hydrogen fuel cell commercial vehicles in China: evaluation of carbon emission reduction and its economic value. Int. J. Hydr. Energy **52**(Jan. Pt.B), 734–749 (2024)
8. Wang, L., Wang, D.H., Xiao, Q.: Carbon emission reduction: contribution of karst carbon sinks and practice in China. China Geol. **6**(4):759–761 (2023)
9. Neha, S., Sharma, R.L., Kundan, Y.: Sustainable development by carbon emission reduction and its quantification: an overview of current methods and best practices. Asian J. Civil Eng.: Build. Hous. **24**(8), 3797–3822 (2023)
10. Zhang, J., Shen, J., Xu, L.Z.Q.: The CO2 emission reduction path towards carbon neutrality in the Chinese steel industry: a review. Environ. Impact Assess. Rev. **99**(Mar.), 107017.1–107017.16 (2023)

Process Simulation and Absorbent Optimization of Marine Carbon Capture Systems Based on Aspen Plus

Youhong Xiao(✉) and Kailong Li

Harbin Engineering University, Harbin 150000, HLJ, China
xiaoyouhong@hrbeu.edu.cn

Abstract. This study aims to advance carbon emission reduction in China's shipping industry by evaluating the effectiveness of three widely used absorbents—Monoethanolamine (MEA), N-Methyldiethanolamine (MDEA), and Diglycolamine (DGA)—in carbon capture processes. Using Aspen Plus software, a carbon capture process model was constructed to investigate the effects of effects of absorbent flow rate and absorbent concentration on carbon capture rate, and the effects of desorption tower top pressure on carbon desorption rate and reboiler heat load. A comparative analysis of the absorption performance, desorption performance, and regeneration energy consumption of the three absorbents was carried out to identify the one with the optimal overall performance. The results indicate that all three absorbents are significantly influenced by variations in flow rate and concentration. MEA exhibited superior absorption performance relative to MDEA and DGA. Under identical simulation conditions, increasing the desorber top pressure resulted in a reduction in reboiler heat duty and a rise in reboiler outlet temperature. In terms of overall performance in absorption, desorption, and regeneration energy consumption, MEA outperformed both MDEA and DGA, suggesting its greater suitability for application in marine carbon capture systems.

Keywords: Marine carbon capture · Aspen Plus simulation · CO_2 absorption and desorption · amine absorbents

1 Introduction

Greenhouse gas emissions cause significant environmental damage. Statistics show that CO_2 emissions from the maritime shipping sector account for 2.8% of global CO_2 emissions, and research into decarbonization technologies in this industry plays a crucial role in mitigating the greenhouse effect [1]. Carbon capture and storage (CCS) is a proven solution to achieving this goal. CO_2 capture technologies are classified into post-combustion, pre-combustion, and oxy-fuel combustion methods [2]. Among these, post-combustion chemical absorption is the most widely studied [3], implemented, and commercialized approach, renowned for its high CO_2 absorption efficiency, stable reaction characteristics, adaptability to low-pressure conditions, and capacity to produce high-purity CO_2[4]. Various CO_2 chemical absorbents have been extensively researched,

D. Han et al. (Eds.): ICEPG 2024, ESE, pp. 381–389, 2025.
https://doi.org/10.1007/978-3-031-86829-0_43

including amine solutions, alkaline aqueous solutions, ionic liquids, and phase-change absorbents [5]. CO_2 absorption using organic amine solutions is a well-established practice, where the amine and hydroxyl groups act as the active sites for CO_2 capture. The amine group offers strong basicity, resulting in a rapid absorption rate, significant absorption capacity, and excellent CO_2 selectivity, making it the key factor driving CO_2 absorption in amine-based methods [6]. Feenstra studied the effect of mass fraction of 30% MEA solution and mass fraction of 30% piperazine aqueous solution (PZ) on carbon capture rate [7]; Casero et al. reported on the application of MDEA in a pilot-scale pre-combustion CO_2 capture and hydrogen production facility [8]; Zahid et al. evaluated the performance of MEA against DGA in natural gas carbon capture [9]. Ammonia (NH_3) served as the solvent for carbon capture in ship-based CO_2 carriers powered by liquefied natural gas (LNG) [10].

This study focuses on three widely used absorbent solutions: MEA, MDEA, and DGA. It employs Aspen Plus chemical process simulation software to conduct an in-depth comparative analysis of the performance characteristics of these amine absorbents in both absorption and desorption processes. The research aims to identify the absorbent with the most superior performance. The findings offer valuable guidance for selecting amine absorbents in marine carbon capture processes and for optimizing absorption and desorption conditions.

2 Model Setup

2.1 Carbon Capture Process

Figure 1 illustrates the process flow for CO_2 capture from ship exhaust using the amine scrubbing method. The high temperature of the exhaust gas and the low CO_2 partial pressure of 5.23% necessitate a greater amount of absorbent to achieve a higher absorption rate. Therefore, the exhaust gas must be cooled and pressurized before entering the absorption tower. The treated gas contacts the lean solution in a counter-current flow at the top of the absorption tower, facilitating gas-liquid heat and mass transfer. The absorbed CO_2, present as ions in the lean solution, flows out from the absorption tower's base as an enriched solution. The enriched solution is pressurized by a pump and sent to the lean-enriched solution heat exchanger. This heat exchanger recovers the heat released by the lean solution during desorption by exchanging heat between the high-temperature lean solution from the desorption tower and the low-temperature lean solution entering the absorption tower. The warmed enriched solution enters the desorption tower's top, where it contacts high-temperature steam from the reboiler in a counter-current flow. Driven by temperature and pressure, CO_2 is desorbed from the enriched solution. The regenerated high-temperature lean solution is cooled through heat exchange with the low-temperature enriched solution from the absorption tower and then flows into the mixer before being sent back to the absorption tower's top. To compensate for solution and water vapor losses, the solution entering the mixer includes absorbent make-up and water, completing the CO_2 capture cycle.

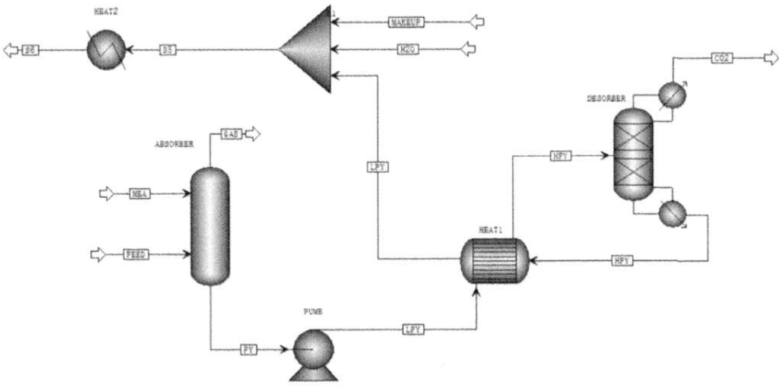

Fig. 1. Sketch of carbon capture process

2.2 Model and Exhaust Parameters

A simulation was conducted to compare the performance of three absorbents using exhaust gas from a marine diesel engine. The engine has a rated power output of 3250 kW, and its emission parameters are detailed in Table 1. The CO_2 capture process from the ship's exhaust was modeled using Aspen Plus V12.0, incorporating an absorption tower, booster pump, lean-rich liquid heat exchanger, desorption tower, and solution mixer. The absorption and desorption towers were modeled using the rate-based approach, while the heat exchanger was modeled as a shell and tube counter-current heat exchanger. The ELECNRTL electrolyte property method was employed to analyze the reactions of MEA, MDEA, and DGA with CO_2, while gas phase components were calculated using Henry's law.

Table 1. Main Exhaust Parameters at 100% load of the Engine

Exhaust properties	Value
Temperature	256 °C
Mass flow rate	7.08 kg/s
Composition (m/m%)	
H_2O	2.37%
CO_2	5.23%
O_2	14.7%
N_2	77.7%

2.3 Performance Evaluation Index

Carbon capture simulations were performed based on the exhaust parameters at the engine's rated operating conditions. The carbon capture rate, denoted as "ψ", reflects the performance of the amine solution in absorbing CO_2. The desorption rate, denoted as "ζ", indicates the efficiency of the amine solution in desorbing CO_2. The regeneration energy consumption, denoted as "q^{evap}", represents the reboiler load required to desorb a unit mass of CO_2, which is influenced by the amount of CO_2 captured and the type of absorbent used. The expressions for these three performance evaluation indicators are as follows:

$$\psi = \frac{m_{CO_2}^{Gas,in} - m_{CO_2}^{Gas,out}}{m_{CO_2}^{Gas,in}} \tag{1}$$

$$\zeta = \frac{x_{rich} - x_{lean}}{x_{rich}} \tag{2}$$

$$q^{evap} = \frac{Q^{evap}}{m_{CO_2}} \tag{3}$$

where $m_{CO_2}^{Gas,in}$ and $m_{CO_2}^{Gas,out}$ is mass flow of CO_2 at the inlet and outlet of the absorption tower; x_{rich} and x_{lean} is the CO_2 load in the rich solution at the inlet of the desorption tower and the CO_2 load in the lean solution at the outlet, respectively; Q^{evap} is the reboiler duty consumed to capture a unit mass of CO_2, GJ/tCO_2; m_{CO_2} is the mass flow of CO_2 at the outlet of the desorption tower.

2.4 Carbon Capture Model Validation

Following the construction of the carbon capture model, validating its accuracy against empirical data from scholarly sources is essential. Reference [11] provides operational data related to the load of a carbon capture facility, including parameters such as flow rate, temperature, and pressure of various process streams. Two specific operating conditions, labeled 19 and 20 in the literature, were selected to simulate the carbon capture process. The boundary and operational parameters were rigorously configured according to the values specified in the academic source, with the results presented in Table 2.

Table 2. The comparison of simulation results with literature results

Operating condition	Amount of CO_2 captured(kg/h)		CO_2 removal rate(%)	
	reference	simulation	reference	simulation
19	82	80.3	95	92.7
20	81	84.9	87	91.2

3 Results and Discussion

3.1 Absorption Performance

The amount and concentration of the absorbent are crucial factors influencing the absorption process. Using the sensitivity analysis tool in Aspen Plus, the effect of absorbent concentration on the capture rate was analyzed by varying its mass fraction. The CO_2

capture rates were determined under varying flow rates (5 to 12 tons per hour) and concentrations (20% to 40%) of different absorbents.

Figure 2 shows that as the absorbent flow rate increases, the CO_2 capture rates of the three absorbents exhibit varying upward trends. A higher flow rate increases the number of amine molecules in the solution, accelerating the reaction with CO_2 and significantly enhancing absorption efficiency. MEA shows the most significant increase, starting at 43.9% and reaching 97.4% at 12 t/h. DGA follows MEA, while MDEA shows the lowest absorption efficiency. This is because MEA is a primary amine, while MDEA and DGA are secondary amines. Primary amines form carbamates, relatively stable intermediates that enhance CO_2 absorption.

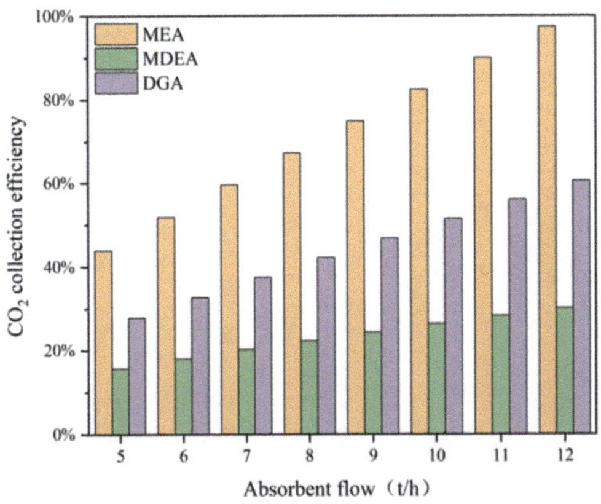

Fig. 2. CO_2 capture rate under different absorbent flow rates

Figure 3 shows that as the absorbent concentration increases, the CO_2 capture rate gradually rises. This increase is due to higher concentrations of key components, which boost the CO_2 capture rate. However, the corrosive nature of the absorbent imposes limitations. This raises the performance standards for the top tower washing unit, so a higher mass concentration is not always better. Current research typically uses a 30% mass fraction for MEA absorbent. In summary, the CO_2 capture efficiency of the amine solutions follows the order: MEA > DGA > MDEA.

3.2 Desorption Performance

Figure 4 depicts the effect of pressure at the top of the desorption tower on CO_2 desorption efficiency. The data indicate a positive correlation between increased desorption tower top pressure and the desorption efficiency for MEA and DGA, while no significant change is observed for MDEA. The MEA process may require less energy during desorption due to the ease of decomposition of the intermediates, thereby reducing the thermal energy demand and enhancing desorption efficiency, which can reach up to 72.3%.

Figure 5 illustrates the relationship between reboiler heat duty and temperature for MEA, MDEA, and DGA under varying pressures at the top of the desorption tower.

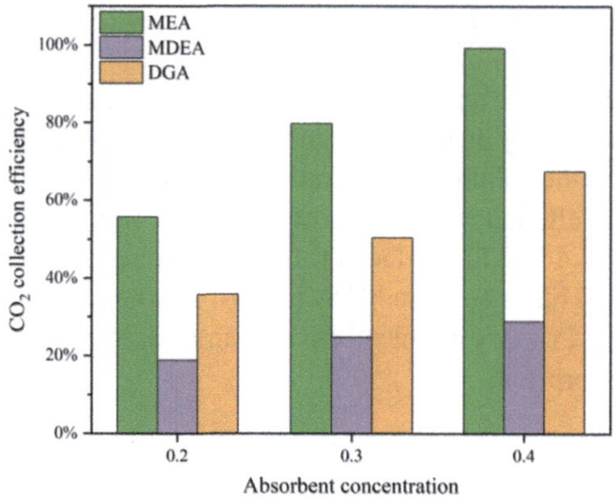

Fig. 3. CO_2 capture rate under different absorbent concentrations

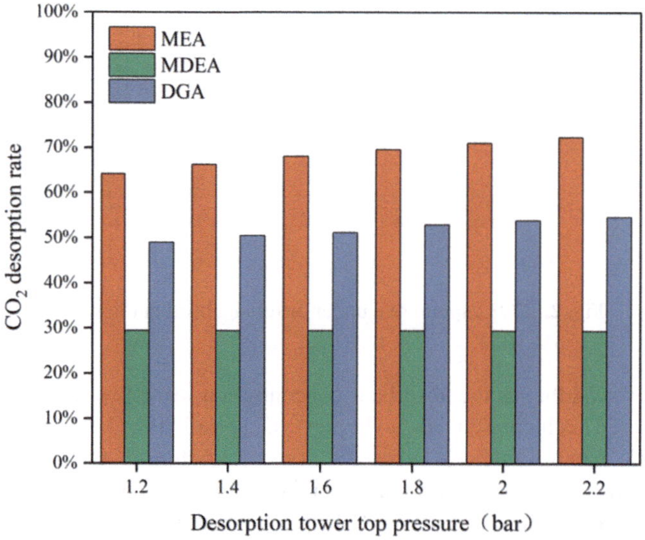

Fig. 4. CO_2 desorption rate under different desorption tower top pressure

As the pressure increases, the required heat duty for the reboiler decreases, while the reboiler temperature correspondingly rises. This phenomenon occurs because, during the desorption process, the transition of CO_2 from the liquid phase to the gas phase elevates the system pressure, which in turn increases the partial pressure of CO_2 in the gas phase. This elevation in partial pressure reduces the chemical potential difference driving CO_2 transfer from the liquid to the gas phase, thereby decreasing the energy required for CO_2 desorption and leading to a lower reboiler heat duty. The increased pressure also reduces the need for cooling water within the tower, enhances CO_2 desorption, and decreases H_2O content. As a result, the condenser duty at the top of the tower is reduced, leading to a continuous rise in reboiler temperature. However, increasing desorption pressure is not universally beneficial, as it is constrained by the solvent's susceptibility to high-temperature degradation and the potential for equipment corrosion. For optimal desorption efficiency, the reboiler temperature should not exceed 125 °C; thus, the maximum allowable pressure for the desorption tower is set at 2 bar, with an operational range between 1.2 and 2 bar.

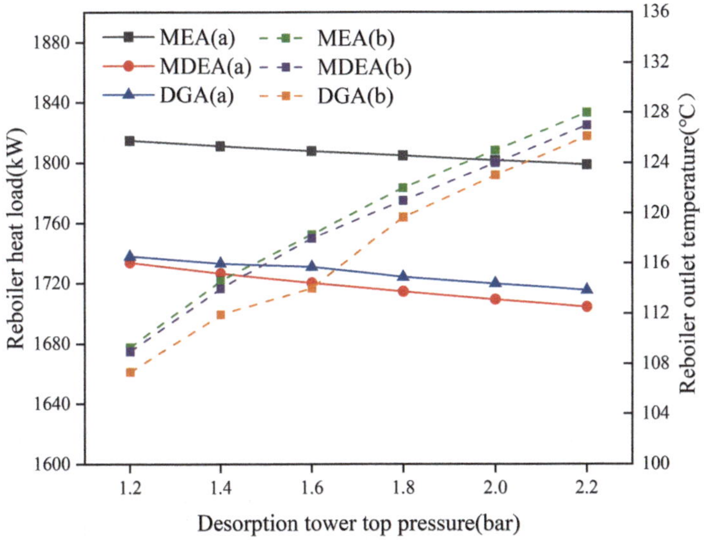

Fig. 5. The performance change of reboiler with different absorbents under different desorption tower top pressure. (a) reboiler heat load; (b) reboiler outlet temperature

In conclusion, the desorption performance of the three amine solutions is ranked as MEA being superior to DGA, which in turn is better than MDEA. When conducting simulation calculations to analyze the desorption performance of carbon, it is essential to consider both the desorption pressure and the reboiler temperature comprehensively.

3.3 Comprehensive Performance Analysis of Carbon Capture

The analysis indicates that MEA exhibits superior absorption efficiency compared to MDEA and DGA. Given the varying regeneration energy requirements of different absorbents, which affect the economic feasibility of the carbon capture system, the effects of different absorbents on the regeneration energy consumption of desorption tower were compared under the same conditions. Simulations were performed under the following conditions: an absorbent flow rate of 12 t/h, an absorbent concentration of 30%, and a desorption tower top pressure of 2 bar. The results, which are presented in Fig. 6, detail the regeneration energy consumption and CO_2 desorption quantities for different absorbents. Figure 6 illustrates that MEA demonstrates the lowest regeneration energy consumption while achieving the highest CO_2 desorption content, followed by DGA, with MDEA proving to be the least effective.

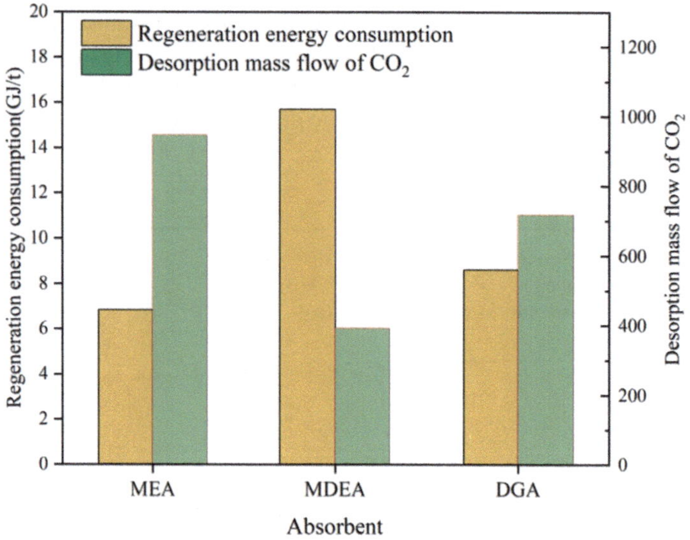

Fig. 6. Regeneration energy consumption and CO_2 desorption amount of different absorbents

Based on a comprehensive evaluation of absorption and desorption performance, as well as regeneration energy consumption, MEA amine absorbent has been identified as superior. In the context of carbon dioxide removal from marine exhaust gases, MEA amine absorbent is more suitable compared to DGA and MDEA amine absorbents.

4 Conclusion

In selecting absorbents for shipboard carbon capture systems, simulations were performed using Aspen Plus software to assess the impact of various amine absorbents (MEA, MDEA, DGA) on CO_2 desorption rates and reboiler loads, focusing on flow rates, mass fractions, and desorption column top pressures. The primary conclusions derived from this study are as follows:

For the three amine absorbents, the absorption performance tends to increase with the flow rate, with the ranking of absorption performance being MEA > DGA > MDEA. Moreover, the concentration of the absorbent should not be too high, with a commonly used absorbent concentration of 30%.

The impact of desorption column top pressure on the desorption performance of the absorbents was studied, revealing that MEA is significantly affected by the desorption column top pressure, while MDEA and DGA are less influenced. MEA exhibits optimal desorption performance at a desorption pressure of 2 bar.

A comprehensive comparison of the absorption and desorption performance as well as the regeneration energy consumption of the three absorbents indicates that the overall CO_2 absorption and desorption performance of the three amine solutions is MEA > DGA > MDEA.

Acknowledgments. This work was supported by the National Key Research and Development Program of China [grant numbers 2022YFE0124300, 2022YFB3504104].

References

1. IEA: Tracking Clean Energy Progress (2023)
2. Mores, P., Scenna, N., Mussati, S.: CO_2 capture using monoethanolamine (MEA) aqueous solution: modeling and optimization of the solvent regeneration and CO_2 desorption process. Energy **45**(1), 1042–1058 (2012)
3. Zhang, Y.D., Hu, D.D., Liu, C., et al.: Research progress of CO_2 capture, utilization and storage (CCUS) technologies in petroleum and petrochemical industry. Oil Gas Stor. Transp. **36**(6), 636–645 (2017)
4. Chai, S.Y.W., Ngu, L.H., How, B.S.: Review of carbon capture absorbents for CO_2 utilization. Greenhouse Gases: Sci. Technol. **12**(3), 394–427 (2022)
5. Zhang, Z.E., Borhani, T.N., Olabi, A.G.: Status and perspective of CO_2 absorption process. Energy **205**, 118057 (2020)
6. Meng, F.Z., Meng, Y., Ju, T.Y., et al.: Research progress of aqueous amine solution for CO_2 capture: a review. Renew. Sustain. Energy Rev. **168**, 112902 (2022)
7. Feenstra, M., Monteiro, J., van den Akker, J.T., et al.: Ship-based carbon capture onboard of diesel or LNG-fuelled ships. Int. J. Greenhouse Gas Control **85**, 1–10 (2019)
8. Casero, P., García-Pena, F., Coca, P.: Elcogas pre-combustion carbon capture pilot real experience of commercial technology. Energy Proc. **37**, 6374–6382(2013)
9. Zahid, U., Al Rowaili, F.N., Ayodeji, M.K., et al.: Simulation and parametric analysis of CO_2 capture from natural gas using diglycolamine. Int. J. Greenhouse Gas Control **57**, 42–51 (2017)
10. Awoyomi, A., Patchigolla, K., Anthony, E.J.: Process and economic evaluation of an onboard capture system for LNG-fueled CO2 carriers. Ind. Eng. Chem. Res. **59**, 6951–6960 (2020)
11. Dugas, R.: Pilot Plant Study of Carbon Dioxide Capture by Aqueous Monoethanolamine. University of Texas at Austin, Austin, Texas, The United States of America (2006)

Numerical Simulation of Low NOx Burner in Lime Rotary Kiln and Analysis of Its Optimization Direction

Chao Bian[1], Tao Zheng[1], Xiaohong Sun[1], and Lianyong Wang[2(✉)]

[1] Benxi Iron and Steel (Group) Mining Co., Ltd., Beitai Iron Mine, Benxi 117017, China
[2] Department of Energy Engineering, Northeastern University, Shenyang 110819, China
wlyneu@126.com

Abstract. This paper presents a numerical simulation study on a low NOx burner for a lime rotary kiln. The simulation employs the Realizable k-ε model for turbulence, the finite rate/eddy-dissipation model for combustion, the P-1 model for radiation, and the extended Zeldovich mechanism and Fenimore mechanism for NO_x formation. The results show that the flame in the kiln exhibits a ring-shaped distribution with a low central temperature and high peripheral temperatures. The combustion process is relatively slow, with a small high-temperature region in the initial stage and a more complete combustion in the later stage. The NO_x concentration at the kiln exit is 369 mg/m^3, exceeding the emission limit set by Liaoning Province. The analysis reveals that the burner does not achieve sufficient mixing of fuel and air, leading to slow combustion, localized high temperatures, and high NO_x emissions. To address these issues, optimization of the burner's structural parameters and the kiln's operating parameters is proposed. Future research will focus on investigating the effects of parameters such as fuel staging, air staging, swirl number, secondary air preheating temperature, secondary air ratio, secondary air supply mode, and flue gas recirculation rate on the combustion process and NO_x emissions.

Keywords: Low NO_x burner · lime rotary kiln · numerical simulation

1 Introduction

Lime rotary kilns are indispensable in the production of quicklime, a fundamental component in steelmaking, construction, and various other industries [1]. However, the combustion of fossil fuels within these kilns leads to the generation of substantial levels of nitrogen oxides (NO_x), which pose significant environmental challenges [2]. NO_x emissions contribute to air pollution, acid rain, and the formation of ground-level ozone, thereby impacting human health and ecosystems. Additionally, NO_x is a potent greenhouse gas, exacerbating global climate change.

This paper presents a numerical simulation study that delves into the combustion characteristics and NO_x emissions of a low NO_x burner specifically designed for lime

D. Han et al. (Eds.): ICEPG 2024, ESE, pp. 390–396, 2025.
https://doi.org/10.1007/978-3-031-86829-0_44

rotary kilns. Utilizing computational fluid dynamics (CFD) techniques, the study investigates the complex interplay between fluid dynamics, combustion chemistry, and NO_x formation within the kiln. The simulation employs the Realizable k-ε model for turbulence, the finite rate/eddy-dissipation model for combustion, the P-1 model for radiation, and the extended Zeldovich mechanism and Fenimore mechanism for NO_x formation. By analyzing the flame shape, temperature distribution, and species concentrations, the study aims to provide valuable insights into the combustion dynamics and identify potential optimization strategies for achieving low NO_x emissions in lime rotary kilns.

2 Numerical Model Establishment

2.1 Basic Governing Equations

(1) The law of conservation of mass means that the increase in the mass of a fluid microelement at the same time is equal to the net mass flowing into the microelement [4, 5]. The mass conservation equation, also called the continuity equation, is expressed as:

$$\frac{\partial \rho}{\partial \tau}+\nabla \cdot \left(\rho \vec{u}\right) = 0 \tag{1}$$

The first term represents the rate of change of fluid density over time, and the second term represents the mass of fluid flow interface entering the micro-element.

(2) The momentum conservation equation is that the sum of external forces acting on an object per unit volume is equal to the sum of the rate of change of the object's momentum with time and the amount of change of the object's momentum [6, 7]. This equation satisfies Newton's second law:

$$\frac{\partial(\rho u)}{\partial \tau} + div\left(\rho u \vec{u}\right) = \frac{\partial(-P+\tau_{xx})}{\partial x} + \frac{\partial \tau_{yx}}{\partial y} + \frac{\partial \tau_{zx}}{\partial z} + F_x \tag{2}$$

$$\frac{\partial(\rho v)}{\partial \tau} + div\left(\rho v \vec{u}\right) = \frac{\partial \tau_{xy}}{\partial x} + \frac{\partial\left(-P+\tau_{yy}\right)}{\partial y} + \frac{\partial \tau_{zy}}{\partial z} + F_y \tag{3}$$

$$\frac{\partial(\rho w)}{\partial \tau} + div\left(\rho w \vec{u}\right) = \frac{\partial \tau_{xz}}{\partial x} + \frac{\partial \tau_{yz}}{\partial y} + \frac{\partial(-P+\tau_{zz})}{\partial z} + F_z \tag{4}$$

(3) The energy conservation equation is that the total amount of energy remains unchanged in the process of transformation, that is, the first law of thermodynamics. This equation analyzes the heat balance of heat transfer. For Newtonian fluid, the specific equation is [8]:

$$\frac{\partial(PT)}{\partial \tau} + div\left(\rho \vec{u} T\right) = div\left(\frac{k}{c_p} gradT\right) + S_T \tag{5}$$

(4) The component mass conservation equation means that the increase in the mass of a component in the system per unit time is equal to the sum of the masses of the component generated through net diffusion or chemical reaction, and is written as:

$$\frac{\partial(\rho c_s)}{\partial t} + div\left(\rho \vec{u} c_s\right) = div(D_s grad(\rho c_s)) + S_s \tag{6}$$

2.2 Gas Phase Turbulent Flow Model

Realizable k-ε model relates Cμ in turbulent viscosity with strain rate, and introduces the related contents of rotation and curvature. It is widely used in rotating uniform shear flow, flow in pipes and flow with separation.

$$\frac{\partial(\rho k)}{\partial t}+\frac{\partial(\rho k u_i)}{\partial x_i}=\frac{\partial}{\partial x_j}\left[\left(\mu+\frac{\mu_t}{\sigma_k}\right)\frac{\partial k}{\partial x_j}\right]+G_k-\rho\varepsilon \tag{7}$$

$$\frac{\partial(\rho\varepsilon)}{\partial t}+\frac{\partial(\rho\varepsilon u_i)}{\partial x_i}=\frac{\partial}{\partial x_j}\left[\left(\mu+\frac{\mu_t}{\sigma_\varepsilon}\right)\frac{\partial\varepsilon}{\partial x_j}\right]+\rho C_1 E\varepsilon-\rho C_2\frac{\varepsilon^2}{k+\sqrt{\nu\varepsilon}} \tag{8}$$

The air flow process in the kiln is complex, and there is an obvious central reflux zone in the furnace due to the existence of rotating jet, so this paper selects the Realizable turbulence model [3] to calculate the air flow.

2.3 Gas Phase Turbulent Combustion Model

In this paper, the finite rate/vortex dissipation model is used to simulate the combustion process, and the reaction rates of Arrhenius formula and vortex dissipation formula are considered at the same time.

Reaction rate corresponding to Arrhenius formula:

$$k_{f,\boldsymbol{r}}=A_{\boldsymbol{r}}T^{\beta_r}e^{-E_rRT} \tag{9}$$

Eddy dissipation formula:

$$R_{i,r}=v'_{i,r}M_{w,r}A\rho\frac{\varepsilon}{k}\min_R\left(\frac{Y_R}{v'_{i,r}M_{w,r}}\right) \tag{10}$$

$$R_{i,r}=v'_{i,r}M_{w,r}A\rho\frac{\varepsilon}{k}\min_R\frac{\sum PY_P}{\sum_j^N v'_{j,r}M_{w,j}} \tag{11}$$

2.4 Radiation Model

For this paper, the P-1 radiation model is selected for calculation by comprehensive calculation accuracy and calculation resources. This model uses orthogonal spherical harmonics to express the radiation intensity, which is the simplest type of P-N model. Its radiant heat flow is expressed as:

$$q_r=-\frac{1}{3(\alpha+\sigma_s)-C\sigma_s}\nabla G \tag{12}$$

3 Model Establishment and Calculation Conditions

3.1 Construction of Geometric Model and Mesh Division

In order to save numerical calculation resources and reasonably simplify the burner and some ancillary equipment as close to the actual operation as possible, the following assumptions are made when constructing the three-dimensional model:

(1) Only the gas combustion and heat transfer phenomenon in the rotary kiln are considered, and the calcination process of the material is not considered;
(2) Ignore the slope of the rotary kiln and consider it to be placed horizontally;
(3) It is considered that the rotary kiln is a thin-walled structure, regardless of the wall thickness of the rotary kiln, the ancillary equipment connecting the kiln head and kiln tail, and the wheel belt outside the kiln body.

The effective inner diameter of the lime rotary kiln used in this paper is 2 m and the length is 45 m. The three-dimensional modeling of the rotary kiln and burner is carried out, as shown in Fig. 1. 25% of the combustion-supporting air is supplied through the burner, which is called primary air. The remaining 75% of the combustion-supporting air is supplied into the kiln from the annular section between the burner and the kiln wall, which is secondary air.

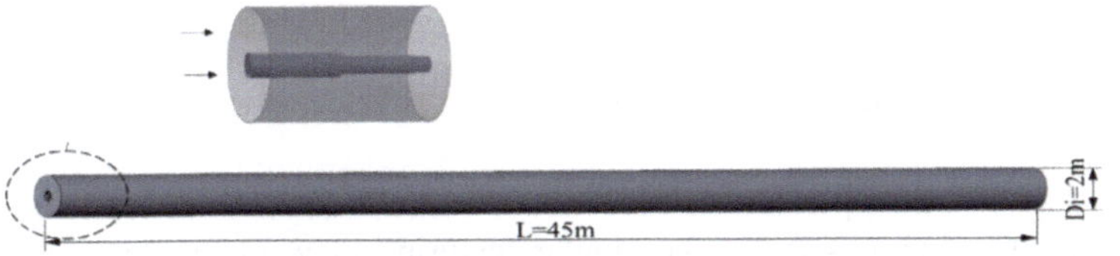

Fig. 1. Schematic diagram of rotary kiln.

In order to reduce the calculation error and improve the calculation speed, local grid encryption is properly carried out in key areas such as burner and kiln head. The quality of the divided mesh is checked, and the quality of the mesh is good and meets the calculation requirements. The divided overall grid of the rotary kiln is shown in Fig. 2.

Fig. 2. Overall grid diagram of rotary kiln.

3.2 Boundary Conditions and Solution Settings

The boundary condition types of the model inlet and outlet are selected as velocity-inlet and pressure-outlet, respectively, and the outlet pressure is set to 0 pa. The wall surface of the rotary kiln is set as a non-slip wall surface, and the wall surface is considered to be a constant temperature wall surface, the temperature is 1150 K, and the emissivity is set to 0.7. The inlet temperature of gas and primary air is constant at 298.15 K, the inlet temperature of secondary air is 698.15 K, the combustion air components are 21% O_2 and 79% N_2, and the air consumption coefficient is 1.05. See Table 1 for specific boundary condition parameters.

Table 1. Boundary conditions.

	Pulverized coal	Primary air			Secondary air	Outlet
		Central wind	Cyclone wind	Axial wind		
Proportion [%]	100	3.375	15	7	75	/
Wind speed[m/s]	30	50	130	200	0.805	/
Types of boundary conditions	Velocity	Velocity	Velocity	Velocity	Velocity	Pressure

4 Analysis of Simulation Results

Figure 3 depicts the temperature distribution across the z-x plane within the furnace. The flame's distribution is observed to be annular, featuring a lower temperature at the center and higher temperatures towards the periphery. Additionally, the combustion process is characterized by an initial slow burn with a small area of high temperature, which intensifies after a certain distance, indicating a more complete burn towards the flame's tail.

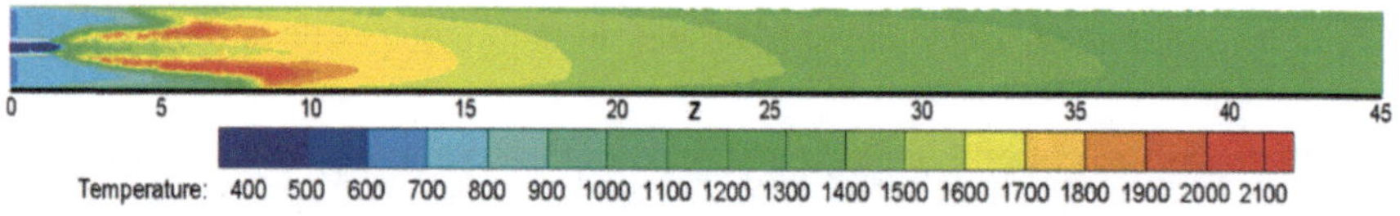

Fig. 3. Cloud diagram of temperature distribution in the z-x plane of the kiln.

Figures 4 and 5 illustrate the distribution patterns of O_2 and NO in the z-x plane within the furnace. Upon examining Fig. 4, it is evident that there exists a significant velocity disparity between the high-temperature secondary air entering the furnace and the swift axial air flow. In the NO distribution map (Fig. 5), a high concentration of NO is observed near the furnace wall, corresponding to the trailing section of the flame front in Fig. 3 and the posterior part of the rapid reaction zone where oxygen participates in combustion, as shown in Fig. 4. This indicates that areas with locally elevated temperatures and oxygen levels contribute to the substantial production of NO. After performing the necessary calculations, the NO_x concentration at the furnace outlet reaches 369 mg/m^3, which significantly surpasses the allowable NO_x emission standards for rotary kilns in Liaoning Province.

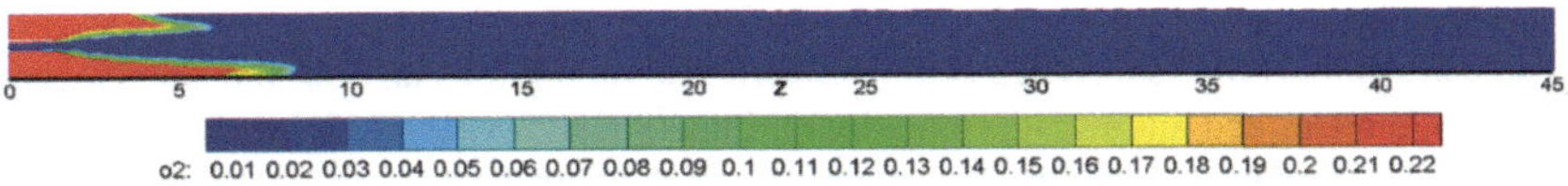

Fig. 4. Cloud map of O_2 distribution in z-x plane in the kiln.

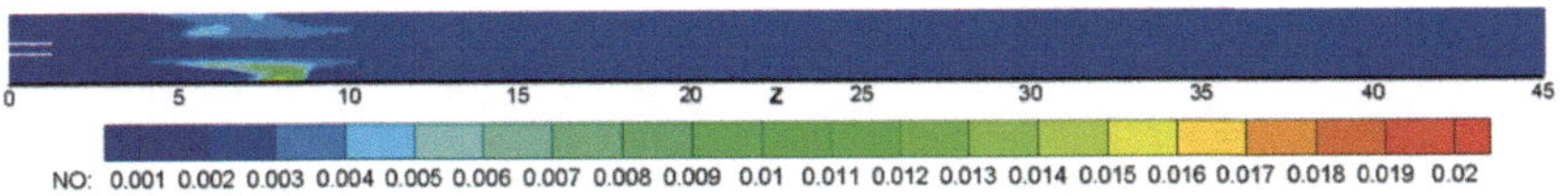

Fig. 5. Cloud map of NO distribution in z-x plane in the kiln.

5 Conclusion

In this paper, the combustion process of low nitrogen burner in lime rotary kiln was numerically simulated. After calculation, it was found that the NO_x concentration at the kiln tail was 369 mg/m^3, far exceeding the NO_x emission limit of relevant rotary kilns in Liaoning Province. Therefore, the improvement direction of the operation parameters of the rotary kiln and the structural parameters of the burner is put forward.

In the future research, according to the structural parameters of burner, the effects of gas classification, air classification ratio and swirl number on the combustion and NO_x generation in furnace can be studied. According to the operating parameters of rotary kiln, the effects of secondary air preheating temperature, secondary air ratio, secondary air supply mode and flue gas recirculation rate on the flame and NO_x emission in the furnace were studied. Through continuous optimization and improvement, we hope to obtain the best working conditions to meet the calcination requirements and NO_x emission requirements of lime rotary kiln.

References

1. Zhou, P., Yang, S., Liu, Q., et al.: Transient simulation of heat and mass transfer in a parallel flow regenerative Lime Kiln with pulverized coal injection. Appl. Therm. Eng. **227**, 120412 (2023)
2. Zhang, F., Bhatia, S.K., Wang, B., et al.: Experimental and numerical study on the kinetics of CO2–N2 Clathrate hydrates formation in Silica gel column with dodecyltrimethylammonium chloride for effective carbon capture. J. Mol. Liq. **363**, 119764 (2022)

3. Farraj, A.R.D., Rajasegar, R., Al-Khateeb, A.N., et al.: Phenomenology of electrostatically manipulated laminar counterflow non-premixed methane flames. J. Energy Eng. **142**(02), 150–162 (2016)
4. Kurose, R., Makino, H., Suzuki, A.: Numerical analysis of pulverized coal combustion characteristics using advanced low-NOx burner. Fuel **83**(6), 693–703 (2004)
5. Wang, Y., Zhou, Y.: Numerical optimization of the influence of multiple deep air-staged combustion on the NOx emission in an opposed firing utility boiler using lean coal. Fuel **269**, 116996 (2020)
6. Duan, E.P., Sun, B.M., Guo, Y.H., et al.: Numerical simulation on an 800 MW swirl-opposed firing boiler retrofitted for low NOx emission. J. Chin. Soc. Power Eng. **32**(11), 825–829 (2012)
7. Zeldvich, Y.B.: The oxidation of nitrogen in combustion explosions. Acta Physicochim. U.S.S.R. **21**, 577–628 (1946)
8. Lfler, G., Sieber, R., Harasek, M., et al.: NOx formation in natural gas combustion—a new simplified reaction scheme for CFD calculations. Fuel **85**(4), 513–523 (2006)

Spatiotemporal Evolutions of Energy Carbon Emissions of Six Metropolitan Areas in the Yangtze River Delta Urban Agglomeration

Yuan Tan, Tianyan Wu(✉), Yuxi Wang, and Yuchen Zhang

College of Urban Construction and Safety Engineering, Shanghai Institute of Technology, Shanghai 201418, China
18512321255@163.com

Abstract. Carbon emissions exert great pressure to the environment, and accurate measurement of carbon emissions is significantly important for regional carbon reduction and sustainable development. This study selects six metropolitan areas (MAs) of the Yangtze River Delta urban agglomeration (YRDUA) to measure the energy carbon emissions (ECE) and energy carbon emissions intensity (ECEI) from 2013 to 2022, depicts the spatiotemporal evolution characteristics of ECEI. The results show that: (1) the ECE of the six MAs remained a stable trend while the ECEI showed a downward trend during the study period, and the ECE are not proportional to the ECEI. (2) The ECEI were continuously decreasing, and the ECEI of each MA were converging towards the low value. High ECEI were concentrated in the middle of the six MAs and dispersed in the edges of the six MAs, while low ECEI were concentrated between the middle and the edges of the six MAs. (3) The variation degree of ECEI in different MAs was large in terms of horizontal comparison, and the variation degree in each MA was small in terms of vertical time change. These findings can provide critical policy implications for regional carbon reductions.

Keywords: Metropolitan Areas · Energy Carbon Emissions · Energy Carbon Emissions Intensity · Spatiotemporal Evolutions

1 Introduction

Rapid urbanization and industrialization have posed enormous environmental challenges over the past decades, with the increase in carbon emissions being the most severe [4]. Carbon emissions are one of the main drivers of global warming, which threatens the sustainability of human societies and even the whole Earth's ecosystem [1]. Therefore, relevant international organizations as well as governments have considered carbon reduction an important measure to cope with global warming [7]. There are many sources of carbon emissions, and the main source remains the consumption of energy. According to the International Energy Agency (IEA), global energy-related CO_2 emissions reached more than 36.8 billion tons in 2022, an increase of 321 million tons or 0.9% from the previous year. Therefore, accurate measurement of energy carbon emissions (ECE) is particularly important for controlling carbon emission reduction.

D. Han et al. (Eds.): ICEPG 2024, ESE, pp. 397–406, 2025.
https://doi.org/10.1007/978-3-031-86829-0_45

Due to regional differences in the distribution of energy consumption, carbon emissions also have regional differences. Based on the calculation of ECE at different levels such as countries, urban agglomerations (UAs), provinces and industries [5], we can clarify the regional differences of carbon emissions and their spatiotemporal evolution, and lay a good foundation for the formulation of carbon emission reduction policies. There are many methods to calculate energy consumption, such as InVEST model and Denitrification Decomposition online model [2], but the most popular method is the carbon emissions coefficient (CEC). This method originated from the IPCC and is more convenient, the carbon emissions caused by fossil fuel use can be directly calculated. For example, Chinese scholar Liu et al. [3] calculated the total carbon emission and carbon emission intensity (CEI) by using the CEC method based on the fossil energy consumption data of the transportation industry in 30 provinces in China, and found that the total carbon emission in the eastern, central and western regions decreased successively, but the CEI was lower in the southeast and higher in the northwest.

Given the remarkable progress, the previous studies on regional differences in ECE mostly took prefecture-level cities as the comparison unit. However, as coordinated regional development becomes increasingly important, metropolitan areas (MAs) and urban agglomerations (UAs) with one or several cities as the core are rising at an accelerated pace. Therefore, it is necessary to take MAs or UAs as the comparison unit to study their carbon emission differences, so as to provide inspiration for policymakers.

This paper selects six MAs of the Yangtze River Delta urban agglomeration (YRDUA) as research objects to measure the ECE and ECEI from 2013 to 2022, depicts the spatiotemporal evolution characteristics of ECEI of the six MAs, through comparative study, we understand the difference of ECE between six MAs, and find the evolution law of ECE and variation degree in each MA.

2 Data and Research Methods

2.1 Study Area and Data Sources

This study selects six MAs of YRDUA as research objects, mainly due to two considerations. First, the YRDUA is the most developed region in China, and six MAs are its core parts (Fig. 1), its energy consumption is especially large. Second, although six MAs (26cities, Table 1) cover the most developed areas in the YRDUA (41cities), there still exist a lot of variations in resource endowment, economic foundation, industrial structure, and environmental protection between six MAs, showing large differences in ECE. Consequently, it is appropriate to conduct comparative studies between six MAs.

Specifically, the data used by this study are obtained from the China Energy Statistical Yearbook (2014-2023), Municipal Statistical Yearbooks (2014-2023) and other sources. And data for missing years were filled in by interpolating indicators from neighbouring years.

Table 1. The composition of six MAs within the YRDUA.

MAs	Prefecture level cities/Municipalities
Shanghai MA	Shanghai, Wuxi, Changzhou, Suzhou, Nantong, Ningbo, Jiaxing, Huzhou, Zhoushan
Hangzhou MA	Hangzhou, Jiaxing, Huzhou, Shaoxing, Quzhou, Huangshan
Nanjing MA	Nanjing, Zhenjiang, Yangzhou, Huai'an, Ma'anshan, Chuzhou, Wuhu, Xuancheng
Hefei MA	Hefei, Ma'anshan, Chuzhou, Wuhu, Huainan, Lu'an, Bengbu
Su-Xi-Chang MA	Suzhou, Wuxi, Changzhou
Ningbo MA	Ningbo, Zhoushan, Taizhou

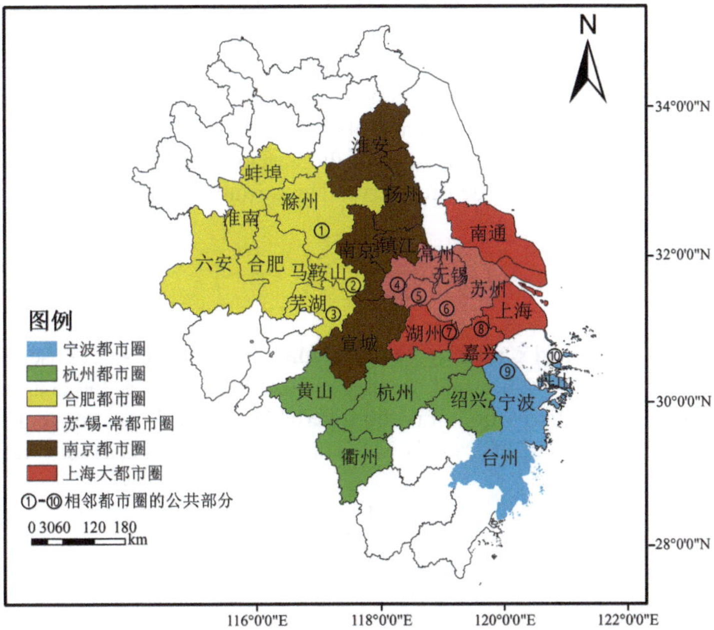

Fig. 1. The geographical locations of six MAs within the YRDUA.

2.2 Research Method

Carbon Emissions Coefficient Method

Based on the energy consumption of the YRDUA and the IPCC method for calculating ECE, this paper selects 10 fossil energy sources, including raw coal, cleaned coal, coke, crude oil, gasoline, kerosene, diesel, fuel oil, natural gas and liquefied petroleum gas, to calculate ECE in six MAs from 2013 to 2022. The calculation formula is expressed as follows [3]:

$$ECE = \sum_{i=0}^{n} E_i \times NCV_i \times CC_i \times OF_i \times 44/12 \tag{1}$$

where ECE is carbon emissions. E_i, NCV_i, CC_i and OF_i refer to consumption of energy i, the average low calorific value of fuel, the carbon content per unit calorific value of fuel, and oxygen content, respectively. 44/12 is the conversion factor of C to CO_2.

Furthermore, considering that more developed regions also have more carbon emissions, the index of carbon emission intensity (carbon emissions per unit GDP) is introduced. The expression is as follows:

$$ECEI = ECE/TGDP \tag{2}$$

where ECEI is carbon emission intensity, TGDP is the gross regional product.

Kernel Density Estimate

Kernel density estimation is a non-parametric method used to assess the spatial distribution disequilibrium by estimating the probability density of a variable and describing the dynamically evolving characteristics of the variable distribution, such as its shape and bandwidth in terms of the density curve. The formula is shown as follows [6]:

$$f(x) = \frac{1}{nh}\sum_{i=1}^{n} K\left(\frac{x_i - \overline{x}}{h}\right) \tag{3}$$

where K, n, h, x_i, and x- refer to the kernel function, the number of observations, the bandwidth, the observed value, and the mean value. In this paper, the Gaussian kernel function with normal distribution of random variable x is used to estimate the dynamic evolution characteristics of ECE of six MAs, and its expression is as follows:

$$K(x) = \frac{1}{\sqrt{2\pi}}\exp\left(-\frac{\overline{x}^2}{2}\right) \tag{4}$$

Coefficient of variation

Coefficient of variation (CV) is a statistic that measures the degree of variation of sequence observations, so CV can be used to measure the difference of ECE in various cities. This study calculates the CV to analyze equilibrium of six MAs, and its expression is as follows [8]:

$$CV = \sqrt{\frac{1}{N}\sum\nolimits_{i=1}^{N}\frac{ECE_i - u}{u}} \tag{5}$$

where N refers to the number of cities, ECE_i refers to ECE for city i, and u refers to the mean ECE for the present year.

3 Results and Discussion

3.1 Overall Change Characteristics of Energy Carbon Emissions

As can be seen from Fig. 2, the ECE value of the six MAs remained a stable trend during the study period, while the ECEI value showed a downward trend, which shows that the carbon reduction policies in recent years have had positive results [9]. For each MA, the Shanghai MA had the highest ECE at approximately 85000 M tons, followed by the Nanjing MA, which accounted for about half of the Shanghai MA, at about 40000 M tons, followed by the Su-Xi-Chang MA at about 28000M tons. Hefei MA and Ningbo MA had no significant difference in ECE, which had the ECE of 20,000 M tons, and the last was the Hangzhou MA, 13,000 M tons or so.

However, the highest ECEI was not Shanghai MA but Ningbo MA, this is because ECE are highly correlated with economic development. For Ningbo MA, ECEI was 1.90 M tons/(100*M yuan) in 2013 and 1.06 M tons/(100*M yuan) in 2022, with an average annual growth rate of -5.67%. The ECEI of Hefei MA and Nanjing MA was similar, Hefei MA was 1.62 in 2013 and 0.93 in 2022, Nanjing MA was 1.67 in 2013 and 0.87 in 2022, and the average annual growth rate was −5.40% and −6.31% respectively. Followed by the Shanghai MA, ECEI was 1.28 in 2013 and 0.66 in 2022, the average annual growth rate was −6.41%. Then, the ECEI of Su-Xi-Chang MA was 1.10 in 2013 and 0.60 in 2022, the average annual growth rate is -5.88%. The last is Hangzhou MA, ECEI was 0.70 in 2013 and 0.35 in 2022, the average annual growth rate was −6.70%.

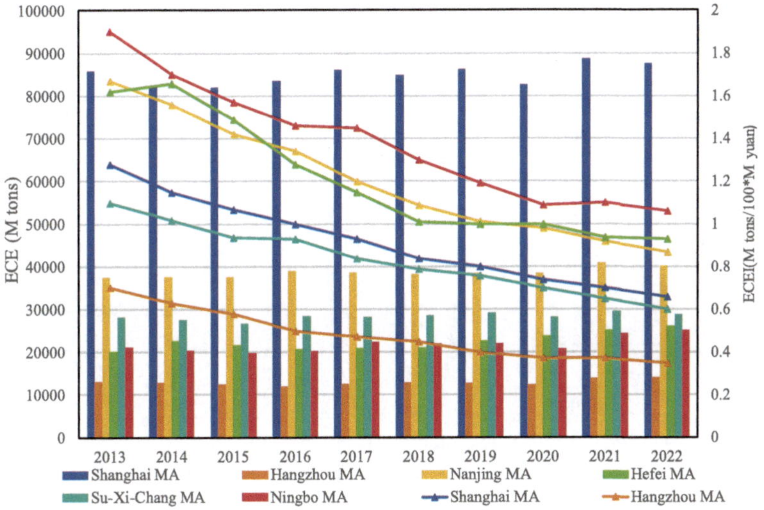

Fig. 2. ECE and ECEI in six MAs during 2013–2022.

3.2 The Spatiotemporal Evolution Characteristics of Energy Carbon Emissions

Considering that ECE of each MA are relatively stable during the study period and ECEI is constantly declining, we select ECEI to characterize the spatiotemporal evolution of the six MAs.

First, to describe the evolution of ECEI in six MAs from temporal perspective, the kernel density of ECEI in six MAs is estimated (Fig. 3). According to Fig. 3, the characteristics of the kernel density curve of ECEI in six MAs are quite different. Therefore, we analyze the temporal evolutions of ECEI from the shape and the bandwidth of the kernel density curves, respectively [6].

Regarding the shape changes of each MA, almost every MA showed obvious single-peak characteristic with a single convergence value during the study period. In 2013, the ECEI of Shanghai MA, Hangzhou MA, Nanjing MA, Hefei MA, Su-Xi-Chang MA, and Ningbo MA was converged 1.05,0.60,1.00,0.80,1.05,1.50, respectively. By 2017, the ECEI was converged 0.75,0.45,0.80,0.95,0.80,1.30, respectively. And by 2022, the ECEI was converged 0.70,0.30,0.60,0.90,0.55,1.05, respectively. In particular, there was an obvious "trailing" phenomenon in Shanghai MA and Nanjing MA, which shows that both Shanghai MA and Nanjing MA had significant low-value convergence and slight high-value convergence. Moreover, except for the Ningbo MA, other MAs' images were

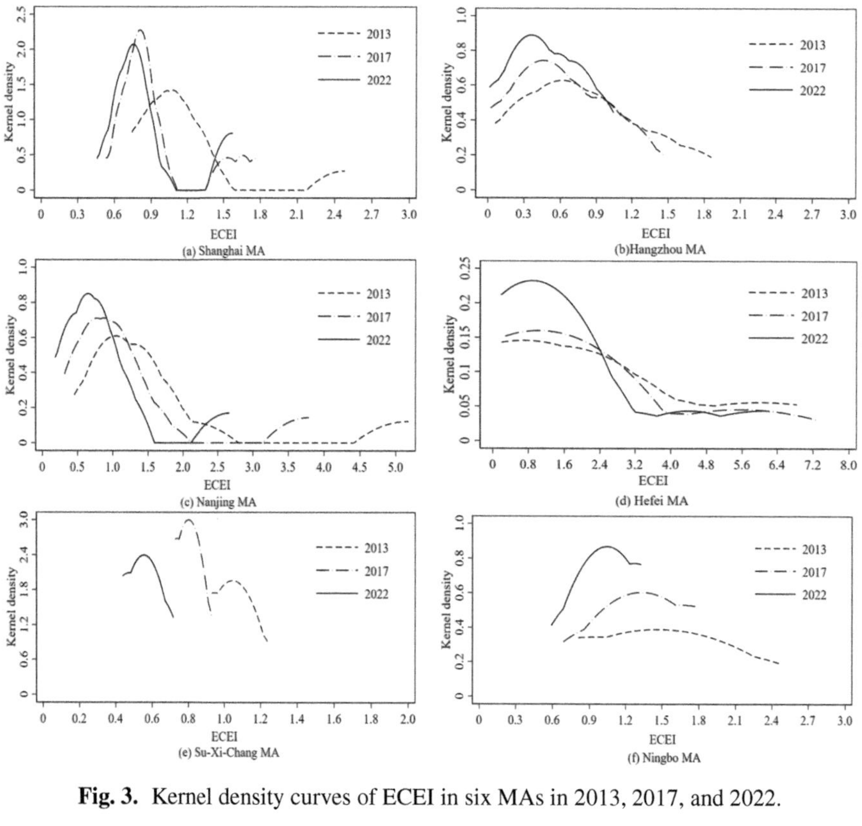

Fig. 3. Kernel density curves of ECEI in six MAs in 2013, 2017, and 2022.

mainly distributed to the right of the convergence value, which showed the mean value of ECEI in MAs was greater than the convergence value. Furthermore, it was obvious that the center of the kernel density curve of the six MAs moved upward to the left, which meant that the ECEI of the six MAs was constantly decreasing and the convergence trend was more obvious.

Regarding the bandwidth changes of each MA, it was obvious that the bandwidth of Hangzhou MA, Hefei MA and Ningbo MA was larger than Shanghai MA, Nanjing MA and Su-Xi-Chang MA, which showed that the inner gap of Hangzhou MA, Hefei MA and Ningbo MA was larger. In addition, during the period, the bandwidth of Hangzhou MA, Nanjing MA, Hefei MA, Su-Xi-Chang MA remained a stable trend while Shanghai MA and Ningbo MA showed a downward trend, this showed the inner gap of Hangzhou MA, Nanjing MA, Su-Xi-Chang MA remained stable while Shanghai MA, Hefei MA and Ningbo MA narrowed.

Second, to describe the evolution of ECEI in six MAs from spatial perspective [10], we plot spatial distributions of ECEI in 2013, 2017, and 2022 (Fig. 4). As can be seen from Fig. 4, high ECEI values were concentrated in the middle of the six MAs and dispersed in the edges of the six MAs during the period, while low ECEI values were concentrated between the middle and the edges of the six MAs. Moreover, the difference of ECEI values between the middle and the edges of the six MAs was the smallest, the

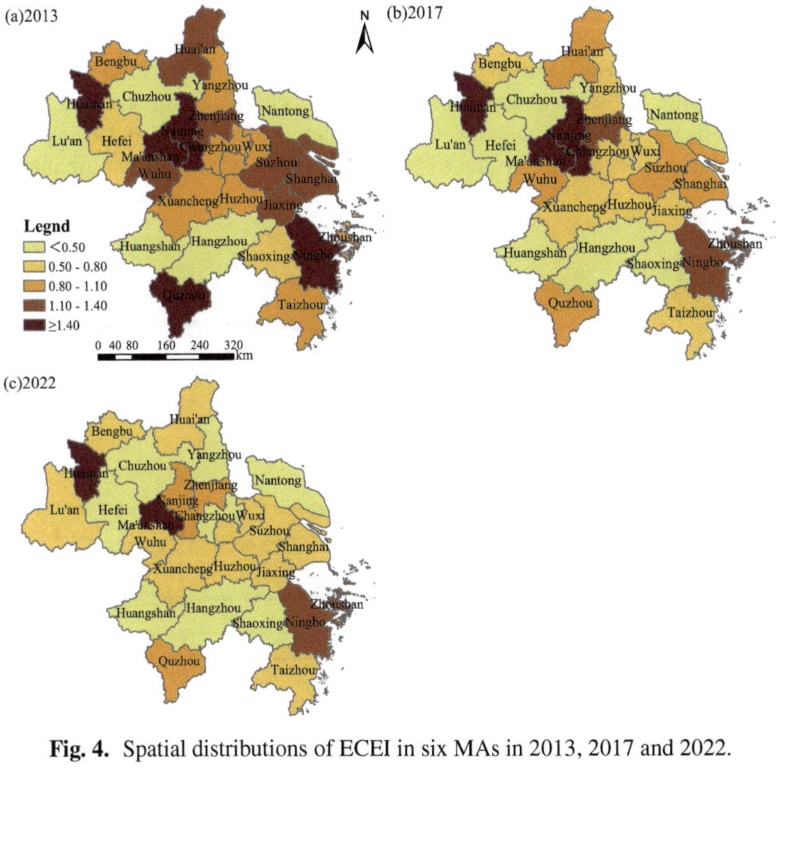

Fig. 4. Spatial distributions of ECEI in six MAs in 2013, 2017 and 2022.

difference of ECEI values in the middle was relatively large, the difference of ECEI values in the edge was the largest.

For each MA, the high ECEI values in 2013 were mainly distributed in Shanghai MA, Ningbo MA, Nanjing MA and Hefei MA, the low ECEI values were mainly distributed in Hefei MA and Hangzhou MA, and the high ECEI values in 2017 and 2022 were mainly distributed in Ningbo MA, Nanjing MA and Hefei MA, the low values were mainly distributed in Hefei MA and Hangzhou MA. Therefore, it can be seen that the ECEI value of Ningbo MA had been at a high level during the study period, while Hangzhou MA had been at a low level during the study period, and Hefei MA has both high and low values during the study period, with the largest internal gap.

3.3 Equilibrium Analysis of Energy Carbon Emissions in Six MAs

According to the previous analysis, we can quantitatively describe the regional difference of ECEI values in the six MAs, but the equilibrium analysis within each MA is insufficient. Therefore, we will further calculate the coefficient of variation (CV) of ECEI in each MA between 2013 and 2022 to analyze the variation of ECEI [8].

According Fig. 5, it can be seen that the CV values of the six MAs differ greatly, but the change trend is not very different. In terms of the size of the values, during the study period, the CV values from high to low are Hefei MA, Hangzhou MA, Nanjing MA, Shanghai MA, Ningbo MA and Su-Xi-Chang MA, respectively. The CV values of Hefei MA were the largest, approximately 2.30, while Su-Xi-Chang MA was the lowest, approximately 0.20, which meant that the internal gap of ECEI in Hefei MA had always been large, while in Su-Xi-Chang MA had always been small. In terms of the change trends, except for the slight increase of CV values in Hefei MA and Hangzhou MA, other MAs have maintained a relatively stable trend, which meant that the internal gap

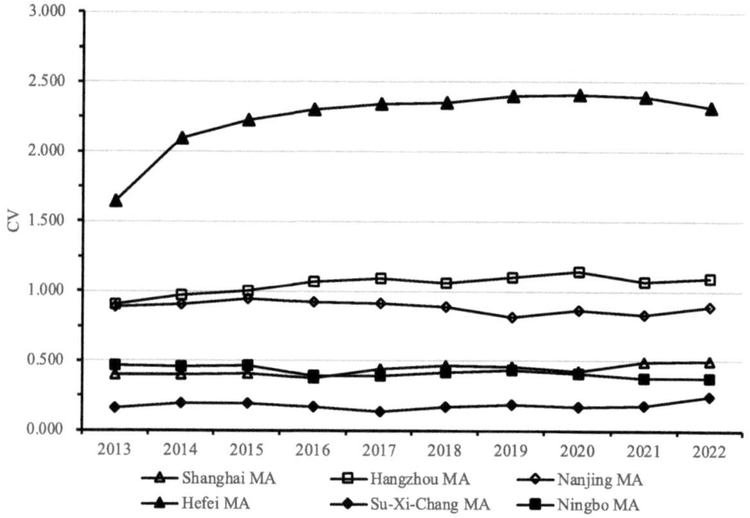

Fig. 5. Variation coefficient of ECEI in six MAs from 2013 to 2022.

between Hefei MA and Hangzhou MA had a slightly expanding trend during the study period.

4 Conclusions

This study measured the ECE and ECEI of six MAs from 2013 to 2022, and analyzed the spatiotemporal evolution characteristics of ECEI, we found that:

(1) the ECE of the six MAs remained a stable trend while the ECEI showed a downward trend during the study period, it shows that carbon reduction policies in recent years have had positive results, and carbon peaking is expected to be achieved ahead of schedule. The ECE are not proportional to the ECEI. For example, the ECE of the Shanghai MA ranks first among the six MAs but the ECEI ranks fourth, which shows that ECE are related to economic development.
(2) from the temporal evolution perspective, the ECEI from 2013 to 2022 were continuously decreasing, and the ECEI of each MA were converging towards the low value. The convergence values of Shanghai MA, Hangzhou MA, Nanjing MA, Hefei MA and Su-Xi-Chang MA were all lower than the average value, and Ningbo MA was higher than the average value. From the spatial evolution perspective, high ECEI values were concentrated in the middle of the six MAs and dispersed in the edges of the six MAs during the period, while low ECEI values were concentrated between the middle and the edges of the six MAs. Furthermore, the ECEI value of Ningbo MA had been at a high level while Hangzhou MA had been at a low level during the study period, and Hefei MA had both high and low values during the study period, with the largest internal gap.
(3) as for internal equilibrium of ECEI in each MA, the variation degree of ECEI in different MAs is large from the perspective of horizontal comparison, and the variation degree in each MA is small from the perspective of vertical time change. Among all MAs, the internal gap of Hefei MA had always been large during the period, while in Su-Xi-Chang MA had always been small. In addition, the internal gap of Hefei MA and Hangzhou MA had a slight enlargement while other MAs remained stable.

References

1. Borozan, D.: Do geopolitical and energy security risks influence carbon dioxide emissions? Empirical evidence from European Union countries. J. Clean. Prod. **439**, 140834 (2024)
2. Fan, P., Qian, X., Wang, J., Yamada, K.: Does ownership structure influence carbon emission? Different roles of institutional investors. J. Clim. Finan. **6**, 100030 (2024)
3. Liu, J., Li, S., Ji, Q.: Regional differences and driving factors analysis of carbon emission intensity from transport sector in China. Energy **224**, 120178 (2021)
4. Nathaniel, S.P., Adeleye, N.: Environmental preservation amidst carbon emissions, energy consumption, and urbanization in selected African countries: implication for sustainability. J. Clean. Prod. **285**, 125409 (2021)
5. Sikder, M., et al.: The integrated impact of GDP growth, industrialization, energy use, and urbanization on CO2 emissions in developing countries: evidence from the panel ARDL approach. Sci. Total Environ. **837**, 155795 (2022)

6. Wang, C., et al.: Is the assessment approach of Sustainable Development Goal 11.3. 1 justified? Evidence from the drivers of future urban land use efficiency. J. Clean. Prod. **444**, 141147 (2024)
7. Wen, W., et al.: Evaluating carbon emissions reduction compliance based on 'dual control' policies of energy consumption and carbon emissions in China. J. Environ. Manage. **367**, 121990 (2024)
8. Wu, X., Huang, X.: Screening of urban environmental vulnerability indicators based on coefficient of variation and anti-image correlation matrix method. Ecol. Ind. **150**, 110196 (2023)
9. Zhang, X., Fan, D.: Research on the synergistic emission reduction effect of carbon emission trading and green financial policy. J. Environ. Manage. **367**, 121924 (2024)
10. Zhou, Y., Lu, Y.: Spatiotemporal evolution and determinants of urban land use efficiency under green development orientation: Insights from 284 cities and eight economic zones in China, 2005–2019. Appl. Geography **161**, 103117 (2023)

Study on the Measurement and Indicators of Carbon Emissions from Major Structures of Highway Bridges

Zhengkai Li[1], Dengyu Rao[1], Mengyue Li[2(✉)], Zhenchun Chen[2], and Guojun Hao[2]

[1] Guangzhou Expressway Company Ltd., Guangzhou 510000, Guangdong, China
[2] Guangdong Provincial Transportation Planning and Research Center, Guangzhou 510101, Guangdong, China
mengyuelee@163.com

Abstract. Bridges are the main source of carbon emissions from highway construction, and the carbon emissions of different types of bridges vary greatly. This study applies the carbon emission accounting theory of life cycle evaluation and the emission factor method to measure the carbon emissions during the construction period of bridges along a highway in Guangzhou, analyses the results and characteristic of carbon emissions of different structures according to the sub-projects, analyses the intensity value of carbon emissions of the major structures of bridges and establishes an estimation model of carbon emissions by taking the quantity of engineering and the length of bridges as the influencing factors. Through the calculation and analysis, it can be seen that the carbon emission of the upper structure of the bridge accounts for the highest proportion, accounting for more than 42%. Calculations show that the carbon emissions per kilometre of cast-in-place concrete steel bridges are much higher than those of other types of bridges, and the carbon emissions per m^3 of different superstructures are the highest for steel box girder structures.

Keywords: Bridges · Emission Factor Approach · Carbon Intensity

1 Introduction

The highway construction process consumes a large amount of materials, fuel and energy, and is a key area of energy conservation and emission reduction in the transport industry. Under the national carbon peak and carbon neutral target, green and low-carbon requirements are put forward through the whole process of highway infrastructure planning, construction, operation and maintenance, and to reduce the energy consumption and carbon emissions in the life cycle. The carbon emission caused by highway construction activities has also been concerned and studied by scholars at home and abroad. In the existing studies, Wang Xianwei and other highway roadbed, pavement, bridge and tunnel projects were respectively calculated for carbon emission, and the conclusion shows that the carbon emission level of bridge and culvert projects is higher than that of tunnel projects and road projects [1]; Liu Yuanyuan et al. pointed out that the ratio of bridge

D. Han et al. (Eds.): ICEPG 2024, ESE, pp. 407–417, 2025.
https://doi.org/10.1007/978-3-031-86829-0_46

and tunnel to tunnel of very large bridges and tunnels is the main factor affecting carbon emissions of road construction activities [2]. The lack of bridges, tunnels and other structures will greatly underestimate the life cycle carbon emissions of the road construction stage [3]. Huangshan Qian et al. pointed out that the carbon emission intensity of typical bridge and culvert projects in the plains and micro-hills area is the highest [4].

Due to the variety of structural forms, loading forms, force states and construction methods of bridges, the carbon emissions of bridges also vary greatly. The calculation methods of carbon emission values for different types of bridges have been studied to cover different cycle stages, and there are discrepancies in the results of carbon emission calculations [5, 6]. This study measures and analyses the carbon emissions of bridge projects (including interchange ramp bridges) of an expressway project under construction in Guangzhou, further studies the carbon emissions of different major structures of bridge projects, compares different types of bridges, and further researches the indicators of carbon emissions of bridge structures, which will help to carry out targeted control of low-carbon construction of bridge projects during the construction process of the project.

2 Measurement Methodology

The theoretical method of life cycle assessment (LCA) is mainly used in the research of carbon emission accounting for highway construction, and with the application of this method in the environmental assessment of road infrastructure such as pavements and roadbeds [7, 8], LCA assessment method is also gradually regarded as an effective means of comprehensively assessing the emissions of complex transport systems [9]. The accounting method adopts the emission factor method of the life cycle. This study mainly considers three phases of material production, transportation and site construction in the construction phase. The accounting boundary includes direct carbon emissions from fossil fuel combustion of machinery, equipment or vehicles used in bridge construction activities and indirect carbon emissions from the production of construction materials, purchased electricity and the materialisation of fossil fuels in construction machinery and equipment.

The bridge structure can be divided into foundation engineering, under structure, upper structure, ancillary structure and other engineering according to the subdivision. The foundation engineering includes pile foundation, bearing platform and tie beam. The lower structure includes abutments, piers, pier cover beam and bearing pads. The upper structure includes pre-stressed concrete girder, pre-stressed concrete T-beam, steel box girder, continuous cast-in-place concrete beam, cast-in-situ continuous concrete girders and cast-in-situ concrete stiffeners and other types. Ancillary structures include bridge bearings, guardrails and nets; other works include bridge drainage system and old bridge treatment. The quantity data of the bridge sub-projects were obtained through the construction project construction drawings and budget documents and combined with the construction site research.

The following methodology for measuring carbon emissions from bridge structures is established according to the division of bridge subcomponents:

$$E = \sum\nolimits_{j=1}^{n} E_j \tag{1}$$

$$E_j = E_{jm} + E_{jy} + E_{jn} \tag{2}$$

where: E is the total bridge carbon emissions; E_j is the carbon emission of bridge sub-projects (by pile foundation, bearing platform, tie girder, etc., a total of n types of structural components); E_{jm}, E_{jy} and E_{jn} are the carbon emissions of the three stages of building materials production, transport and on-site construction corresponding to the bridge sub-projects, respectively.

(1) Building materials production phase

$$E_{jm} = \sum\nolimits_{i=1}^{n} M_{jmi} = \sum_{i=1}^{n} (m_i \times EF_i) \tag{3}$$

where: M_{jmi} is the carbon emission in the production stage of the ith building material; m_i is the amount of the ith material used; EF_i is the carbon emission factor of the ith material.

(2) Transport phase

$$E_{jy} = \sum\nolimits_{i=1}^{n} Y_{jyi} = \sum\nolimits_{i=1}^{n} (y_i \times q_i \times EF_i) \tag{4}$$

where: Y_{jyi} is the carbon emission of the transport equipment corresponding to the ith material; y_i is the transport mileage of the transport equipment corresponding to the ith material; q_i is the energy consumption per unit mile of the transport equipment corresponding to the ith material; EF_i The carbon emission factor of fuel or electricity consumed by the transport equipment corresponding to the i-th material.

(3) On-site construction phase

$$E_{jn} = \sum\nolimits_{i=1}^{n} N_{jni} = \sum\nolimits_{i=1}^{n} (n_i \times s_i \times EF_i) \tag{5}$$

where: N_{jni} is the carbon emission of the ith type of construction machinery and equipment; n_i is the number of workbench shifts of the ith type of construction machinery and equipment; s_i The energy consumption per workbench shift of the ith type of construction machinery and equipment; EF_i The carbon emission factor of fuel or electricity consumed by the ith type of construction machinery and equipment.

The carbon emission factor of corresponding energy includes fossil fuels such as diesel, petrol, heavy oil, natural gas and electricity; the carbon emission factor of materials includes existing road construction materials such as steel, cement, sand and gravel, as well as some self-researched materials such as concrete, mortar and asphalt, which are calculated according to the proportion of raw materials. The carbon emission factors of construction materials are based on the carbon emission databases available at home and abroad and the carbon emission factors disclosed in the relevant standards and norms.

3 Measurement Results

In this study, 105 Bridges of an highway project in Guangzhou were collected. Prestressed concrete small box girder bridges were the main types of Bridges. The total length of the bridge is 39191.2 m, involving 124 kinds of road construction materials and 80 kinds of mechanical equipment, and establishing a total of 204 carbon emission factor lists.

3.1 Total Carbon Emissions

The total carbon emission of the bridge project is calculated to be 1.4097 million tons. From the perspective of bridge structure composition, the carbon emission ratio of upper structure is the highest, up to 42.63%; The proportion of foundation engineering and substructure is 23.91% and 23.69% respectively. Affiliate structure and others accounted for 9.72% (see Table 1).

Table 1. Total and percentage composition of carbon emissions from bridges.

Bridge structures		Unit (of measure)	Project volume	Total carbon emissions (t)	Percentage
Foundation engineering	Foundation for bridge abutment cast piles	m^3	30374.7	18405.5	23.91
	Bridge pier grouted pile foundation	m^3	366836.3	271919.4	
	Bridge deck abutment	m^3	3066.9	1858.1	
	Piers abutment	m^3	27136.0	23702.7	
	Linchpin	m^3	27809.7	21192.0	
Substructure	Bridge platform	m^3	14738.6	13988.5	23.69
	Piers	m^3	187703.7	183442.9	
	Abutment cover girder	m^3	110715.6	133047.7	
	Bearing cushion	m^3	1391.0	3464.7	
Superstructure	Prestressed concrete box girder	m^3	333056.6	465691.6	42.63
	Prestressed concrete hollow core slab	m^3	2044.7	216.5	
	Prestressed concrete T-beam	m^3	377.2	553.9	
	Cast-in-place continuous concrete Beams	m^3	6054.2	8968.4	
	Steel box girder	m^3	7495.9	33718.7	
	Cast-in-place concrete rigid structures	m^3	45951.4	91817.2	
Affiliate structure	Upper main structure ancillary	m^3	81166.4	76359.6	9.72
	Bridge support	classifier for individual items	21581.0	697.1	
	Fences	m	107107.7	60028.2	
Other engineering	Bridge drainage system	m	54314.6	661.4	
	Disposal of old bridges	m	340.3	1.7	

The highest carbon emission by component of the structure is pre-stressed concrete small box girders, accounting for 33% of the total ratio. The following are the pile foundation, bridge pier, bridge pier cover beam and cast-in-place concrete steel structure, accounting for 19.3%, 13%, 9.4%, 6.5% of the total. See Fig. 1 for specific values.

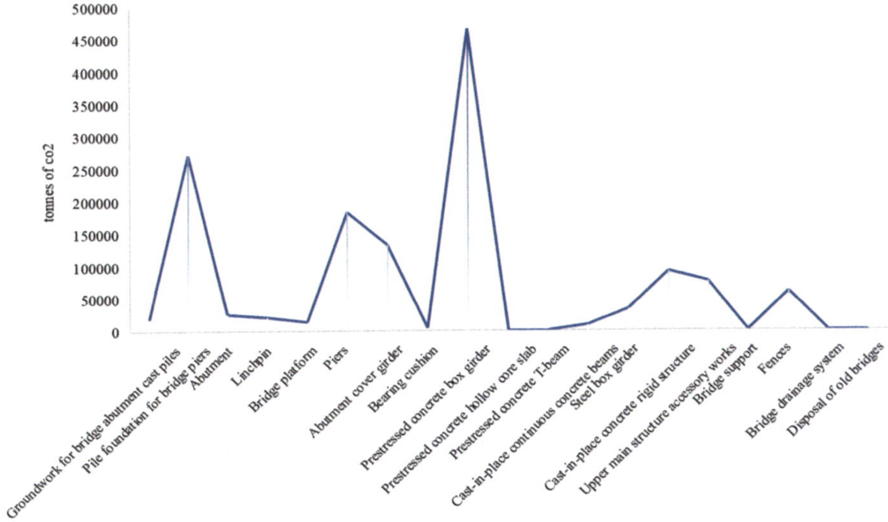

Fig. 1. Carbon emissions from bridge structure components.

3.2 Carbon Emission Characteristics of Single Bridge

In addition to calculating and analysing the total indicators, the carbon emission characteristics of single bridges and their structures are analysed by taking the carbon emission per unit mileage or quantity of engineering on single bridges or structures as intensity indicators. The carbon emission per unit mileage among 105 bridges is 35970.77 kg/km, with the range distributing between 3828.98–177550.43 kg/km, and the intensity value of the carbon emission of single bridges varies greatly (see Fig. 2).

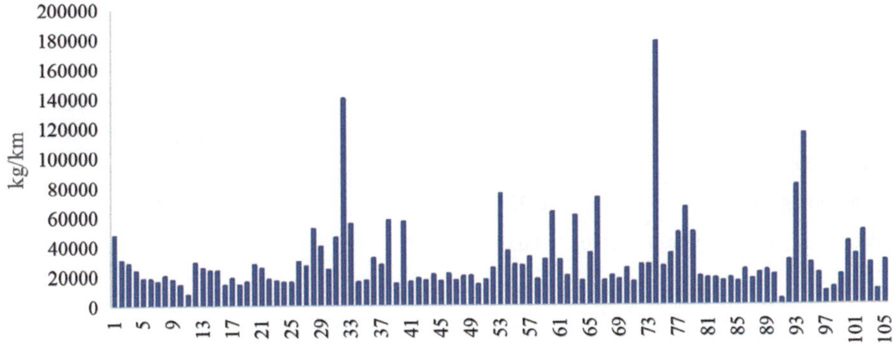

Fig. 2. Carbon emission intensity of single bridge in the whole bridge project.

According to different bridge types to further analyse the carbon emission characteristics, to bridge superstructure prestressed concrete small box girder bridge total carbon emissions accounted for 58.36%, followed by prestressed concrete and steel box girder bridge type carbon emissions accounted for 24.21%, cast-in-place concrete steel bridge accounted for 12.17%. Through the carbon emission per unit mileage of different types of bridges, the carbon emission intensity of cast-in-place concrete steel bridge is the highest, and the carbon emission per kilometre of the bridge reaches 110020.83 kg/km; the bridge type of the combination of prestressed concrete small box girder and steel box girder, the carbon emission per kilometre reaches 48270.74 kg/km, and the difference of the carbon emission intensity of the bridge of the prestressed concrete small box girder and the steel box girder is smaller and higher than that of the cast-in-place continuous concrete girder (see Table 2).

Table 2. Carbon emission share and intensity values of different types of bridges.

Type of superstructure	Share of total carbon emissions (%)	Carbon intensity (kg/km)
Prestressed concrete box girder	58.36	33050.81
Prestressed concrete small box girders, prestressed hollow core slabs	0.25	29685.43
Prestressed concrete small box girders, prestressed concrete T-beams	0.18	23300.64
Prestressed concrete small box girders, cast-in-place continuous concrete girders	4.55	40011.77
Prestressed concrete small box girder, steel box girder	24.21	48270.74
Cast-in-place concrete steel structure	12.17	110020.83
Cast-in-place continuous concrete beams	0.03	16174.30
Steel box girder	0.22	32177.34
Prestressed concrete hollow core slab	0.03	8710.14

The carbon emission intensity of different superstructure types is analysed, comparing the carbon emission intensity values of several typical superstructure forms involved in the project, among which the steel box girder has the highest carbon emission intensity value, with a range of 4386.29–4938.48 kg/m^3; followed by cast-in-place concrete steel structure, with an intensity value ranging from 1950.77–2059.65 kg/m^3; cast-in-place concrete continuous girder strength value ranges from 1499.22–1647.89 kg/m^3 is slightly higher than that of prestressed concrete box girder, and the difference between the two is relatively small; prestressed concrete T-beam has only one bridge in this project, and its value is 1468.48 kg/m^3 (see Fig. 3).

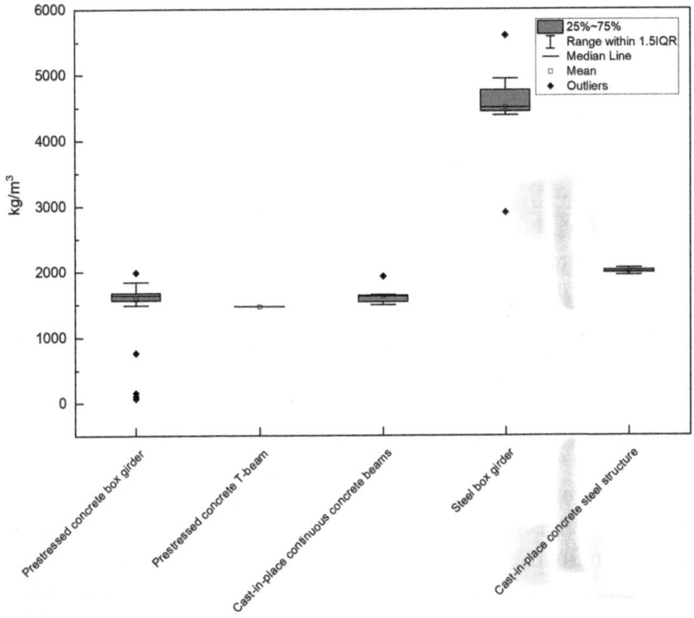

Fig. 3. Range of carbon emission intensity for different bridge superstructure forms.

Analyzing the carbon intensity of 6 types of main structures of prestressed concrete small box girder bridges, which account for a large number of samples and a high percentage of the total number of bridge types (see Fig. 4).

The carbon intensity of pile foundation ranges from 666.90 to 912.32 kg/m^3, the carbon intensity of prestressed concrete box girder ranges from 1489.15 to 1847.95 kg/m^3, the carbon intensity of abutment ranges from 782.69 to 985.85 kg/m^3, and the carbon intensity of abutment cover girder ranges from 1083.49 to 1244.64 kg/m^3. The carbon emission intensity of the upper main structure ancillary engineering ranges from 857.87 to 1022.89 kg/m^3, and the carbon emission intensity of the parapets and guardrail network ranges from 485.11 to 581.8 kg/m^3.

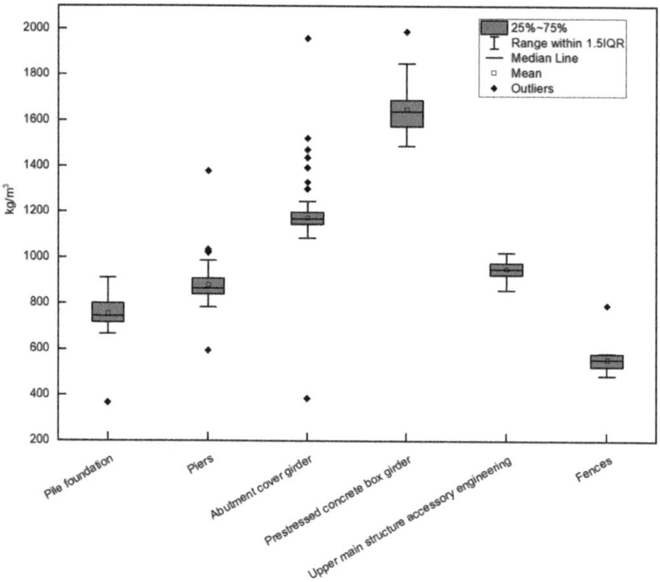

Fig. 4. Carbon intensity range of major structures of prestressed concrete small box girder bridges.

4 Estimation Methods of Carbon Emissions from Major Structures of Bridges

4.1 Correlation Analysis of Carbon Emissions from Bridges

The amount of pile foundation engineering, bearing platform engineering, tie beam engineering, abutment engineering, pier engineering, pier cover girder engineering, superstructure engineering and bridge length are selected as the possible factors affecting the carbon emission of the bridge. SPSS software was used to analyze the correlation between various influencing factors and bridge carbon emissions (see Table 3). Except for the bridge abutment quantity, all other factors are highly correlated, and the correlation between the quantity of each major structure and carbon emission is higher than that of bridge length. Among them, the quantity of superstructure engineering has the highest correlation with the carbon emission of the bridge, with a pearson's correlation coefficient of 0.975, followed by the quantity of pile foundation engineering, with a pearson's correlation coefficient of 0.955.

4.2 Regression Analysis of Carbon Emissions from Bridges

Influencing factors with correlation coefficients higher than 0.9 were selected as independent variables, and the fitting formula of bridge carbon emission was obtained through one-way linear regression analysis. The carbon emission of the bridge can be estimated by the engineering quantity of the main structures, and the length of the bridge can also be used for estimation when the basic data of engineering quantity is insufficient (see Table 4).

Table 3. Correlation analysis between potential influencing factors and bridge carbon emissions.

Relevant factor	Pearson correlation	Significance
Pile foundation engineering quantity	0.955	<0.001
quantity of bearing platforms	0.723	<0.001
Tie beam quantity	0.847	<0.001
Quantity of bridge abutment engineering	0.309	<0.001
Quantity of piers	0.925	<0.001
Quantity of abutment cover girder	0.934	<0.001
Superstructure engineering	0.975	<0.001
Bridge length	0.823	<0.001

Table 4. Equation for estimating carbon emissions from bridges.

Carbon footprint	Independent variable	Fitting formula	R^2
Total bridge carbon emissions	Pile foundation engineering quantity	$y = 3622.9x - 279252.115$	0.912
	Quantity of piers	$y = 6178.506x + 2381007.933$	0.856
	Quantity of abutment cover girder	$y = 10443.019x - 2414559.809$	0.871
	Superstructure engineering	$y = 3818.361x - 937543.693$	0.951
	Bridge length	$y = 53906.913x - 6694669.255$	0.675

By correlation analysis engineering quantity is the key factor affecting carbon emission. Therefore, for the carbon emission of each main structure of the bridge, the amount of structural engineering is chosen as the main influence factor, and the fitting formula of the carbon emission of each structure is obtained by one-dimensional linear regression analysis (see Table 5).

Table 5. Equation for estimating carbon emissions from major structures of bridges.

Carbon footprint		Independent variable	Fitting formula	R^2
Foundation engineering	Pile foundation	Pile foundation engineering quantity	y = 712.407x + 70012.725	0.973
	Abutment	Quantity of bearing platforms	y = 890.554x – 12729.087	0.992
	Linchpin	Tie beam quantity	y = 697.671x + 21566.67	0.982
Substructure	Bridge platform	Quantity of bridge abutment engineering	y = 825.751x + 19761.542	0.926
	Piers	Quantity of piers	y = 1013.969x – 65551.630	0.945
	Abutment cover girder	Quantity of abutment cover girder	y = 1236.803x – 37006.359	0.993
	Bearing cushion	Quantity of bearing cushion stone	y = 2498.105x – 95.751	0.999
Superstructure		Superstructure engineering	y = 1729.094x – 780867.766	0.911
Ancillary engineering	Superstructure ancillary engineering	Quantity of superstructure ancillary	y = 932.985x + 6202.752	0.999
	Fences	Quantity of fencing and netting	y = 557.979x + 2643.707	0.998

5 Conclusion

In this paper, the carbon emission of 105 bridges along a highway in Guangzhou City is studied and analysed by using the carbon emission accounting theory of the life cycle assessment. The bridges are divided into different structures according to the sub-projects, and the carbon emission measurement and analysis of the bridges according to the structural components are used to obtain the carbon emission intensity indexes of the main structures and the carbon emission estimation formula based on the engineering volume. The main conclusions are as follows:

The carbon emissions of bridge superstructures accounted for the highest proportion of 42.63%; followed by foundation works and lower structures accounting for 23.91% and 23.69%, respectively; and bridge ancillary structures accounting for 9.72%. The carbon emission per unit mileage in bridges is 35970.77 kg/km, and the carbon emission intensity value of a single bridge varies greatly, with the range distributed between 3828.98–177550.43 kg/km.

Calculating and analysing the carbon emission intensity of bridges with different superstructures, the carbon emission per kilometre of cast-in-place concrete steel bridges is as high as 110020.83 kg, while pre-stressed concrete small box girder and steel box girder bridges are closer to 33050.81 kg/km and 32177.34 kg/km, respectively, and cast-in-place concrete continuous girder bridges are 16174.3 kg/km.

Calculations were made to analyse the range of carbon intensity values (kg/m^3) for different superstructure forms, in descending order of steel box girder > cast-in-place concrete steel structure > cast-in-place continuous concrete girder > prestressed concrete small box girder > prestressed concrete T girder.

Analysing more types of prestressed concrete small box girder bridges in the project, the carbon intensity values of major structures such as pile foundations and piers were calculated, of which the carbon intensity of prestressed concrete small box girder ranges from 1489.15 to 1847.95 kg/m^3, and that of pile foundations ranges from 666.90 to 912.32 kg/m^3.

Through the correlation analysis, it is found that the engineering quantity is the main factor affecting carbon emission, and the correlation coefficient is higher than that of bridge length, and the fitting formula of the engineering quantity and carbon emission of the bridge and each structure was established as a method of estimating the carbon emission of each structural component of the bridge.

References

1. Wang, X., Wu, L., Yang, D.: Calculation and analysis of CO2 emissions from highway construction. Highw. Traffic Sci. Technol. 150–158 (2014)
2. Liu, Y., Wang, Y., An, D.: Life-cycle CO2 emissions and influential factors for asphalt highway construction and maintenance activities in China. Int. J. Sustain. Transp. 497–509 (2018)
3. Liu, Y., Wang, Y., Li, D., et al.: Life cycle assessment for carbon dioxide emissions from freeway construction in mountainous area: primary source, cut-off determination of system boundary. Resour. Conserv. Recycl. **140**, 36–44 (2019)
4. Huang, S., Huang, X., Gao, S., et al.: Carbon emission accounting for the whole process of highway construction based on LCA. Transp. Res. 72–80 (2022)
5. Liu, M., Dan, O.: Calculation method of life cycle carbon emission of bridge engineering. J. Civ. Archit. Environ. Eng. 125–129 (2011)
6. He, J., Chen, Y., Dong, X., et al.: Whole-life carbon emission characterisation of very large bridges based on LCA. Transp. Energy Conserv. Environ. Prot. https://kns.cnki.net/kcms/detail/10.1261.u.20221112.1521.008.html
7. Fernandez-Sanchez, G., Berzosa, A., Barandica, J., et al.: Opportunities for GHG emissions reduction in road projects: a comparative evaluation of emissions scenarios using CO2NSTRUCT. J. Clean. Prod. **104**, 156–167 (2015)
8. Gschosser, F., Wallbaum, H., Adey, B.T.: Environmental analysis of new construction and maintenance processes of road pavements in Switzerland. Struct. Infrastruct. Eng. **10**, 1–24 (2014)
9. Li, H., Peng, X., Zhang, J.: Highway life cycle carbon emission assessment and its sensitivity analysis: Highw. Eng. 132–137 (2021)

Study on Indoor Ventilation and Wind HCHO Emission Characteristics Based on Multi-zone Model and CFD

Huande Yang[1], Meng Liu[2], Jingjing Hua[1], and Jie Sun[2](✉)

[1] China Construction Eighth Bureau Decoration Engineering CO., LTD, Shanghai, China
[2] Shanghai Urban Construction Vocational College, Shanghai, China
13761710710@139.com

Abstract. The problem of formaldehyde (HCHO) pollution is particularly prominent in indoor environmental pollution, which is mainly due to the materials used in decoration. With the increasing pursuit of air quality, the problem of formaldehyde pollution has become more serious. In the field of decoration engineering, there are still defects in the prediction, simulation and evaluation of indoor air quality pollution. In this study, CONTAM software and CFD simulation technology were used to simulate indoor ventilation and formaldehyde distribution, aiming to enhance the effectiveness of pollution prevention and control. Through these simulation techniques, pollution problems can be identified and dealt with earlier, while reducing energy consumption and ensuring effective control of pollutants.

Keywords: IAQ · Formaldehyde · CONTAM

1 Introduction

The problem of indoor pollution caused by home decoration is becoming more and more serious, which has an important impact on human health and life. With the development of economy and the improvement of living standards, people's demand for indoor air quality is getting higher and higher. Research shows that modern people spend about 80% -90% of their time indoors [1–3]. How to control indoor pollutants and improve indoor air quality has become more and more popular. Traditional decoration projects often become an important source of indoor pollution [4]. Especially in the process of indoor decoration of residential and office space, the materials containing volatile organic compounds (VOCs) such as formaldehyde, such as synthetic resin, oil paint and adhesive, are widely used. Textiles commonly used in decoration, such as plywood, fiberboard, preservatives, flame retardants, and various types of carpets, may release formaldehyde for a long time after decoration, sometimes for decades [5]. It is also urgent to propose corresponding solutions to these problems.

Indoor air quality (IAQ) has an important impact on people's life and health. As one of the main sources of indoor pollution, formaldehyde has attracted people's attention. Indoor formaldehyde pollution is caused by the combined effect of outdoor air pollution and indoor air pollution. Formaldehyde not only has a stimulating effect on human

D. Han et al. (Eds.): ICEPG 2024, ESE, pp. 418–430, 2025.
https://doi.org/10.1007/978-3-031-86829-0_47

organs, skin mucosa and respiratory tract, but also causes allergic reactions in some people. It has been identified as carcinogenic by International Agency for Research on Cancer, IARC, and other organizations [6–8]. Many countries and environmental protection organizations have set strict formaldehyde content standards to control environmental exposure. The corresponding formaldehyde concentration limits are shown in Table 1:

Table 1. Indoor HCHO concentration limits specified by various countries and organizations.

Countries and organizations	limit	Assessment basis
GB/T 18883-2022	Civil Architectural Project Type I: 0.08 mg/m^3	Average of 1 h or 8 h
	Civil Architectural Project Type II: 0.12 mg/m^3	Average of 1 h or 8 h
GB50325-2010	Project acceptance	0.08 mg/m^3
Japan	0.10 mg/m^3	JIS A1901-2015 Average of 30 min
Entrustable Professional Activities (EPA)	$\leq$0.080 mg/m^3,	Based on 8-h time-weighted average concentration
WHO	0.10 mg/m^3	Total population, 30 min guidance limit
EU	E1 $\leq$ 0.124 mg/m^3	EN717-1:2004
	E2 $\leq$ 1.2 mg/m^3	Climate box method

Indoor formaldehyde mainly comes from wood-based panels, floor tiles, wallpaper and other decoration materials with urea-formaldehyde resin as adhesive. In addition, daily necessities such as fabrics release formaldehyde over time, which also poses a threat to indoor air quality. With the improvement of living standards, it is increasingly common for families to purchase new houses or renovate old houses. Plywood, carpets, thermal insulation materials, wood products and other formaldehyde-containing materials are widely used in decoration projects [5]. Due to the continuous release of formaldehyde during the use of these phenolic resin materials, the problem of indoor environmental pollution is aggravated. Hun et al. [9] analyzed many residential data and found that the influence of indoor environment on the initial formaldehyde concentration was as high as 90%, which was significantly higher than that of outdoor environment.

In the field of decoration engineering, there are still many deficiencies in the industry. Firstly, it is in response to the prediction, simulation and evaluation of indoor air quality pollution before decoration. Secondly, the problem of insufficient simulation and prediction of single area and multi area is particularly obvious. It may lead to insufficient monitoring and control of indoor air quality in the construction stage, and it is impossible to find and deal with the pollution sources introduced by the construction site in time, which leads to the lack of necessary evidence support when facing air pollution problems after the handover of the project. CONTAM is the most widely used and validated multi-zone model software for studying airflow and pollutant distribution in various buildings with different ventilation strategies. It is used to solve multi-regional pollutants in this article [10].

In this paper, CFD simulation method is used to simulate and predict an indoor area. CONTAM software is used to simulate ventilation and pollutants in multi-area. Multi-area model is selected to simulate the distribution of indoor pollutant concentration represented by formaldehyde. In the calculation method of natural ventilation, it includes simplified algebraic algorithm, computational fluid dynamics (CFD) method and multi-area network model method. Compared with other methods, CONTAM has the calculation ability of fast simulation, simple setting and easy calculation. Using simulation methods to monitor, predict and evaluate indoor air quality can detect and deal with pollution problems in a timely manner. Improve the active prevention effect in decoration construction, while reducing energy consumption, so as to achieve effective treatment of formaldehyde and indoor air pollutants.

2 Mathematical or Physical Models Involved in the Simulation Process

2.1 Diffusion Model of CONTAM Airflow and Air Pollutants

In the multi-zone model software CONTAM, there are many element models, including the infiltration, leakage and room airflow in the mechanically driven building system, the wind pressure acting on the outside of the building, and so on. Also contains pollutant concentrations, the diffusion of air pollutants transported by these airflows; it is generated by various source mechanisms through various process transformations, including chemical and radiochemical transformations, adsorption and desorption of building materials, filtration and deposition to the building surface, etc. [11].

Airflow Element Model

In the application of CONTAM software, air flow channels such as doors and windows are often defined as "Airflow path", A list of available airflow elements or mathematical models that provide the relationship between airflow and pressure difference for airflow paths. CONTAM performs airflow calculations using the mass flow powerlaw formula,

$$F = C\Delta P^{n} \tag{1}$$

for the following types of airflow elements. Where F is the mass flow rate, P is the pressure difference across a flow path, C is the flow coefficient and n is the flow exponent.

One-way Flow Using Powerlaw Models includes Orifice Area Data and Leakage Area Data. In fluid mechanics, the relationship between the orifice area and the flow rate can usually be described by the orifice flow equation. For Newtonian fluids, this relationship can be derived from the Bernoulli equation and the continuity equation of orifice flow [12].

$$\dot{V} = C_d A\sqrt{\frac{2}{\rho}}\Delta P^{m} \tag{2}$$

where $C_d \in (0,1)$is the dimensionless flow coefficient, and ρ is the density of the medium flowing through the orifice. The characteristic of the large opening is that m is very close to 0.5, while the value of the small crack opening is close to 0.65.

Michael Wetter et al. [13] converted the effective air leakage area into the orifice equation as follows. Let L denote the effective air leakage area in m2. The area A in the orifice equation is expressed by

$$\dot{V} = C_{d,r} L \sqrt{\frac{2\Delta P_r}{\rho}} \tag{3}$$

from which follows that

$$A = \frac{C_d}{C_{d,r}} L \Delta P_r^{0.5-m} \tag{4}$$

if the m value is not reported in the test results, Dols and Walton [14] recommend m = 0.6 to m = 0.7.

Assuming that the air density of each room is constant, the hydrostatic equation is used to correlate the pressure at different heights in each room:

$$P_{0j} = P_j + \rho_j g \left(h_j - h_0\right) \text{ and } P_{0i} = P_i + \rho_i g (h_i - h_0) \tag{5}$$

$$P_j(y) = P_{0j} - \rho_j g y \text{ and } P_i(y) = P_{0i} - \rho_i g y \tag{6}$$

According to Brown and Solvason [15], assuming that the airflow velocity is a function of height, given by the orifice equation:

$$V(y) = C_d \left(\frac{2(P_j(y) - P_i(y))}{\rho}\right)^{\frac{1}{2}} \tag{7}$$

Two Opening in Two-way Flow Models can construct an approximate model by considering the opening between two regions, which have no flow path to any other region. Then the flow in each direction must be equal and $P_{0j} = P_{0i}$. Using Eq. (6) to integrate Eq. (7) from y = 0 to y = H/2 to calculate ΔP, the flow through the upper half of the opening is given.

This is equivalent to a simple power-law model.

$$C_0 = C_d \sqrt{2} \left(\frac{WH}{2}\right) \tag{8}$$

where P_{0j}, P_{0i} are the pressure in the j and i regions when y = 0, the reference elevation of the opening, ρ_j, ρ_i are the air density in the j and i regions, P_j, P_i are the reference pressure in the j and i regions, h_j, h_i are the reference elevation in the j and i regions, and the h_0 elevation in the center of the opening, C_d is the flow coefficient, ρ is the air density through the opening.

CONTAM provides a lot of source models, such as constant coefficient model, pressure driven model, cut-off concentration model, attenuation source model, power law model and so on. In this study, the source model uses constant coefficient model. The constant coefficient or general source/sink model uses the following equation:

$$S_\alpha(t) = G_\alpha - R_\alpha C_\alpha(t) \tag{9}$$

where G_α is the generation, R_α is the removal, S_α is called the contaminant α "source strength".

2.2 Numerical Simulation Model of Indoor Formaldehyde

Gaseous pollutants are generally emitted from the surface or splicing of furniture or decorative materials, and then spread to the breathing area where the person is located, thereby reducing the empty quality and affecting the human body. The total amount of pollutants released from the interior of the decoration materials is mainly determined by the internal diffusion of the materials and the convective diffusion of the gas-solid interface.

Theoretical Model

The theoretical model is based on the mass transfer theory in physics, and the diffusion mechanism is clear. The pressure gradient, temperature gradient and concentration gradient are used as the driving force for the transfer of pollutants inside the material. Tichenor et al. [16] first proposed the VB model determined by the vapor pressure of the volatile organic compound VOC emitted by the pollution source and the boundary layer of the air on the surface:

$$R = h_{\mathrm{D}}\left(\frac{C_{V_0}M}{M_0} - C\right) \tag{10}$$

where R is the emission rate of pollutants (mg/m^2·h), h_{D} is the convective mass transfer coefficient, M and M_0 are the residual emission and initial emission of the pollution source VOC, respectively, and C_{V_0} is the initial vapor pressure.

The VB model mainly describes the emission process of 'wet material' dominated by evaporation in the short term, and has a large error in the description of the emission process of 'dry material'. Yang et al. [17] studied the VOC emitted from the surface of painted coatings and proposed a four-layer mass transfer model:

$$\frac{\partial(\rho c)}{\partial t} + \frac{\partial(\rho u_j c)}{\partial x_j} = \frac{\partial}{\partial x_j}\left(\frac{\mu}{S_c}\frac{\partial c}{\partial x_j}\right) + \rho S_V \tag{11}$$

In the equation: ρ is the air density, c is the concentration of the target pollutant, u_j is the velocity component, and S_c is the dimensionless mass transfer Schmidt number. ρc is the mass concentration of the target pollutant, and D is the diffusion coefficient of the pollutant. The diffusion coefficients are also quite different for different types of target pollutants.

Basic Control Equation

Mass conservation equation. The mass conservation equation of three-dimensional compressible fluid can be obtained by the law of conservation of mass:

$$\frac{\partial\rho}{\partial t} + \mathrm{div}(\rho U) = 0 \tag{12}$$

The momentum conservation equation can be obtained by the law of conservation of momentum:

$$\frac{\partial(\rho i)}{\partial t} + \mathrm{div}(\rho UU) = \mathrm{div}(\mu \cdot gradU) + S_n - \frac{\partial\rho}{\partial \boldsymbol{n}} \tag{13}$$

This equation is referred to as the momentum equation, also known as the Navier-Stokes equation.

Energy conservation equation:

$$\frac{\partial(\rho i)}{\partial t} + \mathrm{div}(\rho \boldsymbol{U} i) = \mathrm{div}(\lambda \cdot gradT) - p \cdot \mathrm{div}(\boldsymbol{U}) + \Phi + S_T \tag{14}$$

Of which:

$$\Phi = \mu \left\{ \begin{array}{c} 2\left[\left(\frac{\partial u}{\partial x}\right)^2 + \left(\frac{\partial v}{\partial y}\right)^2 + \left(\frac{\partial w}{\partial z}\right)^2 \right] \\ + \left(\frac{\partial u}{\partial y} + \frac{\partial v}{\partial x}\right)^2 + \left(\frac{\partial u}{\partial z} + \frac{\partial w}{\partial x}\right)^2 + \left(\frac{\partial v}{\partial z} + \frac{\partial w}{\partial y}\right)^2 \end{array} \right\} + \lambda(\mathrm{div}(U))^2 \tag{15}$$

Fluid state equation:

$$p = p(\rho, T); \; i = i(\rho, T) \tag{16}$$

In the formula: $U = u\mathbf{i} + v\mathbf{j} + w\mathbf{k}$ u, v, w are the components of the flow velocity in the x, y, z coordinate direction respectively; ρ is the fluid density; μ is the dynamic viscosity of the fluid (pa·s); i is the internal energy of fluid; λ is the thermal conductivity; p is the pressure acting on the fluid element; S_n is the source of fluid in all directions; S_T is the heat source.

The component mass conservation equation is used to study the transport and combustion of chemical components. According to the law of conservation of component mass, the mass conservation expression of component i is as follows:

$$\frac{\partial(\rho c_i)}{\partial t} + \mathrm{div}(\rho U c_i) = \mathrm{div}(D_i \cdot grad(\rho c_i)) + S_i \tag{17}$$

where: c_i is the volume concentration of i component; ρc_i is the mass concentration of component i (kg/m^3); D_i is the diffusion coefficient of i component; S_i is the production rate of i component in the system (kg/m^3).

3 Research Object and Method

3.1 CONTAM Physical Model Selection

This study combines CONTAM, a multi-area network software program developed by the National Institute of Standards and Technology (NIST). CONTAM is a popular tool for determining building air infiltration, exudation, and room-to-room airflow driven by external wind pressure on the building, buoyancy effects associated with indoor and outdoor air temperature differences, and mechanical ventilation [18]. It can also predict the diffusion of air pollutants and can be used to calculate the degree of personal exposure to pollutants.

In the CONTAM software, the room of an apartment in Nanjing area is selected as the simulation object to simulate the wind pressure. The room plan is shown in the picture (see Fig. 1), and the room size parameters are shown in the Table 2:

In the wind pressure simulation, the area definition is first carried out, the window is set to the Two Opening original in the Two-way Flow Models, the door uses the Leakage Area Data, and the external wind pressure uses the typical setting Tall building roof, surface averaged, L/W = 1/2 in the database provided by the CONTAM official.

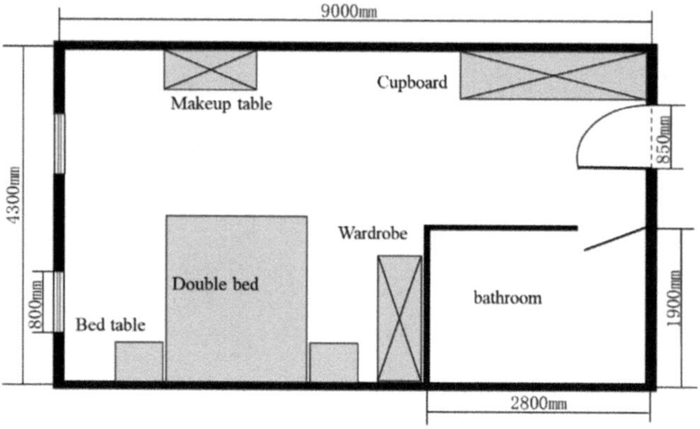

Fig. 1. Plan of an apartment in Nanjing.

Table 2. Furniture facilities of an apartment in Nanjing.

Furniture parameters that may release formaldehyde	
Furniture facilities in the room	Size: length × width × height (mm)
Double bed	2000 × 1800 × 180
Makeup table	800 × 400 × 160
Bed table	460 × 400 × 320
Wardrobe	1600 × 500 × 1900
Cupboard	1600 × 350 × 800
Note: Room area is 40 m^2; the room height is 2900 mm	

The wind pressure coefficient corresponds to the time of 24 h without considering the influence of the weather and the influence of the mechanical exhaust of the toilet.

The civilian area map (with a total area of 143 m^2) of a project in a certain area of Shanghai was selected as the simulation object to simulate the release of pollutants. The room plan is displayed in the CONTAM model as shown in Fig. 2, and the room definition area is shown in the Table 3:

When conducting multi-area pollutant simulation, the region definition is first performed, and the window is set to the original Orifice Area Data. The door uses the Leakage Area Data, and the external wind pressure uses the typical setting in the database in CONTAM. The weather effect is not considered and the external wind pressure is not set. The simulation time is 24 h.

In order to simplify the model, CONTAM has the following ideal assumptions. The air in each room is fully mixed; the convection and diffusion of air between rooms adopt one-dimensional model. When performing steady-state simulations, the mass of air in each room is conserved; on the ventilation path, the relationship between air flow rate and pressure difference adopts exponential law or quadratic relationship.

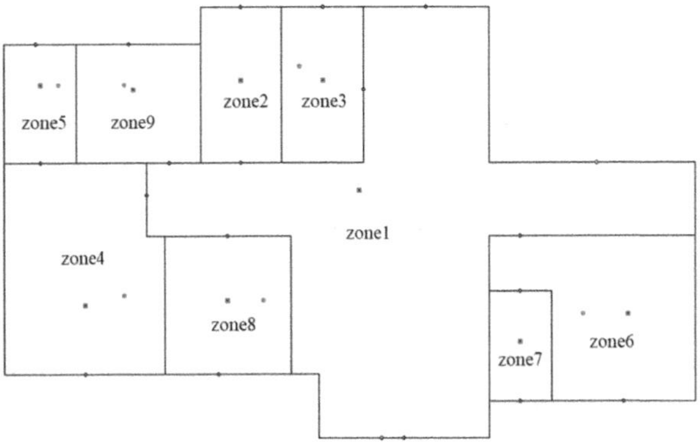

Fig. 2. The room Model in CONTAM

Table 3. Zone number and room

zone1	zone2	zone3	zone4	zone5
Living room, etc.	toilet	kitchen	Master bedroom	Master bedroom toilet
zone6	zone7	zone8	zone9	
Second bedroom	Second bedroom bathroom	Children's room	Study room	

3.2 Numerical Simulation of Physical Model

The three-dimensional physical model of numerical simulation is shown (see Fig. 3). It is north-south, the x-axis direction points to the north, the z-axis positive direction points to the east, and the Y-axis positive direction is the height direction. The cabinet placed inside is the main source of formaldehyde emission. The indoor pressure is a standard atmospheric pressure (101,325 Pa); the background concentration of indoor formaldehyde is zero; in addition to formaldehyde, there is air as the background gas around the walls, ceilings, doors and windows for the insulation wall, ignoring the outdoor through the envelope structure to the indoor heat transfer.

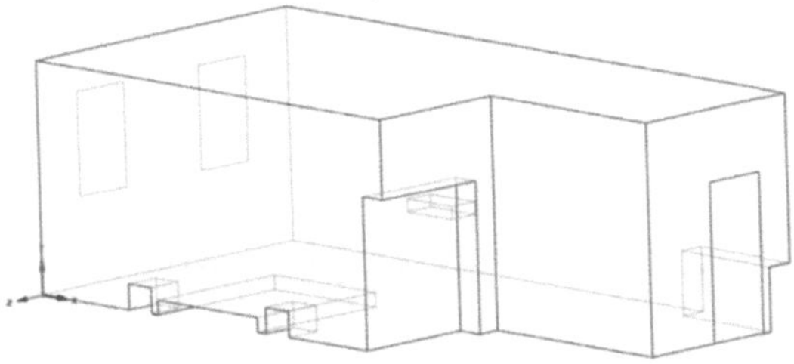

Fig. 3. The room Model in CFD

The release amount of formaldehyde chooses a constant value. The simulation involves a multi-component transport model of single-phase fluid, so the diffusion model of formaldehyde selects the turbulence model (standard k-ε two-equation model) and the species transport model without chemical reaction. The indoor mixed gas adopts incompressible ideal gas, and the density conversion of incompressible ideal fluid is calculated by ideal relationship. However, the flow of the fluid is still solved by the incompressible fluid calculation method, and the transient simulation is carried out by the pressure solver. Gravity acceleration is selected as -9.81 m/s^2 of Y axis. • Transient simulation for the time scale requirements, the number of iterations set to 500 steps to complete the convergence can ensure good convergence. The material properties of formaldehyde in Fluent: the relative molecular mass is 30.06 kg/kgmol, the enthalpy value is -1.159148e08 J/kgmol in the standard state, and the entropy value is 218593.6 J/(kgmol·K) in the standard state.

4 Result and Discussion

4.1 Analysis of CONTAM Results

As shown in the figure (see Fig. 4), the non-steady-state formaldehyde concentration release was simulated for 24 h. After 4 h, the formaldehyde concentration tended to be stable. The toilet, study and children's room were higher than 0.11 mg/m^3, and the specific result was 0.11875 mg/m^3. The formaldehyde concentration in the master bedroom and living room was higher than 0.08 mg/m^3, and the specific result was 0.092 mg/m^3. The concentration of the master bedroom toilet and the second bedroom toilet was less than 0.3 mg/m^3, which may be not connected to the living room corridor, and the formaldehyde concentration in the kitchen was lower than 0.2 mg/m^3.

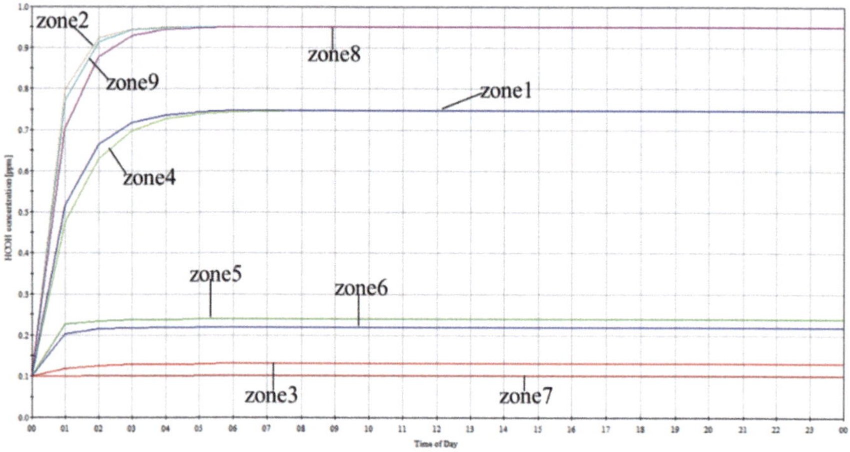

Fig. 4. Simulation results of formaldehyde concentration in CONTAM

The air flow path at the gap is selected from the effective penetration area model. According to the field measurement, the window height is 1.5 m, the width is 0.8 m, and the height of the single open door is 1.95 m, and the width of the door is 0.8 m. The width of the gap at the top and bottom of the door is about 1.25 cm, and the unit penetration c length parameter at the gap of the door frame is calculated to be 0.1 m/m. After 24 h of unsteady simulation, the CONTAM calculation results are shown in the figure (see Fig. 5). The red line represents the pressure difference, the longer the line, the greater the pressure difference; the line extending to the area indicates negative pressure. The green line indicates airflow, and the longer the line, the higher the airflow. Case1 is the wind pressure state without mechanical ventilation in the toilet, case2 is the wind pressure state with mechanical exhaust in the toilet, and the wind speed is set to a constant 1.5 m/s to simulate the wind pressure results.

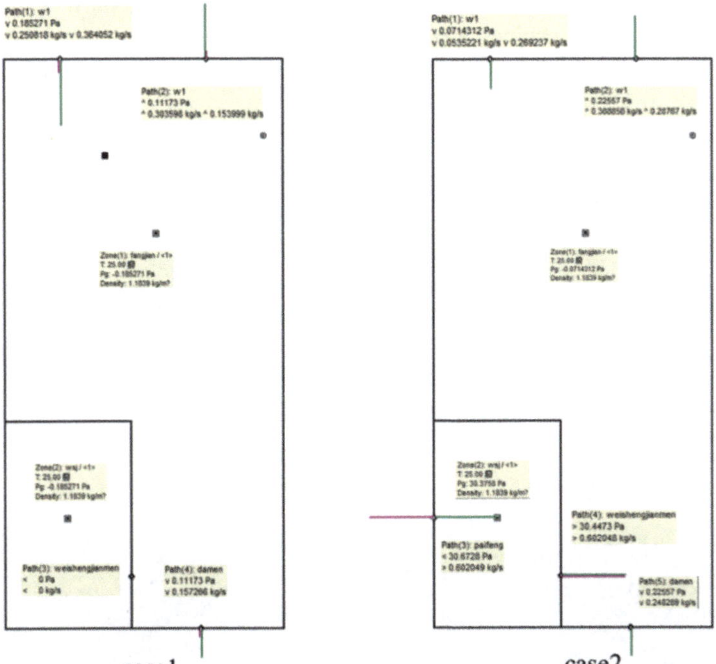

Fig. 5. Room pressure and wind direction simulation results in CONTAM

When the zone2 toilet does not set the exhaust air, the original door of the zone2 has no positive and negative pressure. When the exhaust air is set, the pressure at the door of the zone2 rises, and the influence of mechanical exhaust on indoor ventilation is more significant. Both cases show the airflow direction and pressure direction under natural ventilation, which is helpful for the selection of sampling points for natural ventilation to dilute air pollutants and formaldehyde detection of air pollutants.

4.2 Numerical Simulation Results Analysis

As shown in the diagram (see Fig. 6), using flunet simulation results, the diagram shows the mass concentration cloud map of formaldehyde, and the mass concentration of the color bar increases from bottom to top. At 0.5 m, it is close to the human respiratory area (0.8–1.5 m). The wardrobe and kitchen cabinets characterize higher mass concentrations (close to 0.14 mg/m^3), 1.5 m–2 m (higher than the human respiratory area), and the formaldehyde mass concentration is lower (less than 0.1 mg/m^3). It is consistent with the above 5-point sampling results. The measurement has reference value for the early monitoring of the decoration process, and has certain guiding significance for the decoration evolution strategy. For example, in the middle stage of decoration, the corresponding aldehyde removal and promotion strategy should be made for the furniture with large surface area of wooden furniture. In the later stage, the purifier should be as close as possible to the cabinet, especially the wardrobe and cabinet, and the corner of the wall with poor air circulation, and should be set within 1.5 m breathing as far as possible.

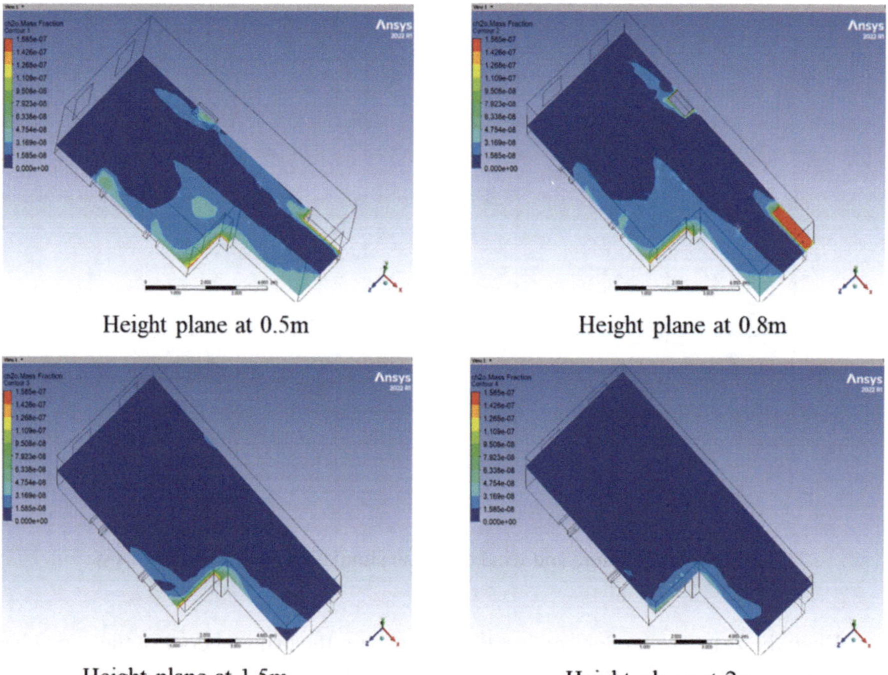

Height plane at 0.5m

Height plane at 0.8m

Height plane at 1.5m

Height plane at 2m

Fig. 6. Formaldehyde concentration mass cloud chart in FLUENT

The simulation results have reference value for the monitoring and distribution of the decoration process in the early stage, and have certain guiding significance for the decoration evolution strategy. For example, in the middle stage of decoration, the corresponding formaldehyde removal and promotion strategy should be made for the furniture with large surface area of wooden furniture. In the later stage, the purifier should be as close as possible to the cabinet body, especially the wardrobe and cabinet, and the corner of the wall with poor air circulation, and should be set within 1.5 m breathing as far as possible.

5 Conclusion

In the process of decoration process, there are still many deficiencies in the prevention and treatment of the influence of indoor air pollutants. In this paper, CONTAM multi-area model software is used to study the ventilation examples. The ventilation examples use two cases to analyze the influence of indoor mechanical exhaust on indoor ventilation. The study of formaldehyde diffusion strength is to select a large multi-room civil residential building sub-region for simulation analysis. The application of numerical simulation is to obtain the cloud map of formaldehyde mass concentration at different heights. From the cloud map, it is known that the formaldehyde mass concentration is higher at the height of 0.8–1.5 m in the case of natural ventilation, and the formaldehyde concentration above 2 m does not show a higher concentration. The simulation results provide a reference value for the decoration process, and the prediction results provide a basis for the treatment of formaldehyde and air pollutants in the whole process in the early, middle and late stages of the decoration process.

References

1. DeCastro, B.R., Sax, S.N., Chillrud, S.N., Kinney, P.L., Spengler, J.D.: Modeling time-location patterns of inner-city high school students in New York and Los Angeles using a longitudinal approach with generalized estimating equations. J. Exposure Sci. Environ. Epidemiol. **17**(3), 233–247 (2007)
2. Hussein, T., Paasonen, P., Kulmala, M.: Activity pattern of a selected group of school occupants and their family members in Helsinki—Finland. Sci. Total. Environ. **425**, 289–292 (2012)
3. Klepeis, N.E., et al.: The National Human Activity Pattern Survey (NHAPS): a resource for assessing exposure to environmental pollutants. J. Exposure Sci. Environ. Epidemiol. **11**(3), 231–252 (2001)
4. Schweizer, C., et al.: Indoor time–microenvironment–activity patterns in seven regions of Europe. J. Exposure Sci. Environ. Epidemiol. **17**(2), 170–181 (2007)
5. Huang, S., Wei, W., et al.: Indoor formaldehyde concentrations in urban China: preliminary study of some important influencing factors. Sci. Total. Environ. **590–591**, 394–405 (2017)
6. IARC Monographs on the Evaluation of Carcinogenic Risks to Humans-Wood Dust and Formaldehyde. IARC Working Group, Lyon (1995)
7. IARC Monographs on the Evaluation of Carcinogenic Risks to Humans-Formaldehyde, 2-Butoxyethanol and 1-tert-Butoxypropan-2-ol. IARC Working Group, Lyon (2006)
8. NTP Response to Issues Raised in the Public Comments for Candidate Substances for the 12th Report on Carcinogens. National Toxicology Program, Bethesda (2011)
9. Hun, D.E., Corsi, R.L., Morandi, M.T., et al.: Formaldehyde in residences: long–term indoor concentrations & influencing factors. Indoor Air **20**(3), 196–203 (2010)
10. Emmerich, S.J.: Validation of multizone IAQ modeling of residential-scale buildings: a review. Trans.-Am. Soc. Heat. Refrig. Air Cond. Eng. **107**(2), 619–628 (2001)

11. National Institute of Standards and Technology U.S Department. www.nist.gov
12. Clarke, J.A.: Energy Simulation in Building Design, 2nd edn. Butterworth-Heinemann, Oxford (2001)
13. Wetter, M.: Multizone Airflow Model in Modelica. The Modelica Association. Modelica, 4th–5th September 2006
14. Dols, W.S., Walton, G.N.: CONTAM 2.0 user manual, multizone airflow and contaminant transport analysis software. Technical Report NISTIR 6921, National Institute of Standards and Technology, November 2002
15. Brown, W.G., Solvason, K.R.: Natural convection through rectangular openings in partitions - 1: vertical partitions. Int. J. Heat Mass Transf. **5**, 859–867 (1962)
16. Tichenor, B.A., Guo, Z., Sparks, L.E.: Fundamental mass transfer model for indoor air emissions from surface coatings. Indoor Air **3**(4), 263–268 (1993)
17. Yang, X., Chen, Q., Zeng, J., et al.: A mass transfer model for simulating volatile organic compound emissions from 'wet' coating materials applied to absorptive substrates. Int. J. Heat Mass Transf. **44**, 1803–1815 (2001)
18. Walton, G.N., Dols, W.S.: CONTAMW 2.4 User Manual, p. 286. National Institute of Standards and Technology, Gaithersburg (2008)

Carbon Reduction Benefit of Green Highway Construction in Dense Urban Areas

Hongjun Li[1](✉), Lei Zeng[2], Wang Tan[2], Jinsheng Fu[1], and Te Song[2]

[1] China Academy of Transportation Sciences, 240 Huixinli, Chaoyang District, Beijing, China
196962282@qq.com

[2] Guangzhou Expressway Co., Ltd., Guangzhou 510290, Guangdong, China

Abstract. In order to improve the level of green construction in urban dense areas, green highway construction is carried out relying on the Nanzhong expressway. Implement the concept of green development into the whole process of highway design, construction and operation. The highway comprehensively develops and applies resource-intensive conservation, intelligent construction, quality engineering construction, and green highway technology to reduce the overall carbon emissions of highway construction, improve the capacity of highway greening and carbon fixation, and provide a reference for the sustainable development of other highly urbanized regions.

Keywords: Urban compact area · Green highway · Nanzhong Expressway · Carbon reduction benefit

1 Introduction

In the face of the grim situation of global climate change, China has formulated and implemented a national strategy to actively respond to climate change and continuously increased its nationally determined contribution [1]. In October 2021, The State Council issued the "Carbon Peak before 2030 action Plan" [2], focusing on the goal of carbon peak proposed ten actions, put forward the "transport green low-carbon action" and "carbon sink capacity consolidation and improvement action" two actions, indicating the goal of green and low-carbon development of highway construction.

As China's urbanization has shifted from rapid development to high-quality development, green infrastructure construction has become an important part of the sustainable development of urban agglomerations. Green highway is a specific starting point for the construction of transportation infrastructure to implement the national dual-carbon strategy, and has been highly valued by the industry. As the province with the largest total mileage of expressways opened to traffic in China, Guangdong Province has always attached importance to the construction of green highways, taking the lead in issuing the Implementation Plan for Promoting the Construction of Green Highways in Guangdong Province and the Technical Guide for the Construction of Green Highways in Guangdong Province (Trial), and has successively set up two batches of typical demonstration

D. Han et al. (Eds.): ICEPG 2024, ESE, pp. 431–438, 2025.
https://doi.org/10.1007/978-3-031-86829-0_48

projects for green highways. Through the construction of various green highway demonstration projects, the green and low-carbon development of transportation infrastructure will be realized, and the goal of carbon neutrality in transportation will be realized.

Nansha to Zhongshan Expressway Project (hereinafter referred to as "Nanzhong Expressway") is an important part of the national expressway network G2518 and Guangdong Expressway S78, with a total length of 32.4 km, located in the core location of Guangdong-Hong Kong-Macao Greater Bay Area, connecting Guangzhou, Shenzhen and Zhongshan. The viaduct of the whole line of the Nanzhong Expressway is a typical representative of the new development mode of the highway of the whole line of the urban agglomeration in the future. Taking the construction of this project as an example, the research sorted out various green construction technologies in the construction of this project, analyzed their carbon reduction benefits, quantified the green construction effects of the construction of engineering highways, in order to provide green and low-carbon development suggestions for the construction of similar environmental highways.

2 The Green Construction Technology

2.1 The Intensive Conservation of Resources Technology

The Nanzhong Expressway makes overall use of resources, makes full use of the space above the municipal road, adopts the large cantilever structure to build the south extension of Waihuan West Road, and saves land covering an area of 310 mu.

The Nanzhong Expressway achieves intensive and economical utilization of land resources by means of route selection scheme comparison, engineering scheme comparison, interworking scheme optimization, construction walkway and material transportation, which reduces the total mileage of the project by 3.102 km and saves an area of 874 mu.

2.2 The Intelligent Construction Technology

The Nanzhong Expressway is a bridge project, consisting of three independent super large Bridges, seven interconnecting, with a total of nearly 13,000 trusses of beams. According to the traditional beam making process, the site is huge and the production cycle is long, which will seriously restrict the construction progress of the project. In order to solve these problems, Nanzhong Expressway established a prefabricated bridge production and development base, developed a prefabricated beam ring production line, and realized the factory assembly line construction of prefabricated beam production, and the bench production capacity was increased from the conventional 10 days/piece to 1 day/piece, greatly improving production efficiency.

2.3 The Quality Engineering Construction Technology

Nanzhong Expressway is located in the estuary of the Pearl River, in the Marine environment of high salinity, high temperature and high humidity, long-term chloride ion erosion of bridge piers will cause steel corrosion, greatly affecting the durability of concrete, resulting in cracking of large-scale concrete in Marine engineering.

According to the characteristics of the Nanzhong Expressway project and the influencing factors of concrete quality, a special study on the technology and quality control of high-performance concrete is carried out. Through research to ensure that the work performance, mechanical properties, durability and appearance quality of concrete meet the design requirements, the formation of high-performance engineering concrete preparation, construction and quality control of a complete set of technologies.

2.4 The Green Highway Technology

The Nanzhong Expressway takes full account of the deep soft soil conditions and climatic conditions deposited by the sea and land in the Greater Bay Area. It crosses the ocean, water network, wetland reserve, etc., and adopts elevated layout with limited landscape greening conditions. However, based on the goal of reducing carbon and increasing sinks, the project carries out landscape greening in the landing green areas such as toll station entrances, service areas, and Spaces under Bridges. Increase the green area as much as possible, effectively improve the overall green amount of roads, increase the potential of road greening to reduce carbon and sequester carbon, and broaden the idea of building low-carbon roads and zero-carbon roads.

3 Energy Saving and Carbon Reduction Benefits

Based on the LCA theory and combined with carbon source analysis, the carbon emission reduction through optimized design of expressways mainly includes three parts: one is to save energy and reduce carbon emission by reducing construction machinery and equipment activities; the other is to reduce carbon emission by reducing road construction material consumption; and the third is to reduce soil organic carbon release by saving land area. The calculation formula is as follows:

$$\Delta C_{OD} = \Delta C_E + \Delta C_M + \Delta C_L \tag{1}$$

ΔC_{OD} - Carbon emission reduced by optimal design.

ΔC_E - Carbon emissions reduced by energy conservation through optimal design. Calculated by statistical summary of each type of energy saving amount multiplied by the corresponding carbon emission factor.

ΔC_M - Reduced carbon emissions through optimized design and material saving. By statistical summary of the saving amount of each type of material multiplied by the corresponding carbon emission factor calculation.

ΔC_L - Reduced carbon footprint through optimized design. The carbon emission reduction of this project is significantly less than the other two projects, so it is ignored in the calculation.

3.1 The Carbon Reduction Benefits of Resource Intensive Saving Technology

The entire line of the project adopts the elevated exhibition line, and the mileage is reduced by 3.102 km through line optimization, reducing the area of about 520.05 mu.

The total carbon emission of a single kilometer viaduct is calculated according to the consumption of machinery, energy and materials of the bridge under construction, and the carbon reduction caused by reducing the road mileage is calculated by analogy. Table 1 is used to calculate the actual activity level of material and energy consumption of engineering construction, and then carbon emissions of different materials are calculated according to the carbon emission factors of different materials and energy sources, so as to obtain the total carbon emissions of bridge construction. The carbon emission factor of fossil fuel was calculated according to the Calculation Standard for Carbon Emission of Buildings [3] and the General rules for calculation of the comprehensive energy consumption [4]. The carbon emission factor of electric power was calculated according to the recommendation factor of Guangdong Provincial Department of Ecology and Environment 0.4512 $kgCO_2e$/kw·h. The carbon emission factors of 453 materials were calculated using the "Building Carbon Emission Calculation Standard", CLCD, the greenhouse gas emission coefficient database of Chinese products throughout their life cycle (2022), the EU ELCD database, and the EFDB database of the UK Department of Environment. Carbon emission factor of machinery and equipment Combined with the "Highway Construction Machinery Shift Cost Quota" [5], the energy consumption amount and fossil fuel carbon emission factor of 972 kinds of machinery and equipment used in the process of highway construction are calculated.

Table 1. The material consumption and energy consumption statistics report.

Construction site	Accomplished amount	Power consumption (kWh)	Fossil fuel (L)	Material 1	Material 2	Material 3

According to the statistical calculation of actual material consumption and energy consumption, the carbon emission factor of electricity, diesel and other energy consumption in the construction of small concrete box girder per square meter is 2554.94 $kgCO_2$, and the carbon emission factor of steel bar, cement, sand and gravel, asphalt and other materials is 43162.35 $kgCO_2$. In the construction process, it is necessary to carry out the construction of the construction road. According to the existing construction specification of the road (4.5 m wide), the carbon emission factor of energy consumption such as electricity and diesel oil per meter of the construction machinery and equipment of the road construction is 16.18 $kgCO_2$, and the carbon emission factor of consumption of cement, sand and gravel is 304.03 $kgCO_2$.

According to the calculation that the length of the expressway route is shortened by 3.102 km, the carbon emission of the main construction is reduced by 141815.03 tCO_2, of which the carbon emission of materials accounts for 94.41%. According to the optimized design of the construction drawing, the reduction of mileage needs to reduce the new construction road by about 6 km, thereby reducing carbon emissions of 1921.26 tCO_2, of which the carbon emissions of material application account for 94.95%. Due to the reduction of mileage, the total carbon emission is reduced by 143736.29 tCO_2, and the carbon emission is estimated to be reduced by 276.39 t/mu according to the land area. According to the data, the total saving area of 874 mu can save carbon emissions of 241564.31 tCO_2.

Using the large cantilever structure to save an area of 310 mu, according to the land area estimated to reduce carbon emissions of about 276.39 t/mu, it can be seen that this measure reduces carbon emissions of about 85680.71 tCO_2.

3.2 The Carbon Reduction Benefits of Smart Construction Technology

The carbon reduction benefit of the precast beam ring production line can be measured by the difference between the carbon emissions corresponding to the actual activity level and the carbon emissions corresponding to the theoretical activity level, in which the actual activity level is derived from the data of the material and energy consumption report of the construction process, and the theoretical activity level is derived from the construction drawing design budget. In order to reduce the difficulty of data acquisition, based on the analysis results of carbon emission contribution characteristics, materials and energy types with the largest carbon emission contribution are selected for comparison. According to the calculation application, the carbon emission of the precast beam ring production line is about 178,400 tCO_2e, and the specific calculation is shown in Table 2.

Table 2. Carbon reduction benefit of precast beam ring production line.

		Materials, energy consumption		Carbon emission factor	Carbon reduction
		Budget	Actual	$kgCO_2e$	tCO_2e
HPB300 rebar	t	710		2340	−1661.4
HRB400 rebar	t	109497	88990	2340	−47986.38
Steel strand	t	21524	16633	2375	−11616.13
Grade 32.5 cement	t	867	0	630	−546.21
Grade 42.5 cement	t	21441	85861	864	55658.88
Grade 52.5 cement	t	229320	61568	981	−164564.71
Sand	m^3	231326	241239	3.765	37.32
Gravel	m^3	50726	370203	3.27	1044.69
Water	m^3	856812	85857	0.168	−129.53
Electricity	Kw • h	23877200	6614474	0.4512	−7788.94
Diesel oil	L	199441	245075	3.77	−843.06
Total					−178395.47

3.3 The Carbon Reduction Benefit of Quality Engineering Construction Technology

Based on the principle of compact particle packing, a new cementing material system of "cement + limestone powder + special active admixture" is proposed based on the raw materials in Guangzhou area and the characteristics of the project in Nanzhong Expressway to carry out relevant research on high-quality and high-performance concrete. The specific parameter ratio is shown in Table 3.

Table 3. High quality and high performance concrete mix ratio (kg/m^3).

ID	Cement	Admixture		Sand	Gravel	Water	Water reducing agent
C1	252	105	63	740	1109	151	4.6
C2	252	84	84	739	1090	151	5.04

Compared with conventional "slag powder + fly ash" concrete (the mix ratio is shown in Table 3), the carbon reduction benefits are analyzed.

Due to the similar production process, the difference in carbon emission between the two is mainly caused by the material production stage. As can be seen from Table 4 and Table 5, concrete "cement + limestone powder + special active admixture" reduces 14.02 $kgCO_2e$ per cubic meter compared with conventional concrete "slag powder + fly ash". About 491,000 m^3 of high quality and high performance concrete is planned to be applied in the whole line of Nanzhong Expressway, and the cumulative carbon emission is reduced by 6883.82 tCO_2e.

Table 4. "Cement + limestone powder + special active admixture" concrete carbon emission.

ID	Material	Unit	Quantity	Carbon emission factor ($kgCO_2e$/kg)	Carbon emission ($kgCO_2e$)
1	Cement	kg	252	0.864	217.73
2	Special active admixture	kg	105		
3	Limestone powder	kg	63	0.00218	0.14
4	Natural river sand	kg	740	0.00251	1.86
5	Gravel	kg	1109	0.00218	2.42
6	Water	kg	151	0.000168	0.03
7	Water reducing agent	kg	4.6	0.02849	0.13
Total					222.30

Table 5. Carbon emission of "slag powder + fly ash" concrete.

ID	Material	Unit	Quantity	Carbon emission factor ($kgCO_2e/kg$)	Carbon emission ($kgCO_2e$)
1	Cement	kg	252	0.864	217.73
2	Slag powder	kg	84	0.0844	7.09
3	Fly ash	kg	84	0.084	7.06
4	Natural river sand	kg	759	0.00251	1.91
5	Gravel	kg	1090	0.00218	2.38
6	Water	kg	151	0.000168	0.03
7	Water reducing agent	kg	5.04	0.02849	0.14
Total					236.32

3.4 The Carbon Reduction Benefit of Green Highway Technology

The carbon sequestration of plants is mainly through photosynthesis, which absorbs carbon dioxide and converts it into organic matter and stores it in the plant body or the soil. In this study, we calculated the carbon stored in plants, i.e. the carbon storage of trees, shrubs and herbaceous communities. At present, the methods of calculating plant carbon storage mainly follow the calculation methods of forestry and grassland system, and the common methods include biomass method [6], flux observation method [7, 8], model simulation method [9, 10] and remote sensing estimation method [11]. However, there are many species of plants in the Nanzhong Expressway, and many landscape plants and seedlings lack calculation models related to carbon sink, so it is difficult to use the existing methods for calculation. Therefore, combined with the existing studies, the carbon density combined with the area is calculated, the carbon density of urban forest is 28.81 t/hm^2 [12], and the carbon density of herbs is 0.36 kg/m^2 [13], and the average annual carbon sequetion rate is about 10% of the carbon density.

The basic data of plant measures were obtained based on construction drawing and field investigation. According to statistical calculation, the construction area of forest and grass vegetation along the whole line of the Nanzhong Expressway is 117.44 hm^2, including 56.21 hm^2 of herbs and 61.23 hm^2 of qiao irrigation. It is calculated that the carbon reserve of green vegetation in the Nanzhong expressway is about 1966.39 t, and the annual carbon sequetion of vegetation is about 196.64 t under the premise that vegetation is not destroyed. Converted to CO_2 about 721.01 t.

4 Conclusions

Adhering to the concept of green construction, Nanzhong Expressway adopts three technologies of resource-intensive conservation, intelligent construction and quality engineering construction to carry out green construction, which can reduce carbon emissions by 512,500 tons compared with traditional construction methods. The green highway technology is adopted to improve the green area of the bridge project, which can absorb and fix about 721.01 t of CO_2 per year.

Acknowledgments. This study was supported by the Study on green highway demonstration project of Nansha to Zhongshan Expressway project. We would like to thank the staff of the Guangzhou Expressway Co., Ltd.

References

1. Zhiguang, L., Haiyan, W., Junxiong, W.: Mechanism, approach and characteristics of carbon sink in soil and water conservation under carbon peaking and carbon neutrality. Bull. Soil Water Conserv. (003), 312–317, 380 (2022)
2. The State Council: The State Council on the issuance of the 2030 Carbon Peak Action Plan Notice, Guofa (2021). Accessed 23–26 Oct 2021
3. GB/T 51366-2019: Building carbon emission calculation standard. China Building Industry Press, Beijing (2019)
4. GB/T 2589-2020: General rules for calculation of the comprehensive energy consumption. Standards Press of China, Beijing (2020)
5. JTG/T 3833-2018: Highway construction machinery platform fee quota. People's Communications Press, Beijing (2018)
6. Fang, J.Y., Guo, Z.D., Piao, S.L., et al.: Estimation of terrestrial vegetation carbon sink in China from 1981 to 2000. Sci. China: Ser. D **37**(6), 804–812 (2007)
7. Clark, D.A., Brown, S., Kicklighter, D.W., et al.: Measuring net primary production in forests: concepts and field measurements. Ecol. Appl. **11**(2), 356–370 (2001)
8. Jiabing, W., Dexin, G., Yushu, Z.: Research methods and progress of CO_2 fluxes in forest ecosystem. J. Northeast For. Univ. **31**(6), 49–51 (2003)
9. Uchijima, Z., Seino, H.: Agroclimatic evaluation of net primary productivity of natural vegetation. J. Agric. Meteorol. **40**(4), 343–352 (1985)
10. Qingfang, S., Liming, J., Yulong, L., et al.: Advances in estimation of forest vegetation and soil carbon storage in China. Environ. Chem. **35**(08), 1741–1744 (2016)
11. Zhizhong, C., Mei, Z., Xuefeng, Y., et al.: Prediction of forest vegetation carbon storage in Xinjiang. Chin. J. Ecol. Environ. **32**(02), 226–234 (2019)
12. Jian, Z., Rongbo, X., Changwei, Z., et al.: The carbon sink of urban forests and efficacy on offsetting energy carbon emissions from city in Guangzhou. Acta Ecol. Sin. **33**(18), 5865–5873 (2013)
13. Feng, Z.: Estimation of carbon pool and greenhouse gas emissions in grassland of China. Lanzhou University (2010)

Analysis of Transmission Path and Potential Sources During Ozone Pollution in Hefei, China

Nina Zheng(✉), Yaqi Xu, Lingyang Pan, and Yuanyuan Fang

School of Urban Construction, Anhui Xinhua University, Hefei 230088, Anhui, China
zhengnina2022@163.com

Abstract. In this paper, based on the hour-by-hour mass concentration of O_3 in Hefei from 2018 to 2021, the backward trajectory analysis, the potential source contribution Function (PSCF) and concentration weighted trajectory (CWT) methods were used to cluster the backward trajectories during O_3 pollution, as well as to analyze the O_3 potential sources simultaneously. The results show that Hefei is dominated by short-range transport airflow, primarily from the east, and that the areas with high WPSCF values are the eastern part of Hubei province (Wuhan), the northern cities of Hefei, the eastern part of Tangshan, and the coastal area of the northern part of Shandong province. The results of the CWT and PSCF assessments are generally consistent. However, the CWT data demonstrate that emissions from Hefei and nearby cities have an impact on O_3 accumulation. To prevent and control O_3 pollution, it is necessary to strictly control the discharge of local pollution sources and strengthen the joint prevention and control of surrounding areas.

Keywords: Ozone Pollution · Backward Trajectories · Potential Sources

1 Introduction

Ozone (O_3) is one of the most powerful oxidants in the atmosphere, playing a significant part in atmospheric chemical reactions. Tropospheric O_3 is mostly produced by its precursor pollutants [1] (e.g., nitrogen oxides (NOx) and volatile organic compounds (VOCs) [2, 3] via a series of complex photochemical reactions in a secondary process. Continuously high concentrations of O_3 in the air can readily generate pollution, such as photochemical haze, which has a major impact on human life and production. High O_3 concentrations can damage the mucosal tissues of the eyes and respiratory system [4–6]. Furthermore, O_3 pollution enhances atmospheric oxidative characteristics, stimulates the formation of secondary organic aerosols, and increases the intensity and probability of heavy pollution weather.

Hefei, located in the central part of Anhui Province, is the center city of the G60 Science and Innovation Corridor, the strategic dual node city of 'One Belt, One Road' and the Yangtze River Economic Belt. As a western city with integrated development in the Yangtze River Delta region, its development has attracted much attention. In the process of economic development, air pollution has become increasingly prominent,

D. Han et al. (Eds.): ICEPG 2024, ESE, pp. 439–445, 2025.
https://doi.org/10.1007/978-3-031-86829-0_49

especially the pollution of O_3, which seriously restricts the improvement of air quality and sustainable economic development in Hefei.

Currently, relevant studies on O_3 pollution in Hefei City focus mostly on the spatial and temporal distribution characteristics of O_3 concentration, as well as their link with precursors and meteorological conditions, with a lack of potential source analysis of O_3 over a long time series. As a result, this study examines the potential sources of high O_3 concentrations in Hefei from 2018 to 2021. It will provide management decision-making reference for joint prevention and control of ozone pollution in Hefei.

2 Data and Research Methodology

The hourly mass concentration data for O_3 in Hefei used in this study were collected from the China Ambient Air Quality Monitoring Platform. The meteorological data utilized for trajectory analysis are the Global Data Assimilation System (GDAS) data given by the National Centers for Environmental Prediction (NCEP) of the United States. The spatial horizontal resolution is $1° \times 1°$.

(1) The backward trajectory analysis model

This work employs the Lagrangian mixed single-particle orbit model to trace the backward trajectories of air mass during O_3 pollution in Hefei. Using the Lagrangian hybrid single-particle trajectory model in the TrajStat plug-in of Meteoinfo software [7] and based on global assimilation meteorological data, the backward trajectories of the air masses arriving at Hefei (31.80°N, 117.31°E) were calculated hour by hour during the period of O_3 pollution in 2018–2021, and the trajectories were delayed backward by 48 h, and the height of the simulated air mass is 500 m. Meanwhile, the Euclidean distance clustering approach is utilized to cluster the trajectories at different times based on the features of the movement speed and transmission direction, resulting in the primary trajectories of the O_3 contaminated air masses.

(2) The Potential Source Contribution Function (PSCF) and Concentration Weighted Trajectory (CWT) Methods

The Potential Source Contribution Function (PSCF) is a method for determining the possible orientation and location of the emission source using the conditional probability function based on the air mass trajectory and pollutant concentration, which is defined as the conditional probability that the air mass passing through the ij-grid reaches the start point when the corresponding pollutant concentration exceeds the threshold. The higher the PSCF value for a grid, the greater its contribution to pollutant concentration at the start point. The weighting function is used to reduce the effects of errors in the conditional probability function caused by the small total sample size [8, 9]. In this article, the PSCF approach establishes an O_3 concentration threshold of 160 $\mu g/m^3$. The CWT method computes the concentration weight value of the trajectories in each grid, indicating the grid's influence on pollutant concentration at the receptor site. By combining the two methods, the amount that the pollutant concentration in a given grid surpasses the threshold at the same PSCF value is produced.

3 Results and Discussion

3.1 Ozone Pollution from 2018–2021

Figure 1 shows that the top primary pollutants from 2018 to 2021 are mostly NO_2, O_3, PM_{10}, and $PM_{2.5}$. O_3 accounted for the largest proportion, with an average of about 40%. Figure 2 shows the time series of O_3 concentration.

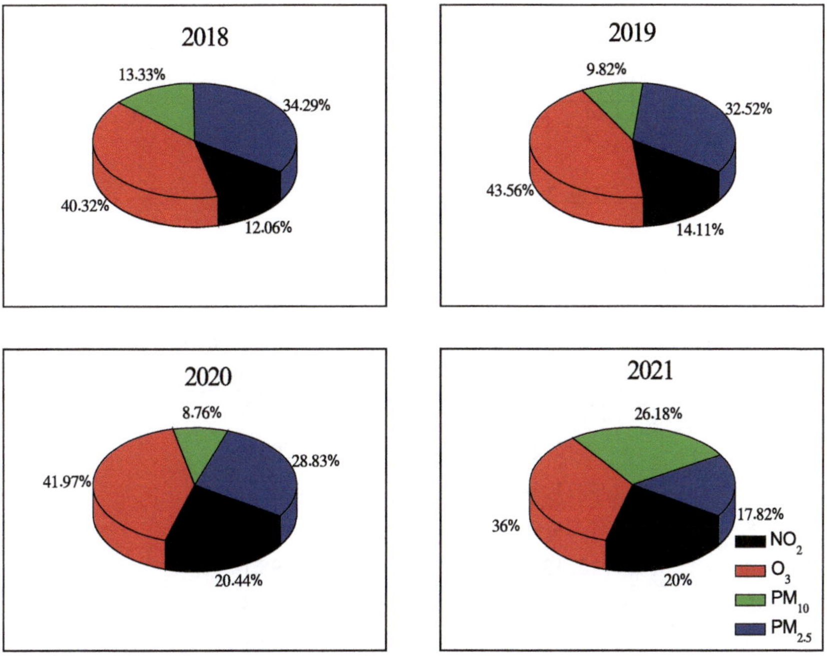

Fig. 1. Percentages of each primary pollutant in Hefei from 2018 to 2021.

The heavy pollution period is defined as an O_3 hourly concentration of more than 200 ug/m^3, as shown in Table 1, and the total number of days with O_3 concentration over the threshold from 2018 to 2021 is 37.

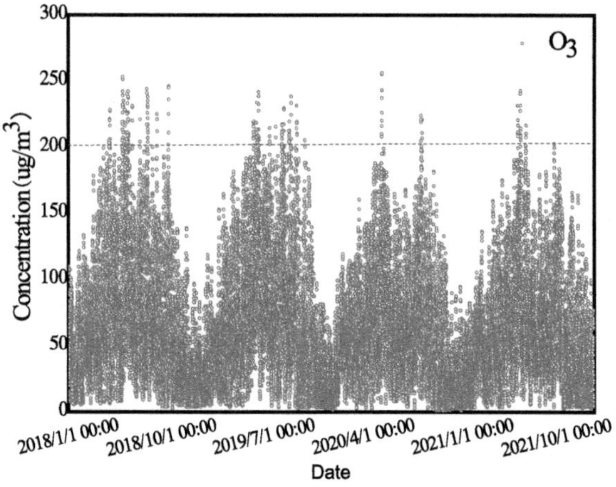

Fig. 2. Time series of O_3 concentration.

Table 1. Ozone pollution period and average concentration.

Year	Month	Day	O_3_8h_max Average concentration (μg/m^3)
2018	4	25, 27, 28	194.5
	6	1, 3, 4, 12, 14, 15, 16, 17, 18, 29	202.3
2019	6	1, 2, 9, 11, 14, 15, 16	199.7
	8	18, 21, 24	201.1
	9	7, 8, 12, 13, 29	198.1
2020	5	20, 21	210.3
	9	6, 7, 9	194.1
2021	6	1, 8, 9, 24	207.8

3.2 O_3 Transport Paths and Potential Source Analysis During the Pollution Period

Because the O_3 concentration is higher during the pollution period, the O_3 transmission path of external sources are more visible. The trajectories of Hefei during the O_3 pollution period from 2018 to 2021 are chosen for clustering and potential source analysis.

During the era of high pollution, the cluster analysis of backward trajectories revealed that Hefei City was dominated by short-range transport airflow, as shown in the Fig. 3, which accounted for 68%. Trajectory 1 mainly comes from the southeast direction, via Shanghai, southeast of Jiangsu Province, and east of Anhui Province to arrive at Hefei, with the highest proportion of 36.07%; Trajectory 2 comes from Hefei and the surrounding western and southern areas, accounting for 32.02%; Trajectory 3 comes from the northeast direction, via the eastern part of Shandong Province, northern Jiangsu

Province, and the central part of the Province to arrive at Hefei, with a high proportion of 20.18%.Trajectory 4 originates in the southeast and travels a relatively long distance to Hefei via northern Zhejiang Province and southeastern Anhui Province, with the lowest share of 11.73%. During the period of O_3 pollution, the air masses in Hefei primarily come from the local area, as well as the southeast and southwest of Hefei, which have a shorter transport distance and faster moving speed, are more likely to arrive in Hefei with precursor pollutants on their paths, and have a greater impact on the concentration of O_3 in Hefei when superimposed on local pollutant emissions.

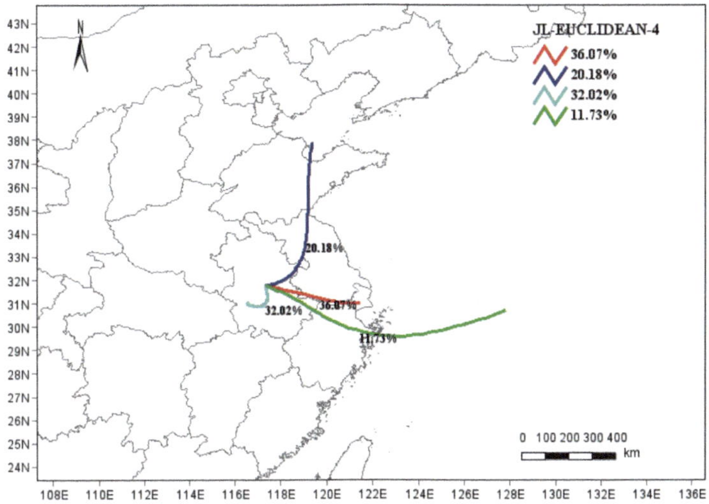

Fig. 3. Cluster analysis of the posterior trajectories during the pollution period.

Figure 4 depicts the findings of the possible source analysis for O_3 pollution in Hefei during the high concentration period. High-value areas (WPSCF > 0.5) were primarily found in the eastern part of Hubei Province (primarily Wuhan and Macheng City) in the southwest of Hefei, the cities of Huaibei, Suzhou, and Fuyang in the north of Hefei City, as well as the eastern part of Tangshan City and the coastal area in the north of Shandong Province. Furthermore, northern Zhejiang Province and eastern Jiangsu Province both have high WPSCF values (WPSCF > 0.4), and their influence cannot be overlooked.

The results of the WCWT and WPSCF analyses were largely consistent. When calculating the contribution factors, WPSCF determines the number of pollutant trajectories, whereas WCWT determines the concentration contribution rate. As shown in Fig. 5, during heavy pollution, the areas contributing to the O_3 concentration (CWT > 100 $\mu g/m^3$) had a wide range and regional contiguous distribution, mainly in the central part of Anhui Province, the northern part of Shandong Province, Jiangsu Province, the northern part of Zhejiang Province, Shanghai, the eastern part of Hubei Province, and the eastern part of Tangshan. The WCWT data demonstrate the impact of emissions from local and neighboring cities in Hefei.

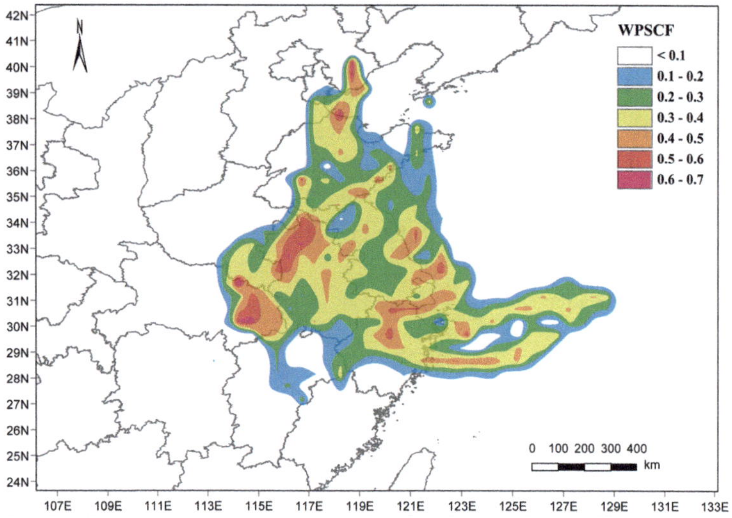

Fig. 4. Distribution of WPSCF during the pollution period.

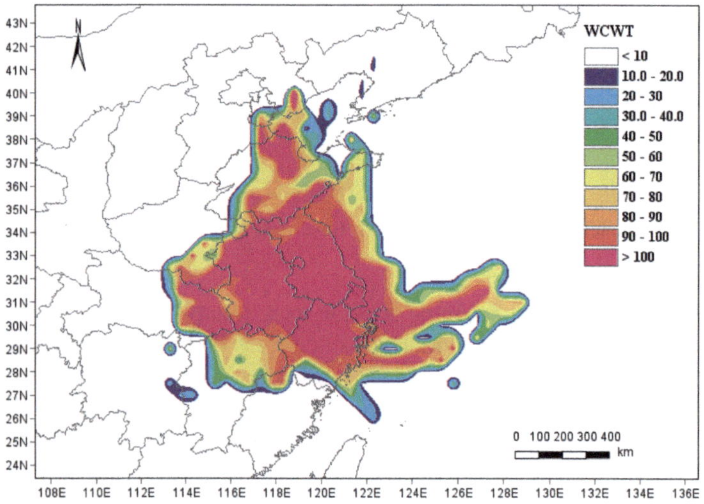

Fig. 5. Distribution of WCWT during the pollution period.

4 Conclusion

The cluster analysis of the backward trajectories during O_3 pollution in Hefei from 2018 to 2021 reveals that short-range transport airflow is dominant. The trajectories are primarily influenced by the eastern airflow. During O_3 pollution, high WPSCF values are observed in the eastern part of Hubei Province (Wuhan), the northern city of Hefei, the eastern portion of Tangshan, and the coastal area in the northern part of Shandong

Province. The results of the concentration weighting analysis demonstrate that locations with higher O_3 concentration contributions are larger in scope and have a regionally contiguous distribution, and the results of the WCWT and WPSCF studies are generally consistent. However, the WCWT data demonstrate that emissions from nearby cities have an impact on O_3 concentration in Hefei. As a result, O_3 pollution prevention and control must be strengthened by coordinated efforts in neighboring areas, with the goal of carefully managing local pollution sources' emissions.

Acknowledgements. The work presented was supported by the Natural Science Research Project in Higher Education Institutions in Anhui Province (Grant No. 2022AH051875, Grant No. 2023AH051812, Grant No. 2024AH050618), the National College Student Innovation Training Program (Grant No. 202312216026) and the Natural Science Research Project of Anhui Xinhua University (Grant No. 2023zr008).

References

1. Bao, J., Li, H., Wu, Z., et al.: Atmospheric carbonyls in a heavy ozone pollution episode at a metropolis in Southwest China: characteristics, health risk assessment, sources analysis. J. Environ. Sci. **113**, 40–54 (2022)
2. Chen, S., Wang, H., Lu, K., et al.: The trend of surface ozone in Beijing from 2013 to 2019: Indications of the persisting strong atmospheric oxidation capacity. Atmos. Environ. **242**, 117801 (2020)
3. Palmisani, J., Nrgaard, A., Kofoed-Srensen, V., et al.: Formation of ozone initiated VOCs and secondary organic aerosol following application of a carpet deodorizer. Atmos. Environ. **222**, 117149 (2020)
4. Boya, L., Lili, W., Lei, Z., et al.: Analysis of severe ozone related human health and weather influence over China in 2019 based on a high resolution data set. Environ. Sci. Pollut. Res. **30**(51), 111536–111551 (2023)
5. Marco, A.D., Garcia-Gomez, H., Collalti, A., et al.: Ozone modelling and mapping for risk assessment: an overview of different approaches for human and ecosystems health. Environ. Res. **211**, 113048 (2022)
6. Lu, J., Yao, L.: Observational evidence for detrimental impact of inhaled ozone on human respiratory system. BMC Public Health **23**(1) (2023)
7. Wang, Y.Q.: MeteoInfo: GIS software for meteorological data visualization and analysis. Meteorol. Appl. **21**(2) (2014)
8. Zhang, H.Y., Cheng, S.Y., Yao, S., et al.: Characteristics and transport pattern of atmospheric particulate matter pollution in Beijing during October-November 2016. Environ. Sci. **40**(5), 1999–2009 (2019)
9. Fu, C.B., Chen, H., Dan, L., et al.: Characteristics and potential sources of four ozone pollution processes in Hainan province in autumn of 2019. Environ. Sci. **43**(11), 5000–5008 (2022)

Research on Emission Characteristics and Test Methods of Brake Wear Particles of Light Vehicles

Le Liu, Zishu Liu(✉), Jiaxing Sun, Chongzhi Zhong, Jingyuan Li, Yi Wen, Yuwei Wang, Zhicheng Ma, Taiyu Zhang, and Zhe Li

China Automotive Technology and Research Center Co., Ltd., Tianjin 300300, China
liuzishu@catarc.ac.cn

Abstract. Brake wear is a significant source of non-exhaust particulate matter (PM) emissions from light-duty vehicles. The $PM_{2.5}$ and PM_{10} particles produced by brake wear pose serious pollution hazards to the atmospheric environment. However, there was a lack of a standardized test method in China to measure the emission characteristics of brake wear particles generated during the actual driving process of vehicles. This paper developed a brake wear particle collection and analysis system based on the chassis dynamometer. The vehicle ran in accordance with the World Light Vehicle Test Cycle (WLTC) on the chassis dynamometer to simulate the actual driving process while measuring brake wear particles. This system could accurately reflect the characteristics of particulate matter emissions during the actual driving process of the vehicle and exhibited high stability and repeatability. The results showed that the normal operation of the fan did not affect the system's performance. In continuous four WLTC cycle tests of the vehicle, the brake wear particles obtained in each cycle tended to be stable, indicating the high long-term stability of the system. Additionally, the change in brake wear particle emissions with WLTC cycle test speed was analyzed, and the particle size distribution of particulate matter was also examined. In brake wear particulate matter, the $PM_{2.5-10}$ fraction was the main component.

Keywords: light-duty vehicle · non-exhaust · braking wear particulate · emission characteristics · test method

1 Introduction

Mobile sources have historically been the primary contributors to particulate matter levels in large and medium-sized urban areas [1–3]. Vehicle emissions, in particular, have long been recognized as significant air pollutants. However, the introduction and enforcement of stricter vehicular emission regulations, coupled with technological advancements such as improved engine efficiency, enhanced fuel quality, and advanced exhaust after-treatment systems [4, 5], have led to a decline in exhaust emissions. Concurrently, non-exhaust emissions—emanating from tire abrasion, brake dust, road degradation, resuspended particles, and general vehicle component wear—have emerged as critical

D. Han et al. (Eds.): ICEPG 2024, ESE, pp. 446–455, 2025.
https://doi.org/10.1007/978-3-031-86829-0_50

contributors to ambient particulate pollution. Notably, in several large and medium-sized cities, these non-exhaust emissions now exceed those from exhaust, positioning them as the predominant source of vehicular particulate matter [6].

In the non-exhaust particulate matter emitted by motor vehicles, brake wear particulate matter constitutes a significant component. These particles originate from the wear of brake pads and discs during deceleration. Approximately 35–55% of these particles diffuse into the air, while the remainder either adheres to the surface of the brake components or accumulates on the road. As vehicles continue to move, those particles that have adhered to surfaces are likely to be re-entrained into the atmosphere [7].

Multiple studies conducted in urban settings have shown that the contribution of brake wear particles to PM_{10} in traffic-related non-exhaust particulates ranges from 16% to 55% [8, 9]. The contribution to traffic-related PM_{10} ranges from 11% to 21% [9, 10], and other relevant studies have indicated that $PM_{2.5}$ in brake wear particulate can account for up to 39% to 63% of emissions in urban areas [11].

As the main component of atmospheric particles, brake wear particles significantly impact environmental pollution. Therefore, it is necessary to study their emission characteristics and provide theoretical guidance for strict control of their emissions. However, the lack of standardized testing methods and processes, as well as differences in testing methods and quantifying standards for particulate emission characteristics, has led to the fact that the particulate emission characteristics obtained from studies cannot be fully compared with each other [12]. This is the primary reason limiting the progress of this research. The research methods for studying brake wear emissions mainly include the pin-on-disk friction meter [13], brake dynamometer [14, 15], chassis dynamometer [16, 17], and actual road tests [18, 19].

At present, the most widely studied method is the method of brake dynamometer, which was independent of the vehicle brake system as a part. It could make the brake system work under a certain working condition, for example, the European seven regulations put forward the brake system to WLTP-B as a working condition, and collect and analyze the particulate matter. The dynamometer test method has the advantages of high test precision and accuracy, but it can only study on brake system and can not completely restore the working behavior in the actual driving process. By contrast, the actual road test has the advantage of reflecting the actual driving behavior of vehicles better. However, many factors, such as differences in environmental background, lead to poor reproducibility of the results. The chassis dynamometer method is a kind of testing method which can make the vehicle drive under the real working condition and can simulate the real braking working condition. Reference to the current domestic and foreign light vehicle brake wear emission characteristics of the mainstream standard test cycle (such as the New European Driving Cycle (NEDC), Worldwide harmonized Light vehicles Test Cycle (WLTC) and China Light-duty vehicle Test Cycle-Passenger (CLTC-P)), chassis dynamometer and WLTC are used to study vehicle braking emissions.

This study introduces a novel testing methodology utilizing a chassis dynamometer to assess brake wear particle emissions from light-duty vehicles. We validate the accuracy and reliability of our test apparatus, thereby contributing valuable theoretical insights and practical direction for future research in brake wear particle analysis.

2 Materials and Methods

2.1 Test System

The experimental setup comprised a conventional fuel-powered test vehicle weighing 1820 kg, a particulate collection apparatus, and an analytical instrument for particle characterization. This setup featured an innovative particulate collection system comprising a gas curtain mechanism, a collection hood, and a transfer conduit. During operation, an annular gas curtain formed between the collection canopy and the vehicle's hub effectively sealed off external contaminants, preventing their ingress into the transfer conduit and thereby preserving particle sample purity. Distinct from Marcel et al.'s [16] approach, our design precluded direct physical interaction between the collection apparatus and the vehicle, mitigating issues associated with rigid couplings such as brake malfunctions, thereby enhancing experimental safety. The collection canopy was positioned externally to the hub, centrally aligned and maintained at a controlled distance, with its opposing end seamlessly integrated with the transfer conduit. Particle analysis was conducted using an Aerodynamic Particle Sizer (APS), which facilitated the assessment of both concentration and size distribution profiles of the particles.

2.2 Test Methods

Figure 1 illustrates the schematic diagram of the brake wear particulate emission test system. The experiment was conducted using a chassis dynamometer. Initially, the test vehicle was securely positioned, and the brake wear particulate collection device was aligned with the vehicle hub. Prior to testing, the air curtain device was activated, setting the air flow rate at the outlet of the air knife to 8 m^3/min, thereby forming an air curtain cover between the collection cover and the wheel hub. This air curtain effectively isolates the external environment, preventing cross-contamination of particles inside and outside the curtain, thus ensuring the efficiency and purity of collected braking wear particles. Upon activation of the sampling fan, brake wear particles generated during vehicle braking are drawn into the conveying pipe. To maximize particle collection, the sampling fan's flow rate was set at 5.76 m^3/s. Throughout the test, the vehicle operated according to the WLTC (Worldwide Light-duty Test Cycle) mode, aligning with the current light-vehicle exhaust emission test protocols and aiding in the enhancement of related brake wear particle emission regulations. An Aerodynamic Particle Sizer (APS) extracted particulate matter from the conveying pipe for emission characteristic analysis. To validate the results, background measurements were performed before each test to eliminate the influence of ambient particulate matter on brake wear emission characteristics. Unless the vehicle remained stationary, both the sampling fan and APS were configured as described, with background measurements spanning one complete WLTC cycle. Each test involved the vehicle completing four consecutive WLTC cycles.

The APS analyzer is employed to assess particles smaller than 20 μm in diameter, detailing their concentration distribution across various size ranges. Operated at a sampling frequency of 1 Hz with a rated flow rate of 5L/min, the device focuses on particulate matter (PM) fractions PM_{10} and $PM_{2.5}$ due to their significant implications for air quality and human health. Notably, PM_{10}, comprising particles up to 10 μm in size, exacerbates air pollution [20], while $PM_{2.5}$, those finer than 2.5 μm, poses heightened risks as they penetrate deeper into the respiratory system, exploiting their minuscule dimensions to reside within lung tissues, thereby amplifying their detrimental effects on health. Consequently, this research concentrates solely on characterizing $PM_{2.5}$ and PM_{10} emissions from brake wear debris. The APS's segmentation criteria for particle sizes below 10 μm are delineated in Table 1.

Table 1. The particle size segmentation standard of APS analyzer

Particle size segment (μm)		
<0.523	1.382	3.786
0.542	1.486	4.068
0.583	1.596	4.371
0.626	1.715	4.698
0.673	1.843	5.048
0.723	1.981	5.425
0.777	2.129	5.829
0.835	2.288	6.264
0.898	2.458	6.732
0.965	2.642	7.234
1.037	2.839	7.774
1.114	3.051	8.354
1.197	3.278	8.977
1.286	3.523	9.647

3 Results and Discussion

3.1 The Stability of the Test System

In chassis dynamometer-based testing, the front-end fan plays a crucial role in heat dissipation, ensuring that the vehicle engine and other critical components maintain an optimal working temperature range. It also simulates actual driving conditions, thereby ensuring that the braking system's temperature remains consistent with real-world scenarios, which is essential for accurate test results. However, because the test system is located outside the wheel hub, it was observed during actual tests that the airflow from the head fan could affect the gas curtain stability. The wind speed of the engine fan varies with the car's speed, necessitating measures to ensure that the test system remains unaffected by the engine fan during the test cycle. To mitigate this issue, a wind direction deviation device (baffle) was installed between the collection device and the head fan, as shown in Fig. 1.

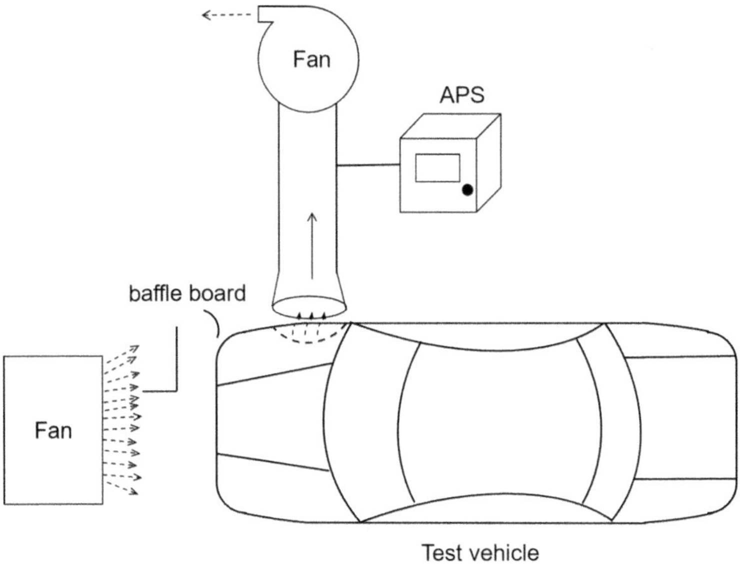

Fig. 1. Schematic diagram of brake wear particulate emission test system

In both testing methods, since the sampling flow rate of the sampling fan and the ASP sampling flow rate are identical, the brake wear particle mass concentration measured by APS can be used as the research subject. The stability of the test system was further analyzed using the mass concentrations of $PM_{2.5}$ and PM_{10} in brake wear particles, with the results illustrated in Fig. 2.

Figure 2 illustrates the variation in mass concentration of $PM_{2.5}$ and PM_{10} particles originating from brake wear during testing. As depicted, the introduction of a baffle into the experimental setup minimally impacts the mass concentrations of these particles, suggesting that the head fan exerts a negligible effect on the measurement system. This observation underscores the robust stability and feasibility of our test methodology. Furthermore, an analysis was conducted to determine the ratio of $PM_{2.5}$ to PM_{10} within two distinct categories of brake wear particles. Under standard testing conditions devoid of a baffle, the $PM_{2.5}$ to PM_{10} ratio stood at 23.82%, while it slightly decreased to 21.91% when a baffle was employed. These findings align with prior research [21, 22], confirming that the predominant size range of brake wear particles falls between $PM_{2.5}$ and PM_{10}.

3.2 The Emission Characteristics of Brake Wear Particles in WLTC Cycle

To further analyze the emission characteristics of brake wear particles during the continuous Worldwide Harmonized Light-Duty Vehicle Test Cycle (WLTC) and to assess the stability of the test system, we conducted four consecutive WLTC cycles under normal test conditions (without a baffle). The particulate matter concentration emission characteristics for each WLTC cycle were examined, as shown in Fig. 3.

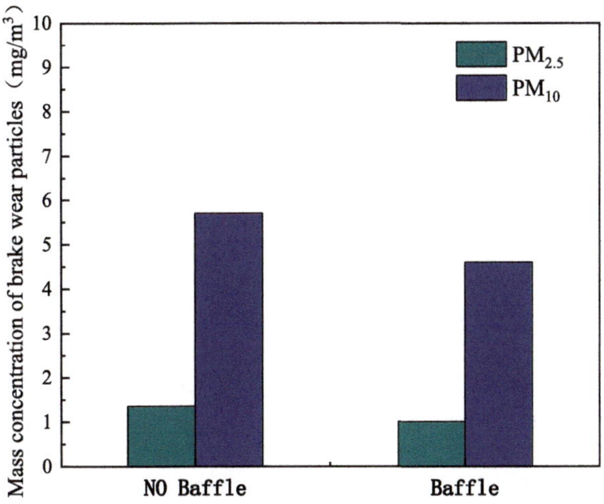

Fig. 2. Mass concentration of brake wear particles

According to Fig. 3, in the four WLTC test cycles, the $PM_{2.5}$ and PM_{10} emissions from vehicles were 1.36 mg/m^3 and 5.71 mg/m^3 respectively in the first WLTC cycle. During the fourth WLTC cycle, the $PM_{2.5}$ and PM_{10} emissions by vehicles were 1.17 mg/m^3 and 5.11 mg/m^3 respectively. By analyzing the change trend of $PM_{2.5}$ and PM_{10} concentrations in brake wear particles, we observed that the differences in $PM_{2.5}$ and PM_{10} concentrations across the WLTC cycles gradually decreased, indicating high feasibility for the test system to operate stably over an extended period.

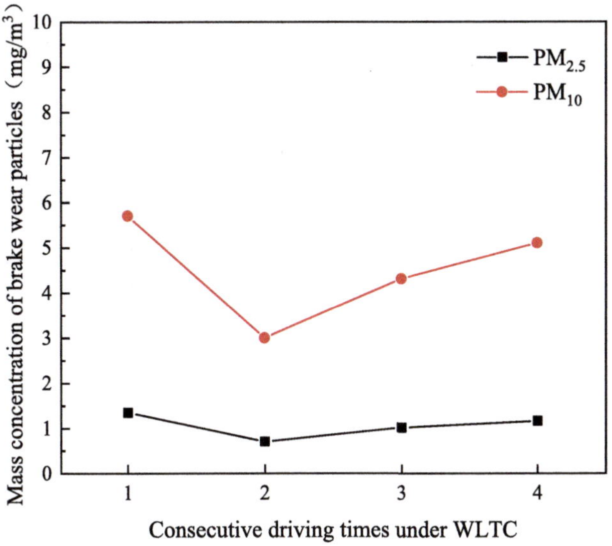

Fig. 3. Change of particle concentration with circulation

3.3 The Characteristics of Particulate Matter Emission Vary with Velocity

To further analyze the change trend of brake wear particle mass concentration emission with varying speed conditions during the vehicle WLTC test cycle, we selected PM_{10} emission characteristics in brake wear particles from the fourth WLTC cycles for analysis based on the stability trend observed in Sect. 2.2. The results are shown in Fig. 4.

According to Fig. 4 (a), the mass concentration of particles in the braking zone exhibits an obvious upward trend throughout the entire WLTC cycle, consistent with actual observations. However, the graph also shows a significant increase in particle emission concentration during some acceleration segments, which may be due to brake drag action [23].

Additionally, we calculated the average mass concentration emissions for four phases: low-speed phase, medium-speed phase, high-speed phase, and extra-high-speed phase. We found that particulate emissions at low speeds were slightly higher than those at medium and high speeds, likely due to the higher braking frequency at this stage. Furthermore, emissions of brake wear particles significantly increased in the ultra-high-speed segment. Studies have shown [24] that an increase in initial braking velocity promotes PM_{10} emissions from brake wear particles. This is likely because higher speeds lead to more kinetic energy being converted into thermal energy during braking, raising the temperature of the braking system. This elevated temperature is a critical factor affecting the emission of brake wear particles [25]. Consequently, the emission of brake wear particles was enhanced, resulting in a significant increase in mass concentration. Additionally, repeated braking actions (braking frequency) contribute to the rise in braking system temperature, further influencing the emission of brake wear particles.

3.4 Characteristics of Particle Size Distribution

To analyze the distribution of brake wear particles with a size less than 10 μm, Aerodynamic Particle Sizer (APS) was employed to examine the mass concentration and particle size distribution under various test conditions. The logarithmic values of particle mass concentration were utilized for a more direct analysis of particle size distribution, as depicted in Fig. 5. Notably, the particle size distribution trends of brake wear particles were consistent across both test conditions. Furthermore, an examination of particle mass concentration revealed that the logarithm of mass concentration was significantly elevated within the particle size range of 2.839 to 6.732 μm. Conversely, particles smaller than 2.5 μm exhibited relatively lower logarithmic mass concentrations, suggesting that the majority of brake wear particles exist as $PM_{2.5-10}$ particles. This finding aligns with the conclusions drawn in Sect. 2.1.

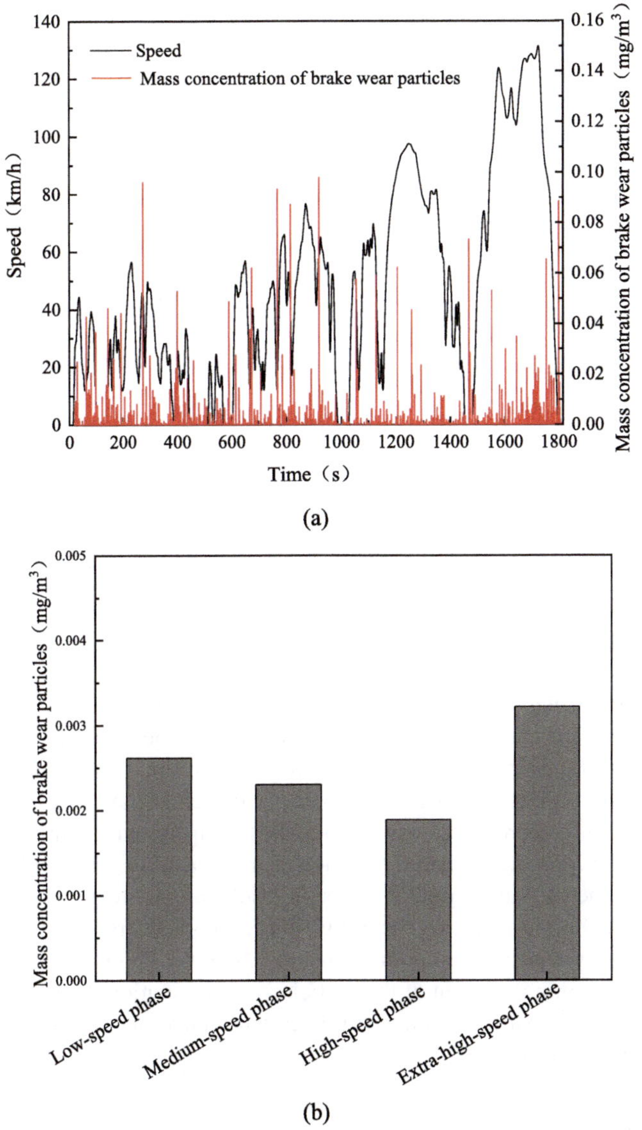

(a)

(b)

Fig. 4. Change of particle concentration with velocity curve

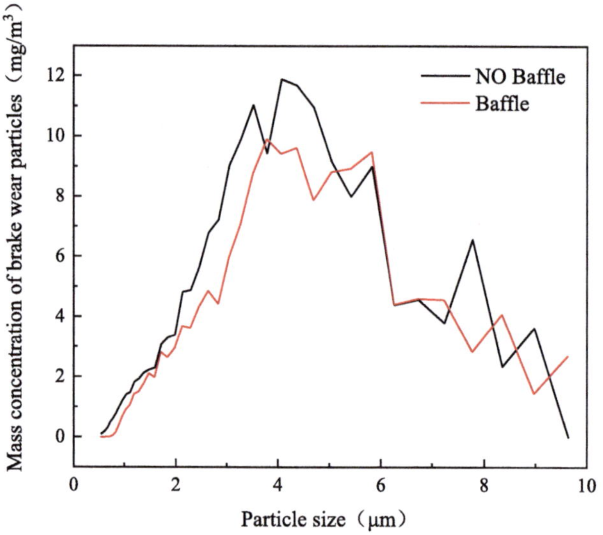

Fig. 5. Characteristics of particle size distribution

4 Conclusion

A method for testing brake wear particles in light-duty vehicles using a chassis dynamometer is introduced, with an investigation into the effect of the head fan on the stability of the testing system. The results indicate that this test method provides a solid foundation for developing brake wear particle emission analysis techniques for light vehicles. The research demonstrates that the testing system exhibits robust stability despite changes in particulate matter emission characteristics. Even after four continuous WLTC cycles, the particulate matter emission results remain relatively stable, suggesting the system's long-term testing capability. Furthermore, the results reveal that particle sizes are primarily concentrated in the range of 2.839–6.732 μm, indicating that $PM_{2.5}$ constitutes only a minor portion of the particle size distribution. In terms of the contribution of brake wear particles to traffic-related particulate matter, their contribution to PM_{10} is more significant.

References

1. Pant, P., Harrison, R.-M.: Estimation of the contribution of road traffic emissions to particulate matter concentrations from field measurements: a review (2013)
2. Moreno-Ríos, A.L., Tejeda-Benítez, L.P., Bustillo-Lecompte, C.F.: Sources, characteristics, toxicity, and control of ultrafine particles: an overview (2022)
3. Karagulian, F., Belis, C.-A., Dora, C.F.-C., et al.: Contributions to cities' ambient particulate matter (PM): a systematic review of local source contributions at global level (2015)
4. Winkler, S., Anderson, J.-E., Garza, L., et al.: Vehicle criteria pollutant (PM, NOx, CO, HCs) emissions: how low should we go? (2018)
5. Harrison, R.M., Allan, J., Carruthers, D., et al.: Non-exhaust vehicle emissions of particulate matter and VOC from road traffic: a review (2021)

6. Piscitello, A., Bianco, C., Casasso, A., et al.: Non-exhaust traffic emissions: sources, characterization, and mitigation measures (2021)
7. Men, Z., Liu, D., Guo, Q.: Chemical component of particulate matters and VOCs characteristics during vehicle brake processes
8. Harrison, R.-M., Jones, A.-M., Gietl, J., et al.: Estimation of the contribution of brake dust, tyre wear and resuspension to non-exhaust traffic particles derived from atmospheric measurements (2012)
9. Lawrence, S., Sokhi, R., Ravindra, K.: Source apportionment of traffic emissions of particulate matter using tunnel measurements
10. Grigoratos, T., Martini, G.: Brake wear particle emissions: a review
11. Ma, Y., Zhang, Z., Liu, X.: Study on particulate emission characteristics of automobile braking wear under different test cycles (2023)
12. Fussell, J.-C., Franklin, M., Green, D.-C., et al.: A review of road traffic-derived non-exhaust particles: emissions, physicochemical characteristics, health risks, and mitigation measures (2022)
13. Wahlström, J., Lyu, Y., Matjeka, V., et al.: A pin-on-disc tribometer study of disc brake contact pairs with respect to wear and airborne particle emissions (2017)
14. Kim, S.-H., Jeong, M.-H., Kim, J., et al.: Dynamometric investigation on airborne particulate matter (PM) from friction materials for automobile: impact of abrasive and lubricant on PM emission factor (2021)
15. Hagino, H., Oyama, M., Sasaki, S.: Laboratory testing of airborne brake wear particle emissions using a dynamometer system under urban city driving cycles (2016)
16. Mathissen, M., Grigoratos, T., Lahde, T., et al.: Brake wear particle emissions of a passenger car measured on a chassis dynamometer (2019)
17. Beji, A., Deboudt, K., Khardi, S., et al.: Non-exhaust particle emissions under various driving conditions: implications for sustainable mobility (2020)
18. Zum Hagen, F.H.F., et al.: Study of brake wear particle emissions: impact of braking and cruising conditions (2019)
19. Oroumiyeh, F., Zhu, Y.: Brake and tire particles measured from on-road vehicles: effects of vehicle mass and braking intensity (2021)
20. Xiao, Y., Cheng, Y., Zhao, H., et al.: Airborne brake wear particle emissions: a review. Tribology **42**(06), 1290–1304 (2022)
21. Iijima, A., et al.: Particle size and composition distribution analysis of automotive brake abrasion dusts for the evaluation of antimony sources of airborne particulate matter (2007)
22. Thorpe, A., Harrison, R.M.: Sources and properties of non-exhaust particulate matter from road traffic: a review (2008)
23. Hagino, H., Oyama, M., Sasaki, S.: Laboratory testing of airborne brake wear particle emissions using a dynamometer system under urban city driving cycles **131**, 269–278 (2016)
24. Liu, Y., et al.: Brake wear induced PM10 emissions during the world harmonised light-duty vehicle test procedure-brake cycle (2022)
25. Feo, M.L., et al.: Laboratory and on-road testing for brake wear particle emissions: a review. Environ. Sci. Pollut. Res. **30**(45), 100282–100300 (2023)

Research on the Impact of Green Technology Progress on Carbon Emission Performance

Zhiyu Li[1(✉)], Jianing Li[1], and Yiran Tao[2]

[1] Hebei University of Economics and Trade, Institute of Economics, Shijiazhuang 050061, China
lizhiyu100@heuet.edu.cn

[2] Department of Finance, Hebei University of Economics and Trade, Shijiazhuang 050061, China

Abstract. The progress of green technology plays a crucial role in improving carbon emission efficiency, while the role of environment-related incentives and constraints policies on green transition cannot be ignored. A threshold effect model is constructed using provincial panel data from 2006 to 2020 to examine the relationship between green technology progress and its impact on carbon emission efficiency. The results show that environmental policies can effectively influence green technological progress, but there is a double threshold effect, and the impact of green technological progress on carbon emission efficiency shows a marginal decline when environmental policies reach a certain threshold. Therefore, utilizing good policy tools at different development stages can effectively enhance green development efficiency, which provides new insights into the green and low-carbon transformation of the regional economy.

Keywords: Green technology progress · environment Regulation · carbon emission reduction

1 Introduction

As the world's second largest economy, China faces significant challenges in achieving carbon neutrality, in particular the ambitious goal of decoupling economic growth from carbon emissions while ensuring stable economic development. Compared to developed countries, China is still advancing in its industrialization and urbanization phases, resulting in a continuous increase in energy consumption. While theoretical analyses suggest that advances in green technologies can successfully reconcile economic development and environmental management [1], current environmental regulatory frameworks often lack the capacity to effectively promote such innovations. In addition, differences in institutional frameworks or sample characteristics may lead to differences in the performance of green technological advances [2, 3]. Research in this area has centered on the cost effects of green innovation and incentives for innovation to explore how green technological advances affect carbon emission efficiency. However, the existence of conflicting theoretical perspectives suggests that there are complex mechanisms and pathways in the green technology innovation and carbon emission sequences that have not yet been explored and are influenced by multiple factors.

D. Han et al. (Eds.): ICEPG 2024, ESE, pp. 456–464, 2025.
https://doi.org/10.1007/978-3-031-86829-0_51

Based on China's development experience, improving ecological conditions in China requires more than the implementation of robust end-of-pipe solutions; it also calls for the establishment of constraints and incentives to improve the efficiency of resource allocation. Regulations concerning the environment play a crucial role in advancing green initiatives. It is important to recognize that there are significant implications between green technology innovation and carbon emission efficiency. Examining the role of environmental policies in improving carbon emission efficiency has important practical implications for achieving sustainable and high-quality development. Such an exploration requires thorough normative assessments and solid empirical investigations to provide informed insights into the issue.

2 Literary Review

The topic of how to promote energy conservation and cut emissions via advancements in environmentally friendly technology has garnered significant interest from scholars. Technological advancements can significantly enhance the efficiency of energy use, consequently leading to a decrease in carbon emissions.Through the development of a simulation system, researchers Turner and Hanley identified that advancements in technology positively influence the rise in carbon emissions, while altering the energy mix can significantly lower CO2 emissions [4]. The indirect effects of advancements in environmentally friendly technology are often deemed to outweigh direct outcomes; The indirect impact of green technology progress is more important than the direct results, green technology spillover capacity increased by 1%, carbon intensity decreased by 0.13% [5]. Some researchers argue that the relationship between technological advancements and carbon emissions lacks clarity. The influence of green technology advancements is complicated by a reliance on established pathways, making it difficult to define the specific nature of innovations stemming from research and development funding [6]. According to the findings of Kim et al. (2015), the relationship between technological advancements and carbon emissions exhibits variability across both space and time, indicating that different geographical areas experience distinct effects during various periods [7]. According to Li Kaijie and Qu Ruxiao (2012), while the effect of technological advancements on reducing carbon emissions may not be markedly observable in the short run, such a mechanism is indeed present over an extended period [8].

In line with the classical neoclassical perspective, constraints imposed by environmental regulations result in increased operational expenses for companies, this can stifle advancements in technology. Nevertheless, A cohort of researchers, including Porter, argues that environmental regulations can lead to an 'innovation compensation effect'. Which can enhance performance related to the environment. Research by BosettiaV (2010) and colleagues utilizing a panel policy performance framework has indicated that non-coercive environmental policies can successfully lead to reductions in CO2 emissions through the encouragement of technological advancements [9]. Scholars in this field have evaluated the effectiveness of these policies, concluding that the integration of environmental regulation with innovation strategies yields optimal outcomes for sustainable development. Fisher.A study by et al. (2017) proposed the integration of mechanisms for reducing carbon emissions with technologies aimed at lowering carbon output to address failures in the innovation market [10].

3 An Overview of the Variables Involved and the Modeling Approach Utilized

3.1 Selection and Assessment of Variables

Describing Variables

Efficiency in Carbon Emissions (ECE). A significant drawback of the traditional Directional Distance Function (DDF) utilized in classical Data Envelopment Analysis (DEA) is its presumption that both the increase of beneficial outputs and the reduction of harmful outputs occur at an equivalent rate. This assumption can introduce inaccuracies in the evaluation of Decision Making Units (DMUs). The Non-Radial Directional Distance Function (NDDF) enables the proportional adjustment of inputs, expected outputs, and undesirable outputs, thereby facilitating a more precise identification of efficiency metrics. This study employs NDDF as a tool to assess the efficiency of carbon emissions. Consider a scenario involving N Decision Making Units (DMUs), all of which utilize capital, labor, and energy as their input measures; Gross Domestic Product (measured in yuan) is identified as the output to be maximized, while carbon dioxide emissions (quantified in units of 10,000 tons) are regarded as the output to be minimized.

Each province's capital consists of its accumulated capital assets, which are estimated utilizing the perpetual inventory technique as outlined by Zhang Jun and colleagues.; The workforce refers to the total number of individuals employed by the end of the fiscal year in each province; This energy metric encompasses the utilization of three fossil fuel categories: coal, petroleum, and natural gas. The consumption of fossil fuels is quantified in terms of standard coal, utilizing the conversion factors outlined in the General Rules for Comprehensive Energy Consumption Calculation. The target GDP is determined based on the actual GDP of each province with the year 2005 serving as the reference point, while the unwanted output pertains to the method of measuring carbon intensity as outlined by the IPCC, which selects the carbon dioxide emissions attributable to the utilization of three fossil fuel types: coal, oil, and natural gas, to signify the carbon emissions in each province.

Using the aforementioned method, the carbon emission effectiveness of thirty Chinese provinces, municipalities, and autonomous regions is assessed, excluding Tibet, Hong Kong, Macao, and Taiwan.

The Primary Explanatory Factors

In terms of selecting alternative indicators for assessing low-carbon technological advancements, many researchers tend to utilize data on environmentally friendly patents or the overall green productivity as a measure of the state of low-carbon tech innovation. The number of granted patents serves as a highly dependable metric and is capable of accurately indicating the degree of innovation due to its consistent criteria for approval and the accessibility of relevant data. The data regarding green patents is sourced from the State Intellectual Property Office (SIPO) patent database, utilizing the list of environmentally friendly patents supplied by the World Intellectual Property Organization (WIPO).

Standard Variables

In regard to environmental regulation, the work of Li Hong and colleagues offers valuable insights., the assessment of environmental control is achieved through a comprehensive index approach that standardizes five separate measures, including the rate of industrial emissions (dust) mitigation, the rate at which industrial SO2 emissions are reduced, overall recycling efficiency of typical industrial waste materials, the rate at which household waste is treated in an environmentally friendly manner, as well as the percentage of sewage effectively processed by treatment facilities, and additionally assessing the significance of the metrics using the entropy approach, followed by assessing the overall index of environmental regulation utilizing the calculated weights and normalized values. Utilizing the established weights alongside the standardized values, the environmental regulation's composite index was derived; a higher index score signifies more stringent environmental regulations imposed by the government.

Control Variable

The variables that serve as controls encompass four aspects: investment in research and development, the level of human resources, market-oriented practices, and the influx of foreign investment. Human capital, represented by the average years of schooling per person, is considered an essential element influencing regional advancements in green innovation. The extent of market development is assessed utilizing indicators from Fan Gang's marketization framework, alongside relevant statistics from each province and municipality for evaluation. The influence of foreign capital inflow on the shift towards sustainable development is significant, and the ratio of such investments to the local province's GDP serves as an approximate indicator of the intensity of these foreign economic contributions.

Data collection: In light of the accessibility and uniformity of statistical quality, this research utilizes samples drawn from 30 provinces across mainland China, excluding Tibet. The foundations data for each indicator is sourced from various publications such as the China Environmental Statistics Yearbook, the China Industrial Statistics Yearbook, the China Statistics Yearbook, and the WIND database from previous years. To convert FDI figures expressed in US dollars into RMB, the middle exchange rate between the US dollar and the RMB is applied. To mitigate the influence of changes in price levels, currency-denominated figures are adjusted to reflect constant prices, using the year 2000 as a reference, based on the corresponding price indices. With the exception of the ratio indicators, other variables are incorporated into the model using their logarithmic values to address issues related to scale and the presence of heteroscedasticity.

3.2 Standard Modeling

The influence of innovation in low-carbon technologies on the efficiency of carbon emissions could be viewed from various dimensions, and the manner in which it affects can vary based on different phases of environmental regulations or geographical contexts, indicating the potential for a nonlinear correlation among the factors. Consequently, the threshold regression approach introduced by Hansen (1999) will be utilized to explore the non-linear connection among the mentioned variables. A model is designed wherein environmental regulation serves as the threshold factor to explore the nonlinear connection between green technology advancements and the efficiency of carbon emissions.

$$CEE_{it} = \beta_0 + \beta_1 Lngti_{it}(Eg \leq \gamma) + \beta_2(Eg \geq \gamma) + \varphi X_{it} + \varepsilon_{it} \tag{1}$$

The variables that provide explanations in the previously mentioned equation pertain to advancements in environmentally friendly technology. The variables that act as thresholds are identified as ecological regulations, and the samples are categorized into distinct groups based on whether these threshold variables exceed their critical values; The notation λXit represents a collection of control factors, comprising aspects such as the intensity of R&D investment, the level of human resources, the extent of marketization, and the influx of foreign direct investment; The term εit represents the stochastic error component.

4 Examination of Observed Outcomes

4.1 Statistical Overview

Table 1 presents the descriptive statistical findings for the primary variables. The lowest observed value for carbon emission efficiency (CEE) is recorded at 0.014, the maxi-mum value is 1.551, with an average value of 0.375, with a standard deviation of 0.248, which reflects an improve-ment in carbon emission efficiency in China, although sig-nificant regional disparities exist; the lowest recorded fig-ure for green technology innovation (Lngti) stands at 2.397, up to 11. 041,The average value is 7.335, with a variation measure of 1.628, and there are significant vari-ations across different regions regarding the extent of in-novation in green technology; Other factors exhibit varying levels of regional disparities.

Table 1. Presents the summary statistics derived from the panel dataset.

Variable	Mean	Standard deviation	Minimum	Max
cee	0.375	0.248	0.014	1.551
lngti	7.335	1.628	2.397	11.041
eg	3.354	2.9851	0.044	24.51
lnrd	14.214	1.473	9.677	17.249
hr	8.834	1.024	6.38	12.78
mi	6.3845	2.1650	0.32	11.1093
fdi	2.293	1.776	0.011	8.078

4.2 Standard Impact Test

This study determines the asymptotic value of the F statistic utilizing threshold effect bootstrap sampling (Bootstrap) testing, and calculates the p value, in addition to examining the threshold value through 300 grid searches and 300 bootstrap samples, respectively.

In the models examining threshold effects, carbon emission efficiency serves as the independent variables, while environmental regulation is treated as the threshold variable. The results of the threshold model are shown in Table 2, the analysis indicates that the F-statistic achieves significance at the 5% level within both the single-threshold and double-threshold frameworks.E.P is less than 0.05, which leads to the conclusion that a dual-threshold exists in the aforementioned two models.

Table 2. Presents the results of self-sampling tests concerning the threshold effect.

explanatory variable	Threshold variables	hypothesis testing	RSS	MSE	F-statistic	P-value
cee	eg	single threshold	4.3613	0.0100	39.53	0.0000
		double threshold	4.1951	0.0096	17.23	0.0233
		triple threshold	4.1199	0.0095	7.94	0.6633

4.3 Examination of the Outcomes from Model Estimations

The samples are classified based on the values of each threshold indicator. The coefficients for the four models have been calculated, the outcomes of the threshold regression analysis can be found in Table 3. From the information presented in Table 3, in cases where the threshold factor pertains to environmental regulation, the impact of different values on carbon emission efficiency has obvious differences. The first column shows that: when $eg \leq 0.0047$, the impact of green technological advancement on the efficiency of carbon emissions is quantified at 0.0258; when the intensity of green technological innovation increases, the regression coefficient is 0.0078 when $0.0047 < eg \leq 0.0056$; in cases where green technological innovation surpasses 0.0056, the regression coefficient registers at 0.0001. This indicates that there exists a non-linear association between green technological innovation and carbon emission efficiency; as the strength of environmental regulations increases and surpasses a specific level, the effect of innovations in green technology on the efficiency of carbon emissions remains considerable, although the degree of influence diminishes, which may be attributed to the fact that many companies are currently at the initial or intermediate phases of their transition towards greener practices, coupled with the rise in operational expenses stemming from environmental

Table 3. Presents the estimates derived from the threshold regression analysis.

	cee	L.cee
independent variable	lngti	lngti
Threshold variables	eg	eg
Eg ≤ 0.0047	0.0258*** (0.0128)	
0.0047 < eg ≤ 0.0056	0.0078 (0.0120)	
Eg > 0.0056	0.0001 (0.0121)	
Eg ≤ 0.0072		0.0272** (0.0130)
0.0072 < eg ≤ 0.0073		0.0095 (0.0122)
Eg > 0.0074		0.0018 (0.0123)
constant	−1.2141*** (0.1812)	−1.1777*** (0.1954)
observations	450	420
R_squared	0.5727	0.5488
F	12.78	11.15

regulations, while the anticipated benefits from innovations in low-carbon technology have yet to manifest.

It is important to note that the standard errors corresponding to the regression coefficients are provided in parentheses, with levels of significance denoted by *, **, and *** for the 10%, 5%, and 1% thresholds, respectively.

4.4 Strength Test

The findings from the threshold model regression indicate that the direction and significance of the results concerning carbon emission efficiency, as well as its lagged effect, remain aligned as explanatory variables. To further validate the reliability of the findings, two meticulous approaches are employed for the robustness assessment. To start with, a sequential approach of gradually incorporating control variables is adopted for multiple rounds of testing. The outcomes indicate that the coefficients and significance levels of all variables remain relatively stable throughout the iterative process. Furthermore, the coefficients related to the primary explanatory variables and threshold indicators align with what is theoretically anticipated. Next, the explanatory variables are substituted. Eco-friendly inventions are primarily categorized as patents for novel ideas, as well as patents for utility models and designs, while patents for green inventions provide the most accurate indication of innovative concepts, therefore, the data pertaining to green invention patents is utilized in place of green patent data to assess the extent of green technological advancement, and perform a robustness analysis, and the findings remain consistent, and the outcomes of the regression analysis align with the hypotheses established by the theoretical framework, which suggests that the conclusions drawn from the research exhibit strong stability.

5 Conclusion and Suggestions

China, being the leading nation in carbon emissions globally, is under significant pressure to take action toward emission reductions. Developing innovative green technologies to mitigate pollution and enhance the effectiveness of carbon reduction is essential for achieving this goal. During this transition, the importance of environmental regulations becomes evident. The study found that there is a nonlinear effect of green technology innovation on carbon emission performance. The results of the threshold model show that as the intensity of environmental regulation increases, the positive promotion effect of green technological innovation on carbon emission efficiency diminishes.

The results mentioned above carry significant implications for policy-making. It is essential to enhance the effectiveness of innovations in sustainable technology to expedite the establishment of a collaborative approach that facilitates both pollution control and carbon emissions reduction. And to fully enable the local area's transition towards a low-carbon economy. It is essential that governmental agencies amplify their assistance towards the advancement of green technological research and development, extend essential assistance to businesses for advancing innovations in technologies aimed at energy efficiency and carbon mitigation, and facilitate the transition of regional economic growth towards a green and low-carbon approach by positioning relevant companies as the 'trailblazers' in the innovation of low-carbon technologies, along with emphasizing the advancement of eco-friendly technologies, establishing an innovation system for green technology that is driven by market demands, and enhancing the efficient use of available resources. There should be initiatives to encourage the adoption of eco-friendly technologies, establish a system focused on innovative advancements in environmentally friendly technologies driven by market principles, enhance the efficiency of resource distribution, and establish the essential framework that supports the sustainable shift towards a greener economy. Implement more policies focused on reducing costs, and focus on alleviating the financial burden on projects aimed at lowering emissions in industries with high pollution levels by offering tax benefits, preferential financing options and supplementary regulations, to foster an advantageous institutional framework for advancing regional low-carbon initiatives. There should be a proactive approach towards government procurement, increase the market value of eco-friendly products, promote the understanding of environmentally friendly consumption, and facilitate the development of an environmentally responsible framework focused on market-driven regulations for sustainable growth.

References

1. Lin, B.: The pursuit of carbon neutrality in China necessitates high-quality economic advancement. Econ. Educ. **57**(01), 56–71 (2022)
2. Shen, H., Zhou, Y.: A study exploring the relationship between environmental oversight and the ecological performance of businesses: insights from quasi-natural experiments involving environmental protection dialogues. Nankai Manag. Rev. **20**(6), 73–82 (2017)
3. Song, D., Zhu, W., Wang, B.: Empirical research at the micro level concerning enterprises involved in China's carbon trading initiatives: the relationship between carbon emissions trading, methods of quota distribution, and advancements in corporate environmental innovation. J. China Popul.-Resour. Environ. **31**(01), 37–47 (2021)
4. The study conducted by Turner and Hanley explores the relationships between energy efficiency, the rebound effect, and the Environmental Kuznets Curve. J. Energy Econ. **33**(5), 709–720 (2011)
5. He, Y., Weidong, W., Lu, W., et al.: In what ways can advancements in low-carbon technology help to reduce carbon emissions? Evidence gathered from China indicates this phenomenon. Manag. Mod. **40**(05), 90–94 (2020)
6. Seoja Choi Jingbo Corporate innovation in green technologies and the development of low-emission urban areas. Ind. Econ. China **2020**(12), 178–196
7. Kim, K.H., Sul, K.H., Szulejko, J.E., et al.: Advancements have been observed in decreasing carbon monoxide concentrations in prominent metropolitan regions across Korea. Environ. Pollut. **207**(7), 420 (2015). Significant advancements have been made
8. Li, K., such as Xiaoqu: An empirical analysis exploring how advancements in technology influence carbon output across provinces. J. Beijing Normal Univ. (Soc. Sci. Ed.) (05), 129–139 (2012)
9. Bosetti, V., Carraro, C., Duval, R., et al.: What can we anticipate from advancements in innovation? An evaluation based on a model regarding the cost effects related to environmental issues and the expenses involved in reducing impacts stemming from climate-focused research and development. Soc. Sci. Electron. Publ. **33**(6), 1313–1320 (2010)
10. Fischer, C., et al.: Options for policy in environmental and technological practices within the electricity industry: are we implementing excessive measures? J. Assoc. Environ. Resour. Econ. **4**(04), 959–984 (2017)

Seasonal Variations of Organic Pollutants in $PM_{2.5}$ Within a Typical Industrial City in Central China

Dayong Tian[1](✉), Xin Han[1], Rongqi Wu[2], Wenyuang Guo[3], Rongrong Zhang[4], and Chao Du[4]

[1] Anyang Institute of Technology, Anyang 455000, Henan, China
astdy@126.com
[2] Anhui University of Technology, Ma'anshan 243000, Anhui, China
[3] Meteorological Administration of Anyang, Anyang National Climate Observatory, Anyang 455000, Henan, China
[4] Licheng Environmental Company, Anyang 455000, Henan, China

Abstract. This study focuses on the seasonal variations of organic pollutants in $PM_{2.5}$ in Anyang, a typical industrial city in central China. $PM_{2.5}$ samples from four seasons were analyzed using GC - MS. The extraction methods for organic chemicals were developed. The results show that acetonitrile has a higher extraction efficiency than dichloromethane for $PM_{2.5}$ samples. Among all the samples, the sample with the most determined types of organic substances contains 24 kinds of organics. Alkanes and phenols are the organic substances with the highest detection frequencies. The concentration of organic pollutants in winter is higher than that in other seasons, indicating a greater harmful for human health. This research provides valuable insights for air pollution control strategies in severely polluted areas in China.

Keywords: Seasonal Variations · $PM_{2.5}$ · organic pollutants · GC-MS

1 Introduction

Air pollution has become a global concern, and particulate matter, especially $PM_{2.5}$, has drawn significant attention due to its adverse effects on human health and the environment [1, 2]. $PM_{2.5}$ might contains a wide variety of bacteria, viruses, heavy metals, and organic pollutants [3, 4]. Among them, organic pollutants are a class of pollutants that pose a threat to human health due to their persistence, toxicity, and hydrophobicity [3]. However, the composition of $PM_{2.5}$ can vary significantly depending on the geographical location, season, and various other factors [5]. Therefore, the research on the variations of organic pollutants in $PM_{2.5}$ across different seasons is of great practical significance.

Gas chromatography - mass spectrometry (GC-MS) is a common analytical technique widely utilized for investigating the organic components in $PM_{2.5}$ [6]. Many literatures reported the variations of organics in $PM_{2.5}$ in cities such as Beijing, Shanghai, and Nanjing [7–10]. However, as an important industrial city mainly engaged in steel

D. Han et al. (Eds.): ICEPG 2024, ESE, pp. 465–470, 2025.
https://doi.org/10.1007/978-3-031-86829-0_52

and chemical industries in the central region of China, the $PM_{2.5}$ pollution situation has long been ranked among the bottom ten in the country. In the air quality ranking of 168 key cities from January to December 2023 announced by the Ministry of Ecology and Environment, Anyang City ranked fourth from the bottom, with the main pollutants being $PM_{2.5}$ and ozone. At present, there is still limited data available on the $PM_{2.5}$ compositions specifically in Anyang City.

This study aims to collect the $PM_{2.5}$ samples in different seasons including spring, summer, autumn and winter, develop a method to extract the organic chemicals, and investigate their seasonal variations. The study will provide an insight into understanding the seasonal variations of organic contaminants in $PM_{2.5}$, which is beneficial to the development of more effective strategies for air pollution control in the severely polluted area.

2 Materials and Methods

2.1 Sampling of $PM_{2.5}$

$PM_{2.5}$ samples were collected using Polytetrafluoroethylene (PTFE) filters based on Thermo2025i Airborne Particulate Sampler (Thermo Fisher Scientific Co., Ltd.). Sampling was located on the roof of Building A05 in the school of chemistry and environmental engineering, Anyang Institute of Technology (36.06°N, 114.35°E).

2.2 Extraction for $PM_{2.5}$ Organic Substances

(1) We first take out two collected $PM_{2.5}$ PTFE filters from the freezer and cut the two filters of the same season into small pieces, followed by transferring them into separate 50 ml centrifuge tubes and adding 15 ml of deionized water to each tube. The tubes are placed in an ultrasonic cleaner, and then sonicated at 28 Hz for 30 min. After the sonication process, the supernatant is filtered and transferred into pre-prepared glass petri dishes. Subsequently, 10 ml of methanol is added. The same processing steps are repeated. Thereafter, another 15 ml of deionized water is added and the identical steps are repeated. The supernatants obtained from the three consecutive ultrasonic treatments are combined into the same glass petri dish.

(2) The glass petri dish is sealed with sealing film and immediately placed in a −80 °C freezer for pre-freezing treatment. After 12 h of pre-freezing, the sample is completely frozen and solidified without any liquid droplets. Subsequently, vacuum freeze-drying is carried out.

(3) After the samples are completely dried (approximately 24 h), the vacuum-dried $PM_{2.5}$ samples are carefully scraped out with a spatula and then weighed.

(4) An equal amount of $PM_{2.5}$ sample is weighed and placed into a 15 ml plastic centrifuge tube. Subsequently, 3 ml of acetonitrile is added to the tube. The tube is then placed in an ultrasonic cleaner, which is immersed in an ice-water bath for 30 min to ensure complete dissolution. A disposable syringe is used to extract 500 μl of the supernatant, and the extracted liquid is filtered through an organic phase filter head. Finally, the filtered solution is transferred into a gas chromatography injection vial. The organics were analyzed using GC-MS (Agilent 8890-5977B).

3 Results and Conclusion

3.1 Analysis of Organic Components in $PM_{2.5}$

Development of the Extraction Method

To compare the extraction efficiencies of different solvents, we compared the results using two different solvent (Fig. 1). It can be found that most of the peaks obtained by extraction with acetonitrile are higher than that with dichloromethane.

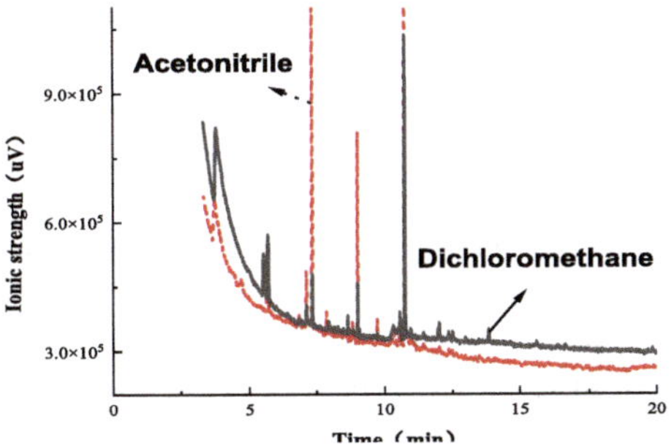

Fig. 1. Comparison of spring $PM_{2.5}$ samples extracted by two different solvents (Apr 8–9 and Apr 13–14, 2019. The red dash line: acetonitrile, black solid line: dichloromethane)

To further investigate the extraction effect, we analyze the peaks with a signal-to-noise ratio greater than 10, and identify the unknown compounds based on the NIST (The National Institute of Standards & Technology) database with a matching score of 95 points. The results are shown as the following figure (see Fig. 2. In this figure, a is the result of organic substances obtained by GC - MS after extraction with acetonitrile as the solvent, and b is the result with dichloromethane).

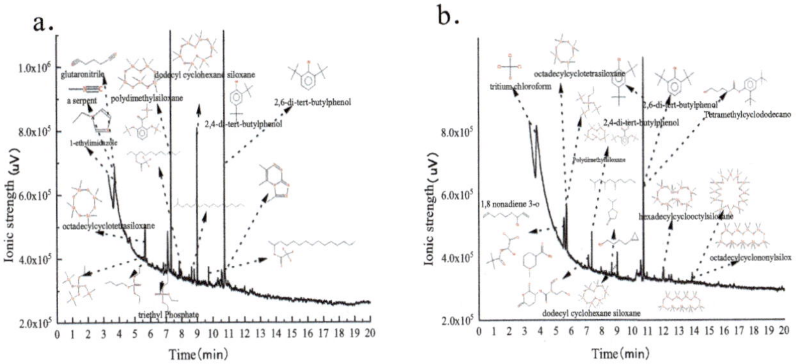

Fig. 2. Organic chemicals extracted from spring $PM_{2.5}$ samples with acetonitrile (a) and dichloromethane (b) on April 8–9, 2019 and April 13–14, 2019.

As can be seen from the figure, after extraction with acetonitrile, 19 main organic substances were determined, including 1 type of salt, 12 types of alkanes, 3 types of phenols, and 3 types of esters. In the sample extracted with dichloromethane as the solvent, 16 main organic substances were determined, including 1 type of alcohol, 2 types of amines, 1 type of salt, 9 types of alkanes, 3 types of phenols, and 5 types of esters. From the perspective of the types of determined organic substances, the number of determined substances in the sample extracted with acetonitrile is greater than that in the sample extracted with dichloromethane using GC-MS. The higher peaks and the more number of compounds in acetonitrile-extracted samples indicating acetonitrile is a more effective solvent for extracting organic substances from $PM_{2.5}$ samples compared to dichloromethane when using GC-MS analysis, which could be ascribed to its superior solubility properties and its capacity to interact with a wider range of organic components existing in $PM_{2.5}$.

Organic Pollutants of $PM_{2.5}$ Samples from Summer, Autumn and Winter

The summer $PM_{2.5}$ samples were collected on July 10–11 and Jul 12–13, 2019. The $PM_{2.5}$ concentrations for these four days were 41, 36, 40, and 35 $\mu g/m^3$ respectively, with an average value of 38 $\mu g/m^3$. The autumn $PM_{2.5}$ samples were collected on Oct 22–23, 2019 and Oct 23–24, 2019. The $PM_{2.5}$ concentrations for these four days were 78, 65, 60, and 69 $\mu g/m^3$ respectively, with an average value of 68 $\mu g/m^3$. The winter $PM_{2.5}$ samples were collected on Jan 10–11 and Jan 12–13, 2020. The $PM_{2.5}$ concentrations for these four days were 238, 172, 148 and 220 $\mu g/m^3$ respectively, with an average value of 195 $\mu g/m^3$ (Fig. 3).

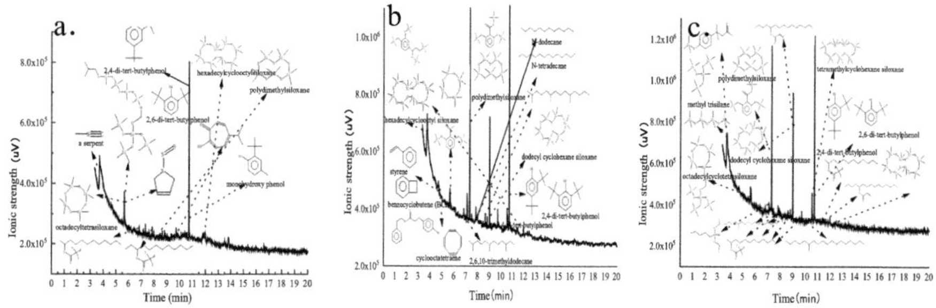

Fig. 3. Organic chemicals extracted from $PM_{2.5}$ samples with acetonitrile during summer (a), autumn (b), winter (c).

As can be seen from the figure, 22 main organic substances were determined, including 15 types of alkanes, 5 types of phenols, and 2 types of esters were determined in summer samples. For samples extracted from autumn $PM_{2.5}$ samples with acetonitrile, 21 main organic substances were determined, including 11 types of alkanes, 7 types of phenols, and 3 types of esters. For samples extracted from winter $PM_{2.5}$ samples with acetonitrile, 21 main organic substances were determined, including 15 types of alkanes, 4 types of phenols, and 2 types of esters. From the perspective of the types of determined organic substances, the number of determined substances in the sample extracted with acetonitrile is greater than that in the sample extracted with dichloromethane using GC-MS.

It can be concluded from above results that, in the atmospheric $PM_{2.5}$ of Anyang City, the sample with the most determined types of organic substances contains 24 kinds of organics. Alkanes and phenols are the organic substances with the highest detection frequencies after extraction with acetonitrile, followed by phenols and esters.

3.2 Comparison of $PM_{2.5}$ Components in Different Seasons

To compare the seasonal variations of the concentration of organic pollutants in $PM_{2.5}$ samples in four seasons, we weighed samples of the same quality from the $PM_{2.5}$ samples extracted in four seasons and compared the concentrations of organics after extraction. The results are shown in Fig. 4.

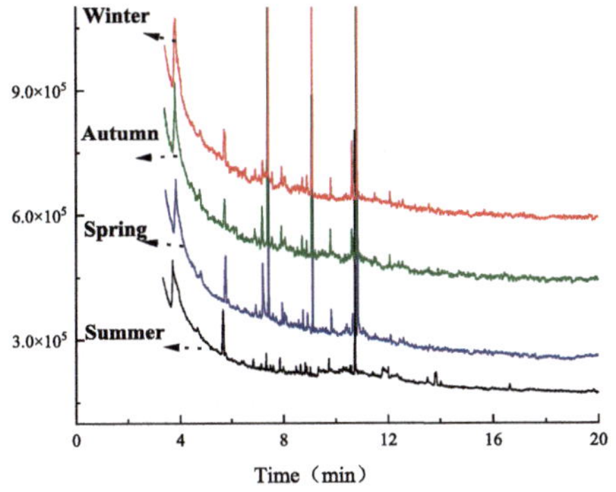

Fig. 4. Comparison of Four $PM_{2.5}$ Samples in Different Seasons

In Fig. 4, winter, autumn, spring, and summer are presented in sequence from top to bottom. It can be seen from the figure that the peaks of pollutants determined in winter are higher than those in other seasons. This indicates that, in $PM_{2.5}$ samples with the same quality, the concentration of organic pollutants in $PM_{2.5}$ per unit mass in winter is higher than that in other seasons. This suggests that $PM_{2.5}$ in winter may pose a greater threat to human health, even with the same $PM_{2.5}$ level.

4 Conclusion

In this study, Anyang City, a typical industrial city in central China, was selected as the research object. $PM_{2.5}$ samples in four seasons (spring, summer, autumn, and winter) were collected to investigate the extraction efficiencies of different solvents. The extraction method and GC - MS determination method were established. The results are as follows:

(1) The extraction efficiency of acetonitrile for $PM_{2.5}$ is higher than that of dichloromethane.

(2) In the atmospheric $PM_{2.5}$ of Anyang City, the sample with the most determined types of organic substances contains 24 kinds of organics. Alkanes and phenols are the organic substances with the highest detection frequencies after extraction with acetonitrile.
(3) The concentration of organic substances in winter $PM_{2.5}$ is higher than that in other seasons, indicating that there are more organic pollutants in winter $PM_{2.5}$.

The results suggests that the winter $PM_{2.5}$ may be more harmful than that of other seasons. This provides an insight into understanding the seasonal variations of organic contaminants in $PM_{2.5}$, which is beneficial to the development of more effective strategies for air pollution control in severely polluted areas.

Acknowledgments. This work was funded by the Open Research Fund of Anyang National Climate Observatory (AYNCOF202306), the Science and Technology Project of Anyang City (2023C01SF217), the Scientific and Technological Project of Henan province (222102320103), Huanshui Scholar Fund Project of AYIT. We are grateful for their financial support for this work.

References

1. Ye, J., Hu, H., Bu, Z., et al.: Spatiotemporal distribution of oxidative potential in $PM_{2.5}$ and its key components across six Chinese cities. J. Hazard. Mater. **476**, 135119 (2024)
2. Park, J., Lee, K.-H., Kim, H., et al.: Analysis of $PM_{2.5}$ inorganic and organic constituents to resolve contributing sources in Seoul, south Korea and Beijing, China and their possible associations with cytokine Il-8. Environ. Res. **243**, 117860 (2024)
3. Lin, X., Cai, M., Pan, J., et al.: $PM_{2.5}$ chemical components are associated with in-hospital case fatality among acute myocardial infarction patients in China. Ecotoxicol. Environ. Saf. **284**, 116898 (2024)
4. Wang, Y., Yuan, Y., Mo, S., et al.: Individual and joint exposures to $PM_{2.5}$ constituents and mortality risk among the oldest-old in China. Sci. China Life Sci. **67**, 2692–2700 (2024)
5. Ran, Z., Wang, X., Yin, X., et al.: Comparison of $PM_{2.5}$ components and secondary formation during the heavily polluted period of two megacities in China. Int. J. Environ. Sci. Technol. **21**(1), 885–894 (2024)
6. Tian, Y., Ma, Y., Wu, J., et al.: Ambient $PM_{2.5}$ chemical composition and cardiovascular disease hospitalizations in China. Environ. Sci. Technol. **58**(37), 16327–16335 (2024)
7. Wen, J., Shi, X., Tian, Y., et al.: Analysis of chemical composition of the fine particulate matter in summer in Tianjin city via a single particle aerosol mass spectrometer (SPAMS). Environ. Sci. **39**(8), 3492–3501 (2018)
8. Yang, X., Huang, S., Li, D., et al.: Water-soluble organic matter with various polarities in $PM_{2.5}$ over Xi'an, China: abundance, functional groups, and light absorption. Particuology **84**, 281–289 (2024)
9. Zhang, J., Li, J., Su, Y., et al.: Interannual evolution of the chemical composition, sources and processes of $PM_{2.5}$ in Chengdu, China: insights from observations in four winters. J. Environ. Sci. **138**, 32–45 (2024)
10. Cai, B., Cheng, H., Kang, T.: Establishing the emission inventory of biogenic volatile organic compounds and quantifying their contributions to o3 and $PM_{2.5}$ in the Beijing-Tianjin-Hebei region. Atmos. Environ. **318**, 120206 (2024)

Assessment of Carbon Emission Reduction Costs in China's Transportation Sector

Jin Tian(✉) and Dong Liang

Beijing Jiaotong University, Beijing 100044, China
19113012@bjtu.edu.cn

Abstract. In the context of intensified global climate change, accurately calculating the carbon emission reduction costs in the transportation sector is of significant practical importance for scientifically advancing carbon reduction efforts and achieving carbon peak and carbon neutrality goals. This paper employs the directional distance function (DDF) to assess the shadow prices of carbon dioxide emissions in the transportation sector across various provinces in China from 2010 to 2020. The results indicate substantial disparities in carbon reduction costs among provinces, with an overall trend of "higher costs in the west and lower costs in the east," placing greater pressure on the western regions. Requiring all regions to achieve the same reduction targets may lead to inequitable outcomes. Based on these findings, this paper suggests formulating more targeted, differentiated carbon reduction policies to promote a green transition in the transportation sector that balances efficiency and equity.

Keywords: Climate Change · Carbon Reduction Costs · Shadow Price

1 Introduction

In the context of global climate change, carbon emission reduction has become a significant topic of concern for governments, businesses, and individuals alike. According to the 2018 report by the Intergovernmental Panel on Climate Change (IPCC), the global average temperature has risen by approximately 1.1 °C since the Industrial Revolution, with greenhouse gas emissions being the primary driving factor behind this change. Carbon dioxide, as the most significant greenhouse gas, has seen a marked increase in emissions over the past few decades, directly contributing to the exacerbation of climate change. The 2023 IPCC report predicts that without proactive reduction measures, the global average temperature could rise by about 3 to 5 °C by the end of this century, resulting in irreversible and severe impacts on ecosystems, economic and social development, and human health. Therefore, carbon reduction efforts are not only a key measure to combat climate change but also an inevitable choice for sustainable development.

The IPCC's Sixth Assessment Report in 2021 highlighted that carbon emissions from the transportation sector grew at an average annual rate of 2% from 2010 to 2019, with emissions in 2019 accounting for 23% of global energy-related carbon emissions. Thus, the transportation sector is a significant source of carbon emissions, and its reduction

D. Han et al. (Eds.): ICEPG 2024, ESE, pp. 471–477, 2025.
https://doi.org/10.1007/978-3-031-86829-0_53

efforts will directly impact the achievement of global climate goals [1, 2]. In February 2024, low-carbon technologies in the transportation sector were incorporated into the national key low-carbon technology promotion plan, and China's transportation sector is actively pursuing carbon reduction. By constructing a scientifically sound carbon reduction cost assessment model and accurately calculating the carbon reduction costs for various provinces, it becomes possible to identify hotspots and key areas for reduction. This provides data support and a decision-making basis for governments to formulate differentiated and precise carbon reduction policies, thus aiding in achieving the goals of carbon peak and carbon neutrality.

In the analysis of carbon reduction costs, shadow prices can help accurately reflect the differentiated carbon reduction costs among regions. Shadow prices comprehensively consider the economic and technological differences between regions by estimating the true economic costs of achieving set reduction targets, enabling a more equitable comparison of the carbon reduction efforts and results across different regions [3]. Overall, using shadow prices can more comprehensively reflect the real economic burdens of carbon reduction, such as the costs of technological transformation and industrial adjustment, revealing disparities in regional reduction costs more authentically and assisting policymakers in formulating fairer and more effective reduction policies.

This paper utilizes the quadratic directional distance function to assess the shadow prices of carbon dioxide emissions in the transportation sector across various provinces in China. This method enhances the accuracy and credibility of the estimated shadow prices, providing a basis for formulating differentiated and precise reduction policies and serving as a reference for promoting fairer global climate governance.

2 Literature Review

In IPCC reports, emission reduction costs are typically defined as the economic costs incurred to reduce greenhouse gas emissions. Specifically, these costs include: (1) Direct costs: Investments required by enterprises or governments to reduce emissions, such as the development and application of clean technologies and improvements in energy efficiency. (2) Opportunity costs: Potential output losses resulting from reduced production and economic activity due to emission reductions. (3) Policy costs: Administrative and compliance costs associated with implementing emission reduction policies (e.g., carbon taxes, emissions trading).

Shadow Price refers to the implicit value of a specific resource or output in the context of incomplete resource allocation or market failure. For carbon dioxide, the shadow price represents the economic cost of reducing one unit of carbon dioxide emissions, reflecting the implied marginal cost. Thus, the shadow price of carbon dioxide can be used to indicate the cost of carbon reduction, quantifying the economic output changes resulting from variations in carbon dioxide emissions. Accurately estimating the shadow price of carbon dioxide not only provides a foundational reference price for carbon trading markets—reflecting the status of factor allocation through discrepancies between marginal reduction costs and actual market prices—but also offers a theoretical basis for governments to formulate carbon tax schemes, thereby promoting reforms in carbon reduction activities. However, studies using shadow prices to calculate reduction costs have predominantly focused on industries such as manufacturing and agriculture, often overlooking the transportation sector.

Since the 1990s, several articles have been published in international environmental, energy, and economic management journals that utilize distance functions to estimate the marginal reduction costs of pollutants [4, 5]. Compared to other methods, the directional distance function has a clear advantage in handling situations involving unwanted outputs, as it can identify effective paths where desired outputs increase while unwanted outputs decrease, aligning with the principles of clean production.

3 Calculation of Carbon Reduction Costs

This study employs a quadratic function form of the directional distance function to estimate the shadow prices of carbon dioxide in the transportation sector. Building upon existing research and considering the actual input and output conditions of the transportation sector, this research selects labor, capital, and energy as input indicators, with the gross value added of the transportation sector as the desired output and the total carbon emissions resulting from transportation operations as the unwanted output. This forms the input-output indicator system for the transportation sector, as shown in Table 1:

Table 1. Input-Output Indicator System for the Transportation Sector

		Indicator Characteristics	Indicator Name
Input-Output Scheme	Input Indicators	Labor	Number of Employees in the Transportation Sector (10,000 person)
		Capital	Fixed Asset Investment in the Transportation Sector (100 million RMB)
		Energy	Energy Consumption in the Transportation Sector (10,000 Tons of Standard Coal)
	Output Indicators	Desired Output	Gross Value Added of the Transportation Sector (100 million RMB)
		Unwanted Output	Carbon Emissions from the Transportation Sector (10,000 Tons)

By standardizing all input and output variables and applying the quadratic directional distance function for linear programming, the shadow prices of carbon dioxide for the transportation sector at the national and provincial levels from 2010 to 2020 were calculated, as shown in Table 2.

Table 2. Carbon Dioxide Shadow Prices in the Transportation Sector by Province (DDF) (Unit: 10,000 RMB/Ton)

	2010	2012	2014	2016	2018	2020
Beijing	1.360	1.466	1.368	1.603	1.842	1.700
Tianjin	1.341	1.370	1.390	1.405	1.332	1.361
Hebei	1.104	1.173	1.166	1.045	1.211	1.415
Shanxi	1.362	1.404	1.359	1.344	1.348	1.642
Inner Mongolia	1.198	1.352	1.409	1.590	1.912	1.647
Liaoning	1.153	1.201	1.237	1.062	1.074	1.095
Jilin	1.476	1.522	1.708	1.617	1.711	1.832
Heilongjiang	1.569	1.574	1.507	1.504	1.620	1.655
Shanghai	0.983	1.021	1.049	1.339	1.239	1.207
Jiangsu	1.205	1.267	1.235	1.156	1.042	1.023
Zhejiang	1.377	1.378	1.474	1.512	1.319	1.441
Anhui	1.505	1.621	1.765	1.734	1.135	1.308
Fujian	1.323	1.533	1.515	1.319	1.327	1.232
Jiangxi	1.500	1.610	1.779	1.546	1.455	1.471
Shandong	0.871	0.783	1.128	0.964	0.845	0.821
Henan	1.363	1.464	1.368	1.198	0.902	1.063
Hubei	1.236	1.285	1.591	1.431	1.163	1.522
Hunan	1.358	1.533	1.470	1.377	1.305	1.338
Guangdong	0.982	1.071	1.712	1.786	0.881	0.972
Guangxi	1.386	1.629	1.504	1.465	1.557	1.961
Hainan	1.501	1.575	1.615	1.675	1.731	1.686
Chongqing	1.444	1.636	1.685	1.584	1.580	1.645
Sichuan	1.479	2.100	2.253	1.956	2.114	2.729
Guizhou	1.481	1.779	2.157	1.623	2.020	1.839
Yunnan	1.543	1.915	3.287	2.442	1.837	2.395
Shanxi	1.387	1.567	1.933	1.613	1.532	1.567
Gansu	1.503	1.590	1.888	1.797	2.134	2.018
Qinghai	1.617	1.891	2.033	1.930	1.828	2.083
Ningxia	1.515	1.558	1.600	1.595	1.666	1.644
Xinjiang	1.544	1.785	1.634	1.667	1.586	1.767
National Total	1.356	1.488	1.627	1.529	1.475	1.569

From a temporal perspective, the estimated shadow prices of carbon dioxide in the national transportation sector show an upward fluctuation, with the annual growth rate of shadow prices across provinces being relatively similar, approximately around 1%. Specifically, the estimated shadow price of carbon dioxide in the national transportation sector in 2020 was 15,700 RMB/ton, reflecting an increase of 15.44% from 13,600 RMB/ton in 2010. As China's transportation sector develops, the economic costs associated with carbon reduction are becoming increasingly significant.

At the provincial level, there is a notable regional disparity in the shadow prices of carbon emissions in the transportation sector across China. Provinces such as Shandong, Jiangsu, Hebei, Henan, Tianjin, and Guangdong exhibit relatively low shadow prices, at 9,500 RMB/ton, 12,100 RMB/ton, 12,100 RMB/ton, 12,500 RMB/ton, 13,100 RMB/ton, and 13,600 RMB/ton, respectively. These provinces are major transportation hubs with favorable production environments, contributing to their well-developed transportation sectors. Conversely, provinces such as Yunnan, Sichuan, Qinghai, Gansu, Guizhou, and Xinjiang have higher shadow prices, at 23,100 RMB/ton, 21,300 RMB/ton, 18,600 RMB/ton, 18,400 RMB/ton, 18,300 RMB/ton, and 16,700 RMB/ton, respectively. Limited by their natural environment and economic development conditions, these regions have relatively underdeveloped transportation sectors, with lower total carbon emissions but higher production costs. Consequently, the economic costs of reducing carbon emissions are higher compared to more developed transportation regions.

Considering regional heterogeneity, the 30 provinces are categorized into four major economic regions: Eastern, Central, Western, and Northeastern China. The shadow prices of carbon dioxide in the transportation sector across these regions are presented in Table 3. Overall, the Western region has the highest shadow prices for carbon dioxide, followed by the Northeastern region, with the Eastern region exhibiting the lowest prices. Compared with other regions, the eastern region has a developed economy, relatively complete transportation infrastructure, and higher energy efficiency, so the carbon emission reduction cost (shadow price) is lower. The Northeastern region, once an industrial base, has a relatively singular economic structure and high levels of industrialization. However, recent economic growth has slowed, creating significant transition pressures, outdated infrastructure and lower technological levels result in high marginal costs for carbon reduction. The Western region, characterized by relatively backward economic conditions and incomplete transportation infrastructure, relies more heavily on traditional energy sources, leading to low energy efficiency and the highest marginal costs for carbon reduction.

Table 3. Shadow Prices of Carbon Dioxide in the Transportation Sector by Region (Unit: 10,000 RMB/Ton)

	2010	2012	2014	2016	2018	2020	Standard Deviation
Eastern	1.205	1.264	1.365	1.380	1.277	1.286	0.096
Central	1.387	1.486	1.555	1.438	1.218	1.391	0.119
Western	1.464	1.709	1.944	1.751	1.797	1.936	0.130
Northeastern	1.399	1.432	1.484	1.395	1.468	1.524	0.040

Overall, during the study period, the disparities in shadow prices of carbon dioxide in the transportation sector both between and within regions have significantly widened. To ensure equitable regional economic development, differentiated emission reduction policies should be implemented based on the distinct endowments of each area.

4 Conclusion and Recommendations

The results indicate that the shadow price of carbon dioxide in China's transportation sector rose by 15.44% from 2010 to 2020, reflecting an overall increase in the economic costs of carbon reduction as the sector develops. Specifically, the shadow prices fluctuated upward from 2010 to 2015, then showed a slight decline before steadily rising from 2015 to 2020. From a heterogeneity perspective, the marginal costs of carbon reduction in the transportation sector exhibit substantial differences both between and within regions, with a trend toward increasing disparity.

Based on the calculation of carbon reduction costs in China's transportation sector, this paper proposes the following policy recommendations aimed at promoting collaboration and balance in emission reduction efforts, ultimately achieving sustainable development in the sector.

Implement Differentiated Carbon Tax Policies to Optimize Emission Reduction Cost Distribution. The cost of carbon reduction in the eastern region is rising rapidly, and market mechanisms should be used to encourage enterprises to increase investment in emission reduction and reduce long-term costs. In the central and northeastern regions, where there are significant differences in emission reduction costs and high-energy, high-emission enterprises, a tiered tax reduction policy should be enacted. Tax incentives can be provided based on the effectiveness of their emission reductions, encouraging enterprises to adopt advanced reduction technologies. In the western region, greater emphasis should be placed on financial support and technological assistance. In addition to providing direct financial subsidies, a platform for carbon reduction technology transfer and sharing should be established to facilitate the transfer of advanced reduction technologies and experiences from the eastern region to the western region.

Enhance Regional Cooperation to Collaboratively Advance Emission Reduction. By closely aligning regional emission reduction costs with technological needs, a cross-regional low-carbon transportation technology research and development fund should be established to encourage enterprises, universities, and research institutions to conduct cutting-edge technology research. By optimizing transportation structures according to regional transportation demands and resource advantages, the shift of freight transportation to low-carbon modes such as rail and waterway should be encouraged, while reducing the proportion of road transport. Additionally, the development of public transportation and shared mobility options can help decrease the frequency of private vehicle use, thus reducing carbon emissions in the transportation sector.

References

1. Meng, C., Du, X., Zhu, M., et al.: The static and dynamic carbon emission efficiency of transport industry in China. Energy **274**, 127297 (2023)

2. Avotra, A.A.R.N., Nawaz, A.: Asymmetric impact of transportation on carbon emissions influencing SDGs of climate change. Chemosphere **324**, 138301 (2023)
3. Zhang, Y., Zhuo, J., Baležentis, T., et al.: Measuring the carbon shadow price of agricultural production: a regional-level nonparametric approach. Environ. Sci. Pollut. Res. **31**(11), 17226–17238 (2024)
4. Wei, C., Löschel, A., Liu, B.: An empirical analysis of the CO2 shadow price in Chinese thermal power enterprises. Energy Econ. **40**, 22–31 (2013)
5. Ma, C., Hailu, A., You, C.: A critical review of distance function based economic research on China's marginal abatement cost of carbon dioxide emissions. Energy Econ. **84**, 104533 (2019)

Study on the Removal Method of Fine Particles from Flue Gas of Iron and Steel Industry Based on Multi-field Synergistic Effect

Haiying Li, Yan Li(✉), Guoliang Li, and Guijie Zhang(✉)

College of Metallurgy and Energy, North China University of Science and Technology, Tangshan 063000, China
15176531006@163.com, zhguijie@vip.sina.com

Abstract. The study of the low efficiency of the treatment of fine dust in the iron and steel industry has selected the sintering flue gas dust as the object of study, and the multi-field synergistic removal of fine particles from the sintering flue gas by chemical agglomeration, phase change agglomeration, and turbulent agglomeration has been carried out. The effects of phase change agglomeration, chemical agglomeration and turbulent agglomeration on the arrangement of turbulence columns, wind speed and other factors on the multi-field synergistic removal of fine particles were investigated through the multi-field synergistic experimental platform. The results showed that the removal efficiencies of PM_{10} were 95.41% and 85.91% with the multi-field synergistic treatment methods of chemical agglomeration, phase change agglomeration and turbulent agglomeration, respectively, and the volume-averaged particle size of the fine dust increased from 19.94 μm initially to 237 μm, which is about 12 times of the increase in particle size. By using Design-Expert software for response surface optimization, the maximum removal efficiencies of $PM_{2.5}$ and PM_{10} were 96.28% and 86.38%, respectively. After adopting the multi-field cooperative agglomeration technology, the volume-averaged particle size of fine dust was significantly improved, and the removal efficiency of fine particles in the sintered flue gas was high, and the results of the study provide important technical support for the ultra-low emission of the sintered flue gas and the multi-field cooperative agglomeration in the future.

Keywords: sintered flue gas · fine particles · multi-field cooperative · agglomeration

1 Introduction

As an important basic industry, traditional industries such as iron and steel not only make an important contribution to economic development, but also have the potential to save energy and reduce emissions. At present, significant results have been achieved in the treatment of large particles, SO_2, NO_X and other pollutants in the flue gas emissions of the iron and steel industry [1], but the flue gas also contains a large number of fine particles, which are also the focus of urgent attention to environmental governance. Fine

D. Han et al. (Eds.): ICEPG 2024, ESE, pp. 478–486, 2025.
https://doi.org/10.1007/978-3-031-86829-0_54

particles in industrial flue gas mainly come from the combustion process as well as ash and other solid impurities in raw materials, and these fine particles, if emitted into the atmosphere, will cause harm to air quality and human health, as well as pollution of the environment. When the diameter of fine dust $\leq$10 μm, such dust in the air in a suspended state, will be inhaled with the air into the lungs, which will lead to a series of lung diseases; fine dust attached to the clothing, and human skin contact, but also trigger allergic reactions, so the fine dust is a serious danger to people's health [2]! To reduce the emission of fine particles in industrial flue gas is of great significance to improve air quality and enhance people's living environment.

Multi-field synergistic removal of fine particles from sintered flue gas is proposed for the characteristics of the steel production process. On the basis of cloud dust removal, the chemical and physical multi-field synergistic method makes the fine particles collide and agglomerate with each other, so that larger particles are formed and then removed, thus improving the removal efficiency of fine particles [3]. The removal efficiency of fine particles is improved.

2 Experimental Method and Setup

2.1 Multi-field Synergistic Treatment of Fine Dust Fundamentals

In the process of using phase change agglomeration, chemical agglomeration, turbulent agglomeration synergistic treatment of fine particulate matter preliminary analysis of fine particles will be subjected to a variety of forces at the same time, including Coulomb force, inertial force, Van der Waals force, surface adsorption force, surface adhesion force, liquid bridge force, fluid trailing force and other forces, which have a greater impact on the fine particles is the liquid bridge force and fluid trailing force.

Phase change agglomeration through the ultrasonic atomizer will be liquid water atomized into micron-sized fog particles, and then build the formation of the relative humidity of the oversaturated environment, fine particles in the oversaturated environment will be collision with the droplets agglomeration, merger and other processes, this process will also release negative ions with the occurrence of electrostatic reaction of fine dust, so that it will be more likely to settle down in the airflow. In the chemical agglomeration process, the sprayed agglomerant droplets collide and adhere with fine dust particles in a very short time, and then the agglomerant droplets or small polymers that have already adsorbed some fine particles and the dust particles will have the same speed and no relative motion, resulting in that some of the fine dust and agglomerant droplets will be discharged directly [4]. The arrangement of the spoiler column further increases the disturbance of the flow field and promotes particle collision [5]. In the turbulence field, the arrangement of the disturbance column, so that the flow field to produce vortex, small particle size dust particles will be in the vortex circling, particle size particles affected by the vortex is small, mainly in the air flow trailing force with the mainstream movement, and through the vortex and the particle size of small particles collision, so as to agglomerate [6]. The basic principle of multi-field synergistic treatment of fine dusts is shown in Fig. 1.

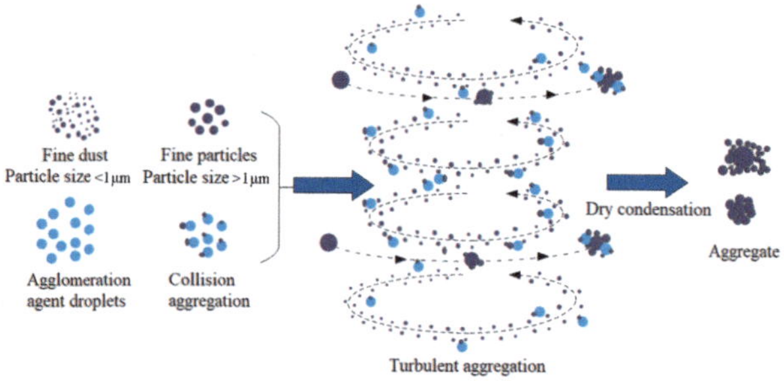

Fig. 1. Basic principle of multi-field synergistic treatment of fine dusts

2.2 Experimental Setup for Synergistic Agglomeration of Fine Particles

Based on the method of multi-field synergistic agglomeration, a multi-field synergistic cloud atomised dust removal test bed was built in the laboratory. The specific device is shown in Fig. 2. The sinter head dust with high concentration of fine dust in the steel production process was selected for the study. The outlet air velocity was set at 3.2 $m \cdot s^{-1}$, and the inlet concentration of dust was kept at 1.4 $g \cdot m^{-3}$, the agglomerating agent was selected as xanthan gum with a mass fraction of 0.04%, and the wetting agent, sodium dodecylbenzene sulphonate (SDBS), was selected with a mass fraction of 0.03%.

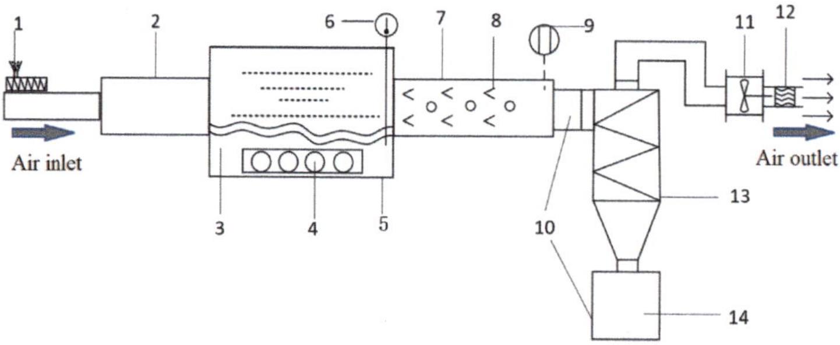

1-vibrating feeder; 2-pre-mixing chamber; 3-water, agglomerant solution; 4-atomiser; 5-atomiser box; 6-thermometer; 7-turbulence Chamber;8 - turbulent plate; 9 - anemometer; 10 - dust particle collection place;11 - induced draft fan; 12 - dehumidifier; 13 - gas-solid separation; 14 -dust collection drum.

Fig. 2. Schematic diagram of multi-field synergistic agglomeration technology

2.3 Effects of Turbulence on Synergistic Agglomeration

Effect of Circular Turbulence Column Diameter and V-Shaped Turbulence Column Edge Length on Co-agglomeration: The side length of the V-shaped turbulent column was selected as 30 mm, the best mass fraction of the above mentioned agglomerating agent and surfactant was selected, and the diameters of the circular turbulent columns were selected as 20 mm, 30 mm, and 40 mm for the experiments, and the transverse distance between the V-shaped turbulent columns was set as 20 mm, and the longitudinal distance between the circular turbulent columns and the V-shape turbulent columns was set as 40 mm, and the experimental measurement of the removal efficiencies of $PM_{2.5}$, PM_{10} were as follows in Fig. 3. From Fig. 3(a), it can be seen that when the circular turbulence column is 30 mm, the dust removal efficiency of fine dust is the highest, and the removal efficiencies of $PM_{2.5}$ and PM_{10} are 92.31% and 85.65%. V-shaped turbulence columns with side lengths of 20 mm, 30 mm, and 40 mm were selected, and a circular turbulence column of 30 mm was selected. The dust removal efficiency of fine dust was obtained, and the results are shown in Fig. 3(b) below. When the side length of the V-shaped turbulence column is 30 mm, the removal rate of sintered fine particles is the highest. The reason for this is that when the turbulence column with a side length of 20 mm is used, because the turbulence column is small in size, the vortex formed in the back of the turbulence column when the airflow flows through the turbulence column is small, and part of the dust-containing airflow flowing through the turbulence column will not be affected by the vortex in the back of the turbulence column.

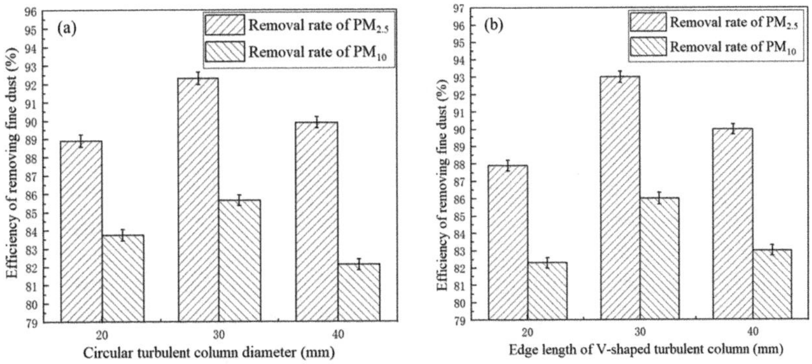

Fig. 3. Effect of turbulent column shape on the removal efficiency of fine dusts

Effect of Transverse and Longitudinal Spacing of Turbulent Columns on Synergistic Agglomeration: Set the lateral distance between the turbulence columns as 15 mm, 25 mm, 35 mm, 45 mm, 55 mm, respectively, because the lateral arrangement of the turbulence columns involves only the V-shaped turbulence columns, so only need to adjust the distance between the V-shaped turbulence columns, resulting in the results of the dust removal efficiency is shown in Fig. 4(a) below, which shows that with the increase of the lateral distance of the turbulence columns, the removal efficiency of fine particles shows a tendency of increasing first and then decreasing.

The longitudinal spacing of the turbulent column was set to be 20 mm, 30 mm, 40 mm, 50 mm and 60 mm for the multi-field collaborative dust removal experiments, and the removal efficiency of fine dust was measured after the experiments as shown in Fig. 4(b) below. From the figure, it can be seen that with the increase of the longitudinal distance of the turbulence column, the dust removal efficiency of fine dust shows a trend of increasing and then decreasing, and the study shows that the vortex formed by the turbulence column will be affected by the neighbouring vortex in addition to the influence of the rear turbulence column.

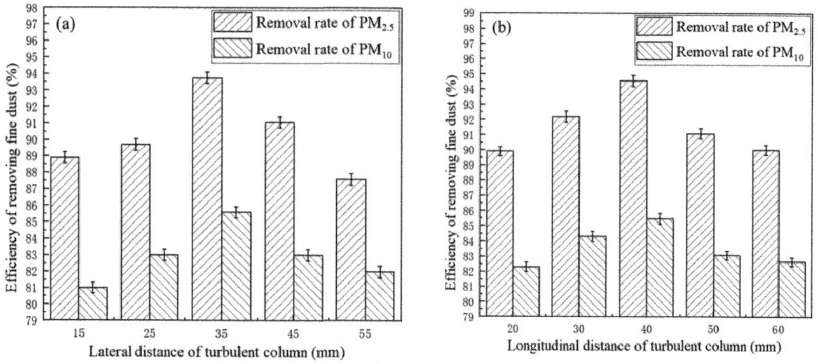

Fig. 4. Influence of turbulence column spacing on the removal efficiency of fine dust

Effect of Flue Gas Flow Rate on Co-agglomeration of Fine Dusts: In the process of turbulent agglomeration, air velocity is an important factor affecting turbulent agglomeration, respectively, set the induced draft fan outlet wind speed of 2 $m{\cdot}s^{-1}$, 3 $m{\cdot}s^{-1}$, 4 $m{\cdot}s^{-1}$, 5 $m{\cdot}s^{-1}$, 6 $m{\cdot}s^{-1}$, and at the same time, set the turbulence column longitudinal spacing is set to 40 mm, and the transverse distance is 35 mm, respectively, to study the effect of chemical agglomeration, phase change agglomeration, turbulent agglomeration on the removal efficiency of fine particles, and the relationship of the removal efficiency and dust removal efficiency of fine particles are shown in Fig. 5 below. The relationship between the removal efficiency and wind speed is shown in Fig. 5 below.

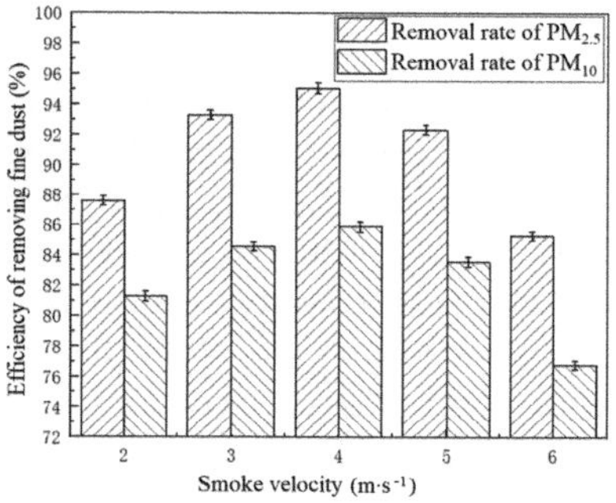

Fig. 5. Effect of flue gas flow rate on removal efficiency of fine dust

From Fig. 5, with the increase of flue gas flow rate, the removal efficiency of fine dust first increases and then decreases, when the flue gas flow rate is 4 $m{\cdot}s^{-1}$, the removal efficiency of fine dust $PM_{2.5}$, PM_{10} is 95.41% and 85.91%, respectively, and when the flue gas flow rate is 6 m·s^{-1}, the removal efficiency of fine dust $PM_{2.5}$, PM_{10} decreases to 88.61% and 79.31%, respectively, and with the increase in the air speed turbulence intensity With the increase of wind speed turbulence intensity increases, vortex intensity increases, so that the degree of diffusion of fine particles increases.

2.4 Comparison of Experiments Under Different Synergistic Agglomerations

A comparison of phase change, chemical co-agglomeration and phase change, chemical, turbulent co-agglomeration particle sizes is shown in Fig. 6 below.

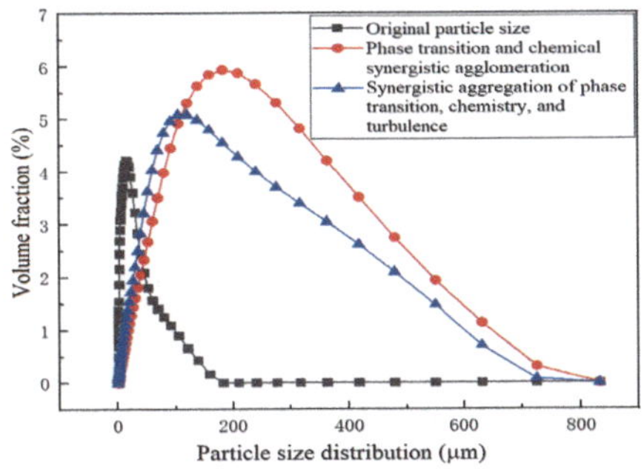

Fig. 6. Comparison of particle size after multi-field synergistic agglomeration

After phase change and chemical synergistic agglomeration, the dust particle size increased to 188.5 μm, and the initial volume-averaged particle size of microfine particles in multi-field synergistic agglomeration after adding turbulent agglomeration increased from 19.94 μm to 237 μm, and the volume-averaged particle size of the dust increased about 12 times, and the removal efficiencies of the chemical agglomeration plus phase change agglomeration of $PM_{2.5}$, and PM_{10} were the highest in removing the microfine particles of $PM_{2.5}$ and PM_{10}, and the removal efficiencies of chemical agglomeration plus phase change agglomeration were the highest in 86.13% and 80.91%, respectively. 80.91%, in the process of chemical agglomeration and phase change agglomeration, the fine particles simultaneously undergo chemical agglomeration dominated by liquid bridge force and phase change agglomeration with nucleation growth. Therefore, chemical and phase change agglomeration occurred simultaneously, and the highest removal efficiency of $PM_{2.5}$ and PM_{10} was 95.41% and 85.91% when chemical agglomeration, phase change agglomeration and turbulent agglomeration were used. Preliminary analyses revealed that chemical agglomeration has a greater influence in the multi-field synergistic agglomeration process, because both phase change agglomeration and turbulent agglomeration involve chemical agglomeration dominated by liquid-bridge forces.

2.5 Experiment on Optimisation of Dust Removal Efficiency by Design-Expert Response Surface

The multi-field synergistic agglomeration technique of phase change, chemistry and turbulence was found to have a high removal efficiency for fine particles through experiments, and in order to further view the factors affecting the agglomeration of fine particles, a mathematical model was developed using Box-Behnken to derive the optimum dust removal process conditions. On the basis of phase change agglomeration the factors that have a greater influence on the dust removal efficiency were selected for response surface optimisation experiments. The response surface experiment was designed as 3 factors and 3 levels, with xanthan gum mass fraction, SDBS mass fraction, and flue gas flow rate as the influencing factors, and the removal efficiencies of $PM_{2.5}$ and PM_{10} as the response values. The mist volume in the experiment was set to 60 g·m^{-3}, and the turbulence column was selected as the best diameter, side length and the best arrangement. The experimental results are shown in Table 1.

Table 1. Box-Behnken test programme and results

serial number	Xanthan gum mass fraction (%)	SDBS mass fraction (%)	flue gas flow rate (m-s^{-1})	$PM_{2.5}$ removal efficiency η_1 (%)	PM_{10} removal efficiency η_2 (%)
1	0.03	0.03	4	94.62	84.61
2	0.03	0.04	5	92.66	81.91
3	0.03	0.05	4	94.64	84.61
4	0.05	0.05	4	93.45	83.71
5	0.04	0.04	4	96.06	85.92
6	0.03	0.04	3	92.17	82.07
7	0.04	0.05	3	94.13	83.91
8	0.04	0.04	4	96.11	86.31
9	0.05	0.03	4	92.48	81.71
10	0.05	0.04	3	91.81	81.39
11	0.04	0.03	5	92.64	81.96
12	0.05	0.04	5	89.17	78.98
13	0.04	0.04	4	96.01	86.21
14	0.04	0.05	5	93.51	83.06
15	0.04	0.03	3	93.89	84.01
16	0.04	0.04	4	95.87	85.89
17	0.04	0.04	4	96.28	86.41

The response surface and contour plots were plotted based on the regression equation using Design-Expert12 software, as shown in Figs. 7 and 8, where it can be seen that the contour plots of xanthan gum mass fraction and flue gas flow rate are elliptical in shape, indicating that the interaction of these two variables is obvious.

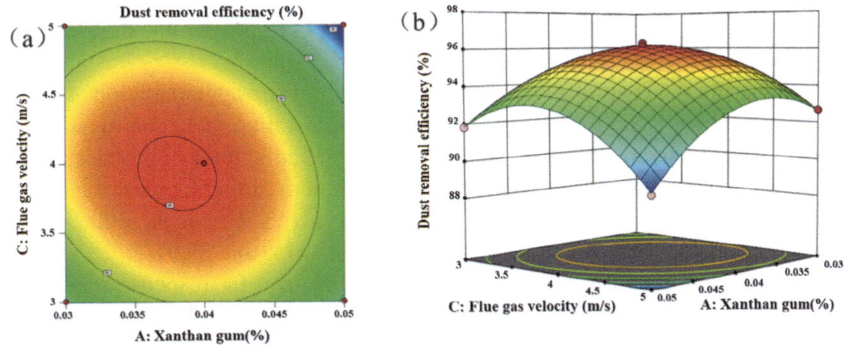

Fig. 7. The interactive effects of xanthan gum mass fraction and flue gas flow rate on $PM_{2.5}$

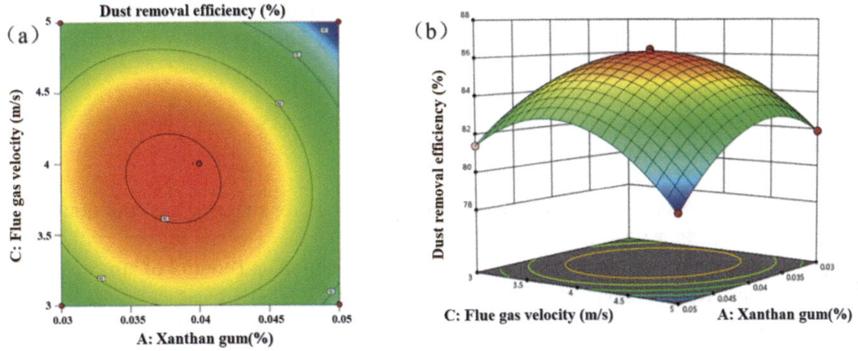

Fig. 8. The interaction between xanthan gum mass fraction and flue gas flow rate on PM_{10}

From the response surface plots, it can be seen that the opening of the surface is downward, indicating that the xanthan gum mass fraction and flue gas flow rate have a parabolic relationship on the removal efficiency of $PM_{2.5}$ and PM_{10}, which means that there is a maximum value in this model. The optimal parameters are: the mass fraction of xanthan gum is 0.04%, the flue gas flow rate is 4 m/s, and the mass fraction of SDBS is 0.05%, the removal efficiencies of $PM_{2.5}$ and PM_{10} are 96.28% and 86.38%, respectively. Based on the derived optimal parameters, it was verified through several experiments, and it was found that the removal efficiencies of $PM_{2.5}$ and PM_{10} could reach 96.28% and 86.38%.

3 Conclusion

The results of the study showed that:

1) After multi-field synergistic agglomeration by phase change, chemical and turbulent agglomeration, the initial volume-averaged particle size of fine particles increased from 19.94 μm to 237 μm, and the volume-averaged particle size of dust increased by about 12 times, and the removal efficiency of chemical agglomeration plus phase change agglomeration of $PM_{2.5}$ and PM_{10} in terms of removing fine dust $PM_{2.5}$ and PM_{10} was the highest at 86.13% and 80.91%.
2) By comparing the chemical and phase change synergistic agglomeration and chemical, phase change and turbulent synergistic agglomeration, it can be found that the chemical agglomeration plus phase change agglomeration has a higher dust removal efficiency, coupled with the turbulent agglomeration to further increase the dust volume average particle size, so that the multi-field synergistic removal of fine particles of $PM_{2.5}$ and PM_{10} has the highest efficiency of 95.41% and 85.91%.
3) A mathematical model was developed by Box-Behnken for response surface optimization, and the highest removal efficiencies of $PM_{2.5}$ and PM_{10} were 96.28% and 86.38%, respectively, when the mass fraction of xanthan gum was 0.04%, the flue gas flow rate was 4 $m \cdot s^{-1}$, and the mass fraction of SDBS was 0.05%.

Acknowledgments. This research was supported by the Natural Science Foundation of Hebei Province (E2022209138). Tangshan Science and Technology Plan Project (No. 21130231C).

References

1. Hu, J.: Exploration on multi-pollutant reduction technology for sintering gas emissions in steel industry. Metall. Power (4) (2022)
2. Wang, Z.: Countermeasures for the comprehensive prevention and control of atmospheric fine dust pollution. Resour. Conserv. Environ. Prot. **08**, 48–52 (2023)
3. Li, H., Hu, Q., Cheng, H., et al.: Effect of chemical agglomeration on dust removal efficiency of sintered fine dust. Mod. Chem. Ind. **40**(08), 139–143+149 (2020)
4. Bin, H., Yang, Y., Lei, Z., et al.: Experimental and DFT studies of PM2.5 removal by chemical agglomeration. Fuel **212**, 27–33 (2018)
5. Li, R., Li, C., Zhuang, J., et al.: Mechanistic influence of chemical agglomeration agents on removal of inhalable particles from coal combustion. ACS Omega **5**(40), 25906–25912 (2020)
6. Afkhami, M., Hassanpour, A., Fairweather, M., et al.: Fully coupled LES-DEM of particle interaction and agglomeration in a turbulent channel flow. Comput. Chem. Eng. **78**, 24–38 (2015)

Synergistic Mitigation of Industrial Pollution and Carbon Emission and Their Drivers in Zhejiang Province, China

Qing Zhang and Hua Zhu(✉)

School of Geomatics, Zhejiang University of Water Resources and Electric Power, Hangzhou 310018, China
zhuhua18@mails.ucas.ac.cn

Abstract. The joint decrease in industrial pollution and CO_2 emissions is crucial for fostering sustainable regional growth. This study investigated the synergistic mitigation of industrial pollution and CO_2 emissions in Zhejiang and the driving forces from 2005 to 2021, utilizing the synergistic control of the cross-elasticity coefficient (S) and the STIRPAT modeling framework. The results indicate that: (1) The industrial air pollution equivalent (LAP) in Zhejiang has continued to decrease from 34.61×10^6 t in 2005 to 8.12×10^6 t in 2020, with a slight rebound to 8.35×10^6 t in 2021. Power plants (PPCB) and other industrial processes (PROT) are the primary contributors to LAP. (2) From 2005 to 2021, the industrial CO_2 emissions in Zhejiang exhibited a fluctuating trend of "increase-decrease-increase". PPCB was the primary contributor to industrial CO_2 emissions, with an average contribution rate of 68.79%. (3) From 2005 to 2021, the synergistic emission reduction levels of industrial LAP and CO_2 in Zhejiang were unstable, and the industrial measures for pollution and CO_2 emissions mitigation had a greater impact on reducing LAP than CO_2 emissions. (4) Population size (PS) is a common key influencing factor for both industrial LAP and CO_2 emissions. On average, a 1% increase in PS leads to a 3.087% decrease in industrial LAP in Zhejiang, while CO_2 emissions increase by 0.405%.

Keywords: Industry · Air Pollution · Carbon Emission · Synergistic Mitigation · Drivers

1 Introduction

China has witnessed substantial improvements in the realm of air pollution control and CO_2 emission reduction in recent years [1], yet its efforts in air pollution prevention still face severe challenges due to the impact of climate change, energy consumption patterns, and industrial layouts [2]. In particular, the increasingly prominent issue of regional composite air pollution, including $PM_{2.5}$ and ozone, has further intensified the difficulty of management [3]. It is estimated that China accounted for 27.3% of global primary energy consumption and 27.7% of worldwide CO_2 emissions in 2022. Moreover, the yearly average $PM_{2.5}$ concentration stood at 29 μg/m^3, substantially higher

D. Han et al. (Eds.): ICEPG 2024, ESE, pp. 487–495, 2025.
https://doi.org/10.1007/978-3-031-86829-0_55

than the World Health Organization's advisable limit of 5 μg/m^3 [4]. The industrial sector is a significant contributor to both atmospheric pollutants and CO_2 emissions in China. According to the China Emission Accounts and Datasets (CEADs) and previous research [5], in 2022, industrial sources were responsible for 38.18% of China's overall carbon emissions, while in 2020, industrial particulate matter emissions constituted 65.6% of the total emissions. Therefore, promoting synergistic governance of CO_2 emissions and air pollution in the industrial sector is crucial for achieving regional sustainability.

Scholars have conducted extensive research on the synergistic effects of air pollution reduction and CO_2 emission mitigation across multiple scales, sectors, and perspectives. This includes scales such as global [6], national [7], provincial [4], and municipal [2]; sectors like energy, transportation [4], and electricity; and various angles such as policy analysis [8], evaluation of the implementation effects of measures [9], comparison of regional differences [10], and exploration of inter-sectoral coordination mechanisms [4]. Nonetheless, there is a scarcity of existing research that delves into the synergistic mitigation of air pollution and CO_2 emissions within the entire industrial sector. Although Kang et al. [11] explored the trends and drivers of industrial air pollution and CO_2 emissions, the industrial emission data utilized in their study were not sourced from the aggregation of sub-sectors, which impedes the comprehensive examination of air pollution reduction and CO_2 emission mitigation efforts within the industrial sector. Additionally, most research has concentrated on identifying factors of the synergy between air pollution and CO_2, with comparatively less attention given to separately analyzing the driving factors for each. For instance, Chen et al. [2] employed a geographic detector model with optimized parameters to investigate the underlying factors influencing the level of synergy between pollution reduction and CO_2 mitigation in the Yangtze River Economic Zone. However, this limits scholars' comprehension of how individual driving factors impact air pollution and CO_2 emissions.

To address the gaps above, this study analyzes the trends in industrial air pollution and CO_2 emissions in Zhejiang, along with the contributions of various industries, based on data from multiple industrial sectors spanning from 2005 to 2021. The study employs the synergistic control of the cross-elasticity coefficient (S) to assess the level of synergistic reduction between industrial air pollution and CO_2 emissions in Zhejiang. Furthermore, we explore the drivers of industrial air pollution and CO_2 emissions using the STIRPAT model, analyzing the mechanisms through which different driving factors influence air pollution and CO_2 emissions.

2 Materials

2.1 Study Area

Zhejiang is situated in the southern Yangtze River Delta (118°01′ E–123°10′ E, 27°02′ N–31°11′ N). In 2020, Zhejiang created the first national pollution and CO_2 reduction innovation zone. According to the research findings of Li et al. [12], in 2021, industrial $PM_{2.5}$ and CO_2 emissions in Zhejiang accounted for 85.70% and 82.61% of the total emissions, respectively. The industrial sector in Zhejiang faces significant challenges, including a large economic scale, high total energy consumption, a substantial proportion of CO_2 emissions, and increasing difficulties in achieving low-carbon transformation.

2.2 Data Sources

The data on industrial air pollutants and CO_2 emissions in Zhejiang are sourced from Li et al. [12]; whereas the data on population, urbanization rate, GDP, and GDP contributed by the secondary industry are derived from the "Zhejiang Statistical Yearbook" covering the years from 2006 to 2022. The proportions of coal fuels in total energy consumption for the years 2006–2020 are sourced directly from the "China Energy Statistical Yearbook". Additionally, the proportions for 2005 and 2021 were calculated based on definitions using respective data on total energy consumption and coal consumption. Given the different emission sources for air pollutants and CO_2 emissions, to facilitate a synergistic analysis, we have selected emission sources from the same industries: industrial boilers (INCB), power plants (PPCB), the cement industry (PRCE), the iron and steel industry (PRIR), and other industrial processes (PROT).

3 Methods

3.1 Industrial Air Pollution and CO_2 Emission Accounting

The emissions of industrial air pollutants and CO_2 in Zhejiang are aggregated from various emission sources across different industries. These air pollutants comprise six types: SO_2, NO_X, CO, VOCs, $PM_{2.5}$ and PM_{10}. To standardize the pollution emission data, Formula 1 is employed to convert the emissions of air pollutants into a common measure, namely the air pollution equivalent (LAP).

$$E_{LAP} = \alpha E_{SO_2} + \beta E_{NO_X} + \gamma E_{CO} + \delta E_{VOCs} + \epsilon E_{PM_{2.5}} + \theta E_{PM_{10}} \quad (1)$$

where E_{LAP} represents the industrial LAP (10^6 t); E_{SO_2} denotes the SO_2 emissions (10^6 t); E_{NO_X} is the NO_X emissions (10^6 t); E_{CO} is the CO emissions (10^6 t); E_{VOCs} is the VOCs emissions (10^6 t); $E_{PM_{2.5}}$ is the $PM_{2.5}$ emissions (10^6 t); and $E_{PM_{10}}$ is the PM_{10} emissions (10^6 t). α, β, γ, δ, ϵ, and θ represent the equivalency coefficients for each pollutant, which are taken as 1/0.95, 1/0.95, 1/16.7, 1/0.95, 1/2.18 and 1/2.18, respectively, referencing the relevant official document.

3.2 Synergistic Control of Cross Elasticity Coefficient

The S can quantitatively describe the co-mitigation and co-emission effects of greenhouse gases and air pollution. This study uses S to quantitatively elucidate the synergistic effects between industrial CO_2 emissions and changes in LAP in Zhejiang. The calculation formula for S is as follows:

$$S = \frac{\Delta E_{CO_2}/E_{CO_2}}{\Delta E_{LAP}/E_{LAP}} \quad (2)$$

where ΔE_{CO_2} denotes the change in CO_2 emissions, E_{CO_2} indicates the amount of CO_2 emissions, and ΔE_{LAP} represents the change in LAP emissions. S represents the synergistic control of the cross-elasticity coefficient, with its meaning detailed in Table 1.

Table 1. The meaning of S.

S	Implication
S > 0 (Both the numerator and the denominator are positive values)	Both LAP and CO_2 emissions have been reduced, showcasing a synergistic effect in emission reduction
S ≤ 0	A reduction has been achieved in either LAP or CO_2 emissions
S = 1	LAP and CO_2 have the same level of emission reduction
0 < S < 1	The degree of emission reduction for LAP is higher than that for CO_2
S > 1	The degree of emission reduction for CO_2 is higher than that for LAP
S > 0 (Both the numerator and the denominator are negative values)	Both LAP and CO_2 emissions are increasing, exhibiting a negative synergy effect

3.3 STIRPAT Model

The STIRPAT model can effectively quantify the independent impacts of changes in factors such as population, economy, and technology on the environment. Due to its strong expandability, the model has become an important method for analyzing the driving factors of CO_2 emissions, which originates from the IPAT equation and overcomes its limitation of assuming equal proportional impacts. The expression is as follows:

$$\mathrm{I} = a \times P^b \times A^c \times T^d \times \mathrm{e} \tag{3}$$

where *a* denotes a constant; *I*, *P*, *A*, and *T* represent environmental impact, population size, affluence, and technology level, respectively.

Based on references [5, 11] and considering the actual situation in Zhejiang, factors influencing industrial pollution and CO_2 emissions were selected to expand Eq. (3):

$$lnI = lna + blnP + clnUR + d\,lnGDP + elnIS + flnES + glnEE + lnh \tag{4}$$

where *I* represent industrial LAP or CO_2 emissions, while *a* and *P* have the same meanings as in Eq. (3), *b* to *h* are regression coefficients. *UR*, *GDP*, *IS*, *ES*, and *EE* stand for urbanization rate, GDP, industrial structure, energy structure, and energy efficiency, respectively. Among these, energy efficiency serves as an indicator of technological level.

4 Results and Discussion

4.1 Changes in LAP and CO_2 Emissions

Figure 1 illustrates the proportion and trends of LAP and CO_2 emissions across various industrial sectors in Zhejiang. From 2005 to 2021, industrial LAP exhibited a trend of continuous decline, followed by a subsequent increase. Specifically, the LAP decreased from 34.61 × 10^6 t in 2005 to 8.12 × 10^6 t in 2020 and then rose to 8.35 × 10^6 t in 2021. This demonstrates that Zhejiang's pollution reduction policies have been highly effective. The outbreak of the COVID-19 pandemic impacted the rollout and implementation of pollution reduction policies, resulting in an increase in LAP in 2021. From 2005 to 2021, the PPCB and PROT sectors contributed to industrial LAP in Zhejiang with fluctuating rates ranging from 21.45% to 57.55% and from 10.58% to 50.71%, respectively. Specifically, PPCB was the main contributing sector to industrial LAP during the earlier period of the study, while PROT became the dominant sector in the later stages. The contribution rates of INCB and PRIR sectors to industrial LAP also showed significant fluctuations, ranging from 4.30% to 19.26% and from 1.83% to 9.87%, respectively. In contrast, the contribution rate of the PRCE sector remained relatively stable, falling between 10.15% and 16.07%.

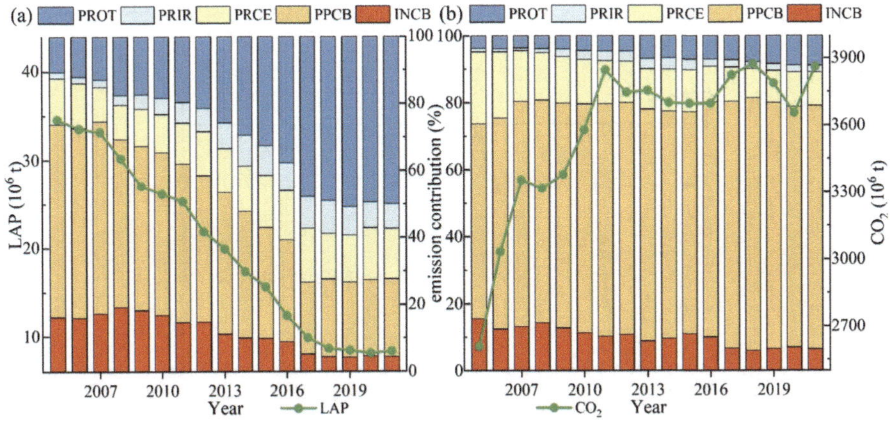

Fig. 1. Changes in LAP and CO_2 emissions from 2005 to 2021 in Zhejiang.

From 2005 to 2021, industrial CO_2 emissions in Zhejiang fluctuated in a cyclical pattern of "increase-decrease-increase". Specifically, emissions reached their lowest point in 2005, totaling 2609.15 × 10^6 t, and climbed to their highest peak in 2018, amounting to 3873.94 × 10^6 t. During the periods of 2005–2007 and 2008–2011, industrial CO_2 emissions experienced rapid increases, with growth rates of 28.60% and 15.90% respectively. In contrast, from 2012 to 2021, industrial CO_2 emissions remained relatively stable. This indicates that the implementation of policies such as the low-carbon pilot program around 2012 and the transition from extensive development to high-quality development in Zhejiang effectively curbed the rapid growth of industrial CO_2 emissions. PPCB is the largest contributor to industrial CO_2 emissions in Zhejiang, with an average contribution rate of 68.79% from 2005 to 2021. In contrast, PRIR contributes the least, averaging only 2.19% over the same period. INCB and PRCE sectors have more pronounced contributions, falling within the ranges of 5.96% to 15.55% and 8.75% to 21.49%, respectively during the specified period. PROT, meanwhile, has a relatively smaller average contribution of 6.00% to industrial CO_2 emissions in Zhejiang.

4.2 Synergistic Effect of Pollution and CO_2 Emission

Based on the emission data of industrial LAP and CO_2 from 2005 to 2021, we utilize Eq. (2) to calculate the S values for industrial LAP and CO_2 emissions in Zhejiang (Fig. 2). The S value for industrial LAP and CO_2 emissions in Zhejiang has shown unstable variations. Specifically, during the periods of 2005–2007, 2009–2011, 2012–2013, and 2015–2018, the S value between industrial LAP and CO_2 emissions was less than 0, indicating that only industrial LAP were reduced after the implementation of emission reduction measures, while industrial CO_2 emissions increased during these years (as illustrated in Figs. 1 and 2). Furthermore, the S values for the periods 2005–2006 and 2006–2007 were −5.52 and −9.96, respectively, suggesting that the increase in industrial CO_2 emissions in Zhejiang from 2005 to 2007 was significantly greater than the reduction in LAP.

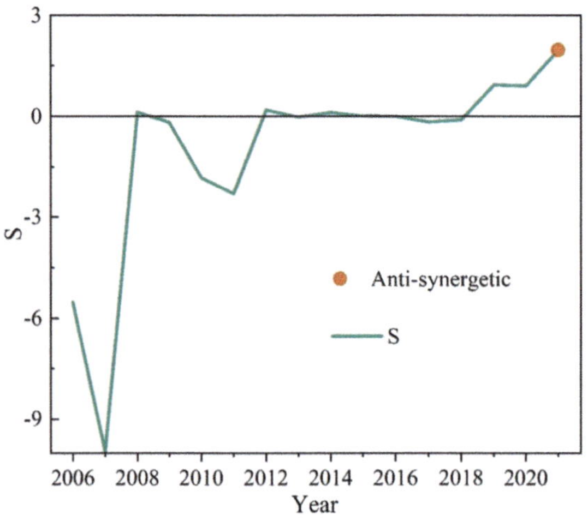

Fig. 2. Changes in S from 2005 to 2021 in Zhejiang.

During the years 2007–2008, 2011–2012, 2013–2015, and 2018–2020, the S values for industrial LAP and CO_2 emissions in Zhejiang fell within the range of 0.01 to 0.93. This signifies that the measures implemented for industrial pollution and CO_2 reduction effectively led to simultaneous reductions in both emissions, albeit with a more pronounced effect on reducing LAP compared to CO_2 emissions. Notably, despite the S being 1.96 for industrial LAP and CO_2 emissions in Zhejiang between 2020 and 2021, both emissions increased, indicating an anti-synergistic state in industrial LAP and CO_2 emissions. The increase in industrial LAP and CO_2 emissions in Zhejiang during the 2020–2021 period was attributed to the impact of the COVID-19 pandemic. To stimulate economic growth, Zhejiang slightly increased its consumption of energy and resources, leading to a rise in both industrial pollution and CO_2 emissions.

4.3 Common Driving Factors of Pollution and CO_2 Emission

Due to the presence of significant multicollinearity among the influencing factors in Eq. (4) (variance inflation factor >10), the ridge regression technique was employed to overcome the adverse effects of multicollinearity on the regression results and to accurately assess the contribution of each influencing factor to the dependent variable. Factors that did not meet the significance threshold were subsequently excluded from the analysis. The influencing factors of industrial pollution and CO_2 emissions in Zhejiang are presented in Table 2. All listed factors have passed the significance test at the 5% level. The regression models constructed for LAP, CO_2, and CO demonstrate significant overall p-values of 0.000, 0.001, and 0.000, respectively, with R^2 of 0.974, 0.811, and 0.876. These results indicate a high degree of model fitting accuracy and effectiveness for the STIRPAT model.

From 2005 to 2021, as shown in Table 2, although PS, UR, GDP, IS, and EE had a negative impact on industrial LAP in Zhejiang, they had a positive impact on CO_2 emissions. However, the direction of influence on industrial CO (LAP component) and CO_2 emissions in Zhejiang is the same for these factors. Additionally, ES has a positive impact on both LAP, CO_2, and CO. Hence, in Zhejiang's industrial sector, effective regulation and management of factors such as PS, UR, GDP, IS, EE, and ES can collectively contribute to reducing pollution and mitigating CO_2 emissions.

In Zhejiang, industrial emissions of LAP, CO_2, and CO exhibit specific trends in response to changes in various factors. Specifically, a 1% average increase in PS leads to a 3.087% decrease in LAP but a 0.405% and 0.419% increase in CO_2 and CO, respectively. Similarly, a 1% rise in UR results in a 0.663% decrease in LAP and a 0.177% and 0.190% increase in CO_2 and CO. GDP growth of 1% causes a slight 0.118% drop in LAP but a minor increase in CO_2 and CO emissions by 0.057% and 0.071%. In contrast, IS growth of 1% boosts LAP by 1.440%. ES growth of 1% leads to increases in LAP, CO_2, and CO emissions by 0.792%, 0.160%, and 0.278%, respectively. Lastly, EE growth of 1% results in a 0.260% decrease in LAP but a 0.185% and 0.233% increase in CO_2 and CO emissions.

Zhejiang, a developed coastal area characterized by its high population density and urbanization levels, is home to enterprises that possess both the financial resources and the drive to invest in environmental technologies aimed at minimizing pollutant emissions. Furthermore, advancements in public transportation networks and infrastructure have significantly reduced the dependence on private automobiles, consequently decreasing the emission of atmospheric pollutants. Nevertheless, the surge in economic growth, population, and urban sprawl has fueled an increase in industrial energy demand, particularly the consumption of fossil fuels, which may have inadvertently led to a rise in industrial CO_2 and CO emissions. In addition, the growth in the proportion of the industrial economy is often accompanied by an expansion in production scale, which in turn boosts resource consumption and pollutant emissions. An increase in the share of coal consumption in total energy will also result in higher emissions of industrial pollutants and CO_2. Notably, the improvement in energy efficiency in Zhejiang from 2005 to 2021 was insufficient to offset the increase in industrial CO and CO_2 emissions caused by rising energy consumption. Instead, it even triggered a rebound effect in consumption, further increasing emissions of industrial CO and CO_2.

Table 2. Synergistic influence factors of industrial pollution and CO_2 emissions.

Factors	Constant	PS	UR	GDP	IS	ES	EE
LAP	24.874	–3.087	–0.663	–0.118	1.440	0.792	–0.260
p	0.000	0.000	0.004	0.001	0.000	0.001	0.036
CO_2	2.364	0.405	0.177	0.057	/	0.160	0.185
p	0.051	0.007	0.003	0.000	/	0.019	0.000
CO	1.461	0.419	0.190	0.071	/	0.278	0.233
p	0.281	0.021	0.011	0.000	/	0.002	0.000

5 Conclusion and Policy Implications

(1) From 2005 to 2021, industrial LAP in Zhejiang showed a continuous decline. However, it rose slightly from 2020 to 2021. During the early stages of the study, PPCB was the main contributor to industrial LAP in Zhejiang, while PROT dominated in the later stages. Overall, industrial CO_2 emissions in Zhejiang fluctuated in a "rise-fall-rise" cycle from 2005 to 2021. They grew rapidly between 2005 and 2007 and again between 2008 and 2011, while remaining relatively stable from 2012 to 2021. PPCB was the main contributor to industrial CO_2 emissions in Zhejiang. Therefore, it is crucial to strengthen pollution control in PROT sectors and CO_2 emission reduction efforts in PPCB sectors in the future.

(2) From 2005 to 2021, the level of synergistic reduction in industrial LAP and CO_2 emissions in Zhejiang was unstable. Moreover, industrial pollution and CO_2 mitigation measures had a greater impact on reducing pollution compared to mitigating CO_2 emissions. The industrial LAP and CO_2 emissions in Zhejiang exhibited an "anti-synergetic" effect from 2020 to 2021. Therefore, it is recommended that Zhejiang focus on strengthening the formulation and implementation of industrial CO_2 reduction policies in the future.

(3) In Zhejiang's industrial sector, policies targeting key indicators like PS, UR, etc., exhibit a synergistic effect in reducing pollution and mitigating CO_2 emissions. Among these, PS is a common key factor influencing both industrial LAP and CO_2 emissions. In the future, Zhejiang should aim to balance population growth, economic development, and urbanization rates, while decreasing coal's share in the energy mix and substantially increasing clean energy utilization. Additionally, research and development of industrial emission reduction technologies must be vigorously promoted.

Funding. This research was supported by the Zhejiang Provincial University Student Science and Technology Innovation Activity Plan (Xinmiao Talent Plan) (grant no. 2024R423A002), Zhejiang University of Water Resources and Electric Power College Students' Innovative Entrepreneurial Training Plan Program (grant no. S202411481039) and the Joint Funds of the Zhejiang Provincial Natural Science Foundation of China (grant no. LZJWY23E090004).

References

1. Song, P., Mao, X., Li, Z., et al.: Study on the optimal policy options for improving energy efficiency and Co-controlling carbon emission and local air pollutants in China. Renew. Sustain. Energy Rev. **175**, 113167 (2023)
2. Chen, S., Tan, Z., Mu, S., et al.: Synergy level of pollution and carbon reduction in the Yangtze River Economic Belt: spatial-temporal evolution characteristics and driving factors. Sustain. Cities Soc. **98**, 104859 (2023)
3. Yang, H., Yao, R., Sun, P., et al.: Spatiotemporal evolution and driving forces of $PM_{2.5}$ in urban agglomerations in China. Int. J. Environ. Res. Public Health **20**(3), 2316 (2023)
4. Qi, Z., Zheng, Y., Feng, Y., et al.: Co-drivers of air pollutant and CO_2 emissions from on-road transportation in China 2010–2020. Environ. Sci. Technol. **57**(50), 20992–21004 (2023)
5. Chen, X., Zhang, J., Tang, X.: Synergistic effect of industrial air pollution and carbon emission reduction in China and influencing mechanism. Resour. Sci. **44**(12), 2387–2398 (2022)
6. Radu, O.B., van den Berg, M., Klimont, Z., et al.: Exploring synergies between climate and air quality policies using long-term global and regional emission scenarios. Atmos. Environ. **140**, 577–591 (2016)
7. Nam, K.-M., Waugh, C.J., Paltsev, S., et al.: Synergy between pollution and carbon emissions control: comparing China and the United States. Energy Econ. **46**, 186–201 (2014)
8. Rive, N.: Climate policy in Western Europe and avoided costs of air pollution control. Econ. Model. **27**(1), 103–115 (2010)
9. Huang, X., Xu, X., Wang, Q., et al.: Assessment of agricultural carbon emissions and their spatiotemporal changes in China, 1997–2016. Int. J. Environ. Res. Public Health **16**(17), 3105 (2019)
10. Chen, Y., Zhao, Z., Yi, W., et al.: Has China achieved synergistic reduction of carbon emissions and air pollution? Evidence from 283 Chinese cities. Environ. Impact Assess. Rev. **103**, 107277 (2023)
11. Kang, Z., Li, W., Liu, W.: Influencing factors and promoting measures of industrial pollution abatement and carbon reduction of the city clusters in the Yellow River basin. China Environ. Sci. **43**(4), 1946–1956 (2023)
12. Li, S., Wang, S., Wu, Q., et al.: Emission trends of air pollutants and CO_2 in China from 2005 to 2021. Earth Syst. Sci. Data **15**(6), 2279–2294 (2023)

Spatio-Temporal Analysis of Tropospheric NO_2 in Northeast Asia Based on Ozone Monitoring Instrument (OMI) Data

Huiqian Wang(✉)

School of Surveying and Land Information Engineering, Henan Polytechnic University, Jiaozuo 454003, Henan, China
whq200300@163.com

Abstract. NO_2 is an important trace gas in atmosphere, influencing air quality and human health. The spatio-temporal distribution of tropospheric NO_2 Vertical Column Density (VCD) in Northeast Asia is analyzed based on OMI data from 2005 to 2016. Results show that average NO_2 VCD in Northeast Asia over these 12 years is 3.97×10^{15} molec/cm^2, with high values mainly in areas with large population density and developed industry. The average NO_2 in Beijing–Tianjin–Hebei region of China reaches 11.41×10^{15} molec/cm^2. Other NO_2 high-value areas include Shandong province, Henan province, Pearl River Delta, and Yangtze River Delta in China, Seoul in Korea, and Tokyo in Japan. The average NO_2 in South Korea is 7.18×10^{15} molec/cm^2, almost the same as that in Pearl River Delta of China. The average NO_2 VCD in Japan is 2.96×10^{15} molec/cm^2. NO_2 is the highest in winter, followed by autumn and spring, and the lowest in summer. The change of NO_2 shows the trend of first increase then decrease, reaching peak in 2011 (4.99×10^{15} molec/cm^2).The average growth rate is 7.9% in 2005–2011, and −7.6% in 2012–2016, with average growth rate of 0.88%. NO_2 increases by 0.6×10^{15} molec/cm^2 in North Korea, increases by 2.45×10^{15} molec/cm^2 in Beijing-Tianjin-Hebei region, whereas it decreases at varying degrees in Japan, Yangtze River Delta, Pearl River Delta, Henan and Anhui provinces. It is necessary for developing regions to transform economic development patterns, optimize industrial structures and promote the use of clean energy.

Keywords: Tropospheric NO_2 Vertical Column Density · Northeast Asia Area · OMI Data · Spatio-temporal Distribution

1 Introduction

Nitrogen dioxide (NO_2) is an important component of troposphere and trace gas influencing the quality of atmosphere, including the formation of haze, acid mist, and acid rain. It is a major air pollutant and an important precursor of volatile organic compounds and aerosol particles, which affects climate change and harms human health [1]. The main NO_2 sources are anthropogenic and natural, with anthropogenic sources comprising of combustion of fossil fuels, industrial production and transportation, while natural sources mainly include soil emissions and lightning generation, with two-thirds of the total emissions attributed to anthropogenic sources [2].

D. Han et al. (Eds.): ICEPG 2024, ESE, pp. 496–502, 2025.
https://doi.org/10.1007/978-3-031-86829-0_56

As the largest country in Asia, China's land area accounts for over 50% of the land area in Northeast Asia, therefore representing a major source of NO_X emissions in Northeast Asia. China's rapid economic growth occurred early on without strict NO_X emission control measures. This created serious pollution characterized by increasingly low air quality, causing atmospheric pollution to evolve from environmental problem into social problems. To remedy severe air pollution, Chinese government adopted a series of policies and measures to strictly control NO_X emissions. These measures directly impacted the distribution of tropospheric NO_2 concentration in Northeast Asia. Therefore, monitoring changes of NO_2 concentration in Northeast Asia remains significant. At present, the two main monitoring methods for trace gases in atmosphere include fixed-point sampling monitoring and spectral monitoring. Compared with ground observation, satellite observation has an advantage due to the wide coverage, strong real-time performance, continuous observation and collection of high-resolution data, and provides important parameters that cannot be obtained by conventional observation methods. Monitoring of global tropospheric NO_2 distribution has been conducted through satellite remote sensing since 1996. Kim et al. used satellite and ground observation data to analyze NO_2 and $PM_{2.5}$ concentrations during the COVID-19 pandemic in South Korea [3]. Xu et al. used GOME and SCIAMACHY satellite data to analyze the distribution of NO_2 density in China from 1996 to 2010, showing that North China Plain, Pearl River Delta and Yangtze River Delta were areas of high NO_2 density and annual emissions growth rates [4]. Zia et al. used OMI satellite data to predict NO_2 concentrations based on machine learning method, and made case studies in major cities [5]. By collecting the annual emission densities of NO_2, SO_2, and PM_{10} from White Paper of Ministry of Environment, Piao et al. found that the NO_2 density in South Korea remained stable in urban areas like Seoul from 2005 to 2010 [6].

2 Study Area and Data Processing

2.1 Study Area

Northeast Asia (105°E–145°E, 20°N–53°N) includes South Korea, North Korea, Japan, Mongolia, and north and east of China (see Fig. 1), covering a land area of more than 16 million km^2 that accounts for about 40% of Asia's total area. The climate is mostly temperate continental, characterized by cold winter and hot summer, with high annual temperature difference, low annual precipitation and four distinct seasons. The rest is temperate and subtropical monsoon climate with cold and dry winter, and summer impacted by marine air masses is hot and rainy. The high NO_2 density areas include Beijing-Tianjin-Hebei region, Shandong province, Yangtze River Delta region and Pearl River Delta region.

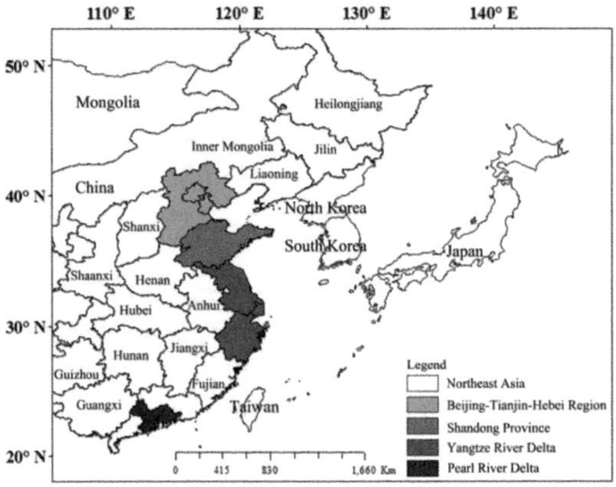

Fig. 1. Location of the study area.

2.2 Data Processing

The sensor OMI is onboard Aura satellite launched in 2004, and it monitors daily global tropospheric trace gases like O_3, NO_2 and SO_2, with spatial resolution of 13 km × 24 km and wavelength of 270–500 nm. The data used in this study is tropospheric NO_2 vertical column density(VCD) of OMI from January 2005 to December 2016, provided by Royal Netherlands Meteorological Institute, DOMINO version 2.0 [7], ESRI grid format and spatial resolution of 0.125° longitude × 0.125° latitude, in units of 10^{15} molec/cm^2. ArcGIS software is mainly used to extract monthly NO_2 VCD for analysis, then seasonal and annual average NO_2 are calculated. Seasons are divided into Winter (December–February), spring (March–May), summer (June–August) and autumn (September–November).

3 Distribution and Change Trend of NO_2 VCD

3.1 Yearly and Seasonal Distribution of NO_2

The spatial distribution of annual average NO_2 VCD in Northeast Asia from 2005 to 2016 is displayed in Fig. 2. The average NO_2 VCD over these 12 years is 3.97×10^{15} molec/cm^2, with high values mainly in areas with large population density and developed industry. For example, the average NO_2 VCD in Beijing–Tianjin–Hebei region of China reaches 11.41×10^{15} molec/cm^2, almost 2.9 times that of Northeast Asia. Other NO_2 high-value areas include Shandong province, Henan province, Pearl River Delta, and Yangtze River Delta in China, Seoul in Korea, and Tokyo in Japan. The average NO_2 VCD in South Korea is 7.18×10^{15} molec/cm^2, almost the same as that in Pearl River Delta of China. The average NO_2 in Japan is 2.96×10^{15} molec/cm^2, and low values is attributed to long-time strict environmental policies. Other low NO_2 areas in Northeast Asia are Mongolia and North Korea, due to low economic development and relatively backward industries involving animal husbandry and mining.

The spatial distribution of seasonal NO_2 VCD in Northeast Asia from 2005 to 2016 is shown in Fig. 3. The low NO_2 region exhibits little change over four seasons, but

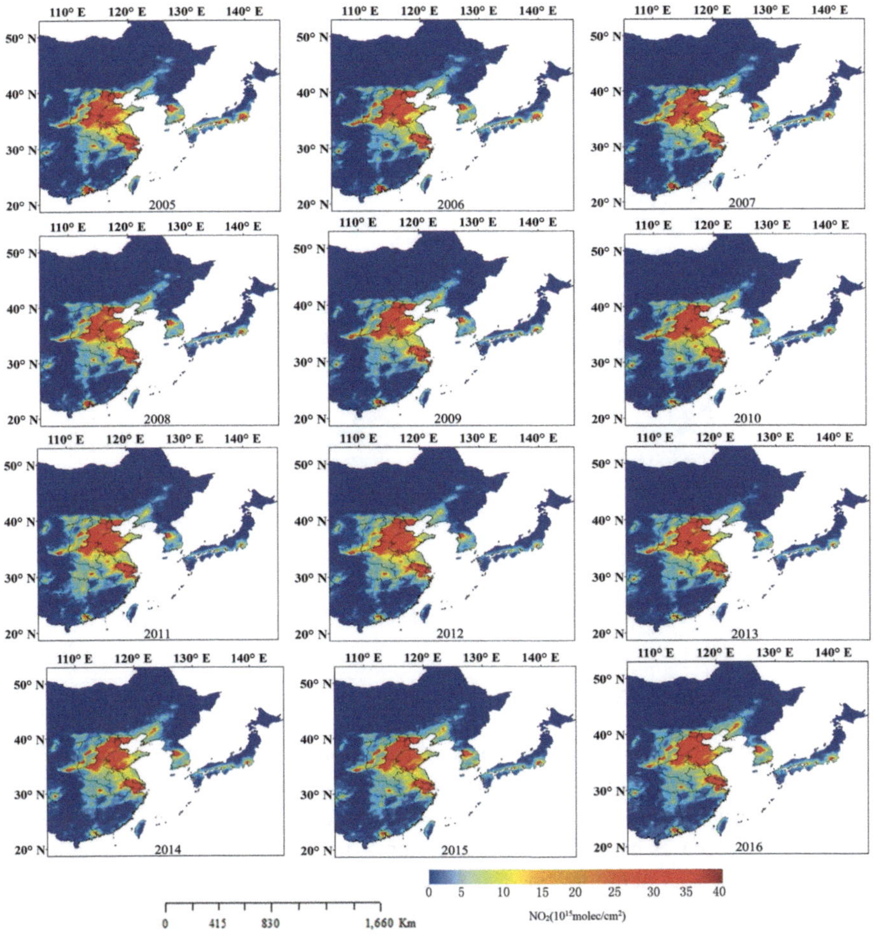

Fig. 2. Spatial distribution of NO_2 VCD in Northeast Asia from 2005 to 2016.

change is obvious in NO_2 high-value region. The NO_2 VCD is the highest in winter, followed by autumn and spring, and the lowest in summer. This pattern is attributed to the burning of fossil fuels in winter to meet heating needs in urban areas, thereby promoting increased NO_2 emissions. In addition, due to weak solar radiation in winter and long lifespan of NO_2, NO_2 VCD is the highest in winter. In summer, the intensity of illumination is relatively high, and under sunlight with wavelength <424 nm, NO_2 participates in photolysis reaction. In addition, Northeast Asia is strongly influenced by oceanic air masses in summer, with prevalent southeast monsoon increasing precipitation and promoting the transmission of NO_2, so that NO_2 is the lowest in summer.

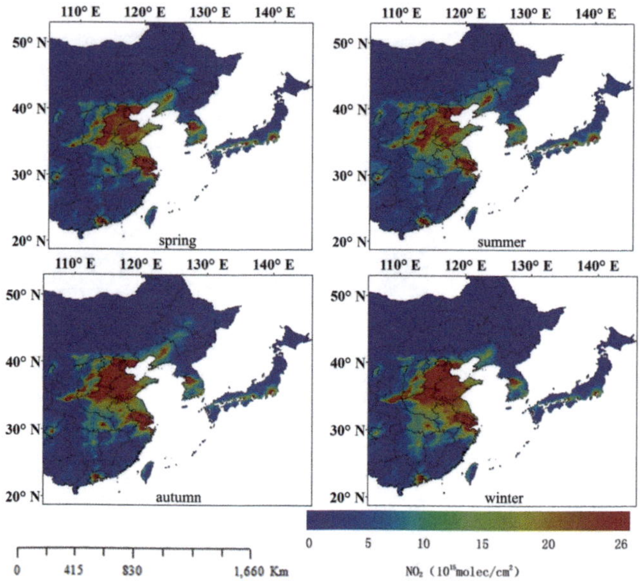

Fig. 3. Spatial distribution of seasonal NO_2 VCD from 2005 to 2016.

3.2 Change Trend of NO_2 VCD

Distribution of NO_2 VCD change in Northeast Asia from 2016 to 2005 is shown in Fig. 4. The overall NO_2 emissions in Mongolia, North Korea, and central and western China are relatively unchanged, except in Seoul. NO_2 increases by 0.6×10^{15} molec/cm^2 in North Korea, increases by 2.45×10^{15} molec/cm^2 in Beijing-Tianjin-Hebei region and also increases in Shandong province, whereas it shows decrease at varying degrees in Yangtze River Delta, Pearl River Delta, Henan province and Anhui province. The NO_2 in Pearl River Delta decreases by 2.22×10^{15}molec/cm^2.

The change of NO_2 VCD in Northeast Asia shows the trend of first increase then decrease, reaching peak in 2011 with a value of 4.99×10^{15} molec/cm^2, which is 1.58 times in 2005 (see Fig. 5). The main reason is Chinese government issued the "12th Five-Year Plan for the Prevention and Control of Air Pollution in Key Areas" in 2012. The growth rate of average annual NO_2 in Northeast Asia is 7.9% in 2005–2011, and -7.6% in 2012–2016, with average growth rate of 0.88%.

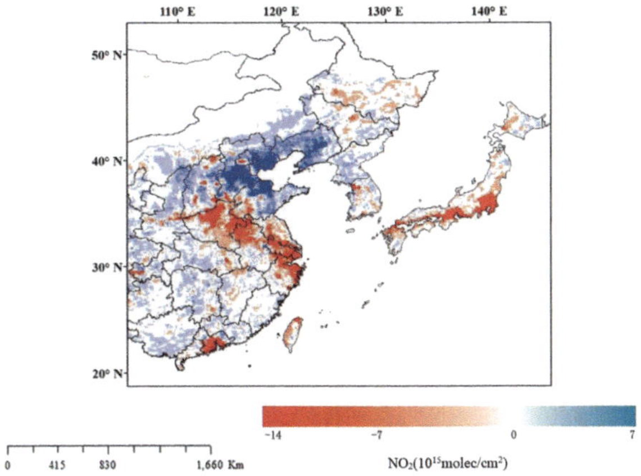

Fig. 4. Distribution of NO_2 VCD Change from 2005 to 2016.

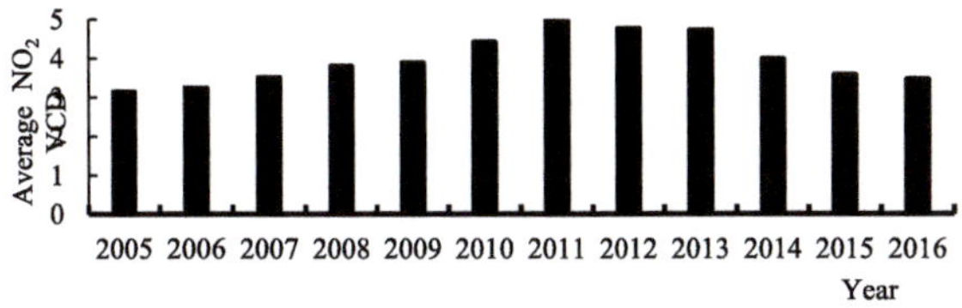

Fig. 5. Change of NO_2 VCD from 2005 to 2016.

4 Conclusions

The NO_2 in Japan exhibits continuous downward trend from 2005 to 2016, which is related to policies by Japanese government for environmental protection. China issued many key projects for air pollution prevention in key areas in Twelfth Five-Year Plan, thus decreasing air pollution. Further works are to establish joint prevention and control mechanism, to establish multi-scale, high-resolution atmospheric emission inventory, and multiple environmental data assessment models, and clean energy use.

References

1. Fan, C., Li, Z., Li, Y., et al.: Variability of NO_2 concentrations over China and effect on air quality derived from satellite and ground-based observations. Atmos. Chem. Phys. **21**(10), 7723–7748 (2021)
2. Chen, Y., Wei, Y., Chang, Z., et al.: Variation characteristics of concentration ratio of NO_2 and CO_2 in 31 cities, China based on remote sensing data and XGBoost algorithm. J. Earth Sci. Environ. **45**(06), 1355–1367 (2023)
3. Kim, D., Lee, J., Kim, J.: Spatiotemporal variations of NO_2 and $PM_{2.5}$ concentrations during the COVID-19 pandemic in South Korea: a satellite and ground-based observation study. Sci. Total Environ. **853**, 158343 (2023)

4. Xu, Y., Wang, D., Wu, Z.: Remote sensing monitoring of temporal and spatial changes of tropospheric NO_2 in China from 1996 to 2010. Remote Sens. Technol. Appl. **28**(5), 898–903 (2013)
5. Zia, U., Salman, T.: Machine learning-based prediction of NO_2 concentrations using satellite data: a case study of major cities. Environ. Res. Lett. **18**(2), 024001 (2023)
6. Piao, C., Liu, G., Long, F., et al.:The situation and management policy of South Korea's air pollution. Environ. Sci. Technol. **36**(6L), 382–385+432 (2013)
7. Van Geffen, J., Eskes, H., Compernolle, S., et al.: Sentinel-5PTROPOMI NO_2 retrieval: impact of version v2.2 improvements and comparisons with OMI and ground-based data. Atmos. Meas. Tech. **15**(7), 2037–2060 (2022)

Projections of Land Use Change and Carbon Emission in Fujian Province Under Multiple SSP-RCP Scenarios

Zhihao Liu, Pangpang Gao(✉), Shiqi Li, Yuanke Sun, and Hejie Zhou

School of Environmental Science and Engineering, Xiamen University of Technology, Xiamen 361024, China
Gaopp@xmut.edu.cn

Abstract. Land use change is the most important factor influencing change in the carbon cycle. Predicting how land use will change in the future can enable stakeholders understand and identify appropriate land use configurations and effectively control carbon emissions. This investigate utilizes the Patch-generating Land Use Simulation (PLUS) model to simulate land use in Fujian province. The model is used to project future land use based on data derived from the LUH2 datasets. Land use change related carbon emission is evaluated by carbon emission factor method. Results indicate that: (i) The Kappa coefficient, OA, and FOM are 0.89, 0.84, and 0.19, indicating that the model performs well; (ii) In the SSP1-2.6 scenario, forest land, construction land, and bareland areas are anticipated to expand, whereas cropland and grassland areas are projected to shrink. Conversely, under the SSP2-4.5 and SSP5-8.5 scenarios, decreases are expected in cropland, grassland, and construction land areas. (iii) The trends of carbon emissions follow a similar pattern across various scenarios. Emissions are projected to increase steadily until around 2050, after which they begin to gradually decline. To achieve the double carbon target, the future land use planning should focus on the protection of cropland, the development of a rational planning strategy for construction land, and the improvement of grassland restoration.

Keywords: Carbon emission · Land use change · PLUS model · SSP-RCP

1 Introduction

Carbon emissions, which directly contributing to global climate change, have emerged as a major challenge of the 21st [1]. This phenomenon has already had adverse impacts on ecological environment and socioeconomics [2]. Predicting future carbon emissions can help in formulating adaptive strategies to address the challenge of climate change and achieve a proper harmonize between economic growth and environmental protection.

Simulating and predicting future land use can effectively support future carbon emission projection. Land use prediction is comprised of two parts: one is the quantitative changes, and the other is the spatial distribution. The LUH2 dataset has already provided data for quantitative change predictions, while methods for spatial prediction mainly

D. Han et al. (Eds.): ICEPG 2024, ESE, pp. 503–509, 2025.
https://doi.org/10.1007/978-3-031-86829-0_57

include FLUS, CA, CLUE, and PLUS models. PLUS has been widely applied in recent studies [3, 4]. Hu et al. 2021; utilized PLUS model to analyze the historical and projected land use changes and spatiotemporal variations in terrestrial ecosystem carbon storage in Jiangxi Province under different development scenarios. Zhang et al.2003; used the PLUS model to identify the key driving factors of land use change in the Yangtze River Delta and simulated various future land [5].

There are several common methods for calculating land use-related greenhouse gas emissions are the carbon stock change method, the IPCC method, the method combining remote sensing and GIS technology, the latter is particularly versatile, as it can be applied to a wide range of land use types and multiple land use change scenarios. Many international organizations provide standardized carbon emission factors, which make the carbon emission factor method highly comparable internationally, contributing to global climate change research and policy formulation.

This study employs land use simulation based on the future land use demand derived from LUH2 dataset, and the PLUS model is employed to simulate the land use pattern of Fujian Province under SSP1-2.6, SSP2-4.5, and SSP5-8.5. The carbon emissions of Fujian Province under multiple scenarios are then calculated using the carbon emission factor method. The adaptive strategies for carbon emission reduction will be designed based on future spatial distribution characteristics of land use in Fujian Province under different scenarios.

2 Description of the Method and Data

2.1 PLUS Model

The PLUS model is primarily constituted of two modules: the Land Expansion Analysis Strategy (LEAS) and the Cellular Automata (CA) Markov module [6]. LEAS: First, land use data from two periods is extracted, and then driving factors influencing LUCC are added to the expanded land use layer. The Random Forest formula is then applied to the land use extension data for calculation. The formula for this is as follows [7]:

$$P\frac{d}{i,k(x)} = \frac{\sum_{n=1}^{M} I[h_n(x) = d]}{M} \tag{1}$$

CARS: CARS utilizes the predicted evolution probabilities of diverse land use categories, as derived from LEAS, to project the prospective distribution of land use patterns [8, 9]. The comprehensive transition probability formula for multiple land use types is presented as follows [7]:

$$p_{o,i,k}^{d=1} = p_{i,k}^{d=1} \times \Omega_{i,k}^{t} \times D_{k}^{t} \tag{2}$$

2.2 Calculation of Carbon Emissions

The Carbon emission factor method is introduced as a means of calculating carbon emissions from land use, these can be further subdivided into two categories: immediate carbon emissions and indirect carbon emissions. The immediate carbon emission calculates the carbon emissions from forest land, grassland, cropland, and bareland. The carbon emission coefficients for cropland, forest land, grassland and bareland are set at 0.0422, −0.05789, −0.0021 and −0.0005, respectively, while construction land has a coefficient of 42.97 [10, 11]. The calculation formulas for direct and indirect carbon emissions are as follows.

$$E_K = \sum e_i = \sum s_i \times \delta_i \quad (3)$$

$$E_t = \sum E_{t,i} = \sum E_{ni} \times \theta_i \times f_i \quad (4)$$

2.3 Model Performance Evaluation

In this study, values of FOM, kappa coefficient and OA values are used to evaluate the performance of PLUS [11]. The specific evaluation metrics are defined by Eqs. (5) to (7).

$$Kappa = \frac{p_o - p_e}{1 - p_e} \quad (5)$$

$$FOM = \frac{B}{A + B + C + D} \quad (6)$$

$$OA = \frac{TP + TN}{TP + TN + FP + FN} \quad (7)$$

3 Study Area and Data Acquisition

3.1 Study Area

Fujian Province has a total area of 124000 km^2, lying between 23°33′ and 28°20′ north latitude and 115°50′ and 120°40′ east longitude (as shown in Fig. 1). Fujian is rich in natural resources, boasting the highest forest coverage rate in the country. The province's coastline is winding, with abundant port resources.

3.2 Data Acquisition

The fundamental data necessary to this study includes land uses and driving factor data, which are listed in Table 1. Slope data is processed using ArcGIS 10.8. River and road data are sourced from Open Street Map, with relevant factor data obtained by calculating Euclidean distances.

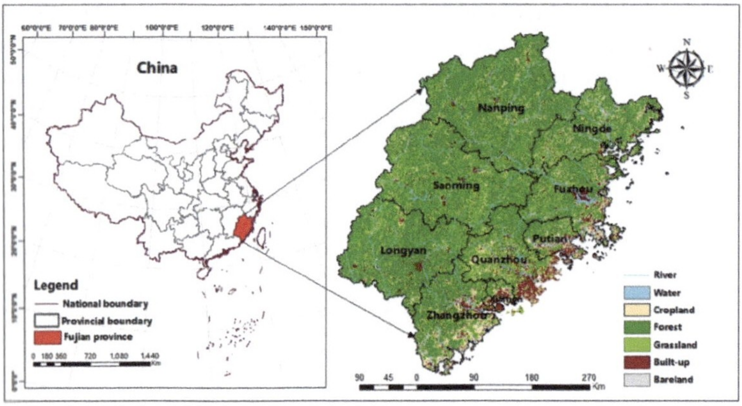

Fig. 1. Study area and the land use distributions

Table 1. List of the date

Date	Resolution	Date resource
Land use, City1, Stream2	1 km	National Catalogue Service for Geographic Information
Driving factor data (Temperature, Dem, Slope, Pop, GDP, NDVI)	1 km	Resource and Environmental Science Date Platform
Road data	30 m	Open Street Map

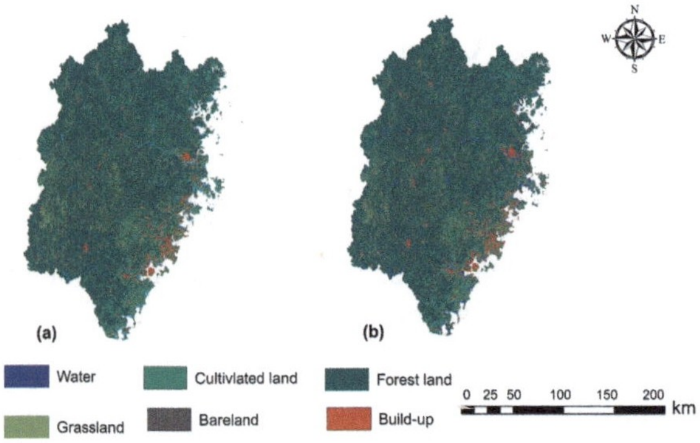

Fig. 2. (a) Simulated result in 2020; (b) Actual result in 2020

4 Result and Discussion

4.1 Evaluation of Model Performance

This study uses Fujian Province's 2000 and 2010 land use date as initial inputs to predict the 2020 land use distribution. Figure 2 demonstrates the model's performance by contrasting the projected land use in 2020 with the actual land use in 2020. The Kappa, OA, and FOM figures are 0.89, 0.84, and 0.19, indicating a good model performance of the model.

4.2 Land Use Patterns Under Multiple Scenarios

The trends for various the different types of land use under the three scenarios simulated by the PLUS model are shown in Fig. 3. Results indicate that the trends are relatively consistent across different scenarios. Under the SSP1-2.6, both forest land and construction land show varying degrees of growth, while cropland first decreases and then increases. The data indicate a trend of initial increase and subsequent decrease in construction land and grassland. Under SSP2-4.5 and SSP5-8.5, the land consistently demonstrates a decline. In the SSP2-4.5 scenario, grassland and construction land exhibit the smallest changes, while cropland shows an overall decreasing trend with a significant reduction in area. By 2060, cropland will have decreased by 4,980.92 km^2.

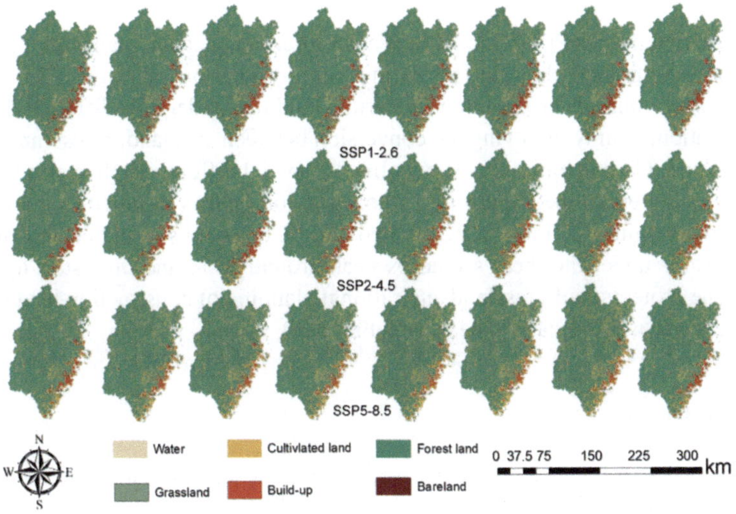

Fig. 3. Land use changes in 2025–2060 under different scenarios

4.3 Carbon Emissions Under Multiple Scenarios

Figure 4 indicates that the trends of carbon emissions across different scenarios. All the scenarios for carbon emissions show a steady increase until 2050, after which a gradual decline is observed. The maximum carbon emissions occur in 2050 under the SSP1-2.6 scenario, which is 1.21×10^{11} t. If the 2030 carbon peak target wants to be achieved, stronger carbon reduction strategies need to be further developed.

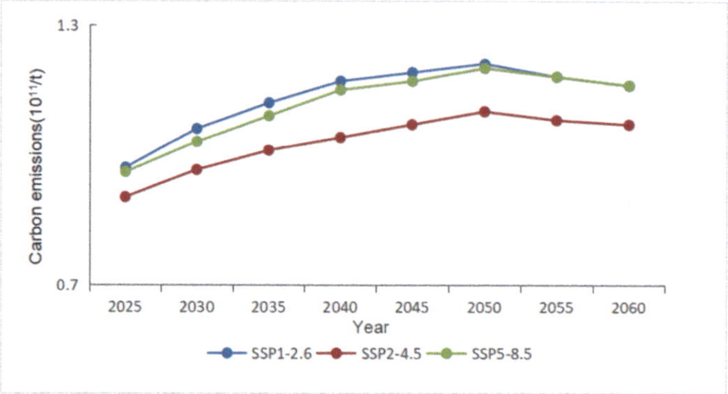

Fig. 4. Carbon emissions in Fujian Province under different scenarios

5 Conclusion

This research employs PLUS model to project future land cover change by considering 14 socioeconomic and climatic factors in a multi-scenario framework. The results of the study show that under the three scenarios, land use undergoes different degrees of transformation, mainly involving the conversion between cropland, forest land and construction land. Carbon emissions show obvious spatial differences, because the coastal economy is better, so the expansion of construction land is more concentrated in the coastal area resulting in more concentrated carbon emissions in the coastal area. Carbon emissions under the three scenarios peak around 2050 and then start to decrease slowly. The protection of grassland and rational planning of construction land should be strengthened to achieve carbon peak target in 2030.

References

1. Awotwi, A., Anornu, G.K.: Water balance responses to land-use/land-cover changes in the Pra River Basin of Ghana, 1986–2025. CATENA **182**, 104129 (2019)
2. Zhang, Y., Lin, J.: Delineation of urban growth boundary based on FLUS model under the perspective of land use evaluation in hilly mountainous areas. J. Mt. Sci. **21**(5), 1647–1662 (2024)
3. Zhang, Z., Wang, X.: Simulating land use change for sustainable land management in rapid urbanization regions: a case study of the Yangtze River Delta region. Landsc. Ecol. **38**, 1807–1830 (2023)
4. Liu, J., Liu, B.: Prediction of land use for the next 30 years using the PLUS model's multi-scenario simulation in Guizhou Province, China. Sci. Rep. **14**, 13143 (2024)

5. Hu, J.X., Le, X.W.: Temporal and spatial evolution and prediction of ecosystem carbon storage in Jiangxi Province based on PLUS-INVEST model. Huan Jing Ke Xue **45**(6), 3284–3296 (2024)
6. Li, S., Cao, Y., Liu, J.: Simulating land use change for sustainable land management in China's coal resource-based cities under different scenarios. Sci. Total Environ. **916**, 170126 (2024)
7. Liang, X., Guan, Q., Liu, S.: Understanding the drivers of sustainable land expansion using a patch-generating land use simulation (PLUS) model: a case study in Wuhan, China. Comput. Environ. Urban Syst. **85**, 101569 (2021)
8. Zhu, Z., Duan, J., Li, R.: Spatial evolution, driving mechanism, and patch prediction of grain-producing cultivated land in China. Agriculture **12**(6), 860 (2022)
9. Li, L., Huang, X.: Scenario-based urban growth simulation by incorporating ecological-agricultural-urban suitability into a Future Land Use Simulation model. Cities **137**, 104334 (2023)
10. Wei, J., Xia, L.: A network-based framework for characterizing urban carbon metabolism associated with land use changes: a case of Beijing city, China. J. Clean. Prod. **371**, 133695 (2022)
11. Yao, Y., Liu, X.: Simulating urban land-use changes at a large scale by integrating dynamic land parcel subdivision and vector-based cellular automata. Int. J. Geogr. Inf. Sci. **31**(12), 2338–2363 (2017)